NEUROMETHODS

Series Editor
Wolfgang Walz
University of Saskatchewan
Saskatoon, Canada

For further volumes:
http://www.springer.com/series/7657

Neuromethods publishes cutting-edge methods and protocols in all areas of neuroscience as well as translational neurological and mental research. Each volume in the series offers tested laboratory protocols, step-by-step methods for reproducible lab experiments and addresses methodological controversies and pitfalls in order to aid neuroscientists in experimentation. *Neuromethods* focuses on traditional and emerging topics with wide-ranging implications to brain function, such as electrophysiology, neuroimaging, behavioral analysis, genomics, neurodegeneration, translational research and clinical trials. *Neuromethods* provides investigators and trainees with highly useful compendiums of key strategies and approaches for successful research in animal and human brain function including translational "bench to bedside" approaches to mental and neurological diseases.

Ketamine

Pain Management, Critical Care, Depression, and Resource-Poor Settings Volume II

Edited by

Sabine Himmelseher

Department of Anesthesiology and Intensive Care, Technical University of Munich – School of Medicine and Health, Munich, Germany

Humana Press

Editor
Sabine Himmelseher
Department of Anesthesiology and
Intensive Care
Technical University of Munich – School of Medicine and Health
Munich, Germany

ISSN 0893-2336 ISSN 1940-6045 (electronic)
Neuromethods
ISBN 978-1-0716-4598-7 ISBN 978-1-0716-4599-4 (eBook)
https://doi.org/10.1007/978-1-0716-4599-4

Image Caption for Volume 2: Fluorescence microscopy photos of cultured mouse cortical stylized neurons which express a genetically encoded calcium indicator during ketamine treatment, overlayed with a human brain-neuron avatar and a stylized neuron generated by artificial intelligence, as well as the molecular structure of ketamine (from public domain). Image designed by Samuel Kohtala and Iina Annala, Helsinki, Finland.

This Humana imprint is published by the registered company Springer Science+Business Media, LLC, part of Springer Nature.
The registered company address is: 1 New York Plaza, New York, NY 10004, U.S.A.

If disposing of this product, please recycle the paper.

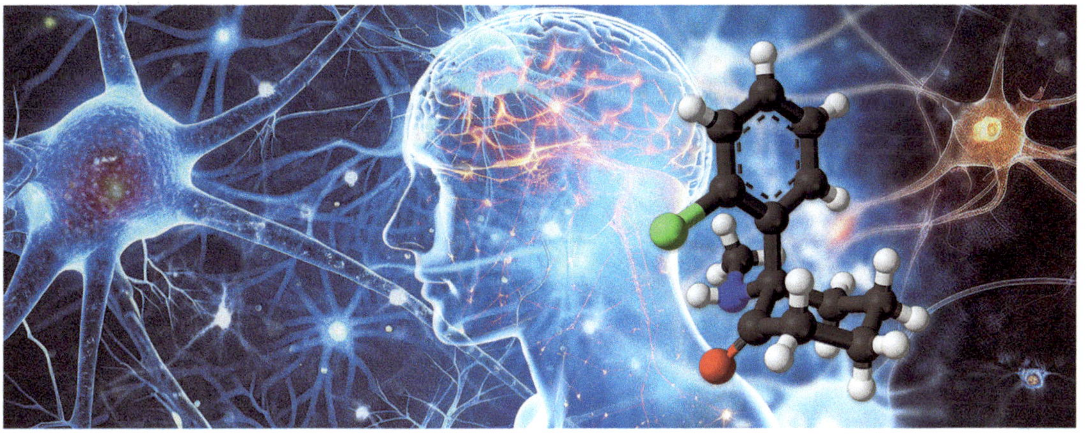

Fluorescence microscopy photos of cultured mouse cortical stylized neurons which express a genetically encoded calcium indicator during ketamine treatment, overlayed with a human brain-neuron avatar and a stylized neuron generated by artificial intelligence, as well as the molecular structure of ketamine (from public domain). Image designed by Samuel Kohtala and Iina Annala, Helsinki, Finland

Preface to the Series

Experimental life sciences have two basic foundations: concepts and tools. The Neuromethods series focuses on the tools and techniques unique to the investigation of the nervous system and excitable cells. It will not, however, shortchange the concept side of things as care has been taken to integrate these tools within the context of the concepts and questions under investigation. In this way, the series is unique in that it not only collects protocols but also includes theoretical background information and critiques which led to the methods and their development. Thus it gives the reader a better understanding of the origin of the techniques and their potential future development. The *Neuromethods* publishing program strikes a balance between recent and exciting developments like those concerning new animal models of disease, imaging, in vivo methods, and more established techniques, including, for example, immunocytochemistry and electrophysiological technologies. New trainees in neurosciences still need a sound footing in these older methods in order to apply a critical approach to their results.

Under the guidance of its founders, Alan Boulton and Glen Baker, the Neuromethods series has been a success since its first volume published through Humana Press in 1985. The series continues to flourish through many changes over the years. It is now published under the umbrella of Springer Protocols. While methods involving brain research have changed a lot since the series started, the publishing environment and technology have changed even more radically. Neuromethods has the distinct layout and style of the Springer Protocols program, designed specifically for readability and ease of reference in a laboratory setting.

The careful application of methods is potentially the most important step in the process of scientific inquiry. In the past, new methodologies led the way in developing new disciplines in the biological and medical sciences. For example, Physiology emerged out of Anatomy in the nineteenth century by harnessing new methods based on the newly discovered phenomenon of electricity. Nowadays, the relationships between disciplines and methods are more complex. Methods are now widely shared between disciplines and research areas. New developments in electronic publishing make it possible for scientists that encounter new methods to quickly find sources of information electronically. The design of individual volumes and chapters in this series takes this new access technology into account. Springer Protocols makes it possible to download single protocols separately. In addition, Springer makes its print-on-demand technology available globally. A print copy can therefore be acquired quickly and for a competitive price anywhere in the world.

Saskatoon, Canada ***Wolfgang Walz***

Preface to Volume II

Ketamine remains an enigmatic drug, even for researchers and clinicians who have worked with this substance for a long time. As a medical student, I was deeply impressed by the pioneering clinician-scientists who studied the effects of intravenous *S*(+)-ketamine (esketamine) in experimental and clinical studies compared to the racemate in anesthesia and analgesia. In 1992, the US Food and Drug Administration had complained that too little attention was directed toward the separation of racemic drugs because this process might be difficult for the pharmaceutical industry and therefore not of interest.

Time has shown that such an endeavor can lead to amazing results: Based on many clinical advantages, intravenous esketamine was approved in 1997 in Germany, and then in other European countries. However, it quickly became clear that even esketamine was also not an "ideal" monoanesthetic and needed to be combined in balanced regimen except for emergencies and other specific indications. Then the growing popularity of illicit use of racemic ketamine as a recreational drug with a hot spot in China and Asia generated much concern about its adverse effects. It is not much different today, with addiction psychiatrists warning of the serious consequences of years of high dose ketamine abuse by young people. To be clear, long-term abuse must not be confused with medical treatment with ketamine. In medical settings, the psychoactive effects of ketamine are clearly dose-dependent, transient, and can be managed with skillful use, patient guidance, or combination therapy with, for example, gamma amino butyric acid (GABA)ergic agents.

In contrast, over the past two decades, the discovery of ketamine's fast-acting, subanesthetic-dose, antidepressant benefits in treatment-resistant depression has spurred global interest not only in new uses for ketamine in psychiatry and many other medical fields but also in neuroscience and neuropharmacology. This development culminated in the official approval of a dedicated intranasal (IN) esketamine device in the United States, Canada, and other countries in 2019 as adjunctive antidepressant for the treatment of major depression, suicidal ideation, and resistant depression treatment. Meanwhile, 6.5 years with flexibly dosed IN esketamine have been observed, showing not only sustained improvements in patients' depressive symptoms but also reduced mortality and hospitalization compared to previous standard of care. In anesthesia and critical care, evidence is accumulating to support expanded indications for esketamine in particular including new uses in perioperative multimodal analgesia, prevention and treatment of perioperative depression, prevention of postoperative neurocognitive disorders, and opioid-free anesthesia. Professional society guidelines and recommendations for sedation and analgesia in pediatric and adult critically ill patients have been updated to include direct advice on the use of ketamine. Research into the effects of ketamine in chronic pain states and other serious mental health and medical conditions has been intensified. This repurposing includes a bundle of innovative approaches ranging from Alzheimer's and Parkinson's disease to substance use disorders, chronic inflammatory bowel syndrome, and even stuttering in children. The second volume of the textbook provides further clarification of ketamine's important role in managing (acute) perioperative and chronic pain, and emphasizes its significance in challenging situations in intensive and neurocritical care, such as difficult sedation in seriously ill COVID-19 patients. The book also highlights the significant improvements in patient outcomes with major depressive disorders after therapy with racemic or es-ketamine, including those desperate situations where patients did not respond to other antidepressant treatments. It also presents practical regimen for healthcare professionals seeking to implement effective antidepressant treatment strategies with racemic or es-ketamine. Ketamine's utility in other severe mental health issues such as substance use disorders is also explored. It is imperative to acknowledge ketamine's distinctive role in healthcare provision in resource-constrained areas and under conditions of conflict and disaster. The book is further enriched by expert- and evidence-based ideas and prospects regarding ketamine's role in enhancing medical care in challenging fields like mountain medicine, cardiac anesthesia, and space travel or long-term missions in microgravity.

Is the textbook format still relevant today? If you were to ask our authors, they would vividly echo a "Yes, of course, it is"! Everyone wants a printed book, something to hold in their hands! And what about the "value" of a textbook's contents in our age of constantly new, fast, often short-lived information? The content value of a textbook written by experts and leaders in their fields is undeniably and unalterably great, because it offers a comprehensive approach to a topic synthesizing evidence- and eminence-based knowledge. Textbook authors typically share tremendous expertise whether or not controlled studies, trials, meta-analyses, and guidelines are available. They provide accurate advice and guidance often long before formal evidence is gained.

Is ketamine a cure? Our international interdisciplinary group of expert authors from all parts over the world allows us to look at this question from very different angles, from the perspective of basic mechanisms to computer simulations, and from clinical trials to everyday practice. Well, it is certainly not a panacea, but it can have beneficial, long-lasting effects. Far beyond that, it is very appealing to read that all of our authors do not only see an important role for ketamine in their respective fields but also potential for many further developments. Without exception, the globally localized authors come to this conclusion, which is remarkable.

In this spirit of unity, I would like to express my personal gratitude to all authors and friends. This textbook would not have been possible without their great dedication and absolute will to teach and educate. It is with enthusiasm that we offer this unique collection of knowledge on ketamine and its isomers. Enjoy, expand, and let our patients benefit from the information.

Ulm, Baden-Württemberg, and München, Bayern, Germany *Sabine Himmelseher*

Contents

Contributors

LOBNA ABDELWAHAB • *Department of Epilepsy, Geisel School of Medicine at Dartmouth, Lebanon, NH, USA*

THEODOROS ASLANIDIS • *Department of Anesthesiology and Intensive Care Unit, Saint Paul ("Agios Pavlos") General Hospital, Thessaloniki, Greece*

ANDREW BAUERSCHMIDT • *Departments of Neurology and Neurosurgery, New York Medical College, Valhalla, NY, USA; Westchester Medical Center, Valhalla, NY, USA*

MARC MORITZ BERGER • *Department of Anesthesiology, Intensive Care Medicine, Emergency Medicine, and Pain Therapy, RKH Hospital, Ludwigsburg, Germany*

CARLOS DARCY ALVES BERSOT • *Department of Anesthesiology, BP Mirante Hospital – Anextesia, São Paulo, Brazil; Paulista School of Medicine, UNIFESP, São Paulo, Brazil*

HELMAR BORNEMANN-CIMENTI • *Department of Anaesthesiology and Intensive Care Medicine, Medical University of Graz, Graz, Austria*

LUDIVINE BOUDIEU • *Department of Psychiatry, CHU Clermont-Ferrand, University of Clermont Auvergne, CNRS, Clermont Auvergne INP, Institut Pascal (UMR 6602), Clermont-Ferrand, France*

J. MICHAEL BREWER • *Nazih Zuhdi Transplant Institute, Specialty Critical Care and Acute Circulatory Support Service, INTEGRIS Health Baptist Medical Center, Oklahoma City, OK, USA*

STUART BRILL • *Westchester Medical Center, Valhalla, NY, USA*

CARSON CHRENEK • *University of Alberta, Edmonton, AB, Canada*

IVANA ĆIRKOVIĆ • *Institute of Microbiology and Immunology, Faculty of Medicine, University of Belgrade, Belgrade, Serbia*

GREG DAL-BO • *French Armed Forces Biomedical Research Institute (IRBA) Toxicology and Chemical Risks Department, Brétigny-sur-Orge, France*

FREDERIC DORANDEU • *French Military Health Academy, Ecole du Val-de-Grace, Paris, France*

MICHAEL EICHLSEDER • *Department of Anaesthesiology and Intensive Care Medicine, Medical University of Graz, Graz, Austria*

MOHAMED ELSAYED • *Department of Psychiatry, Giesel School of Medicine at Dartmouth, Concord, NH, USA*

THIAGO RAMOS GRIGIO • *Department of Anesthesiology, Pain Centre, University of Groningen, University Medical Centre Groningen, Groningen, The Netherlands; Postgraduate Program of Anesthesiology, Surgical Sciences and Perioperative Medicine, Faculdade de Medicina da Universidade de São Paulo (USP), São Paulo, Brazil*

KENJI HASHIMOTO • *Chiba University Center for Forensic Mental Health, Chiba, Japan*

SABINE HIMMELSEHER • *Department of Anesthesiology and Intensive Care, Technical University of Munich – School of Medicine and Health, Munich, Germany*

JOCHEN HINKELBEIN • *University Department of Anaesthesiology, Intensive Care Medicine, and Emergency Medicine, Johannes Wesling Klinikum Minden, University Hospital Ruhr-University Bochum, Minden, Germany; German Society of Aerospace Medicine (DGLRM), Munich, Germany*

MICHAEL G. IRWIN • *Department of Anaesthesiology, University of Hong Kong, Hong Kong, People's Republic of China*

ANA JOTIĆ • *Clinic for Otorhinolaryngology and Maxillofacial Surgery Clinical Center of Serbia, Belgrade, Serbia*

HASSAN SAYED KANDIL • *Department of Anesthesiology, Critical Care and Pain Management, Faculty of Medicine, Ain Shams University Hospitals, Cairo, Egypt*

DANIA IMTIYAZ KHAN • *Faculty of Medicine, Ain Shams University, Cairo, Egypt*

PATRYCJA KLECZKOWSKA • *Maria Sklodowska-Curie Medical Academy in Warsaw, Warsaw, Poland; Department of Biomedical Research, National Medicines Institute, Warsaw, Poland*

JÜRGEN KNAPP • *Department of Anaesthesiology and Pain Medicine, Inselspital, Bern University Hospital, University of Bern, Bern, Switzerland*

LYDIA JOHNSON KOLAPARAMBIL VARGHESE • *University Department of Anaesthesiology, Intensive Care Medicine, and Emergency Medicine, Johannes Wesling Klinikum Minden, University Hospital Ruhr-University Bochum, Minden, Germany; German Society of Aerospace Medicine (DGLRM), Munich, Germany*

SIMON KUNG • *Department of Psychiatry and Psychology, Mayo Clinic, Rochester, Minnesota, USA*

HÜLYA KURSUN • *Department of Anesthesiology, Intensive Care Medicine, Emergency Medicine, and Pain Therapy, RKH Hospital, Ludwigsburg, Germany*

SEBASTIAN LABENBACHER • *Department of Anaesthesiology and Intensive Care Medicine, Medical University of Graz, Graz, Austria*

KAVEH LATIFI • *Department of Anesthesiology and Pain Medicine, School of Medicine, Iran University of Medical Sciences, Tehran, Iran*

QIAN LI • *Department of Anesthesiology, Laboratory of Mitochondrial Metabolism and Perioperative Medicine, National-Local Joint Engineering Research Centre of Translational Medicine of Anesthesiology, West China Hospital of Sichuan University, Chengdu, China*

TAO LI • *Department of Anesthesiology, Laboratory of Mitochondrial Metabolism and Perioperative Medicine, National-Local Joint Engineering Research Centre of Translational Medicine of Anesthesiology, West China Hospital of Sichuan University, Chengdu, China*

PIERRE-MICHEL LLORCA • *Department of Psychiatry, CHU Clermont-Ferrand, University of Clermont Auvergne, CNRS, Clermont Auvergne INP, Institut Pascal (UMR 6602), Clermont-Ferrand, France*

WEI MA • *Department of Anesthesiology, Laboratory of Mitochondrial Metabolism and Perioperative Medicine, National-Local Joint Engineering Research Centre of Translational Medicine of Anesthesiology, West China Hospital of Sichuan University, Chengdu, China*

CHRISTOPHER V. MAANI • *Department of Anesthesia and PeriOperative Services, Brooke Army Medical Center, Fort Sam Houston, San Antonio, TX, USA*

MARC O. MAYBAUER • *Department of Anesthesiology and Intensive Care Medicine, Philipps University, Marburg, Germany; Critical Care Research Group, The Prince Charles Hospital, University of Queensland, Brisbane, QLD, Australia; Department of Anesthesiology, Division of Critical Care Medicine, University of Florida, Gainesville, FL, USA*

BRANISLAVA MEDIĆ • *Department of Pharmacology, Clinical Pharmacology and Toxicology, Faculty of Medicine, University of Belgrade, Belgrade, Serbia*

JANNICKE MELLIN-OLSEN • *Department of Anaesthesiology and Intensive Care Medicine, Baerum Hospital, Oslo, Norway*

CHRISTINE H. MEYER-FRIEßEM • *Department of Anesthesiology, Intensive Care and Pain Medicine, KLW St. Paulus GmbH, St. Marien Hospital, Luenen, University Teaching Hospital, Luenen, Germany*
ALETTA M. E. MILLEN • *Integrated Molecular Physiology Research Initiative, School of Physiology, Faculty of Health Sciences, University of the Witwatersrand, Johannesburg, South Africa*
GEORGES MION • *Department of Anesthesia, Intensive Care, and Perioperative Medicine, Assistance Publique – Hôpitaux de Paris, Cochin Hospital, Paris, France*
MARINA RAMZY MOURID • *Faculty of Medicine, Alexandria University, Alexandria, Egypt*
CLAIRE C. NESTOR • *Department of Anaesthesiology, Tallaght University Hospital, Dublin, Ireland*
CHIMAOBI TIM NNAJI • *Department of Anaesthesia, Federal Medical Centre Owerri, Owerri, Nigeria; Department of Anaesthesia, Federal Teaching Hospital Owerri, Owerri, Nigeria*
JOSE EDUARDO GUIMARAES PEREIRA • *Department of Anesthesiology, Hospital Unimed Volta Redonda, Rio de Janeiro, Brazil; Department of Anesthesiology, Hospital Central do Exé rcito, Rio de Janeiro, Brazil*
MAXWELL Z. PRICE • *Department of Psychiatry, Hackensack Meridian School of Medicine, Nutley, NJ, USA*
RICHARD L. PRICE • *Department of Psychiatry, Yale School of Medicine, New Haven, CT, USA*
POUYA MOVAHED RAD • *Department of Clinical Sciences, Division of Adult Psychiatry Faculty of Medicine, Lund University, Lund, Sweden*
JON ROSENBERG • *Departments of Neurology and Neurosurgery, New York Medical College, Valhalla, NY, USA; Westchester Medical Center, Valhalla, NY, USA*
LUDOVIC SAMALIN • *Department of Psychiatry, CHU Clermont-Ferrand, University of Clermont Auvergne, CNRS, Clermont Auvergne INP, Institut Pascal (UMR 6602), Clermont-Ferrand, France*
ISLAM MOHAMMAD SHEHATA • *Anesthesia Department, Faculty of Medicine, Ain Shams University, Cairo, Egypt*
BALWINDER SINGH • *Department of Psychiatry and Psychology, Mayo Clinic, Rochester, Minnesota, USA*
RICHARD STEFFEN • *Department of Anaesthesiology and Pain Medicine, Inselspital, Bern University Hospital, University of Bern, Bern, Switzerland*
PAUL A. STEWART • *Department of Anaesthesiology, Tallaght University Hospital, Dublin, Ireland*
JENNIFER SWAINSON • *University of Alberta, Edmonton, AB, Canada*
AKIHIRO TAKAMIYA • *Department of Neurosciences, Neuropsychiatry, Leuven Brain Institute, KULeuven, Belgium; Hills Joint Research Laboratory for Future Preventive Medicine and Wellness, Keio University School of Medicine, Tokyo, Japan*
KARINE THIBAULT • *French Armed Forces Biomedical Research Institute (IRBA) Toxicology and Chemical Risks Department, Brétigny-sur-Orge, France*
HELMUT TRIMMEL • *Department of Anaesthesia, Emergency Medicine and Intensive Care, and Karl Landsteiner Institute of Emergency Medicine, University Hospital Wiener Neustadt, Wiener Neustadt, Austria; Faculty of Medicine/Dentistry, Department of Medicine, Danube Private University, Krems, Austria*
JENNIFER L. VANDE VOORT • *Department of Psychiatry and Psychology, Mayo Clinic, Rochester, Minnesota, USA*

KATARINA SAVIĆ VUJOVIĆ • *Department of Pharmacology, Clinical Pharmacology and Toxicology, Faculty of Medicine, University of Belgrade, Belgrade, Serbia*
XIAYUN WAN • *Chiba University Center for Forensic Mental Health, Chiba, Japan*
TOBIAS WARNECKE • *University Clinic of Anaesthesiology, Intensive Care, Emergency Medicine, and Pain Therapy, Klinikum Oldenburg, Oldenburg, Germany*
YUMING WU • *Department of Critical Care Medicine, Union Hospital, Tongji Medical College, Huazhong University of Science and Technology, Wuhan, People's Republic of China*
HUIJING XIE • *Department of Anesthesiology, Laboratory of Mitochondrial Metabolism and Perioperative Medicine, National-Local Joint Engineering Research Centre of Translational Medicine of Anesthesiology, West China Hospital of Sichuan University, Chengdu, China*
MIN XIE • *Department of Anesthesiology, Sichuan Academy of Medical Sciences & Sichuan Provincial People's Hospital, Chengdu, China*
SHIYING YUAN • *Department of Critical Care Medicine, Union Hospital, Tongji Medical College, Huazhong University of Science and Technology, Wuhan, People's Republic of China*
MALGORZATA ZAREMBA • *Department of Experimental and Clinical Pharmacology, Centre for Preclinical Research (CePT), Medical University of Warsaw, Warsaw, Poland*
JIANCHENG ZHANG • *Department of Critical Care Medicine, Union Hospital, Tongji Medical College, Huazhong University of Science and Technology, Wuhan, People's Republic of China*

Part I

Pain Management

Chapter 1

Ketamine for Acute Postoperative Pain Management and Prevention of Chronic Postsurgical Pain: What Do Recent Meta-Analysis Teach?

Sebastian Labenbacher, Michael Eichlseder, and Helmar Bornemann-Cimenti

Abstract

Perioperative pain management is a fundamental aspect of anesthesia care. Immediate postoperative pain management is a challenging area with huge variability in pain influenced by the surgical procedure, preoperative use of opioids, preexisting disease, and many more factors. Especially orthopedic, gynecological, and thoracic operations prove to be highly painful. Postoperative pain results in complications including prolonged hospitalization, risk of pain chronification, and a higher rate of cardiovascular, pulmonary, and gastrointestinal complications. Despite the mounting evidence in favor of multimodal analgesia, the majority of drugs employed for postoperative pain management are opioids. Ketamine could therefore be a valuable adjunct for perioperative analgesia. We performed a narrative review of systematic reviews published between January 1, 1990, and February 29, 2024, on perioperative ketamine for postsurgical pain published in Embase (via Ovid), Ovid MEDLINE, and Cochrane Database of Systematic Reviews. The evidence on perioperative use of ketamine shows a relevant pain reduction extending beyond the immediate postoperative phase. Especially patients with high risk for severe, prolonged postoperative pain may benefit from perioperative ketamine. With vast dosing and application differences between studies, further research is required to find optimized use schedules.

Key words Postoperative pain, Perioperative ketamine, Chronic postsurgical pain, Pain prevention

1 Introduction

In their 2023 report, the perioperative quality improvement program conducted in the United Kingdom with 135 hospitals and nearly 45,000 patients states that up to 26% of patients still suffer from severe pain after surgery stratified to the procedure [1, 2]. The highest incidence was found after orthopedic (26%), spinal (23%), gynecological (20%), and thoracic surgery (20%). The problem is a global health issue: In a cross-sectional study of 122 Chinese hospitals from 2019 to 2021, nearly half of all patients (49%)

Sabine Himmelseher (ed.), *Ketamine: Pain Management, Critical Care, Depression, and Resource-Poor Settings Volume II*, Neuromethods, vol. 225, https://doi.org/10.1007/978-1-0716-4599-4_1,

experienced pain rated as 4 or above on a visual analogue scale (VAS) on the first postoperative day [3]. More than 10 years ago, a prospective cohort study in 105 German hospitals identified most painful procedures with regard to the first postoperative day (POD): In orthopedics, open calcaneus reduction, dorsal spinal fusion, complex spinal reconstruction, arthrodesis in the foot or hand joints, and shoulder joint replacement; in gynecology and obstetrics, myomectomy and caesarean section; and, in general surgery, proctocolectomy were associated with the highest intensity of pain [4]. Yet, with great disappointment, it has to be stated that acute postoperative pain is still grossly undertreated.

During postoperative care, severe pain impacts patients in all aspects of recovery, may cause extra health challenges, worsens prognosis, and reduces quality of life. Complications from inadequate postoperative pain therapy have high-cost implications to health and social care, such as prolonged hospitalization and unanticipated hospital readmission [5].

High-intensity pain conditions are associated with serious pathophysiological sequelae, including over-activation of the autonomic nervous system with numerous adverse effects: For example, in the cardiovascular system, pain increases heart rate and blood pressure and may enhance the cardiac workload with the risk of a supply and demand mismatch in vulnerable patients. As a consequence, myocardial ischemia and even (silent) infarction may develop. In the respiratory system, pain may impair breathing mechanics, reduce vital capacity, and cause atelectasis with the result of hypoventilation and hypoxemia. This is especially valid after thoracic procedures. In the gastrointestinal tract, pain may increase secretions via sympathetic activation on the one hand and reduce motility on the other hand. From a mental health perspective, severe postoperative pain may cause sleep disturbance, anxiety, fear, and depression. Early convalescence may be protracted and daunting, reducing all kinds of patient-centered outcomes, well-being, and satisfaction.

A major complication of acute postoperative pain is the development of chronic postsurgical pain (CPSP). In addition to the severity of acute postoperative pain, its duration and trajectory have been identified as important contributors [6]. The ICD-11 definition highlights CPSP as pain that develops after surgery or tissue injury and persists beyond the healing process, with a pain persistence of at least 3 months from surgery plus distribution relatable to the performed surgery. Other possible causes such as infection, malignancy, or loose hardware have to be ruled out [7].

However, this definition does not answer all questions related to the development of CPSP, especially those surrounding the multidimensional feature of chronic pain. Recent research has identified several risk factors for CPSP in various surgical procedures: These include younger age, female sex, previous cardiac surgery, and presurgical depression and anxiety [8, 9]. For risk assessment and to predict CPSP development, a score has been generated [10]. It is based on values of patient self-reported visual and numerical pain rating on the 14th POD, preoperative opioid use, bone surgery, and the presence of painful cold sensations in the operated area. To take these important factors into analytic account, core patient-centered outcome domains need to be addressed, including physical functioning, emotional status, and patient ratings of improvement and satisfaction with treatment.

In trials evaluating strategies to reduce postoperative pain, it is therefore of utmost importance to recognize the ubiquity of pain after surgery, individual patient- and procedure-related conditions, and modifiable "factors" in new trial designs. However, there is still a paucity of firm guidance with regard to pharmaceutical interventions that may improve pain care.

For decades, ketamine has been and is commonly used as potent analgesic agent in pre- and in-hospital emergencies, austere environments, disasters, and military operations, and anesthesia induction in the critically ill. Nonetheless, routine use for pain therapy in hospital is still often shied away from because of concerns about unwanted psychoactive impact.

This chapter examines present systematic reviews and meta-analysis on postoperative pain with perioperative ketamine as therapy including trials which aimed to prevent CPSP development. Our work will include evaluation and reporting of preexisting risk factors for severe pain, assessment of acute pain intensity, and patient-centered core outcomes if described. To increase our knowledge of clinical evidence, trials with improved benefit-risk ratio, long-term effects, and adverse psychic phenomenon will briefly be discussed.

2 Search Strategy

Our objective was to map, present, and synthesize contemporary meta-analytic evidence of primary clinical trials about perioperative ketamine as a therapy in multimodal analgesia for better postoperative pain care and prevention of CPSP. We developed our search strategy according to the Peer Review of Electronic Search Strategy (PRESS) recommendations with adaption for a narrative review of the presently available systematic reviews published from January 1, 1990, and February 29, 2024. We searched:

- Ovid MEDLINE.
- Embase (via Ovid).
- Cochrane Database of Systematic Reviews.

We included analyses assessing the effects of intraoperative, postoperative, and perioperative interventions with ketamine in adult patients undergoing surgery with general or regional anesthesia (RA). Eligible comparators were placebo, other analgesics, or no intervention. Analyses had to provide pooled results of at least one patient-centered outcome with regard to change in postoperative pain after ketamine or difference to the comparators at any length of follow-up. Only systematic reviews in the English or German language were eligible. To give a short overview, we summarized the included systematic reviews and their main results of interest in Table 1.

3 The Perioperative Period

3.1 Acute Postoperative Pain and Ketamine

In one of the largest existing reviews published in 2018, Brinck et al. [11] evaluated the effects of perioperative intravenous (IV) ketamine for acute postoperative pain in adults. They included 130 (120 racemic and 10 esketamine) trials (from 1996 to 2017) with 8341 patients and reported a significant reduction in pain of 5 on a 100 mm visual analogue scale (VAS) with an overall pain median of 25 mm. Racemic ketamine dosing schedules were highly variable with perioperative boluses ranging from <0.25 up to 1 mg/kg or infusion rates from 0.7 to 42 μg/kg/min. Trials using esketamine had an equally high dosing variability, applying boluses from 0.075 up to 0.5 mg/kg or infusion rates from 0.25 to 6.7 μg/kg/min. In addition to pain reduction, the time to first analgesic request was noted to be more than doubled with ketamine, averaging 93 min compared to 39 min in controls. The amount of morphine equivalents was slightly reduced by 8 mg during the POD. Importantly, Brinck et al. stratified their results according to the type of surgery, such as thoracotomy, major orthopedic or abdominal surgery, laparoscopy, and total abdominal hysterectomy. While a reduction in opioid needs was found throughout the first 48 h after all kinds of procedures except for laparoscopy, the decrease in pain at rest and movement at 24 and 48 h was not consistent and/or insignificant despite ketamine. After total abdominal hysterectomy, no decrease in opioid requests was observed, although there was less pain at rest 24 h after surgery. In laparoscopic procedures, ketamine did not induce a detectable decrease in pain [11].

Another systematic review focused on perioperative IV esketamine for acute postoperative pain in adults. It examined 12 studies with 905 patients published between 2002 and 2020, revealing a significant pain reduction. The highest reduction in pain was reported at 4 h after surgery and decreased to non-significance at 48 h. Similar to patient pain ratings, morphine consumption was reduced in the first 4 and 12 h following surgery but not at 24 and

Table 1
Systematic reviews on perioperative ketamine/esketamine administration and their main results

Systematic review	Number of RCTs/ patients	Time span of publications	Main results of interest
Brinck et al. [11]	130/8341	1996–2017	Perioperative ketamine reduced postoperative analgesic consumption and pain
Wang et al. [12]	12/905	2002–2020	Esketamine improved resting pain scores at 4, 12, and 24 h after surgery compared to placebo but did not affect pain scores during movement
Wang et al. [14]	20/1737	1997–2017	Ketamine reduced postoperative pain and analgesic requirements and prolonged the time to first analgesic request
Caouch et al. [17]	7/412	2009–2021	Ketamine reduced opioid consumption at 24 h and pain at 4 h post-surgery
Bi et al. [18]	13/1182	2008–2020	Ketamine reduced wound pain at 6 and 24 h post-surgery Ketamine lowered the incidence of PMPS
Viderman et al. [20]	8/896	2008–2017	Ketamine did not lower the incidence of postoperative delirium
Fellous et al. [21]	14/1618	2008–2021	Ketamine did not lower the incidence of postoperative delirium or postoperative cognitive dysfunction
Zhou et al. [23]	9/573	2013–2023	Ketamine does not change the incidence of postoperative cognitive dysfunction during the first 7 days after surgery
Meyer-Frießem et al. [24]	9/802	2007–2019	Ketamine reduced pain at movement during the first 48 h after surgery Ketamine reduced opioid requirements during the first 48 h post-surgery
Guo et al. [26]	15/3159	2002–2022	Ketamine reduced postoperative depression during the first 7 days and beyond
Ma et al. [27]	11/1855	2021–2023	Esketamine reduced the incidence of postpartum depression in the first 7 days after cesarean section
Xiang et al. [28]	20/1011	1996–2023	Ketamine as adjuvant to local anesthetics can prolong the duration of analgesia
Abouarab et al. [29]	36/3572	2001–2021	Ketamine may reduce the incidence of chronic postsurgical neuropathic pain at 3 months after surgery
Doleman et al. [30]	132/23902	1997–2021	Ketamine reduced the incidence of chronic postsurgical pain at 6 months post-surgery
Sun et al. [31]	20/1561	2001–2021	Ketamine reduced the incidence of chronic postsurgical pain at 3 and 6 months after surgery

PMPS postmastectomy pain syndrome, *RCT* randomized controlled trial

48 h [12]. Without surprise, a sub-analysis comparing intraoperative only with intra- and postoperative use revealed lower pain scores during the first POD after intra- and postoperative application.

3.1.1 Obstetric Surgery

As noted above, obstetric surgery has been associated with intense, acute (postoperative) pain. For example, in neuraxial anesthesia, it occurs more often than previously assumed: The American Society of Anesthesiologists points out that pain is prevalent during caesarean section in 12–22% of patients, with an up to 5% necessity of conversion to general anesthesia due to either insufficient spinal, epidural, or combined spinal-epidural anesthesia [13]. In previous years, ketamine has often been used as analgesic adjunct during obstetric anesthesia. Wang et al. conducted a systematic review on the impact of ketamine used during cesarean section involving 20 randomized controlled trials (RCT)s published until 2018 with 1737 patients. When ketamine was administered IV or together with bupivacaine in spinal anesthesia (SPA), a significant pain reduction and longer time until analgesic request was noted [14]. Ketamine doses ranged from 0.15 to 1 mg/kg with different use periods. However, in view of the neurodevelopmental toxicity debate, it is not clear what kind of (lasting) effects such an exposure to ketamine may have on fetal brain development. This will be discussed in a separate chapter in the book. For the time being, it appears to be prudent not to apply a higher than low subanesthetic ketamine dose in this setting. This is also valid with regard to the NIH lactation database report about racemic ketamine, stating that with appropriate time delay after one time use, the risk of harm for a baby drinking milk from a breastfeeding mother appears to be small given ketamine's low oral bioavailability [15]. In addition, even ketamine drug preparations without neurotoxic preservatives (including esketamine) are not approved for intrathecal use because of concerns of spinal toxicity. Therefore, intrathecal application should only be performed in selected cases, for example, in palliative care.

3.1.2 Bariatric Surgery

Bariatric patients have an elevated risk of postoperative complications. Relevant aspects include obstructive sleep apnea and obesity hypoventilation syndrome, which even increase the risk for perioperative mortality (4% vs 0.2% in controls) in open bariatric surgery [16]. Pain management is challenging, and maintaining respiratory drive is of utmost importance. Ketamine offers the advantage of high analgesic efficacy and minimal effects on respiration. A systematic review on ketamine as a component of multimodal analgesia for pain management in bariatric surgery included seven RCTs with 412 patients published between 2009 and 2021. It demonstrated not only an early reduction of pain scores 4 and 8 h after surgery but also a decrease in postoperative opioid need and shorter hospital stay [17]. There was no difference in postoperative nausea and vomiting (PONV), dizziness, hallucinations, and euphoria.

3.1.3 Breast Surgery

Breast surgery is a procedure where in particular better pain management is urgently needed. Postmastectomy pain syndrome (PMPS) is a complication that affects 20–68% of patients after breast surgery often involving nerve injury. It may result in chronic (neuropathic) pain localized in and around the surgical wound. A systematic review on ketamine use for acute and chronic wound pain in breast surgery included 13 RCTs with 1182 patients published between 2008 and 2020. Adding ketamine to local anesthetics or as IV adjunct to general anesthesia reduced pain at 6 and 24 h after surgery, opioid consumption, and postoperative depression at 1 week up to 1 month. Pooled data of five trials on PMPS showed a significant reduction after perioperative ketamine at 3 months and to some degree at 1 and 6 months after surgery [18]. This indicates a relatively long-lasting improvement in a patient-centered outcome variable.

3.2 Postoperative Neurocognitive Dysfunction and Delirium

Postoperative neurocognitive dysfunction (PND) and delirium have a major impact on patients, relatives, and caregivers. Pathogenesis is multifactorial, and confounders are difficult to differentiate in effect, including preexisting frailty, surgical trauma, reperfusion injury, inflammatory response, hemodynamic instability, inevitable use of deliriogenic drugs, and inappropriate anesthetic depth [19–21]. The psychoactive impact of ketamine is assumably dose-dependent, and the use of GABA-ergic substances during anesthesia reduces or prevents negative ketamine action. However, when ketamine is applied intraoperatively to prevent postoperative pain, it has to be stopped or decreased in dose early enough before the end of surgery. Otherwise, delayed awakening and unwanted psychic effects may occur and increase neurocognitive impairment during and after emergence.

A systematic review of 14 RCTs with 1628 patients published between 2008 and 2021 found no difference in the incidence of PND or delirium with perioperative ketamine. A sensitivity analyses adjusting for ketamine dose as well as bolus application versus continuous infusion revealed no difference as compared to placebo. The evaluation of other patient-centered outcomes, such as acute pain, respiratory depression, PONV, psychic adverse effects, length of stay, and mortality, did not detect differences between ketamine use and not [21]. In a letter to the editor, Hung et al. performed a trial sequential analysis showing that in order to detect or reject a 20% relative risk reduction, almost 6500 patients would have been necessary [22].

Viderman et al. conducted a systematic review with the title "the effect of ketamine on postoperative neurocognitive disorders" which included eight trials and 896 patients published between

2008 and 2017. The primary findings indicated no difference in the incidence of postoperative delirium but an increase in hallucinations with ketamine, which was primarily derived from one study. As observed in other analyses, less morphine was consumed with adjunctive ketamine. A possible limitation of this review, however, is the low number of RCTs assessed [20].

A recently published systematic review by Zhou et al. on the efficacy of subanesthetic doses of ketamine/esketamine for the prevention of postoperative cognitive dysfunction included nine RCTs published between 2013 and 2023. It did not show significant differences with respect to postoperative cognitive dysfunction in adults between days 1 and 7. A subgroup analysis of patients older than 60 years did not find significant results either [23].

3.3 Ketamine in Patients with Long-Term Opioid Use

Patients who have received preoperative opioids are of special concern in perioperative pain therapy especially after longer time opioid use. Opioid tolerance is a well-known problem with major perioperative implications because sufficient analgesia is often hard to accomplish. Meyer-Frießem et al. systematically reviewed RCTs studying perioperative ketamine for postoperative pain in patients with preoperative opioid intake. Nine RCTs published between 2007 and 2019 with 802 patients undergoing spine or orthopedic lower limb surgery who received opioids for at least 2 weeks up to several years before the operation were included. Although ketamine infusions affected pain intensity only minimally, considerable opioid sparing was observed when ketamine was administered as a bolus and continuously infused for the first POD. During this time, a mean reduction of 97.3 mg oral morphine was observed, which amounted to a reduction of 186 mg in 48 h. This was related to ketamine's anti-hyperalgesic properties [24]. In the eight RCTs reporting less postoperative opioid consumption with ketamine, four found no difference in the first 2 h after surgery and in the stay at the postoperative care unit. Nevertheless, the results are associated with an uncertain risk of bias for selective outcome reporting with high heterogeneity in all trials included.

3.4 Ketamine and Postoperative Depression

Depression is a disease with great relevance in perioperative care. A Canadian questionnaire study among 1061 patients revealed that females with preoperative depression had a higher risk for postoperative complications, extended hospitalization, and hospital readmission than those without [25]. Meanwhile, ketamine is established as effective antidepressant worldwide and has recently been used for this purpose in perioperative care. Guo et al. systematically reviewed the efficacy and safety of perioperative ketamine for postoperative depression including 15 RCTs with 3159 patients. IV ketamine was efficacious as antidepressant at the first, third, and seventh POD, as well as at long-term follow-up until the

42nd day after general or SPA. Postoperative pain was rated as less on the first, but not third, day after surgery. There was more PONV, headaches, hallucinations, and dizziness with ketamine. This may partially be attributed to ketamine's dosage which was mostly in the higher subanesthetic range around and above 0.5 mg/kg, with bolus doses from 0.1 to 1 mg/kg [26]. In addition, Ma et al. performed a systematic review on postpartum depression in patients undergoing caesarean section delivery who were randomized to IV intraoperative esketamine therapy. They included 11 trials published between 2021 and 2023 and found a significant risk reduction for postpartum depression in patients who received i.v. esketamine until the seventh POD [27]. The review by Bi et al. assessed three trials which looked at postoperative depression as an outcome after breast surgery: With ketamine, there was less depression during the first week and 1 month after surgery, but this must be interpreted with caution given the high trial heterogeneity [18].

4 Ketamine as Adjunct in Regional Anesthesia

Regional anesthesia may offer a solution to the challenge of providing effective analgesia while simultaneously reducing opioid intake. A disadvantage of single-shot RA is the limited duration of effect. Different adjuncts to local anesthetics have therefore been investigated: A systematic review with the title "efficacy and safety of ketamine as an adjuvant to RA" examined 20 trials which combined (es-)ketamine with local anesthetics in SPA (11 trials) or peripheral nerve block (nine trials). Ketamine doses ranged from 0.1 to 1 mg/kg, causing a significant effect prolongation in SPA (58 min) as well as in peripheral RA (367 min). However, there was considerable heterogeneity between the different RA techniques. In peripheral RA, motor blocks were not significantly extended, but sensory blocks lasted for approximately 29 min longer with ketamine. No differences were found between ketamine and placebo in onset of motor and sensory block. Dizziness was the only adverse event occurring more often after ketamine, while the incidence of PONV, nausea, vomiting, headache, or hallucinations did not differ. However, the overall quality of evidence was low because of high risk of publication bias and study heterogeneity [28].

5 Unwanted Ketamine Effects

A more in-depth examination of the systematic reviews discussed uncovers that unwanted (adverse) effects have not consistently been reported. This lack of uniformity or paying attention at all makes it difficult or rather impossible to evaluate the incidence and severity of adverse patient outcomes. Overall, this may result in an enormously distorted perspective.

The symptoms that were most frequently reported from ketamine's psychoactive action were hallucinations, nightmares, visual disturbances, sedation, dizziness, drowsiness, and confusion. PONV was reported most often as physical adverse postoperative outcome.

In a recent systematic review on the efficacy and safety of perioperative ketamine for the prevention of CPSP, Abouarab and co-workers tried to comprehensively examine adverse psychoactive effects as reported in 36 trials with a total of 3572 patients. They did not find a difference in the incidence of postoperative hallucinations in a comparison of ketamine with placebo, no treatment, or lidocaine. No differences were also reported with regard to the incidence of postoperative dizziness, sedation, and bad dreams. However, after ketamine, postoperative nystagmus and visual disturbances were more frequent [29]. In their review from 2018, Brinck et al. combined all psychoactive effects reported in the 130 analyzed trials on postoperative pain and found a 5% incidence after ketamine versus a 4% incidence in the controls (187/3614 versus 122/2924 patients) [11]. Wang and co-workers also reported no significant differences with regard to CNS effects in their review of esketamine for acute postoperative pain in adults [12]. As outlined above, Guo et al. related the relatively high incidence for hallucinations (relative risk (RR): 15.35, 95% confidence interval (CI): 6.24–37.34), dizziness (RR: 3.48, 95% CI: 2.68–4.50), and headache (RR: 2.47, 95% CI: 1.42–4.32) to the relatively high ketamine doses administered: Boluses up to 1 mg/kg were applied [26]. Among adverse physical phenomenon, PONV is a common burden in general anesthesia. Comparing adjunctive ketamine versus control in 95 studies revealed that 761 of 3263 patients receiving ketamine (23%) and 731 of 2702 patients receiving placebo (27%) experienced PONV (RR: 0.88, 95% CI: 0.81–0.96) [11]. Wang et al. and Fellous et al. were unable to reproduce the reduction in PONV observed by others with the use of esketamine [12, 21]. In fact, the meta-analysis by Guo et al. found a higher incidence of PONV [26]. Further side effects that were inconsistently mentioned include respiratory depression and various psychological phenomena. For both, Fellous et al. did not find a different incidence [21] as well as Meyer-Frießem and co-workers focusing on respiratory depression [24].

6 Perioperative Ketamine for Prevention of Chronic Postoperative Pain

A recent systematic review of non-opioid analgesics for the prevention of CPSP included 132 trials with 23,902 patients, reporting that ketamine was one of the most studied substances [30]. The work shows that ketamine was associated with "second-best" effects in the reduction of CPSP for up to 6 months after surgery (only lidocaine was associated with more pain relief). Administration schedules varied considerably, with dosage ranging from 2 to 500 μg/kg/h over a time span "just covering surgery" up to 96 h after the procedure. While some trials started ketamine with a bolus, others directly applied a continuous infusion. Four studies assessed long-term ketamine effects but could not find reduced CPSP beyond 6 months. There was no decrease in opioid need during this time [30]. The review also identified a variable baseline risk as a potential violation to the network meta-analysis transitivity assumption. This caused authors to report from a fixed value for risk, which results in less confidence in the findings because of possible bias and imprecision. In the review by Sun et al. with 20 trials and 1561 patients, CPSP was significantly reduced for 3–6 months after perioperative ketamine or esketamine across multiple surgical procedures. There was no increase in the incidence of PONV and a trend toward more hallucinations [31]. However, Abouarab et al.'s work could not show a difference in CPSP at 3, 6, or 12 months after surgery when all selected ketamine trials were analyzed. Nevertheless, the development of chronic neuropathic postsurgical pain was reduced with perioperative ketamine up to a maximum of 3 months after surgery [29].

7 Limitations

First of all, we did not undertake a formal risk-of-bias analysis of our scoping description of systematic reviews and meta-analysis, and we relied on the narrative conclusions of the authors. Secondly, the reviews mapped were focused on the outcome measure "acute or chronic pain" but often did not pay attention to the multidimensional character of pain. Our work reveals a substantial degree of variability in ketamine regimen used in clinical different conditions and circumstances. To demonstrate the immense range of doses reported, a diagram (Fig. 1) indicates the various doses that have been applied to a patient with 80 kg body weight as a representative example. To prevent confusion, the distinction between racemic ketamine and its isomers with regard to dosage needs to be clear: Approximately twice the analgesic effects can be expected with dextrorotatory esketamine at the same "drug quantity" as racemic ketamine, because it offers the advantage of being twice as potent as analgesic and anesthetic as the racemate [32]. Levorotatory arketamine, which is half as potent as analgesic as racemic ketamine, has rarely been examined in humans so far.

Overall, only a few trials were multicentered, all were with a very variable duration of follow-up, and several did not report a registration or protocol publication. The efficacy of pain therapy

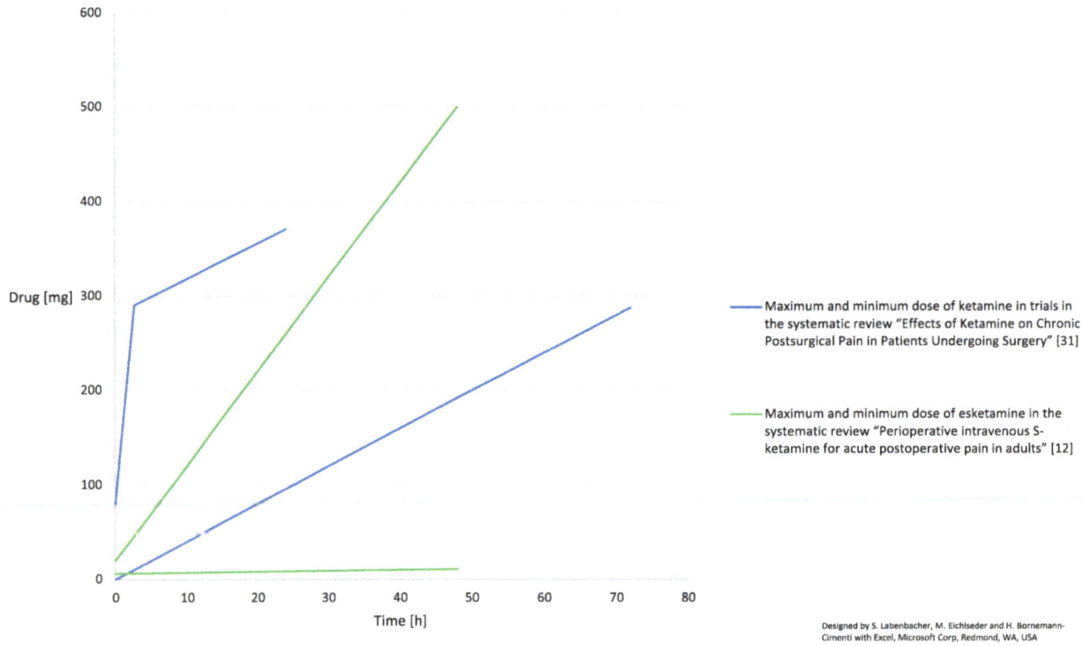

Fig. 1 The immense range of perioperative ketamine doses used for postoperative pain reduction and prevention of CPSP in RCTs. For postoperative pain, maximum and minimum doses (mg) of IV esketamine are indicated as green lines [12], and for prevention of CPSP, maximum and minimum doses of IV racemic ketamine are shown as blue lines [31]

was typically measured with subjective patient pain ratings and opioid consumption. However, these methods have many pitfalls: For example, rating of pain in different scales leads to inconsistency in systematic reviews' calculation. As an "objective" surrogate for pain, the opioid need quantified in morphine equivalents has often been used. Opioid consumption, however, exhibits a non-Gaussian distribution with parametric tests being inapplicable [33]. Sometimes, placebo-treated patients received (rescue) medication for pain relief. Although this is important in clinical quality care, it may diminish and mask effects of study drugs, as does RA or multimodal analgesia. It has also not been clarified whether the timepoint of ketamine application, i.e., before versus after opioid administration, has an effect on its analgesic potency. Ketamine's perioperative effects appear to wear off variably, and maintenance of antinociceptive treatment gain is only possible with repeated or prolonged use in perioperative care [32]. Nevertheless, we could not clarify which criteria for use were applied in longer-term administration schedules.

The incidence and severity of the psychopathological phenomena reported did not seem to be different in patients receiving ketamine as compared to those without ketamine. However, no review clearly differentiated between arousal from anesthesia and effects from the other drug combinations in comparison to possible ketamine-related psychoactive impact. There was no measurement of drug plasma concentrations or those of ketamine's metabolites, although there are multiple active metabolites with their own (analgesic) action.

Finally, our review identified a substantial knowledge gap or "no-reporting" of patient-centered outcome domains such as patient satisfaction, well-being, quality of life, or functional capabilities in meta-analysis. Nevertheless, most recent original work reports improved quality of recovery scores and other specific outcome measure domains such as patient satisfaction with perioperative esketamine, including better analgesia [34–36].

8 Conclusion

Our work demonstrates perioperative racemic and esketamine's therapeutic potential for reduction of acute and long-term postsurgical pain after multiple surgical procedures. Ketamine may successfully be used not only as an analgesic adjunct in general anesthesia at subanesthetic dose but also as a "complementary measure" in RA. To improve ketamine's benefits, preoperative risk factors and constellations for expectedly higher postoperative pain, such as individual patient disposition and surgical procedure-related conditions, have to be considered as well. According to the pathophysiological cascade that develops after severe tissue injury with nociceptive stimuli lasting over longer time [37] and the clear dose-effect relationship in acute pain, ketamine protocols that cover intraoperative and at least early postoperative time periods will likely be more effective. Regimen in the low subanesthetic dose range will not induce typical ketamine-related psychoactive effects but may be insufficient in terms of pain reduction. Regimen in the higher subanesthetic dose range may be associated with improved pain relief but more psychoactive impact. This has to be managed with "skillful" ketamine use (such as including a GABA agonist in awake patients). Future trials need to incorporate patient-centered outcomes to better evaluate patient and health system relevancy of ketamine. The ongoing Australian ROCKet trial examines perioperative ketamine in a study population of 4000 surgical patients with use regimen of up to 72 h after surgery. The results of this study will hopefully bring further insights regarding effects and optimization for ketamine use strategies.

Disclaimer

After completing the manuscript, AI-assisted language improvement for better readability was performed with DeepL (DeepL SE, Cologne, Germany).

Conflicts of Interest Disclosure

SL, ME, and HBC report no conflict of interest.

References

1. Perioperative Quality Improvement Programme (2023) Report 4 July 2021 to March 2023. https://pqip.org.uk/FilesUploaded/PQIP-Report2023_Finalversion_140623.pdf. Accessed 10 May 2024
2. Moonesinghe SR, McGuckin D, Martin P et al (2022) The perioperative quality improvement programme (PQIP patient study): protocol for a UK multicentre, prospective cohort study to measure quality of care and outcomes after major surgery. Perioper Med 11:1–10
3. Liu Y, Xiao S, Yang H et al (2023) Postoperative pain-related outcomes and perioperative pain management in China: a population-based study. Lancet Reg Heal – West Pacific 39:1–12
4. Gerbershagen HJ, Aduckathil S, van Wijck AJM, Peelen LM, Kalkman CJ, Meissner W (2013) Pain intensity on the first day after surgery. Anesthesiology 118:934–944
5. Joshi GP, Ogunnaike BO (2005) Consequences of inadequate postoperative pain relief and chronic persistent postoperative pain. Anesthesiol Clin North Am 23:21–36
6. Rosenberger DC, Pogatzki-Zahn EM (2022) Chronic post-surgical pain – update on incidence, risk factors and preventive treatment options. BJA Educ 22:190–196
7. World Health Organization (2022) International classification of diseases eleventh revision (ICD-11). World Health Organization, Geneva
8. Xiao MZX, Khan JS, Dana E et al (2023) Prevalence and risk factors for chronic postsurgical pain after cardiac surgery: a single-center prospective cohort study. Anesthesiology 139:309–320
9. Zhang Y, Zhou R, Hou B et al (2022) Incidence and risk factors for chronic postsurgical pain following video-assisted thoracoscopic surgery: a retrospective study. BMC Surg 22:1–8
10. van Driel MEC, van Dijk JFM, Baart SJ, Meissner W, Huygen FJPM, Rijsdijk M (2022) Development and validation of a multivariable prediction model for early prediction of chronic postsurgical pain in adults: a prospective cohort study. Br J Anaesth 129:407–415
11. Brinck ECV, Tiippana E, Heesen M et al (2018) Perioperative intravenous ketamine for acute postoperative pain in adults. Cochrane Database Syst Rev 12:CD012033
12. Wang X, Lin C, Lan L, Liu J (2021) Perioperative intravenous S-ketamine for acute postoperative pain in adults: a systematic review and meta-analysis. J Clin Anesth 68:110071
13. American Society of Anesthesiologists (2023) Statement on pain during cesarean delivery. Edited by Committee on Obstetric Anesthesia. American Society of Anesthesiologists. https://www.asahq.org/standards-and-practice-parameters/statement-onpain-duringCesareandelivery. Accessed 30 June 2024
14. Wang J, Xu Z, Feng Z, Ma R, Zhang X (2020) Impact of ketamine on pain management in cesarean section: a systematic review and meta-analysis. Pain Physician 23:135–148
15. Ketamine (2023) Drugs lact database. https://www.ncbi.nlm.nih.gov/books/NBK500566/. Accessed 27 Mar 2024
16. Chau EHL, Mokhlesi B, Chung F (2013) Obesity hypoventilation syndrome and anesthesia. Sleep Med Clin 8:135–147
17. Chaouch MAMA, Daghmouri MA, Boutron MC et al (2022) Ketamine as a component of multimodal analgesia for pain management in bariatric surgery: a systematic review and meta-analysis of randomized controlled trials. Ann Med Surg 78:103783
18. Bi Y, Ye Y, Zhu Y, Ma J, Zhang X, Liu B (2021) The effect of ketamine on acute and chronic wound pain in patients undergoing breast

surgery: a meta-analysis and systematic review. Pain Pract 21:316–332

19. Pérez-Otal B, Aragón-Benedí C, Pascual-Bellosta A et al (2022) Neuromonitoring depth of anesthesia and its association with postoperative delirium. Sci Rep 12:1–9
20. Viderman D, Aubakirova M, Nabidollayeva F et al (2023) Effect of ketamine on postoperative neurocognitive disorders: a systematic review and meta-analysis. J Clin Med 12:4314
21. Fellous S, Dubost B, Cambriel A, Bonnet MP, Verdonk F (2023) Perioperative ketamine administration to prevent delirium and neurocognitive disorders after surgery: a systematic review and meta-analysis. Int J Surg 109:3555–3565
22. Hung KC, Chen IW, Liu PH (2024) Impact of ketamine on postoperative delirium: insights from trial sequential analysis. Int J Surg 110: 1834–1835
23. Zhou N, Chen C, Liu Y, Yu Z, Chen Y (2023) Efficacy of intraoperative subanesthetic dose of ketamine/esketamine in preventing postoperative cognitive dysfunction: a systematic review and meta-analysis. Ther Adv Psychopharmacol 13:1–12
24. Meyer-Frießem CH, Lipke E, Weibel S et al (2022) Perioperative ketamine for postoperative pain management in patients with preoperative opioid intake: a systematic review and meta-analysis. J Clin Anesth 78:110652
25. Geoffrion R, Koenig NA, Zheng M et al (2021) Preoperative depression and anxiety impact on inpatient surgery outcomes: a prospective cohort study. Ann Surg Open 2:e049
26. Guo J, Qiu D, Gu H et al (2023) Efficacy and safety of perioperative application of ketamine on postoperative depression: a meta-analysis of randomized controlled studies. Mol Psychiatry 28:2266–2276
27. Ma S, Dou Y, Wang W et al (2024) Association between esketamine interventions and postpartum depression and analgesia following cesarean delivery: a systematic review and meta-analysis. Am J Obstet Gynecol MFM 6:101241
28. Xiang J, Cao C, Chen J et al (2024) Efficacy and safety of ketamine as an adjuvant to regional anesthesia: a systematic review and meta-analysis of randomized controlled trials. J Clin Anesth 94:111415
29. Abouarab AH, Brülle R, Aboukilila MY, Weibel S, Schnabel A (2024) Efficacy and safety of perioperative ketamine for the prevention of chronic postsurgical pain: a meta-analysis. Pain Pract 24:553–566
30. Doleman B, Mathiesen O, Sutton AJ, Cooper NJ, Lund JN, Williams JP (2023) Non-opioid analgesics for the prevention of chronic postsurgical pain: a systematic review and network meta-analysis. Br J Anaesth 130:719–728
31. Sun W, Zhou Y, Wang J et al (2023) Effects of ketamine on chronic postsurgical pain in patients undergoing surgery: a systematic review and meta-analysis. Pain Physician 26: E111–E122
32. Mion G, Himmelseher S (2024) Esketamine: less drowsiness, more analgesia. Anesth Analg 139:78–91
33. Moore RA, Ní Mhuircheartaigh RJ, Derry S, McQuay HJ (2011) Mean analgesic consumption is inappropriate for testing analgesic efficacy in post-operative pain: analysis and alternative suggestion. Eur J Anaesthesiol 28: 427–432
34. Hung KC, Kao CL, Ho CN et al (2024) The impact of perioperative ketamine or esketamine on the subjective quality of recovery after surgery: a meta-analysis of randomised controlled trials. Br J Anaesth 132(6):1293–1303
35. Huang Z, Liu N, Hu S, Ju X, Xu S, Wang S (2023) Effect of Dexmedetomidine and two different doses of Esketamine combined infusion on the quality of recovery in patients undergoing modified radical mastectomy for breast cancer – a randomised controlled study. Drug Des Devel Ther 17:2613–2621
36. Qian XL, Li P, Chen YJ, Xu SQ, Wang X, Feng SW (2023) Opioid free total intravenous anesthesia with Dexmedetomidine-Esketamine-lidocaine for patients undergoing lumpectomy. J Clin Med Res 15:415–422
37. Himmelseher S, Durieux ME (2005) Ketamine for perioperative pain management. Anesthesiology 102:211–220

Chapter 2

Perioperative Esketamine and Pain Relief: Endeavors to Optimize Use Schedules, Combination Regimen, Opioid-Sparing, and the Strive Not to Negatively Impact Upon Natural Postsurgical Healing Processes

Huijing Xie, Min Xie, Wei Ma, Qian Li, and Tao Li

Abstract

Esketamine, a potent N-methyl-D-aspartate (NMDA) receptor antagonist with a higher affinity than worldwide used racemic ketamine, has emerged as an important analgesic agent in perioperative pain management. Especially when the use of traditional analgesics is challenging, such as in care of pediatric patients and conscious analgesia and sedation, esketamine is of great value. This chapter reviews the pharmacological profile of esketamine, focusing on its efficacy to provide improved pain relief; the opportunity of administration via diverse routes including intravenous, intranasal, and intrathecal application; and its potential to reduce opioid consumption. Esketamine has little impact on respiratory function, which makes it a preferable option in perioperative pain therapy, enhancing patient comfort without prolonging hospital stay. It is helpful for reducing the incidence of opioid-induced adverse effects and may alleviate chronic pain exacerbated by opioid dependence and hyperalgesia. However, despite all benefits, esketamine's clinical use must be carefully managed due to possible adverse psychoactive effects. The chapter also highlights the importance of continued study of esketamine in large-scale, multicenter trials to elucidate regimen with best possible analgesic efficacy and safety. This may significantly improve patient outcomes.

Key words Esketamine, Perioperative pain, Combination regimen, Opioid-sparing, Postoperative healing

1 Introduction

(S)-Ketamine and (R)-ketamine are optical enantiomers of the equimolar isomer mixture of racemic ketamine [1]. Esketamine (ESKET), the dextrorotatory enantiomer, acts as an N-methyl-D-aspartate (NMDA) receptor antagonist while also interacting with opioid, muscarinic, and monoaminergic receptors. ESKET exhibits an approximately three- to fourfold greater affinity for NMDA

The author Huijing Xie and Min Xie have equally contributed to this chapter.

Sabine Himmelseher (ed.), *Ketamine: Pain Management, Critical Care, Depression, and Resource-Poor Settings Volume II*, Neuromethods, vol. 225, https://doi.org/10.1007/978-1-0716-4599-4_2,

receptors compared to racemic ketamine: This is associated with an approximately two times greater analgesic and anesthetic potency. ESKET is metabolized more rapidly than the racemate which leads to a shorter (postoperative) recovery time, emergence from the drug, and a shorter return of neurocognition. ESKET causes less impairment of logical functions than the racemate and induces less unpleasant psychoactive effects at half-dose racemic ketamine. It provides analgesia without or minimal cardiorespiratory depression. This pharmacological profile makes the substance very suitable for minor surgery, for example, in pediatric patients, and for procedures with conscious analgesia and sedation.

ESKET is also effective in reducing opioid-induced hyperalgesia (such as with remifentanil) and perioperative opioid consumption. Opioids are commonly employed for perioperative analgesia. However, opioid use may result in a range of adverse effects, including respiratory depression, constipation, and tolerance and addiction with longer-term use. Prolonged opioid use after surgery has been associated with a 64% increase in all-cause mortality in the United Kingdom (hazard ratio: 1.64) [2]. If skillfully applied, ESKET reduces perioperative opioid requirements, thereby decreasing opioid-related adverse effects. It may also prevent the development of chronic postsurgical pain, which has often led to opioid dependence.

ESKET is versatile in its administration, and employable via intravenous (IV), intramuscular, oral, intranasal (IN), intrathecal, and intrarectal routes. To achieve best possible antinociception with few side effects, different doses and use regimen are explored. When administered perioperatively, IV subanesthetic low-dose ESKET is usually applied as an adjunct to a multimodal analgesia regimen within general anesthesia to improve intra- and postoperative pain management. A combination of ESKET and other analgesic and sedative drugs, such as dexmedetomidine, ropivacaine, and pregabalin, can reduce perioperative pain, better maintain hemodynamic stability, and reduce opioid needs.

A patented new device for an IN ESKET preparation has recently been approved for use as antidepressant in several countries. It is considered as attractive alternative to IV administration. Its application allows for little invasive drug administration with rapid drug absorption at a bioavailability of slightly more than 50%. However, as with every IN solution, a part of the rest gets uncontrollably swallowed. This part is subject to first-pass hepatic metabolism, so that approximately 20% will reach the systemic circulation [3]. In addition, in view of the fact that some experts are of the opinion that different metabolites of ketamine and ESKET will be formed after IV versus IN use with various pharmacologic actions of their own and that there are no valid pharmacokinetic/pharmacodynamic models that could simulate such scenarios, it is at present not possible to directly compare an IV racemic ketamine dose with an IN ESKET dose.

Inflammation is a critical homeostatic mechanism used by the human body to heal tissue injury. Some analgesics, such as nonsteroidal anti-inflammatory drugs (NSAIDs) and opioids, can disrupt the inflammatory balance and may lead to delayed healing or healing with injury such as scars [4, 5]. Beneficial anti-inflammatory effects of ketamine have been reported which may regulate inflammatory and immune disbalances: Ketamine inhibits the production of immune reaction-induced transcription factors and proinflammatory cytokines, including nuclear factor κB, and decreases blood levels, for example, of tumor necrosis factor-α, interleukin 6, C-reactive protein, and inducible nitric oxide synthase [6–8]. Notably, ketamine's anti-inflammatory effects occur in the presence of an existing immune stimulus maintaining an appropriate inflammatory response. This feature suggests that ketamine/ESKET may not negatively impact upon natural postsurgical healing processes. For the purpose of this review, the term "ESKET" will refer to the dextrorotatory isomer S(+)-ketamine and "ketamine" to the racemic ketamine mixture.

2 Pain Relief

2.1 Analgesic Mechanisms

ESKET's analgesic mechanisms of action are akin to that of ketamine. The well-characterized analgesic effects are primarily attributed to NMDA receptor inhibition [9]. Ketamine interacts with dopamine, serotonin, sigma, opioid, and cholinergic receptors [8]. It has been reported that ESKET affects connectivity in brain areas involved in endogenous pain inhibition [10]. Various channels, such as Ca^{2+}-activated K^{+} channels, voltage-activated K^{+} channels, and hyperpolarization-activated cyclic nucleotide-gated channels, are also implicated in its analgesic actions [11]. As outlined, ESKET is a more potent NMDA receptor antagonist than its stereoisomer arketamine or the racemate [12], with strong analgesic effects and weak respiratory depression.

2.2 Perioperative Analgesia

2.2.1 Pain Management in Pediatric Patients

Intranasal Administration

The delivery of analgesics may sometimes prove challenging due to problems with establishing an IV access in pediatric patients. In 2019, an IN ESKET spray was approved in the USA, Canada, and Europe for major depressive disorder and treatment-resistant depression [13, 14]. IN administration also is an excellent route for ESKET use in children offering a high bioavailability [8]. After tonsillectomy with low-temperature plasma ablation, nasal ESKET drops, i.e., two drops (~1.25 mg/drop) per nostril at a time, once in the morning and once in the evening on each postoperative day (POD), have been found to significantly reduce pain at POD 1–3 in children at 3–10 years of age [15]. In addition, a case series has reported positive experiences with IN ESKET for prehospital analgesia in nine especially challenging cases including six children.

According to prehospital healthcare professionals' impression, IN ESKET at a total maximum dose of 1.5 mg/kg body weight (BW) offered a fast, easy, and essentially noninvasive opportunity to reduce acute pain after traumatic injuries without observable unwanted effects. Nevertheless, this was not a randomized controlled trial (RCT), and therefore the preliminary findings cannot be generalized [16]. However, they represent a starting point for a clinical study.

Intravenous Administration

Most commonly, ESKET is administered via the IV route. The combined use of low-dose 0.3 mg/kg ESKET during anesthesia induction and in the postoperative period via patient-controlled IV analgesia at 1 mg/kg together with hydromorphone was reported to effectively alleviate postoperative pain after pediatric urological procedures in 3–6.5-year-old children. The study also reported that ESKET significantly reduced the incidence of adverse effects, such as nausea and vomiting [17]. However, when a single bolus dose of 0.46 mg/kg ESKET at average was administered at anesthesia induction in 2–7-year-old preschool children undergoing minor surgical procedures, no significant differences in pain scores versus placebo were noticed. Moreover, ESKET increased the incidence of emergence delirium [18]. This comes without surprise: In short-term surgery, acute effects of this relatively high subanesthetic dose will not have dissipated until the end of the procedure. Children were additionally allowed to receive dezocine as analgesic, which may have masked positive ESKET effects. For all beneficial uses of ESKET and especially in pediatric patients, it is an absolute necessity that its dose and administration regimen is adapted to the respective scenario, conditions, circumstances, and individual patient dispositions. Ketamine effects are always context- and dose-dependent.

2.2.2 Pain Management in Adult Patients

Intravenous Administration

ESKET counters opioid-induced respiratory depression [19], and ESKET and ketamine exert respiratory-stimulating effects [20]. ESKET is therefore used in analgesia and sedation regimen without intubation and maintained spontaneous breathing. For example, propofol combined with ESKET reduces pain during magnetic resonance imaging-guided high-intensity focused ultrasound ablation for uterine fibroids [21]. In acute ischemic stroke patients receiving endovascular stroke treatment, the combination of propofol and esketamine was proven to be feasible and safe without causing delay in periinterventional management [22]. In women undergoing cesarean delivery with epidural anesthesia, 0.25 mg/kg IV ESKET administered before incision provided transient analgesia and sedation but did not induce significant neonatal depression [23]. In patients undergoing painless gastroscopy, ESKET at half-dose racemic ketamine showed the same efficacy as the racemate but allowed for shorter recovery and orientation time in routine care [24]. In this setting, the use of one single dose of 0.5 mg/kg ESKET was generally safe and tolerable.

In 2021, a meta-analysis of 12 trials came to the conclusion that IV ESKET improved pain scores at rest as assessed with the visual analogue scale or numerical rating scale at 4, 12, and 24 h after surgery (4 h: standardized mean difference (SMD) −1.11; 12 h: SMD −0.88; 24 h: SMD −0.39). Administration times were crucial variables in the effect evaluation: A subgroup analysis of intra- and postoperative administration showed significantly reduced pain scores at 24 h after surgery at rest but an increased risk of psychotomimetic adverse events compared with intraoperative use only [25]. In another RCT, continuous ESKET infusion at 0.3 mg/kg/h during laparoscopic cholecystectomy provided better pain relief at 4 and 12 h after surgery than placebo [26]. These results show that skillful use of ESKET is beneficial in the management of acute postoperative pain. In addition, in patients with chronic pain and opioid dependence undergoing lumbar fusion surgery, intraoperative ESKET (0.5 mg/kg bolus and 0.25 mg/kg/h infusion) reduced chronic pain at 6 months after surgery [27]. However, in opioid-naïve patients undergoing major lumbar fusion surgery, intraoperative ESKET used at a comparable dose to the above-cited study had no significant effect on postoperative pain at 48 h after surgery [28]. This difference may in part be explained by ESKET's beneficial effects on opioid-induced tolerance and dependency but also by lack of longer-term study observation time and/or lack of patient-individualized approaches before comparing opioid-dependent versus opioid-naïve patients. In various patient-controlled IV analgesia settings, a drug combination regimen consisting of low-dose ESKET and an opioid provided enhanced analgesic effects [29, 30].

Intrathecal and Epidural Administration

Research on intrathecal and epidural administration of ESKET is limited. In patients undergoing video-assisted thoracoscopic surgery in opioid-free anesthesia, intraoperative replacement of opioids by IV ESKET injection and epidural infusion combined with postoperative epidural infusion reduced the incidence of mild chronic postoperative pain at 6 months after surgery and unwanted postoperative effects [31]. Another study reported that the 90% effective dose (ED90) value of epidural ESKET co-administered with 0.075% ropivacaine for labor analgesia in nulliparous parturients was ~1.0 mg/mL. However, the study results also showed that epidural ESKET was dose-dependently associated with unwanted effects such as mild dizziness, especially at doses up to 1.0 mg/mL [32]. Nonetheless, ESKET has not officially been approved for clinical use at the epidural or caudal let alone intrathecal route because of serious concerns about spinal neurotoxicity. While neurotoxic effects in various animal models vary, even one epidural injection of preservative-free ESKET in a dog model showed severe neurotoxic effects [33]. It is therefore prudent not to use epidural ESKET outside of palliative care settings, limited patient life expectancy, and desperate situations.

3 Opioid-Sparing

Long-term harms from prescription opioids, including addiction, misuse, and increased mortality, have become much more obvious in recent years [34]. Opioid-sparing anesthesia (OSA) strategies have emerged as a realistic alternative to opioid-based anesthesia. OSA can enhance postoperative recovery by reducing serious opioid-related side effects and improves patient satisfaction. Numerous studies have applied low-dose ESKET in perioperative settings with OSA, demonstrating the opportunity of reducing opioid requirements. In the already mentioned meta-analysis, consumption of morphine was significantly reduced after perioperative ESKET at 4 h (SMD −0.98) and at 12 h after surgery (SMD −1.36) [25]. In opioid-dependent patients, intraoperative ESKET significantly decreased morphine needs in patient-controlled analgesia until 24 h after lumbar fusion surgery [27].

Further adverse effects of opioids in perioperative pain management include opioid-induced hyperalgesia and tolerance phenomenon. In addition, patients who present for surgery and have used opioids for a longer time before are likely to have aberrant somatosensory responses to painful stimuli [34]. A meta-analysis including nine RCTs with 802 patients with at least 2 weeks of opioid intake before surgery suggested that ketamine might be a useful anti-hyperalgesic adjuvant in these patients' perioperative care. A clinically relevant opioid-sparing effect was evident associated with a reduced risk of postoperative sedation and no increased risk for harm. However, given the considerable clinical heterogeneity of the RCT included and the different pharmacological effects of ketamine and ESKET, it is not yet possible to propose a specific ESKET use protocol to most effectively approach preoperative opioid consumers [35].

In human volunteers, remifentanil-induced post-infusion hyperalgesia was abolished by ESKET, suggesting NMDA receptor-related downstream mechanisms [36]. ESKET may also be useful for reduction of opioid tolerance in desperate settings. In a case report, intrathecal ESKET in a chronic pain patient with morphine tolerance led to significant pain relief and, consecutively, to an enormous, possible reduction of morphine requirements. After cessation of 24-day ESKET administration, the gain in pain relief persisted for weeks [37], and the patient could later be treated with neuromodulation modalities.

4 Endeavors for Optimizing Use Schedules

4.1 Dosage and Route of Administration

As outlined, ESKET can be administered IV, intramuscularly, subcutaneously, orally, nasally, caudally, or even rectally. Unfortunately, even preservative-free solutions may rather not be considered for epidural or intrathecal administration because of concerns of spinal neurotoxicity. The dose required depends on the therapeutic effect that shall be achieved and the route of administration. A double-blind, crossover, three-way placebo-controlled study found that IV ESKET was approximately twice as potent as IV racemate in inhibiting temporal and spatial summation of pain at a plasma concentration of 350 ng/mL racemate versus 180 ng/mL esketamine [38]. In acute pain after orofacial surgery, esketamine plasma concentration and postoperative pain reduction strictly correlated. The analgesic effect was strongest when the plasma concentration reached about 0.5 μM/L [39].

4.2 The Definition of Low-Dose Esketamine

At first, ketamine was used as mono-anesthetic for general anesthesia at anesthetic dose also inducing analgesia. However, over the years, it became clear that ketamine exerts analgesic effects at lower, subanesthetic dose as well. Joel Katz et al. therefore proposed that a distinction must be made between the use of high-dose ketamine as an anesthetic agent and the use of lower-dose ketamine for antinociceptive, analgesic, or anti-hyperalgesic purposes [40]. A single IV ESKET dose < 0.5 mg/kg BW can be defined as subanesthetic dose, and < 0.3 mg/kg BW as low dose, with sole anti-hyperalgesic doses being even smaller [1].

4.3 The Administration of Low-Dose Esketamine

Primarily, the analgesic effects of low- and subanesthetic-dose ESKET are related to drug dose, i.e., bolus or load dose, and infusion rate/dose. In addition, its analgesic efficacy must always be evaluated within the context of administration, i.e., for example, painful versus less painful surgery, duration of procedure, individual (genetic) patient disposition, patient comorbidities and age, preexisting risk factors for enhanced pain, use of other analgesics and/or multimodal analgesia, combination regimen with other sedatives and anesthetics, heterogenous study design, rationale for dose selection, and many more influencing factors. Some of the various doses investigated in diverse patient trials and settings are summarized in Table 1.

All of these diverse components affect the analgesic efficacy of ketamine and ESKET not only with respect to pain but also with regard to general patient outcomes from surgery. For example, only a small part of the trials report intraoperative anesthesia monitoring although it is well known that, for example, inhalational anesthetics and opioids can intraoperatively be reduced when low-dose ESKET is applied as analgesic adjunct in general anesthesia. Moreover, most

Table 1
Uses and dosages of esketamine

Purpose	Setting	Design	Sample size	Dosing	References
Pain management in pediatric patients	Tonsillectomy, postoperative analgesia: IN	RCT	76	Two drops ESKET per nostril at a time, once in the morning and once in the evening daily	[15]
	Mountain rescue, prehospital analgesia: IN	Case series	9	0.5 mg/kg ESKET as first dose, thereafter additional doses as needed; total dose limited to 1.5 mg/kg in children	[16]
	Urological surgery, postoperative analgesia: IV	RCT	80	Anesthesia induction: 0.3 mg/kg ESKET as analgesic adjuvant. Postoperative: PCA with 1.0 mg/kg ESKET and 0.1 mg/kg hydromorphone	[17]
	Minor surgery, postoperative analgesia: IV	Prospective observational study	230	Mean ESKET dose used: 0.46–1 mg/kg	[18]
Pain management in adult patients	Uterine fibroid surgery, intraoperative analgesia: IV	Pilot study	20	Before start of operation: Single 0.25 mg/kg ESKET bolus then via infusion: 5 mg/kg/h propofol and 0.2 mg/kg/h ESKET	[21]
	Cesarean section, perioperative analgesia: IV	RCT	600	0.25 mg/kg ESKET injected over 1 min	[23]
	Gastroscopy, intraprocedural analgesia: IV	RCT	32	Equipotent comparison: 0.5 mg/kg ESKET versus 1 mg/kg racemic ketamine	[24]
	General anesthesia, postoperative analgesia: IV	Systematic review and meta-analysis of RCTs	905	Dose range of bolus: 0.075–0.5 mg/kg Dose range of infusion: 0.075–0.6 mg/kg/h ESKET	[25]
	Laparoscopic cholecystectomy, postoperative analgesia: IV	RCT	48	Intraoperative infusion: 0.3 mg/kg/h ESKET	[26]
	Spinal fusion, surgery, postoperative analgesia: IV	RCT	147	0.5 mg/kg bolus ESKET injected after anesthesia induction and then 0.25 mg/kg/h ESKET infusion	[27]
	Lumbar fusion surgery, postoperative analgesia: IV	RCT	189	0.5 mg/kg bolus ESKET injected before incision followed by intraoperative infusion of 0.12 or 0.6 mg/kg/h ESKET	[28]

	Thoracoscopic lung surgery, postoperative analgesia: IV	RCT	242	Postoperative PCA for 48 h: 0.1 mg/kg hydromorphone and 0.5–2 mg/kg ESKET	[29]
	Cesarean section, postoperative analgesia: IV	RCT	140	Postoperative PCA: 0.5 mg/kg ESKET, 150 μg sufentanil, and 4 mg ondansetron	[30]
	Thoracoscopic surgery, intra- and postoperative analgesia: IV and epidural	RCT	159	Anesthesia induction: 0.125 mg/kg IV ESKET as adjuvant and then additional 0.125 mg/kg ESKET during surgery as needed. Before incision, 10 mL 0.2% ropivacaine plus 0.25 mg/kg EPI ESKET Postoperative, EPI PCA: 0.15% ropivacaine with 25 mg ESKET	[31]
	Labor analgesia, postoperative analgesia: Epidural	RCT	65	ED90 for EPI ESKET combined with 0.075% ropivacaine for labor analgesia in nulliparous parturients: ~ 1.0 mg/mL	[32]

mg/kg body weight, *ED90* 90% effective dose, *EPI* epidural, *ESKET* esketamine, *IN* intranasal administration, *IV* intravenous, *RCT* randomized controlled trial, *PCA* patient-controlled analgesia commercially available anesthesia depth monitors are not suited for correctly reflecting ketamine's sedative effects. Therefore, well-designed, high-quality studies are still needed to elucidate best possible drug dosing schedules and administration regimen.

5 Patient Comfort and Adverse Events

A recent meta-analysis evaluated the impact of ketamine on post-operative subjective quality of recovery (QoR) scores including 18 RCTs with more than 1500 patients. The analysis demonstrated that ketamine/ESKET induced substantial improvements in patient experiences with respect to emotional state, physical comfort, physical independence, pain management, and psychological support until 2 days after surgery. Subgroup analyses revealed that improved QoR scores were linked to ESKET use, but ESKET did not reduce length of hospital stay. In addition, ESKET was associated with a lower risk of postoperative nausea and vomiting. The incidence of dizziness and drowsiness was similar between ESKET and the controls. Importantly, the effect size of IV ESKET when used at a range from 0.2 to 0.6 mg/kg BW did not significantly correlate with its exact dose [41]. However, when ESKET doses of 1 or 2 mg/kg/48 h were combined with hydromorphone in patient-controlled IV analgesia, QoR scores were significantly higher than with lower-dose regimen. Notably, 2 mg/kg/48 h ESKET had a similar analgesic but deeper sedative effect compared to 1 mg/kg/48 h ESKET [29]. Context dependent, repeated ESKET doses, and condition and circumstance appropriate usage within the subanesthetic dose range of ~0.5 mg/kg or slightly more in adult patients may therefore contribute to reduce unwanted events after surgery and anesthesia and may increase patient comfort. Nonetheless, it has to be clearly stated that there is no simple recipe with "cutoff lines," and it is unlikely that there will ever be a "one-size-fits-all" concept.

6 Combination Regimens

- Propofol: An ESKET-propofol mixture reduced the incidence of propofol-induced injection pain in patients undergoing general anesthesia [42]. In gastroscopy, the combination of propofol with 0.2 mg/kg ESKET reduced the total amount of propofol required, provided more hemodynamic stability, and did not affect recovery time [43]. In addition, combining ESKET with propofol significantly reduced propofol-related circulatory and respiratory depression and various other unwanted effects [44].

- Benzodiazepines: Remimazolam besylate combined with ESKET shows a similar efficacy as sedation and analgesia regimen as propofol and ESKET during outpatient colonoscopy and surgical abortion [45, 46].
- Opioids: As described above, a combination of ESKET and opioids can be very useful to improve intra- and postoperative pain management and may reduce opioid needs. In addition, a RCT demonstrated that ESKET effectively counteracts remifentanil-induced respiratory depression [19].
- Dexmedetomidine: Dexmedetomidine plus low-dose ESKET improved analgesia in modified radical mastectomy and scoliosis correction surgery [47, 48].
- Pregabalin: ESKET (0.5 mg/kg intraoperative bolus and 0.12 mg/kg/h postoperative infusion for 48 h) combined with oral pregabalin (75–150 mg/kg per day) could alleviate moderate-to-severe acute postsurgical pain after spinal surgery. However, there was a slightly increased risk of mild dissociative symptoms after surgery with this strategy [49]. Another RCT reported that perioperative use of oral pregabalin (150 mg 2 h before surgery and twice daily for 7 days) combined with an IV patient-controlled analgesia regimen consisting of 100 μg sufentanil, 1.25 mg/kg ESKET, and 4 mg tropisetron could improve acute postoperative pain, reduce postoperative opioid consumption, and prevent chronic pain after breast cancer surgery [50].

The combination regimens of the milestone articles we selected to show the bandwidth of ESKET use in clinical practice are summarized in Table 2.

7 The Strive Not to Negatively Impact on Natural Postsurgical Healing Processes

The four phases of healing are hemostasis, inflammation, proliferation, and remodeling. Inflammation fights infection and promotes wound tissue healing if proinflammatory and anti-inflammatory mechanisms are well balanced. However, both excessive and insufficient inflammations are harmful to the human body. An excessive inflammatory response interferes with homeostasis and may affect survival; conversely, excessive immunosuppression increases mortality and complication rates. Maintenance of a proper balance between pro- and anti-inflammatory effects is therefore essential for postsurgical healing and recovery [6].

Analgesic agents that inhibit inflammation, such as NSAIDs, interfere with natural recovery processes [51, 52]. To address these unwanted effects, some researchers have proposed limiting the use of NSAIDs at the time of their greatest negative impact, that is, during the first days after injury [53]. For example, Sagi and

Table 2
Drug combination regimen with esketamine

Setting	Combination regimen	Design	Sample size	References
General anesthesia	Mixture of 1% propofol (20 mL) and ESKET 4/12/20 mg (diluted with 0.9% normal saline, 1 mL)	RCT	226	[42]
Gastroscopy	Low-dose propofol and 0.2 mg/kg ESKET	RCT	20	[43]
Sedation and analgesia, spontaneous ventilation	Low-dose propofol and subanesthetic doses of ESKET	Systematic review and meta-analysis	14 articles	[44]
Outpatient colonoscopy	0.15 mg/kg IV ESKET injected over 30 s, 0.15–0.25 mg/kg IV remimazolam besylate injected at a rate of 1 mL/s, and then 1–2 mg/kg/h remimazolam besylate infusion	RCT	150	[45]
Surgical abortion	0.3 mg/kg remimazolam and 0.3 mg/kg ESKET compared with 2 mg/kg propofol and 0.3 mg/kg ESKET	RCT	200	[46]
Modified radical mastectomy	0.5 μg/kg dexmedetomidine as loading dose and then 0.4 μg/kg/h infusion and 0.5 mg/kg ESKET bolus and 0.12–0.24 mg/kg/h infusion	RCT	135	[47]
Scoliosis correction surgery	Postoperative PCA, mean infusion rates: 5.5 μg/kg/h ESKET and 0.02 μg/kg/h dexmedetomidine	RCT	199	[48]
Resection of spinal neoplasms	0.5 mg/kg IV ESKET bolus after anesthesia induction and postoperatively: 0.12 mg/kg/h ESKET infusion for 48 hours combined with oral pregabalin 75–150 mg twice per day Pregabalin: First intake 2 h before surgery until 2 weeks after surgery	RCT	90	[49]
Breast cancer surgery	2 h before surgery 150 mg oral pregabalin Postoperatively: IV PCA with 100 μg sufentanil, 1.25 mg/kg ESKET, and 4 mg tropisetron in 100 mL saline solution	RCT	90	[50]

mg/kg mg/kg body weight, *μg/kg* μg/kg body weight, *RCT* randomized controlled trial, *PCA* patient-controlled analgesia co-authors recommended to cut down on the use of NSAIDs to not more than 1 week after acetabular fracture surgery. This approach would reduce the risk of heterotopic ossification without increasing the odds of nonunion healing [54]. More restrictive advice was given by Marquez-Lara et al., who recommended to replace NSAIDs with opioids for analgesic therapy [55]. Several groups examined the local application of opioids as an alternative to systemic application in painful wounds, such as skin ulcers, burns, or knee surgery [56, 57]. However, potential use of opioids in wound healing deserves careful considerations: It has been reported that chronic opioid application results in delayed healing and complications, particularly in infected wounds [58–60]. In mammals, growing evidence shows that opioids direct tissue repair toward scar healing with tissue damage, instead of allowing natural regenerative processes that lead to recovery of morphology and function [61]. There are also scientists who believe that the effects of opioids on wound healing are related to different opioid receptor types. For example, mu opioid receptors play a critical role during proliferation with formation of granulation tissue, whereas kappa receptors may not be involved in wound healing [62]. High-dose opioids and repeated opioid administration also induce neuroinflammation which exacerbates and maintains post-injury pain [6]. Much more comprehensive considerations and careful choices should therefore be made when using opioids in everyday practice and future studies.

In view of ESKET's weak agonistic effect at opioid receptors, its use may be very valuable in wound healing. As yet, research on the effect of ESKET on wound healing is limited, although the benefits of ketamine and ESKET on overshooting inflammatory responses have consistently been shown, such as in on-pump cardiac surgery patients, septic patients, and COVID-19 patients. Experimental data point to the fact that a ketamine-related decrease of high proinflammatory cytokines positively affects, for example, the kynurenine pathway with the result of less neurodegenerative effects and depression [63]. Moreover, recent evidence has been gained in RCT of patients receiving opioid-free anesthesia with ESKET and ultrasound-guided regional anesthesia such as pectoral nerve block II in modified mastectomy for breast cancer surgery or quadratus lumborum block in lower abdominal or pelvic surgery: It has been found that ESKET use resulted in a smaller increase in proinflammatory cytokines and higher patient QoR scores with less anxiety and postoperative depression [64, 65]. Furthermore, in a retrospective evaluation of 77 consecutive patients undergoing pancreatic resection with opioid-free anesthesia, those receiving intraoperative infusion of dexmedetomidine, lidocaine, and ESKET had a significantly lower rate of all kinds of complications including a lower incidence of pancreatic fistula and less gastrointestinal transit time [66].

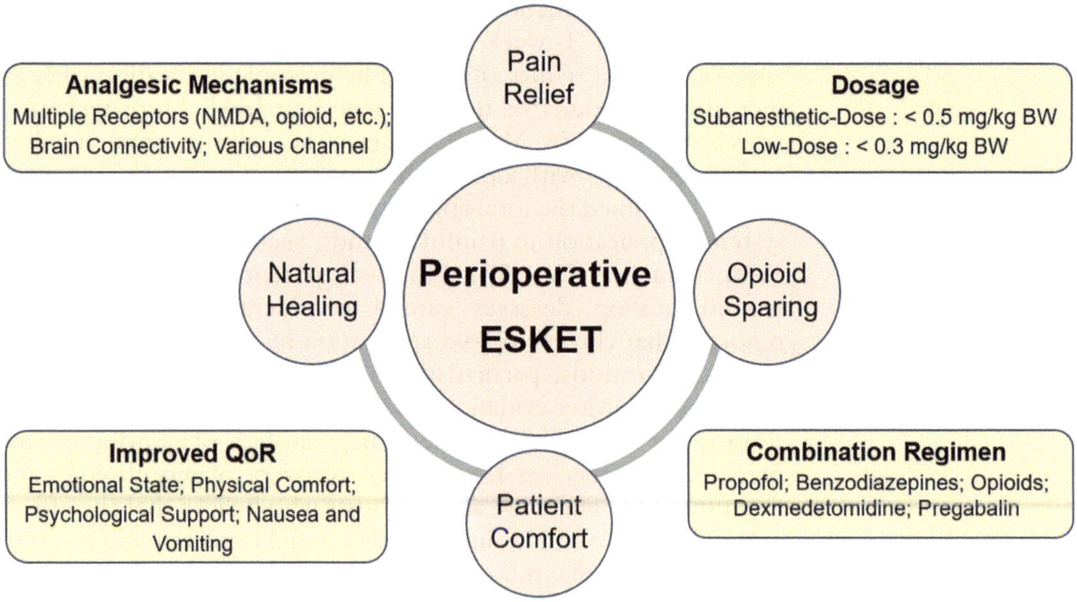

Fig. 1 Decisive effects and factors for beneficial use of perioperative esketamine at low - subanesthetic dose. *ESKET* esketamine, *NMDA* N-methyl-D-aspartate, *BW* body weight, *QoR* quality of recovery (created by the authors in PowerPoint, Microsoft Corp., Redmond, WA, USA)

Taken together, ketamine and ESKET appear to be unique "homeostatic regulators" of disbalanced inflammatory reactions and immune disturbances. The substances prevent exacerbation and extension of local tissue and general inflammation without blunting natural local processes or delaying inflammatory resolution. ESKET's therapeutic potential for supporting natural wound healing processes warrants further investigation.

Use patterns and potential benefits of ESKET in perioperative analgesia are shown in Fig. 1.

8 Conclusion

Substantial evidence has shown that low or subanesthetic doses of ESKET as adjunctive analgesic can help reduce postoperative pain and opioid use and improve patient comfort. Despite these positive effects, it may be too early to recommend ESKET as routine medication for everyday practice. Many physicians lack sufficient clinical experience with its use, and possible interactions with other medications as well as patient heterogeneity have to be considered. Nevertheless, especially in combination with opioids, esketamine has enormous beneficial potential. IN administration of ESKET is presently approved for the treatment of depression, but its use in analgesia would benefit from better pharmacokinetic models. Nonetheless, clinical trials have been started. Far beyond, a deeper investigation of the potentially positive effects of ESKET on wound healing may provide novel, important insights and indications for use of ESKET. In addition to the need for more high-quality trials, it is therefore now a priority to share practical expertise and transfer know-how about ESKET uses from studies to practical care. This is especially valid for education and training programs of younger physicians.

Acknowledgments

This work was supported by Sichuan Science and Technology Program (2023ZYD0168).

Conflict of Interest

The authors declare no conflict of interests.

References

1. Mion G, Himmelseher S (2024) Esketamine: less drowsiness, more analgesia. Anesth Analg 139(1):78–91
2. Srivastava D et al (2021) Surgery and opioids: evidence-based expert consensus guidelines on the perioperative use of opioids in the United Kingdom. Br J Anaesth 126(6):1208–1216
3. Perez-Ruixo C et al (2021) Population pharmacokinetics of Esketamine nasal spray and its metabolite Noresketamine in healthy subjects and patients with treatment-resistant depression. Clin Pharmacokinet 60(4):501–516
4. Dahners LE, Mullis BH (2004) Effects of nonsteroidal anti-inflammatory drugs on bone formation and soft-tissue healing. J Am Acad Orthop Surg 12(3):139–143
5. Rook JM, Hasan W, McCarson KE (2008) Temporal effects of topical morphine application on cutaneous wound healing. Anesthesiology 109(1):130–136
6. De Kock M, Loix S, Lavand'homme P (2013) Ketamine and peripheral inflammation. CNS Neurosci Ther 19(6):403–410
7. Beilin B et al (2007) Low-dose ketamine affects immune responses in humans during the early postoperative period. Br J Anaesth 99(4): 522–527
8. Zanos P et al (2018) Ketamine and ketamine metabolite pharmacology: insights into therapeutic mechanisms. Pharmacol Rev 70(3): 621–660
9. Franks NP, Lieb WR (1994) Molecular and cellular mechanisms of general anaesthesia. Nature 367(6464):607–614
10. Niesters M et al (2012) Effect of subanesthetic ketamine on intrinsic functional brain connectivity: a placebo-controlled functional magnetic resonance imaging study in healthy male volunteers. Anesthesiology 117(4):868–877
11. Hayashi Y et al (2011) Microglial Ca(2+)-activated K(+) channels are possible molecular targets for the analgesic effects of S-ketamine on neuropathic pain. J Neurosci 31(48): 17370–17382
12. Johnston JN, Henter ID, Zarate CA Jr (2023) The antidepressant actions of ketamine and its enantiomers. Pharmacol Ther 246:108431
13. Kim J et al (2019) Esketamine for treatment-resistant depression – first FDA-approved antidepressant in a new class. N Engl J Med 381(1):1–4
14. Mahase E (2019) Esketamine is approved in Europe for treating resistant major depressive disorder. BMJ 367:l7069
15. Qian Q, Liu HX, Li YQ (2023) Effect of esketamine nasal drops on pain in children after tonsillectomy using low temperature plasma ablation. Front Pediatr 11:1110632
16. Johansson J et al (2013) Prehospital analgesia using nasal administration of S-ketamine – a case series. Scand J Trauma Resusc Emerg Med 21:38
17. Xie Y et al (2024) Effectiveness of perioperative low-dose esketamine infusion for postoperative pain management in pediatric urological surgery: a prospective clinical trial. BMC Anesthesiol 24(1):65

18. Chen S et al (2023) Risk of esketamine anesthesia on the emergence delirium in preschool children after minor surgery: a prospective observational clinical study. Eur Arch Psychiatry Clin Neurosci 274:767
19. Jonkman K et al (2018) Esketamine counters opioid-induced respiratory depression. Br J Anaesth 120(5):1117–1127
20. Suleiman A et al (2023) Effects of ketamine infusion on breathing and encephalography in spontaneously breathing ICU patients. J Intensive Care Med 38(3):299–306
21. Vaessen HHB et al (2017) Moderate-to-deep sedation technique, using propofol and ketamine, allowing synchronised breathing for magnetic resonance high-intensity focused ultrasound (MR-HIFU) treatment for uterine fibroids: a pilot study. J Ther Ultrasound 5:8
22. Schonenberger S et al (2019) The keep simplest study: improving in-house delays and periinterventional management in stroke thrombectomy-a matched pair analysis. Neurocrit Care 31(1):46–55
23. Xu LL et al (2023) Efficacy and safety of esketamine for supplemental analgesia during elective cesarean delivery: a randomized clinical trial. JAMA Netw Open 6(4):e239321
24. Wang J et al (2019) Pharmacokinetics and safety of esketamine in Chinese patients undergoing painless gastroscopy in comparison with ketamine: a randomized, open-label clinical study. Drug Des Devel Ther 13:4135–4144
25. Wang X et al (2021) Perioperative intravenous S-ketamine for acute postoperative pain in adults: a systematic review and meta-analysis. J Clin Anesth 68:110071
26. Miziara LE et al (2016) Efficacy of continuous S(+)-ketamine infusion for postoperative pain control: a randomized placebo-controlled trial. Anesthesiol Res Pract 2016:6918327
27. Nielsen RV et al (2017) Intraoperative ketamine reduces immediate postoperative opioid consumption after spinal fusion surgery in chronic pain patients with opioid dependency: a randomized, blinded trial. Pain 158(3): 463–470
28. Brinck ECV et al (2021) Analgesic effect of intraoperative intravenous S-ketamine in opioid-naive patients after major lumbar fusion surgery is temporary and not dose-dependent: a randomized, double-blind, placebo-controlled clinical trial. Anesth Analg 132(1): 69–79
29. Zhang A et al (2024) Effects of S-ketamine added to patient-controlled analgesia on early postoperative pain and recovery in patients undergoing thoracoscopic lung surgery: a randomized double-blinded controlled trial. J Clin Anesth 92:111299
30. Han T et al (2023) Low-dose esketamine with sufentanil for postcesarean analgesia in women with gestational diabetes mellitus: a prospective, randomized, double-blind study. Front Endocrinol (Lausanne) 14:1202734
31. Yan H et al (2023) Opioid-free versus opioid-based anesthesia on postoperative pain after thoracoscopic surgery: the use of intravenous and epidural Esketamine. Anesth Analg 137(2):399–408
32. Lou S et al (2023) ED(90) of epidural esketamine with 0.075% ropivacaine for labor analgesia in nulliparous parturients: a prospective, randomized and dose-finding study. Front Pharmacol 14:1169415
33. Gomes LM et al (2011) Neurotoxicity of subarachnoid preservative-free S(+)-ketamine in dogs. Pain Physician 14(1):83–90
34. Colvin LA, Bull F, Hales TG (2019) Perioperative opioid analgesia-when is enough too much? A review of opioid-induced tolerance and hyperalgesia. Lancet 393(10180): 1558–1568
35. Meyer-Friessem CH et al (2022) Perioperative ketamine for postoperative pain management in patients with preoperative opioid intake: a systematic review and meta-analysis. J Clin Anesth 78:110652
36. Koppert W et al (2003) Differential modulation of remifentanil-induced analgesia and postinfusion hyperalgesia by S-ketamine and clonidine in humans. Anesthesiology 99(1): 152–159
37. Sator-Katzenschlager S et al (2001) The long-term antinociceptive effect of intrathecal S(+)-ketamine in a patient with established morphine tolerance. Anesth Analg 93(4): 1032–1034
38. Arendt-Nielsen L et al (1996) Effect of racemic mixture and the (S+)-isomer of ketamine on temporal and spatial summation of pain. Br J Anaesth 77(5):625–631
39. Mathisen LC et al (1995) Effect of ketamine, an NMDA receptor inhibitor, in acute and chronic orofacial pain. Pain 61(2):215–220
40. Schmid RL, Sandler AN, Katz J (1999) Use and efficacy of low-dose ketamine in the management of acute postoperative pain: a review of current techniques and outcomes. Pain 82(2):111–125
41. Hung KC et al (2024) The impact of perioperative ketamine or esketamine on the subjective quality of recovery after surgery: a meta-analysis of randomised controlled trials. Br J Anaesth 132(6):1293–1303

42. Xu C et al (2022) Esketamine prevents propofol-induced injection pain: randomized controlled trial. Front Pharmacol 13:991559
43. Liu X, Xiao Q, Zhuang S (2023) Comparison of propofol-esketamine versus propofol for anesthesia in gastroscopy: a double-blind, randomized controlled clinical trial. Front Med (Lausanne) 10:1184709
44. Chen H et al (2023) Analysis of the efficacy of subclinical doses of esketamine in combination with propofol in non-intubated general anesthesia procedures – a systematic review and meta-analysis. BMC Anesthesiol 23(1):245
45. Li W et al (2023) The efficacy and safety of remimazolam besylate combined with esketamine for outpatient colonoscopy: a prospective, randomized, controlled clinical trial. Drug Des Devel Ther 17:2875–2887
46. Yue L et al (2023) Remimazolam versus propofol in combination with esketamine for surgical abortion: a double-blind randomized controlled trial. Clin Transl Sci 16(9): 1606–1616
47. Huang Z et al (2023) Effect of dexmedetomidine and two different doses of esketamine combined infusion on the quality of recovery in patients undergoing modified radical mastectomy for breast cancer – a randomised controlled study. Drug Des Devel Ther 17:2613–2621
48. Zhang Y et al (2023) Mini-dose esketamine-dexmedetomidine combination to supplement analgesia for patients after scoliosis correction surgery: a double-blind randomised trial. Br J Anaesth 131(2):385–396
49. Zhou Y et al (2024) Effect of esketamine combined with pregabalin on acute postsurgical pain in patients who underwent resection of spinal neoplasms: a randomized controlled trial. Pain 165:e96
50. Wang M et al (2023) Perioperative administration of pregabalin and esketamine to prevent chronic pain after breast cancer surgery: a randomized controlled trial. Drug Des Devel Ther 17:1699–1706
51. Delande S, Lavand'homme P (2023) Acute pain management and long term outcomes. Curr Opin Anaesthesiol 36(2):222–227
52. Chen MR, Dragoo JL (2013) The effect of nonsteroidal anti-inflammatory drugs on tissue healing. Knee Surg Sports Traumatol Arthrosc 21(3):540–549
53. Lisowska B, Kosson D, Domaracka K (2018) Positives and negatives of nonsteroidal anti-inflammatory drugs in bone healing: the effects of these drugs on bone repair. Drug Des Devel Ther 12:1809–1814
54. Sagi HC et al (2014) Indomethacin prophylaxis for heterotopic ossification after acetabular fracture surgery increases the risk for nonunion of the posterior wall. J Orthop Trauma 28(7):377–383
55. Marquez-Lara A et al (2016) Nonsteroidal anti-inflammatory drugs and bone-healing: a systematic review of research quality. JBJS Rev 4(3)
56. Stein C et al (1991) Analgesic effect of intraarticular morphine after arthroscopic knee surgery. N Engl J Med 325(16):1123–1126
57. Farley P (2011) Should topical opioid analgesics be regarded as effective and safe when applied to chronic cutaneous lesions? J Pharm Pharmacol 63(6):747–756
58. Lam CF et al (2008) Prolonged use of high-dose morphine impairs angiogenesis and mobilization of endothelial progenitor cells in mice. Anesth Analg 107(2):686–692
59. Brack A, Rittner HL, Stein C (2011) Immunosuppressive effects of opioids – clinical relevance. J Neuroimmune Pharmacol 6(4): 490–502
60. Martin JL et al (2010) Chronic morphine administration delays wound healing by inhibiting immune cell recruitment to the wound site. Am J Pathol 176(2):786–799
61. Berthezene CD et al (2021) Tissue regeneration: the dark side of opioids. Int J Mol Sci 22(14):7336
62. Wang Y et al (2017) Opioids and opioid receptors orchestrate wound repair. Transl Res 185: 13–23
63. Jozwiak-Bebenista M et al (2024) Ketamine – a new antidepressant drug with anti-inflammatory properties. J Pharmacol Exp Ther 388(1):134–144
64. Yu L, Zhou Q, Li W et al (2022) Effects of esketamine combined with ultrasound-guided pectoral nerve block type II on the quality of early postoperative recovery in patients undergoing a modified radical mastectomy for breast cancer: a randomized controlled trial. J Pain Res 15:3157–3169
65. Dai J, Li S, Zheng R, Li J (2022) Effects of esketamine on inflammatory factors in opioid-free anesthesia based on quadratus lumborum block: a randomized trial. Medicine (Baltimore) 102:e34975
66. Hublet S, Galland M, Navez J et al (2022) Opioid-free versus opioid-based anesthesia in pancreatic surgery. BMC Anesthesiol 22:9

Chapter 3

Ketamine as an Adjuvant to Local and Regional Anesthesia? Efficacy and Safety Considerations

Paul A. Stewart, Claire C. Nestor, and Michael G. Irwin

Abstract

This chapter explores adjunctive ketamine use in regional/local analgesia and general anesthesia. It highlights the inconsistent study results of ketamine's efficacy and safety. Local anesthesia, of limited duration, is seen with subcutaneous infiltration. Better pain relief occurs in combinations with bupivacaine. Topical ketamine can reduce allodynia in complex regional pain syndrome and induces similar analgesia to EMLA® in venipuncture. Evidence of ketamine mouthwashes for mucositis and stomatitis pain is limited. Intrathecal ketamine can shorten spinal anesthesia onset with mixed effects on block quality and duration, but there are significant concerns of neurotoxicity. Epidural ketamine is promising in pediatric caudal anesthesia but requires further evidence of safety. In peripheral nerve blocks, ketamine use shows inconsistent evidence, with both improved analgesia and increased side effects. Intra-articular ketamine exerts mixed results on pain after arthroscopic knee surgery. Local toxicity and chondrocyte damage might be possible. Ketamine can be combined with propofol in general anesthesia, offering hemodynamic stability and reduced propofol requirements. However, fixed drug mixtures limit individual dose titration. The efficacy, benefits, and risks of ketamine as a supplement to local anesthesia remain uncertain, emphasizing the necessity for further high-quality research.

Key words Anesthesia, General, Regional, Analgesia, Local anesthesia, Ketamine, Efficacy, Safety

1 Introduction

Ketamine modulates neuronal excitability and influences electrochemical signaling within the nervous system by blocking multiple channels including those targeted by local anesthetics (LAs). Its effects extend well beyond these ion channels, secondarily impacting a multitude of neurotransmitter systems which contribute to its anesthetic and psychotropic effects. Consequently, ketamine has been investigated either in place of or in addition to traditional LA drugs (Fig. 1).

Sabine Himmelseher (ed.), *Ketamine: Pain Management, Critical Care, Depression, and Resource-Poor Settings Volume II*, Neuromethods, vol. 225, https://doi.org/10.1007/978-1-0716-4599-4_3,

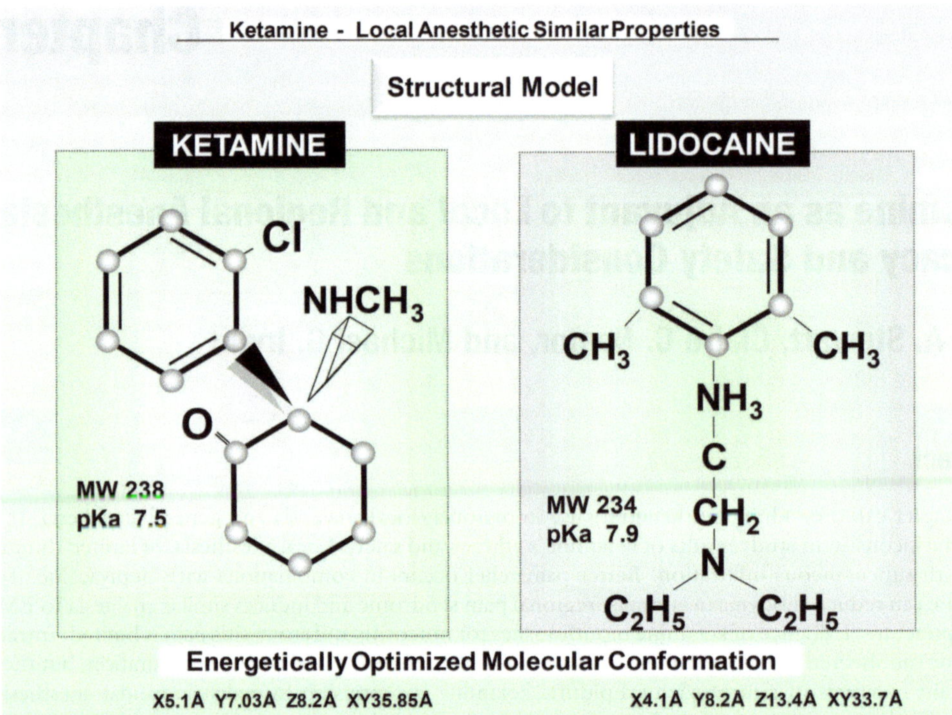

Fig. 1 Structural similarities between ketamine and lidocaine. *MW* molecular weight, *pKa* acid dissociation constant at logarithmic scale (designed by S. Himmelseher in Power Point, Microsoft Corp., Redmond, WA, USA)

2 Ketamine in Local Analgesia/Anesthesia

2.1 Pharmacological Background

Ketamine may act as a LA by blocking voltage-gated sodium channels (VGSCs) in a similar fashion to commonly used clinical LAs. It also shares a binding site with these anesthetics. Mechanistically, ketamine exhibits both tonic (reducing sodium channel current) and phasic blockade (use-dependent inhibition). In comparison with lidocaine, ketamine shows greater tonic inhibition but less phasic blockade. It also affects voltage-gated potassium channels (VGPCs) by inducing a concentration-dependent hyperpolarizing shift in the activation curve of these channels without altering their steady-state activation [1].

2.2 Clinical Use: Local Ketamine as Analgesic/Anesthetic Agent

In volunteers, subcutaneous (SC) infiltration of 3 mg/mL ketamine has been shown to produce LA effects, albeit very short-lived [2]. In patients undergoing minor dermatological surgery, skin infiltration with 12.5–30 mg ketamine and adrenaline 10 mcg/mL produced anesthesia but lasting only 11 min [3]. The technique in this study was limited by injection pain, skin erythema, and a 56% incidence of lightheadedness and dizziness with ketamine. This may be explained by the relatively high bioavailability of SC ketamine at 64% [4]. In contrast to the limited duration of action seen in earlier studies, a recent trial found that 1 mg/kg SC ketamine either alone or in combination with 20 mL 0.5% bupivacaine provided better postoperative pain relief and reduced postoperative opioid consumption when compared to the use of bupivacaine alone at 12 h post-caesarean section [5]. Unfortunately, a lack of data on potential side effects and transfer to breast milk limits the clinical application of this finding. Ketamine for topical use is a specially compounded product that is not available commercially but is under investigation in various preparations. It is made by mixing a powder form of ketamine with a gel or cream. It is used to reduce allodynia in complex regional pain syndrome (CRPS) [6] but has also been applied for analgesia prior to venipuncture. In comparison to EMLA® (lidocaine and prilocaine) cream, 2 g of 10% topical ketamine cream produced similar analgesia, with minimal side effects and a lower incidence of itch or skin irritation [7].

Ketamine mouthwashes have been used for managing the pain of mucositis and stomatitis [8, 9], but supporting evidence is limited [10].

2.3 Safety Considerations: Combining Drugs with Local Anesthetics

If considering mixing additional drugs such as ketamine into solution with LAs, it is essential to consider physicochemical compatibility and potential pharmacokinetic and pharmacodynamic consequences. Both lidocaine and bupivacaine have been shown to be physically compatible in solution with ketamine without precipitation [11]. LAs are weak bases, existing in both ionized and unionized forms. The proportion of drug existing in each state is dependent on the drug pKa and environmental pH. At a pH below the pKa, the majority of drug exists in the ionized form, with the unionized proportion increasing as pH approaches the pKa. The primary action of LAs is blockade of neuronal transmission by binding of neuronal voltage-gated sodium channels in the active state. These channels span the cell membrane, but the binding site of LAs is on the intracellular side. Only unionized LAs can cross the cell membrane (Fig. 2), and, therefore, the proportion of drug existing in the unionized state (in addition to the molecular weight) has a major impact on the time to onset. Drugs with a lower pKa (and those with a lower molecular weight) such as lidocaine (7.8) or prilocaine (7.7) have a faster onset than those with higher pKa such as bupivacaine (8.1), as more of the drug will exist in the unionized form at physiological pH.

Commercially available lidocaine preparations have a pH of 4 or 6 with and without adrenaline, respectively [12]. Bupivacaine and levobupivacaine are commonly buffered to a pH between 4 and 6.5. Racemic ketamine solutions have a pH between 3.5 and 5.5 [13]. The addition of ketamine to LA solutions will reduce the

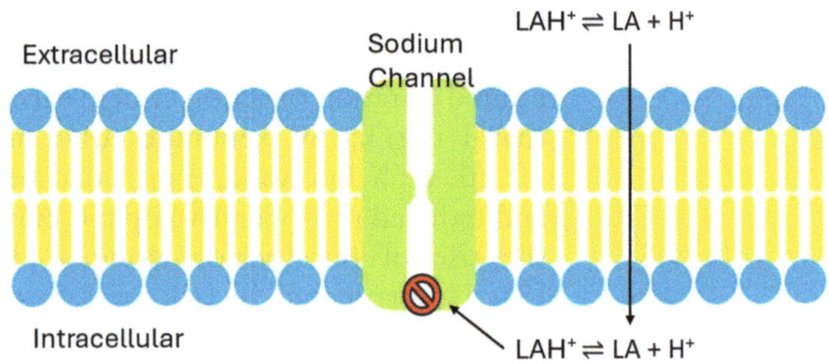

Fig. 2 Mechanism of action of local anesthetics. *LA* local anesthetic in non-ionized form, *LAH*$^+$ local anesthetic in ionized form, *H*$^+$ hydrogen ion (designed by the authors in Microsoft PowerPoint, Microsoft Corp., Redmond, WA, USA)

overall pH of the solution. Theoretically, this has potential to delay onset, but, as will be discussed below, it does not necessarily have a clinical impact.

2.4 Ketamine as Adjunct to Local Anesthetics in Clinical Application

2.4.1 Intrathecal Use

The addition of S(+) ketamine to intrathecal (IT) bupivacaine has been demonstrated to shorten the onset of spinal anesthesia [14, 15]. The effects of ketamine admixture on the quality and the duration of spinal anesthesia remain uncertain. Potential effects range from reduced motor blockade and shorter duration of action with 0.1 mg/kg S(+) ketamine added to 7.5 or 10 mg bupivacaine [14] to increased segmental spread with no impact on block duration with 0.05 mg/kg S(+) ketamine added to 10 mg bupivacaine [15] or even no effect on segmental spread or duration with 25 mg racemic ketamine added to 7.5 mg bupivacaine compared to 10 mg bupivacaine alone [16]. It is possible that these differences may be accounted for by difference in admixture technique. Those studies demonstrating increased spread added ketamine to a fixed dose of bupivacaine, thereby increasing the total drug volume administered. In contrast, the studies demonstrating shorter duration of action replaced a portion of the local anesthetic dose with ketamine, keeping volume constant but reducing the local anesthetic dose. Discrepancies in dosing and specific drug choice may also account for the differences in systemic side effects. Adding 25 mg of racemic ketamine to IT bupivacaine significantly increased patient dizziness and postoperative nausea and vomiting to the extent that the trial was abandoned and unblinded [16]. Conversely, the addition of 0.1 mg/kg S(+) ketamine to IT bupivacaine did not increase adverse systemic effects [14]. Addition of 0.1 mg/kg IT racemic ketamine to 0.3 mg morphine and 10 mg hyperbaric bupivacaine reduced pain after major abdominal cancer surgery compared to morphine and bupivacaine alone, at the expense of an increase in sedation [17]. None of the studies reported any dissociative effects. However, there is a significant safety concern regarding IT ketamine that cannot be overlooked: Autopsy studies have reported histopathological lesions of the spinal cord following IT administration of ketamine in both animals [18] and humans [19]. We are convinced that possible risks of central nervous system damage outweigh potential ketamine benefits. We cannot therefore recommend IT ketamine use without further safety data.

2.4.2 Epidural Use

The epidural (EPI) administration of ketamine in LA admixtures has been studied extensively, especially in pediatric caudal anesthesia. Systemic bioavailability of ketamine after EPI administration is 77% [20]. Individual studies have demonstrated that pediatric caudal 0.5 mg/kg ketamine admixture prolongs postoperative analgesia with 0.25% bupivacaine 1 mL/kg [21] and ropivacaine [22] with minimal side effects. More recent studies have suggested an increase in postoperative sedation, albeit of only 30-min duration with 0.5 mg/kg racemic ketamine added to 0.25% bupivacaine 0.75 mL/kg [23]. These findings are supported by systematic review data, with a mean increase in duration of analgesia by 5.6 h but with a trend toward increased sedation and hallucinations which were, albeit, not statistically significant [24]. A concern in this cohort is potential neurotoxicity, especially with respect to local developing neurons and neurocognitive developmental issues from systemic ketamine absorption. However, ketamine has been shown to exert both neuroprotective and neurotoxic effects, dependent on dose, frequency of use, and setting. For example, under conditions and circumstances of painful (surgical) stimulation with increased inflammatory signaling, NMDA receptor sensitization, and windup of signal transmission, skilled use of ketamine exerts neuroprotection. Without treatment, severe pain otherwise causes neurodegeneration. Nonetheless, the same ketamine dose may induce neurotoxic effects on developing brain in settings without noxious stimulation [25].

Trials examining the effectiveness of epidural ketamine admixture in adults have been heterogenous in both design and results, limiting the potential for meta-analysis [26]. In patients undergoing major upper abdominal surgery, the addition of 0.2 mg/kg ketamine to a 0.0625% bupivacaine and 0.05 mg/mL morphine patient-controlled epidural analgesia (PCEA) strategy reduced total opioid consumption and pain scores at both movement and rest with minimal side effects [27]. In one study, a single dose of 0.5 mg/kg EPI racemic ketamine combined with 0.5% bupivacaine did not improve pain control post-total knee arthroplasty in comparison to epidural 0.5% bupivacaine alone. However, patients received a continuous 5 mL/h infusion of 0.1% bupivacaine and 1 μg/mL fentanyl for postoperative pain therapy so that beneficial effects of ketamine may have been masked [28]. In contrast, another study investigating the addition of epidural S(+) ketamine 0.25 mg/kg to epidural ropivacaine in patients undergoing total knee arthroplasty reported a reduction in VAS pain scores at 24 and 48 h post-op, when compared to ropivacaine alone [29]. PCEA bolus demands were higher in the ropivacaine only group at 48 h post-op. No difference in the incidence of adverse effects was observed. A recent randomized controlled trial examined the use of 0.25 mg/kg EPI S(+) ketamine and 10 mL 0.2% ropivacaine admixture for opioid-free anesthesia (OFA) administered before surgical incision in video-assisted thoracoscopic surgery [30]. In comparison with 2 mg EPI morphine with 10 mL 0.2% ropivacaine, OFA resulted in a reduced incidence of mild chronic pain (numerical rating scale 1–3) at 3 and 6 months post-surgery. A better side effect profile with EPI S(+) ketamine and ropivacaine as compared to an anesthesia regimen with IV opioids, EPI morphine, and ropivacaine was reported. Data on psychomimetic effects were not described. However, it is difficult to draw firm conclusions from this study with regard to EPI S(+) ketamine, because 0.125 mg/kg IV S(+) ketamine could intraoperatively be administered at the discretion of the anesthetist in the OFA group or 1 μg/kg fentanyl as deemed necessary in those patients receiving opioids. One potential novel use of EPI ketamine admixture is the treatment of refractory pain in postherpetic neuralgia [31].

Adjunctive use of ketamine in EPI anesthesia/analgesia appears to have potential for improved pain management with minimal systemic side effects, but serious concerns relating to potential local neurotoxicity continue to remain. High-quality evidence of both efficacy and safety is required. EPI use of ketamine preparations with toxic preservative is unethical, because, for example, benzethonium chloride may account for 30% of racemic ketamine's neurotoxicity. In addition, EPI ketamine is not officially approved for clinical use in many countries, and liability issues have not been clarified.

2.4.3 Use in Peripheral Nerve Blocks

The adjunct use of ketamine in peripheral nerve blocks has conflicting evidence. The addition of 30 mg ketamine to 30 mL 0.5% ropivacaine for interscalene block [32] or 75 mg S(+) ketamine to 30 mL 0.5% bupivacaine for inguinal field block [33] did not improve analgesia but resulted in a significant increase in side effects. This contrasts with a previously reported improvement in duration of analgesia and increase in pain threshold when 0.3% ketamine was added to 0.5% bupivacaine for wound infiltration post-inguinal hernia repair. The addition of 1 mg/kg ketamine to 0.1% ropivacaine applied as continuous infusion for femoral nerve block in patients having anterior cruciate ligament repair did not improve pain control versus 0.2% ropivacaine at equivalent volume [34]. Conversely, a separate a study compared combined perineural administration of 40 mg ketamine and 40 mL 0.375% ropivacaine, to 40 mL perineural 0.375% ropivacaine alone, or

40 mg IV ketamine and 40 mL perineural 0.375% ropivacaine in combined femoral and sciatic nerve block. Perineural combined ketamine and ropivacaine resulted in better pain scores, prolonged analgesia, and improved patient satisfaction [35]. Hallucinations were noted in the IV ketamine group, but there was no significant difference in adverse effects between plain ropivacaine and perineural ketamine groups. A meta-analysis published in 2021 examined the effects of perioperative ketamine on acute and chronic wound pain from 13 randomized controlled trials incorporating 1182 patients. It was shown that IV ketamine or the addition of ketamine to bupivacaine in thoracic paravertebral analgesia improved early postoperative pain relief and opioid needs [36]. Perioperative ketamine reduced postoperative depression symptoms and showed long-term benefit on postmastectomy syndrome. Local infiltration with a mixture of 1 mL ketamine (25 mg/mL) and 2 mL 0.5% bupivacaine also demonstrated improved pain scores in comparison to bupivacaine alone or placebo after anal fistula repair [37]. Taken together, the overall evidence for admixture of ketamine with LA is still of low quality and inconsistent. However, a meta-analysis published in June 2024 reviewed the effects of ketamine as adjuvant to regional anesthesia in 21 RCT on 1011 patients undergoing diverse surgical procedures and regional anesthesia techniques [38]. The authors came to the conclusion that ketamine admixture prolongs the duration of regional anesthesia especially with regional nerve blocks and may prolong sensory blocks without significantly increasing motor blockade and ketamine-related adverse effects. Nevertheless, there was trend ($p = 0.08$) toward more adverse effects with ketamine.

2.4.4 Intra-articular Ketamine

An additional area of interest is the admixture of ketamine with LA drugs for intra-articular (IA) administration. Early studies demonstrated no improvement in postoperative pain scores following arthroscopic knee surgery with IA ketamine compared to 0.9% saline, either as a bolus of 0.5 mg/kg [39] or an infusion of 10 mg via catheter [40]. A subsequent study comparing 0.5 mg/kg IA ketamine, neostigmine, or bupivacaine to placebo did demonstrate improved analgesia with all three drugs [41]. Notably, duration of analgesia was longest with ketamine, but pain scores were lowest with bupivacaine. In patients undergoing knee arthroscopy for meniscal repair, pain scores were higher after 1 mg/kg IA ketamine compared to 20 mL 0.25% bupivacaine, and there was no improvement when ketamine was added to bupivacaine alone [42]. In contrast, a later study using 0.25 mg/kg IA S(+) ketamine found that postoperative pain scores were lower after IA S(+) ketamine than 10 mL 0.25% bupivacaine [43]. The discovery that 1 mg/kg IA ketamine can enhance pain relief post-arthroscopic meniscectomy compared to a placebo was further investigated:

Postoperative pain scores were improved when a combination of 0.5 mg/kg ketamine and 50 mg levobupivacaine were compared with 1 mg/kg IA ketamine or placebo [44]. Additionally, a study using the same dose of IA ketamine showed better pain management after meniscectomy when compared with IA magnesium or IA tramadol [45]. However, there was a 30% incidence of hallucinations among patients who received 1 mg/kg ketamine. These adverse effects were not present in a subsequent study where doses of 0.5 mg/kg and 1 mg/kg of IA ketamine were compared [46]. While most data regarding IA ketamine injections pertains to knee arthroscopy, their potential in rotator cuff surgeries has also been explored. A mix of 50 mg IA ketamine and 10 mg morphine injected prior to surgical incision did not show a statistically significant difference in post-surgical pain when compared to 20 mg morphine alone; however, both were found to be more effective than placebo [47].

As with IT delivery, there is a concern for local toxicity with IA ketamine administration. Laboratory studies have observed a dose-dependent cytotoxic effect on chondrocytes from humans [48] and animals [49]. On the other hand, a study conducted in live animals suggested that ketamine might play an anti-inflammatory role and could mitigate the development of osteoarthritis [50].

3 Ketamine as Admixture, Adjunct, or Induction Agent in General Anesthesia

A further application of ketamine admixture to consider is in combination with propofol, commonly named "ketofol." Physiochemical compatibility has been extensively tested. Combinations of 10 mg/mL ketamine and 10 mg propofol are physically and chemically compatible and stable at both 50:50 and 30:70 ratios [51]. Each drug retained 97% of its initial concentration after 3 h of storage in polypropylene syringes with exposure to light. A further study demonstrated the stability of various propofol and ketamine admixtures for up to 6 h at room temperature [52]. It should be noted, however, that even where mixtures are physiochemically compatible, homogeneity of the solution cannot be guaranteed, even if standardized mixing techniques are applied [53]. When any drug admixture is administered as an IV infusion, differing drug pharmacokinetics and pharmacodynamics must be accounted for. Ketamine has a faster onset than propofol and a longer duration of action. The context-sensitive half-life of ketamine increases with prolonged infusions, resulting in delayed emergence and recovery from anesthesia if admixture is administered.

Significant concerns have been raised regarding the admixture of additional drugs with propofol [54]. Infusion rates and dosing are often guided by one drug, frequently propofol, with the resultant potential for over- or underdosing of the other drug in solution. As the relative proportions of each drug are fixed, the ability to individually titrate the dose and resultant pharmacodynamic effects of each drug is lost. Consequently, many of the advantages of total intravenous anesthesia, such as the ability to individually control anesthesia and analgesia, and improved emergence characteristics may be lost if drug admixture is performed. If ketamine is administered as an adjunct to propofol-based total intravenous anesthesia, a significant reduction in propofol requirement is seen of approximately 20–40%. This reduction can confer hemodynamic stability advantages. The recent National Audit Project (NAP) 7 recommended that ketamine should have been used in place of propofol or other anesthetic agents in 12 cases of perioperative cardiac arrest due to its better hemodynamic profile [55]. "Ketodex" (ketamine and dexmedetomidine) is promoted as a "sedo-analgesic" agent. The combination is purported to have benefits in terms of hemodynamic stability, absence of respiratory depression, postoperative analgesia, and recovery. The most common regimen seems to be the use of 1 μg/kg dexmedetomidine and 2 mg/kg ketamine given as a bolus over 5 min. Dexmedetomidine has very different pharmacodynamic effects from ketamine and propofol. Rapid infusion of dexmedetomidine is likely to cause hypertension and reflex bradycardia. Arterial blood pressure and heart rate increase after ketamine, even at sub-anesthetic doses, and decrease after propofol. In addition, co-administration of anesthetics, sedatives, hypnotics, or opioids with dexmedetomidine is likely to lead to an enhancement of their effects, requiring a reduction in dose. As outlined, this cannot be controlled with a fixed mixture.

An additional consideration is the effect of ketamine on intraoperative depth of anesthesia monitoring. Processed EEG monitors use various proprietary algorithms and techniques to generate a dimensionless index value from a raw EEG trace to represent depth of anesthesia. The widely available Bispectral Index (BIS™) uses a combination of EEG suppression, beta ratio, and bispectral frequency analysis to estimate anesthetic depth [56]. Propofol anesthesia characteristically results in an alpha and delta wave predominant EEG pattern. Ketamine administration has markedly different effects on the EEG trace, both when given in isolation, or in combination with propofol. Isolated ketamine administration produces a "gamma-burst" EEG pattern with alternating slow-delta and high-frequency gamma waves at higher doses [57] or a stable beta/gamma predominant pattern at lower doses [58]. When administered in combination with propofol, ketamine shifts the prevalent alpha oscillations into the higher frequency beta

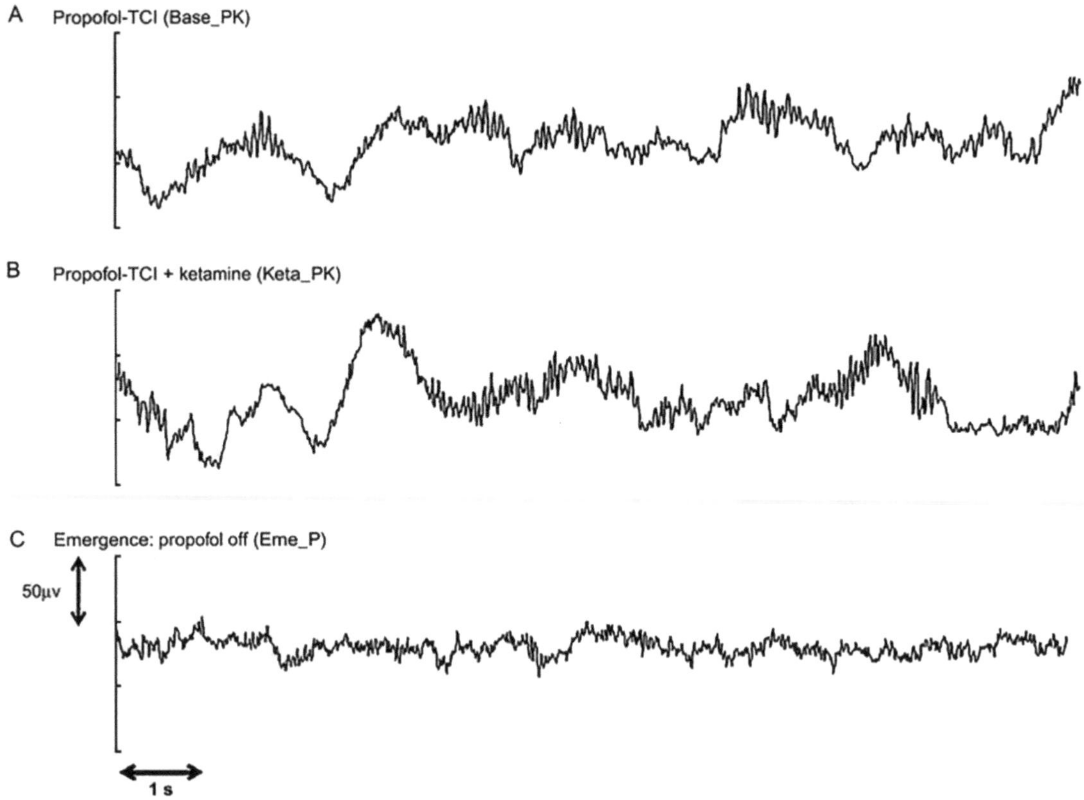

Fig. 3 Raw electroencephalography spindle oscillations under anesthesia with (**a**) TCI-propofol only (Base_PK), (**b**) TCI-propofol combined with IV ketamine (Keta_PK), and (**c**) during emergence from propofol (Eme_P). Ketamine increased the frequencies of an alpha-spindle wave induced by propofol and did not block their formation. During emergence, the alpha waves vanished. *TCI* target-controlled infusion (from Tsuda et al. [59] with permission from John Wiley and Sons conveyed via Copyright Clearance Center, Inc.)

range [59] (Figs. 3 and 4) resulting in an increase in both calculated spectral edge frequency (SEF) and numerical BIS values [60].

In the authors' experience, this increase in beta activity and consequently depth of anesthesia index values is transient, lasting approximately 15–20 min after addition of adjunctive ketamine to a steady-state propofol infusion.

4 Conclusions

Despite nearly 30 years of research, significant uncertainty remains about the efficacy and safety of ketamine as a supplement to LA. Most research consists of small studies with inconsistent application methods, which largely impedes more advanced analyses. A selection of relevant meta-analyses [36, 38, 61–65] is summarized

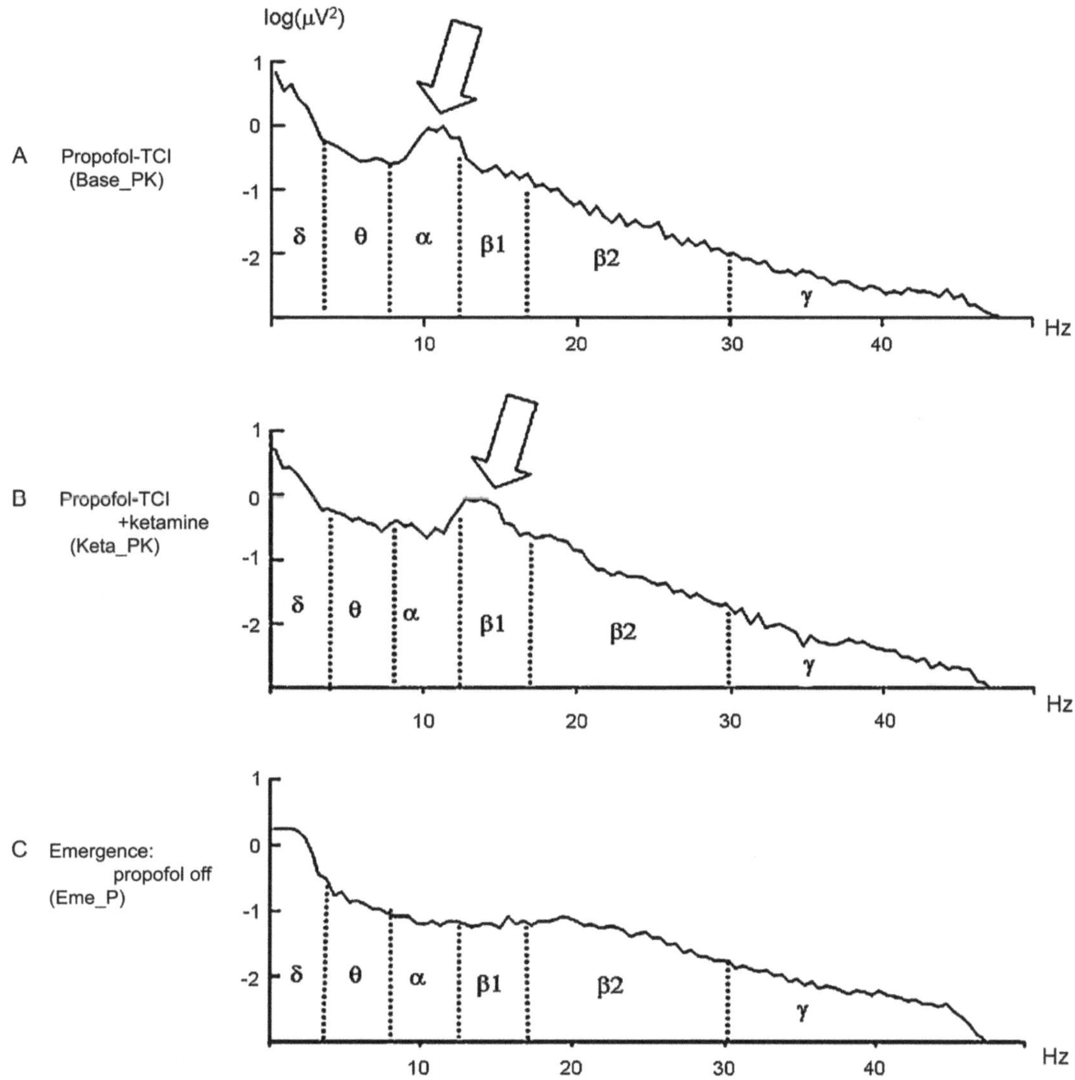

Fig. 4 Electroencephalography power spectra under anesthesia with (**a**) TCI-propofol only (Base_PK), (**b**) TCI-propofol combined with IV ketamine (Keta_PK), and (**c**) during emergence from propofol (Eme_P). Propofol alone caused the alpha-peaks of the power spectra to occur at an average frequency of 10.4 +/− 0.9 Hz; the addition of ketamine shifted the peaks to higher frequencies of 15.1 +/− 1.4 Hz ($P < 0.05$). On the other hand, when the EEG was activated by discontinuation of propofol, the corresponding alpha-peaks disappeared. *TCI* target-controlled infusion (from Tsuda et al. [59] with permission from John Wiley and Sons conveyed via Copyright Clearance Center, Inc.)

in Table 1. Consequently, the risk-to-benefit ratio of combinations of ketamine and LA remains undefined. Due to variable results, combined with common systemic adverse effects and possible cytotoxicity, it is difficult to endorse such combinations. There is stronger evidence of ketamine's efficacy when used as a supplement to

Table 1
Relevant systematic reviews of ketamine as an adjunct to various regional anesthesia techniques

Systematic review	Number of RCT/patients	Application route	Outcome: ketamine efficacy	Outcome: ketamine-related safety
Xiang et al. [38]	21 studies/ 1011 patients	Ketamine as adjunct to various regional anesthesia techniques	Adjuvant ketamine produces: Longer time to first analgesic rescue in spinal anesthesia (58 mins $P = 0.005$) and peripheral nerve blocks (366 mins, $P = 0.0007$) No significantly longer motor block in spinal anesthesia or peripheral nerve blocks (6.94 min, $P = 0.16$) Longer sensory block (30 mins, $P = 0.003$) in overall analysis No significant difference in spinal and peripheral nerve subgroup analyses ($P = 0.22$ spinal, $P = 0.25$ peripheral) No significant difference in block onset time (−1.17 min, $P = 0.13$)	Overall trend toward increased adverse effects but without significance (OR 1.97, $P = 0.08$) Ketamine increases incidence of dizziness (OR 19.17, $P = 0.01$) No significant increase in hallucinations (OR 4.04, $P = 0.22$) but only assessed in four trials with 250 patients, three events total No significant difference in nausea and vomiting (OR 1.08, $P = 0.83$) No significant difference in headache (OR 0.33, $P = 0.51$)
Du et al. [61]	11 studies/ 532 patients	Ketamine added to lidocaine in IV regional anesthesia Upper limb surgery	IV ketamine added to IV lidocaine: Shortens onset time of sensory block (MD 2.04 mins, $P < 0.00001$) Shortens onset time of motor block (MD 3.7 mins, $P < 0.00001$) Prolongs the time to sensory (MD −4.88 mins, $P = 0.001$) and motor (MD −2.56 mins, $P = 0.006$) recovery Reduces postoperative fentanyl requirement (MD 59.68 mcg, $P < 0.00001$)	No significant difference in overall incidence of complications (OR 1.21, $P = 0.64$) 92 adverse events in total across both treatment groups in 472 surgeries, but nature not specified

			Prolongs time to tourniquet pain onset (MD −24.9 mins, $P < 0.00001$) Reduces VAS scores after tourniquet inflation and deflation, $P < 0.05$ from 10 mins post-inflation to 4 h post-deflation	
Sohnen et al. [62]	11 studies/ 750 patients	Single-dose intrathecal ketamine added to intrathecal bupivacaine. Caesarean sections, lower limb, urological and lower abdominal surgery	Addition of ketamine to intrathecal bupivacaine: Prolongs time to first analgesic request (58.23 mins, $P < 0.0001$) Shortens the onset of sensory blockade (−0.87 mins, $P = 0.0005$) Shortens the onset of motor blockade (−0.88 mins, $P = 0.05$) Prolongs duration of sensory blockade (39.73 mins, $P = 0.001$) Prolongs duration of motor blockade (4.02 mins, $P < 0.0001$)	No statistical analysis of safety data or adverse effects One study using 50 mg intrathecal ketamine reported emergence delirium in 2 patients Authors suggest limiting dose to 0.7 mg/kg or less to avoid neuropsychiatric effects Three studies reported APGAR scores following Caesarean section with no impact noted
Shah et al. [63]	Total 87 studies/ 5285 patients Ketamine 16 studies/ 852 patients	Caudal ketamine and other agents as adjuvants Pediatric patients undergoing infraumbilical surgery	Ketamine as an adjunct to caudal epidural analgesia: Prolongs duration of analgesia (network estimate weighted mean difference = 297 mins) Reduces number of analgesia doses in 24 hrs post-op (network estimate weighted mean difference = −1.2 doses) Does not significantly reduce total paracetamol dose in 24 h post-op	Adverse effects not included in analysis Included studies did not evaluate long-term neurological safety

(continued)

Table 1 (continued)

Systematic review	Number of RCT/patients	Application route	Outcome: ketamine efficacy	Outcome: ketamine-related safety
Bi et al. [36]	13 studies/ 1182 patients	IV ketamine (nine studies) or adjunctive ketamine to paravertebral block (three studies) or pectoral nerve block (one study) Breast surgery	IV ketamine: Reduces postoperative pain intensity at 6 h (WMD −0.83, $P = 0.048$) Reduces postoperative pain intensity at 24 h (WMD −0.65, $P < 0.001$) Reduces morphine consumption in the first 24 h post-op (WMD −4.14 mg, $P = 0.035$) Adding ketamine to paravertebral blocks: Reduces postoperative pain intensity at 6 h (WMD −0.59, $P = 0.014$) Reduces postoperative pain intensity at 24 h (WMD −0.9, $P < 0.001$) Reduces morphine consumption in the first 24 h post-op (WMD −4.59 mg, $P < 0.001$) Adding ketamine to pectoral nerve blocks: Does not reduce postoperative pain intensity at 24 h (WMD −0.01, $P = 0.866$) Reduces morphine consumption in the first 24 h post-op (WMD −6.36 mg, $P < 0.001$) Perioperative ketamine use: Reduces incidence of postoperative depression Reduces incidence of postmastectomy pain syndrome at 3 months post-op	Adverse effects reported including psychomimetic effects, drowsiness, hallucinations, nightmares, anxiety, dizziness, headache, and PONV, but no statistical significance was reached for either GI or CNS side effects

			(RR 0.79, $P = 0.04$) with a nonsignificant tendency to reduce its incidence at 1 month (RR 0.83, $P = 0.107$) and 6 months (RR 0.83, $P = 0.348$)	
Covington et al. [64]	17 studies/ 1039 patients	Intra-articular ketamine Predominantly knee surgery	Intra-articular ketamine may result in lower VAS pain scores during the first 4 h post-op Onset of analgesia between 10 min and 8 h Duration of analgesia between 2 and 48 h Dose of ketamine varied between 0.25 mg/kg and 2 mg/kg No meta-analysis of outcome measures due to significant heterogeneity in study design, dosing strategy, and measurement techniques	No adverse effects reported in 10/17 studies. One study did not include any data on adverse effects Reduction in opiate-related side effects when sparing effect seen IA ketamine associated with fewer adverse effects than IM ketamine in one small study Large number of patients in one study reported hallucinations after 1 mg/kg ketamine dose Longer-term safety concerns, e.g., potential chondrocyte toxicity not examined/included
Bai et al. [65]	Total 89 studies, various adjuncts Ketamine 12 studies/ 889 patients	Ketamine and other agents as adjuncts to local anesthetic wound infiltration	Ketamine wound infiltration has overall weak analgesic benefit Modest improvement in opioid consumption, duration of analgesia, and pain scores when compared to saline infiltration Minimal to no benefit when compared to local anesthetic infiltration, either alone or as an adjunct Variable results when compared to systemic ketamine administration	Most studies report minimal or no hemodynamic or psychomimetic symptoms One study reported 10% incidence of hallucinations One study reported increased sedation Overall benefits do not outweigh potential side effects, and authors recommend against use in this context

APGAR appearance, pulse, grimace, activity, reflex (score assessing newborn health), *CNS* central nervous system, *GI* gastrointestinal, *IA* intra-articular, *IM* intramuscular, *MD* mean difference, *OR* odds ratio, *PONV* postoperative nausea and vomiting, *RR* relative risk, *VAS* visual analogue scale for pain assessment, *WMD* weighted mean difference pediatric caudal anesthesia, yet concerns over possible neurotoxicity remain [24]. Definitive answers will ultimately require large, high-quality studies, including those that examine long-term outcomes. Ketamine admixture in total intravenous anesthesia, although technically feasible, has significant clinical limitations. There is certainly a role for co-administration of ketamine and propofol but not necessarily in the same syringe.

Conflicts of Interest

MGI: Editor, *Anaesthesia*; Perioperative Medicine; Expert Opinion Pharmacotherapy; Journal of Clinical Medicine; Hong Kong Medical Journal.
PS and CN declare no conflicts of interest.

References

1. Larry K, John YJ (2001) Ketamine blockade of voltage-gated sodium channels. Anesthesiology 95(6):1406–1413. https://doi.org/10.1097/00000542-200112000-00020
2. Tverskoy M, Oren M, Vaskovich M, Dashkovsky I, Kissin I (1996) Ketamine enhances local anesthetic and analgesic effects of bupivacaine by peripheral mechanism: a study in postoperative patients. Neurosci Lett 215(1):5–8. https://doi.org/10.1016/S0304-3940(96)12922-0
3. Saarelainen IO, Larni HM, Kiistala U (2012) Ketamine and lidocaine as local anesthetics in minor dermatological surgery. Prehosp Disaster Med 13(S1):S18. https://doi.org/10.1017/s1049023x00038371
4. Abuhelwa AY, Somogyi AA, Loo CK, Glue P, Barratt DT, Foster DJR (2022) Population pharmacokinetics and pharmacodynamics of the therapeutic and adverse effects of ketamine in patients with treatment-refractory depression. Clin Pharmacol Ther 112(3):720–729. https://doi.org/10.1002/cpt.2640
5. Aksoy H, Gökahmetoğlu G, Ak M, Aksoy Ü (2023) Subcutaneous wound infiltration of ketamine is superior to bupivacaine in terms of pain perception and opioid consumption after cesarean section: a double-blinded randomized placebo-controlled clinical trial. Eur Rev Med Pharmacol Sci 27(18):8860–8867. https://doi.org/10.26355/eurrev_202309_33806
6. Maloney J, Pew S, Wie C, Gupta R, Freeman J, Strand N (2021) Comprehensive review of topical analgesics for chronic pain. Curr Pain Headache Rep 25(2):7. https://doi.org/10.1007/s11916-020-00923-2
7. Heydari F, Khalilian S, Golshani K, Majidinejad S, Masoumi B, Massoumi A (2021) Topical ketamine as a local anesthetic agent in reducing venipuncture pain: a randomized controlled trial. Am J Emerg Med 48: 48–53. https://doi.org/10.1016/j.ajem.2021.03.055
8. Shillingburg A, Kanate AS, Hamadani M, Wen S, Craig M, Cumpston A (2017) Treatment of severe mucositis pain with oral ketamine mouthwash. Support Care Cancer 25(7): 2215–2219. https://doi.org/10.1007/s00520-017-3627-6
9. Higgins EA, West JA (2021) A novel use of topical ketamine for the treatment of oral pemphigus: a case report. J Palliat Care 36(3): 146–147. https://doi.org/10.1177/0825859720946504
10. Prakash S, Meena JP, Gupta AK et al (2020) Ketamine mouthwash versus placebo in the treatment of severe oral mucositis pain in children with cancer: a randomized double-blind placebo-controlled trial. Pediatr Blood Cancer 67(9):e28573. https://doi.org/10.1002/pbc.28573
11. Christie JM, Jones CW, Markowsky SJ (1992) Chemical compatibility of regional anesthetic drug combinations. Ann Pharmacother 26(9): 1078–1080. https://doi.org/10.1177/106002809202600904
12. Frank SG, Lalonde DH (2012) How acidic is the lidocaine we are injecting, and how much bicarbonate should we add? Can J Plastic Surg

20(2):71–73. https://doi.org/10.1177/229255031202000207

13. Pai A, Heining M (2007) Ketamine. BJA Educ 7(2):59–63. https://doi.org/10.1093/bjaceaccp/mkm008
14. Togal T, Demirbilek S, Koroglu A, Yapici E, Ersoy O (2004) Effects of S(+) ketamine added to bupivacaine for spinal anaesthesia for prostate surgery in elderly patients. Eur J Anaesthesiol 21(3):193–197. https://doi.org/10.1017/s0265021504003059
15. Unlugenc H, Ozalevli M, Gunes Y et al (2006) A double-blind comparison of intrathecal S(+) ketamine and fentanyl combined with bupivacaine 0.5% for caesarean delivery. Eur J Anaesthesiol 23(12):1018–1024. https://doi.org/10.1017/s0265021506000950
16. Kathirvel S, Sadhasivam S, Saxena A, Kannan TR, Ganjoo P (2000) Effects of intrathecal ketamine added to bupivacaine for spinal anaesthesia. Anaesthesia 55(9):899–904. https://doi.org/10.1046/j.1365-2044.2000.01472.x
17. Abd El-Rahman AM, Mohamed AA, Mohamed SA, Mostafa MAM (2018) Effect of Intrathecally administered ketamine, morphine, and their combination added to bupivacaine in patients undergoing major abdominal cancer surgery a randomized, double-blind study. Pain Med 19(3):561–568. https://doi.org/10.1093/pm/pnx105
18. Jan TD, De Haan P et al (2006) Severe toxic damage to the rabbit spinal cord after intrathecal Administration of Preservative-free S(+)-ketamine. Anesthesiology 105(4):813–818. https://doi.org/10.1097/00000542-200610000-00028
19. Vranken JH, Troost D, Wegener JT, Kruis MR, van der Vegt MH (2005) Neuropathological findings after continuous intrathecal administration of S(+)-ketamine for the management of neuropathic cancer pain. Pain 117(1):231–235. https://doi.org/10.1016/j.pain.2005.06.014
20. Pedraz JL, Lanao JM, Calvo MB, Muriel C, Hernández-Arbeiza J, Domínguez-Gil A (1987) Pharmacokinetic and clinical evaluation of ketamine administered by i.v. and epidural routes. Int J Clin Pharmacol Ther Toxicol 25(2):77–80
21. Cook B, Grubb DJ, Aldridge LA, Doyle E (1995) Comparison of the effects of adrenaline, clonidine and ketamine on the duration of caudal analgesia produced by bupivacaine in children. Br J Anaesth 75(6):698–701. https://doi.org/10.1093/bja/75.6.698
22. Wang H-Z, Wang L-Y, Liang H-H, Fan Y-T, Song X-R, She Y-J (2020) Effect of caudal ketamine on minimum local anesthetic concentration of ropivacaine in children: a prospective randomized trial. BMC Anesthesiol 20(1):144. https://doi.org/10.1186/s12871-020-01058-y
23. Aliena SP, Lini C, Chirayath JJ (2018) Comparison of postoperative analgesic effect of caudal bupivacaine with and without ketamine in pediatric subumbilical surgeries. J Anaesthesiol Clin Pharmacol 34(3):324–327. https://doi.org/10.4103/joacp.JOACP_60_17
24. Schnabel A, Poepping DM, Kranke P, Zahn PK, Pogatzki-Zahn EM (2011) Efficacy and adverse effects of ketamine as an additive for paediatric caudal anaesthesia: a quantitative systematic review of randomized controlled trials. Br J Anaesth 107(4):601–611. https://doi.org/10.1093/bja/aer258
25. Yan J, Jiang H (2014) Dual effects of ketamine: neurotoxicity versus neuroprotection in anesthesia for the developing brain. J Neurosurg Anesthesiol 26(2):155–160. https://doi.org/10.1097/ana.0000000000000027
26. Elia N, Tramèr MR (2005) Ketamine and postoperative pain – a quantitative systematic review of randomised trials. Pain 113(1):61–70. https://doi.org/10.1016/j.pain.2004.09.036
27. Sethi M, Sethi N, Jain P, Sood J (2011) Role of epidural ketamine for postoperative analgesia after upper abdominal surgery. Indian J Anaesth 55(2):141–145. https://doi.org/10.4103/0019-5049.79894
28. Mathew LM, Mathew L, Cherian V, David A (2022) Effect of epidural ketamine on pain after total knee arthroplasty: a randomized double- blind placebo controlled clinical trial. Christian J Global Health 9(1):77–86. https://doi.org/10.15566/cjgh.v9i1.641
29. Himmelseher S, Ziegler-Pithamitsis D, Argiriadou H, Martin J, Jelen-Esselborn S, Kochs E (2001) Small-dose S(+)-ketamine reduces postoperative pain when applied with Ropivacaine in epidural anesthesia for total knee arthroplasty. Anesth Analg 92(5):1290–1295. https://doi.org/10.1097/00000539-200105000-00040
30. Yan H, Chen W, Chen Y et al (2023) Opioid-free versus opioid-based anesthesia on postoperative pain after thoracoscopic surgery: the use of intravenous and epidural esketamine. Anesth Analg 137(2):399–408. https://doi.org/10.1213/ANE.0000000000006547
31. Lee JY, Sim WS, Kim KM, Oh MS, Lee JE (2014) The effect of ketamine as an additive in epidural block on the intractable herpetic

neuralgia: a case report. Korean J Anesthesiol 66(1):64. https://doi.org/10.4097/kjae.2014.66.1.64

32. Lee I-O, Kim W-K, Kong M-H et al (2002) No enhancement of sensory and motor blockade by ketamine added to ropivacaine interscalene brachial plexus blockade. Acta Anaesthesiol Scand 46(7):821–826. https://doi.org/10.1034/j.1399-6576.2002.460711.x
33. Clerc S, Vuilleumier H, Frascarolo P, Spahn DR, Gardaz J-P (2005) Is the effect of inguinal field block with 0.5% bupivacaine on postoperative pain after hernia repair enhanced by addition of ketorolac or S(+) ketamine? Clin J Pain 21(1):101–105
34. Rahimzadeh P, Faiz SH, Ziyaeifard M, Niknam K (2013) Effectiveness of adding ketamine to ropivacaine infusion via femoral nerve catheter after knee anterior cruciate ligament repair. J Res Med Sci 18(8):632–636
35. Zhu T, Gao Y, Xu X, Fu S, Lin W, Sun J (2020) Effect of ketamine added to ropivacaine in nerve block for postoperative pain management in patients undergoing anterior cruciate ligament reconstruction: a randomized trial. Clin Ther 42(5):882–891. https://doi.org/10.1016/j.clinthera.2020.03.004
36. Bi Y, Ye Y, Zhu Y, Ma J, Zhang X, Liu B (2021) The effect of ketamine on acute and chronic wound pain in patients undergoing breast surgery: a meta-analysis and systematic review. Pain Pract 21(3):316–332. https://doi.org/10.1111/papr.12961
37. Kazemeini A, Rahimi M, Fazeli MS et al (2014) The effect of local injections of bupivacaine plus ketamine, bupivacaine alone, and placebo on reducing postoperative anal fistula pain: a randomized clinical trial. Sci World J 2014:424152. https://doi.org/10.1155/2014/424152
38. Xiang J, Cao C, Chen J et al (2024) Efficacy and safety of ketamine as an adjuvant to regional anesthesia: a systematic review and meta-analysis of randomized controlled trials. J Clin Anesth 94:111415. https://doi.org/10.1016/j.jclinane.2024.111415
39. Huang GS, Yeh CC, Kong SS, Lin TC, Ho ST, Wong CS (2000) Intra-articular ketamine for pain control following arthroscopic knee surgery. Acta Anaesthesiol Sin 38(3):131–136
40. Rosseland LA, Stubhaug A, Sandberg L, Breivik H (2003) Intra-articular (IA) catheter administration of postoperative analgesics. A new trial design allows evaluation of baseline pain, demonstrates large variation in need of analgesics, and finds no analgesic effect of IA ketamine compared with IA saline. Pain 104(1):25
41. Dal D, Tetik O, Altunkaya H, Tetik Ö, Doral MN (2004) The efficacy of intra-articular ketamine for postoperative analgesia in outpatient arthroscopic surgery. Arthrosc J Arthrosc Relat Surg 20(3):300–305. https://doi.org/10.1016/j.arthro.2003.11.038
42. Kumar Batra Y, Mahajan R, Kumar Bangalia S, Nath Nagi O, Singh DM (2005) Bupivacaine/ketamine is superior to intra-articular ketamine analgesia following arthroscopic knee surgery. Can J Anesth/J Can Anesth 52(8):832–836. https://doi.org/10.1007/bf03021778
43. Borner M, Bürkle H, Trojan S, Horoshun G, Riewendt HD, Wappler F (2007) Intraartikuläre Ketamingabe bei arthroskopischen Knieoperationen. Anaesthesist 56(11):1120–1127. https://doi.org/10.1007/s00101-007-1249-y
44. Isik C, Demirhan A, Yetis T et al (2015) Efficacy of intraarticular application of ketamine or ketamine-levobupivacaine combination on post-operative pain after arthroscopic meniscectomy. Knee Surg Sports Traumatol Arthrosc 23(9):2721–2726. https://doi.org/10.1007/s00167-014-2962-0
45. Salman N, Olgunkeleş B, Bektaş U et al (2019) Effects of intraarticular tramadol, magnesium and ketamine on postoperative pain in arthroscopic meniscectomy. Braz J Anesthesiol 69(1):35–41. https://doi.org/10.1016/j.bjane.2018.09.002
46. Sağır Ö (2020) Effects of intraarticular ketamine combined with periarticular bupivacaine on postoperative pain after arthroscopic meniscectomy. Joint Dis Relat Surg 31(3):589–596. https://doi.org/10.5606/ehc.2020.75274
47. Khashan M, Dolkart O, Amar E et al (2016) Effect of preemptive intra-articular morphine and ketamine on pain after arthroscopic rotator cuff repair: a prospective, double-blind, randomized controlled study. Arch Orthop Trauma Surg 136(2):233–239. https://doi.org/10.1007/s00402-015-2346-z
48. Stueber T, Karsten J, Stoetzer C, Leffler A (2014) Differential cytotoxic properties of drugs used for intra-articular injection on human chondrocytes: an experimental in-vitro study. Eur J Anaesthesiol 31(11):640–645. https://doi.org/10.1097/EJA.0000000000000121
49. Ozturk AM, Ergun MA, Demir T, Gungor I, Yilmaz A, Kaya K (2014) Ketamine is toxic to chondrocyte cell cultures. Bone Joint J 96(7):989–994. https://doi.org/10.1302/0301-620x.96b7.33005
50. Lu W, Wang L, Wo C, Yao J (2016) Ketamine attenuates osteoarthritis of the knee via modulation of inflammatory responses in a rabbit

model. Mol Med Rep 13(6):5013–5020. https://doi.org/10.3892/mmr.2016.5164
51. Donnelly R, Willman E, Andolfatto G (2008) Stability of Ketamine–propofol mixtures for procedural sedation and analgesia in the emergency department. Can J Hosp Pharm 61:426
52. Izgi M, Basaran B, Muderrisoglu A, Ankay Yilbas A, Uluer MS, Celebioglu B (2018) Evaluation of the stability and stratification of propofol and ketamine mixtures for pediatric anesthesia. Paediatr Anaesth 28(3):275–280. https://doi.org/10.1111/pan.13318
53. Pauline T, Lucie B, Pascal B (2018) Validation of a procedure to mix homogenous solutions in bags and syringes. Eur J Hosp Pharm 25(e2): e126–e1e9. https://doi.org/10.1136/ejhpharm-2017-001284
54. Nestor CC, Sepulveda VP, Irwin MG (2022) Drug mixtures and infusion technology. MasterChef or hell's kitchen? Anaesthesia 77(3): 260–263. https://doi.org/10.1111/anae.15561
55. Kursumovic E, Cook TM, Lucas DN et al (2023) The 7th National Audit Project (NAP7) baseline survey of individual anaesthetists: preparedness for and experiences of perioperative cardiac arrest. Anaesthesia 78(12): 1453–1464. https://doi.org/10.1111/anae.16154
56. Rampil IJ (1998) A primer for EEG signal processing in anesthesia. Anesthesiology 89(4):980–1002. https://doi.org/10.1097/00000542-199810000-00023
57. Akeju O, Song AH, Hamilos AE et al (2016) Electroencephalogram signatures of ketamine anesthesia-induced unconsciousness. Clin Neurophysiol 127(6):2414–2422. https://doi.org/10.1016/j.clinph.2016.03.005
58. Purdon PL, Sampson A, Pavone KJ, Brown EN (2015) Clinical electroencephalography for anesthesiologists: part I: background and basic signatures. Anesthesiology 123(4): 937–960. https://doi.org/10.1097/aln.0000000000000841
59. Tsuda N, Hayashi K, Hagihira S, Sawa T (2007) Ketamine, an NMDA-antagonist, increases the oscillatory frequencies of alpha-peaks on the electroencephalographic power spectrum. Acta Anaesthesiol Scand 51(4): 472–481. https://doi.org/10.1111/j.1399-6576.2006.01246.x
60. Hirota K, Kubota T, Ishihara H, Matsuki A (1999) The effects of nitrous oxide and ketamine on the bispectral index and 95% spectral edge frequency during propofol-fentanyl anaesthesia. Eur J Anaesthesiol 16(11): 779–783. https://doi.org/10.1046/j.1365-2346.1999.00585.x
61. Du J, Li C, Zhang W, Lu X, Xia Y, Qin X (2024) Efficacy and safety of adding ketamine to lidocaine in intravenous regional anesthesia: a meta-analysis of randomized controlled trials. J Perianesth Nurs 40:195. https://doi.org/10.1016/j.jopan.2024.03.013
62. Sohnen S, Dowling O, Shore-Lesserson L (2021) Single dose perioperative intrathecal ketamine as an adjuvant to intrathecal bupivacaine: a systematic review and meta-analysis of adult human randomized controlled trials. J Clin Anesth 73:110331. https://doi.org/10.1016/j.jclinane.2021.110331
63. Shah UJ, Karuppiah N, Karapetyan H, Martin J, Sehmbi H (2022) Analgesic efficacy of adjuvant medications in the pediatric caudal block for Infraumbilical surgery: a network meta-analysis of randomized controlled trials. Cureus 14:e28582. https://doi.org/10.7759/cureus.28582
64. Covington SM, D'Souza RS, Erwin PJ, Qu W (2020) A systematic review of intra-articular ketamine for postoperative analgesia. Am J Phys Med Rehabil 99(11):1039–1047. https://doi.org/10.1097/PHM.0000000000001483
65. Bai JW, An D, Perlas A, Chan V (2020) Adjuncts to local anesthetic wound infiltration for postoperative analgesia: a systematic review. Reg Anesth Pain Med 45(8):645–655. https://doi.org/10.1136/rapm-2020-101593

Chapter 4

Relief from Chronic Neuropathic Pain with Ketamine but Increased Burden of Psychedelic Adverse Effects?

Jose Eduardo Guimaraes Pereira, Carlos Darcy Alves Bersot, Thiago Ramos Grigio, Theodoros Aslanidis, and Sabine Himmelseher

Abstract

Neuropathic pain (NP) is a frequently undertreated component of chronic pain, posing challenges due to its complexity. This has led to interest in N-methyl-D-aspartate receptor (NMDAR) antagonists like ketamine. NP is linked to maladaptive processes, such as NMDAR-related windup and sensitization. While results from ketamine trials on chronic NP have been inconsistent, perioperative use has shown potential to reduce chronic postsurgical pain and NP. For chronic NP, intravenous racemic ketamine at doses of 0.5–1.5 mg/kg/day for 4–5 days is recommended as an initial hospital-based approach. When effective, this provides pain relief lasting 1–3 months. Non-IV administration with de-escalating doses can then be tailored to individual patients, and infusions may be repeated quarterly if necessary. However, neurocognitive side effects can cause discomfort or treatment discontinuation. Psychological guidance during treatment helps patients remain engaged, and using gamma aminobutyric acid (GABA)-ergic agents beforehand can reduce adverse effects. Despite some challenges, the potential benefits of ketamine in chronic NP management often outweigh the risks. Further high-quality trials are urgently needed to refine ketamine use for NP.

Key words Ketamine, Chronic pain, Neuropathy, Psychoactive events, Chronic postsurgical pain prevention, Use schedule, Comorbidity, Depression, Anxiety, Stress disorder

1 Introduction

The International Association for the Study of Pain defines neuropathic pain (NP) as severe pain caused by a lesion or disease of the somatosensory nervous system (SNS) that is limited to a "neuroanatomically plausible" distribution. There may be a loss of sensation in the territory innervated by injured nerves that corresponds to a spinal or brain region with direct or indirect damage [1]. A feature of NP is the paradox of sensory loss and pain which may coexist with or without sensory hypersensitivity from aberrant nerve signaling. Table 1 presents a classification of the various forms of NP.

Sabine Himmelseher (ed.), *Ketamine: Pain Management, Critical Care, Depression, and Resource-Poor Settings Volume II*, Neuromethods, vol. 225, https://doi.org/10.1007/978-1-0716-4599-4_4,

Table 1
Neuropathic pain: classification including overlapping entities

Peripheral focal	Symmetric	Central neuropathic pain
Post-traumatic neuropathy	Polyneuropathy (PNP)	Brain infarct
Diabetic mononeuropathy	Diabetic PNP	Thalamus, brainstem
Postherpetic neuralgia	Chemotherapy-related	After brain trauma/class
Trigeminal neuralgia	PNP	After spinal cord injury
Radiculopathy	Toxic PNP (alcohol solvents, lead)	Neuroinflammation
Sciatica		Myelitis, multiple sclerosis
Carpal tunnel syndrome	Vit. B12 deficiency	
Small fiber neuropathy	Small fiber neuropathy	
Phantom limb pain		Phantom limb pain
Vasculitis	Vasculitis	Vasculitis
Gluten neuropathy	Gluten neuropathy	
COVID-19-related acute phase pain and long COVID-19-related pain		

The prevalence of pain varies according to the specific type of neuropathy. It is important to differentiate the etiology, for example, in the case of painful trigeminal disorder, where classical neuralgia requires a different treatment plan than trigeminal neuropathy resulting from a herpetic zoster infection. *COVID-19* coronavirus disease-2019, *PNP* polyneuropathy, *Vit.* vitamin

While NP and nociplastic forms of pain are frequently observed concurrently, the former arises from nociceptive stimuli and functional changes in the SNS, without evidence of structural damage.

The World Health Organization (WHO) has estimated that 22% of patients receiving primary care suffer from chronic pain that has not been adequately treated and is a significant source of distress. Chronic pain is a disease that needs to be addressed by all health professionals [2–5]. In the absence of a "gold standard" for diagnosis of chronic NP, the accuracy of prevalence data is uncertain. However, evidence suggests that up to 10% of the global population may be affected by chronic NP [4–6].

In addition to an evaluation of a patient's symptoms and medical history, the diagnostic workup of a pain patient is frequently complex including the examination of location and quality of pain [7]. Clinical NP manifests as ongoing or intermittent spontaneous pain, described as burning, shooting, pricking, pins and needles, squeezing, freezing, itching pain or as repeated paroxysmal pain akin to that experienced in a series of electric shocks [8]. There may be dysesthesia (painful abnormal sensations) or paresthesia (abnormal not so unpleasant sensations). Evoked pain often occurs in addition to spontaneous pain and includes touch- or cold-evoked pain, allodynia, hyperalgesia, and after-sensations. For initial detection, "screening" questionnaires may be useful: The Douleur Neuropathique en 4 Questions (DN4) and the Leeds Assessment of Neuropathic Symptoms and Signs (LANSS) have recently received a strong, and the PainDETECT and quantitative sensory and evoked potential tests a weak recommendation with regard to diagnostic utility [7, 9]. Spontaneous and evoked pain may be caused by different or overlapping mechanisms due to preserved or destroyed afferent nerve fibers. Hyperpathia with an explosive painful reaction may occur after repeated stimuli in affected areas with increased pain threshold. Although there is a weak correlation between patient-reported pain and quantitative sensory testing, the early onset of evoked pain or postoperative hypersensitivity has been linked to the development of chronic NP [7, 8, 10]. A discriminative feature of NP is the insufficient response to conventional analgesics such as nonsteroidal anti-inflammatory drugs (NSAIDs).

Medications effective in other disorders may exert a partial effect in the context of NP. Guidelines recommend tricyclic antidepressants, serotonin-norepinephrine uptake inhibitors, gabapentinoids, anticonvulsants, or local lidocaine or capsaicin applications [11–13]. But prognosis for NP is poor, with a <50% response to clinically available drugs.

Animal models show a multitude of pathophysiological cascades involved in NP development and maintenance [1, 8]. This may in part explain the high incidence of comorbidities such as depression [2]. However, underlying etiologies are not completely understood so that better management of symptoms often becomes focus of therapy.

In human pain models and patient neuroimaging, converging evidence with clinical relevance shows that ionotropic glutamate (Glu) NMDARs are important in pain transmission, neuroplasticity, and sensitization [14]. Targeting altered NMDAR function and maladaptive system change in NP may therefore be a more mechanistic-oriented approach to NP management. Indications for the NMDAR antagonist ketamine have recently been expanded to use as a rapid antidepressant, anti-hyperalgesic, and analgesic adjunct in perioperative multimodal regimen and as a potential remedy in NP [15–18]. This chapter aims to elucidate the pathophysiological background and significant developments in the use of ketamine in NP to inform the most appropriate therapeutic approaches.

2 Pathophysiology of Neuropathic Pain

2.1 Peripheral Mechanisms

Peripheral mechanisms contribute to transition from acute to chronic pain [1, 8]. In active settings, nociceptive neurons that have been stimulated convey high-threshold noxious stimuli to the central nervous system (CNS). This includes ascending nociceptive neural pathways which can develop hyperexcitability and ectopic pacemakers after peripheral sensitization in SNS injury. Severe and prolonged painful stimulation upregulates physiological receptor signals and transcription and manifests nociception-increasing mediators (e.g., through release of vesicles containing pro-inflammatory cytokines and growth factors [19]). The process of inflammation is propagated and matched by cytokines directly released from all kinds of injured cells which bind to neural and glial receptors, triggering enzymatic cascades and transcriptional response.

Further downstream, phosphorylation processes may result in the insertion of new channels into neuronal membrane, thereby increasing sensitivity and enhancing the action potential. Subsequently, responding cells may migrate or otherwise disseminate signals that further sensitize or activate surrounding nociceptors originally outside of the primary nociceptive field. Increased transcription of nociceptive mediators also occurs after ligand binding, finally resulting in the formation of new protein complexes which are transported to nociceptive nerve terminals. Upon release, an increased expression of (voltage-gated) cation channels and ligand-gated receptors associated with ectopic discharge occurs [20]. In addition, SNS damage results in the activation of glial cell plasticity including nociceptor expression [21]. The production of neuropeptides including cyto- and chemokines, bradykinin, histamine, and prostaglandins alters the focal neurochemistry at peripheral nociceptor terminals. This augments the already increased sensitivity to stimuli and enhances ectopic activity [22]. Neuroinflammation is considered as some kind of self-reinforcing phenomenon: The vast majority of nociceptive nerve endings possess receptors for the molecules they are capable of releasing, so that neurons entirely unaffected by the original stimulus can be activated. All of these processes may underlie the emergence and maintenance of NP [1, 8, 10].

2.2 Nerve, Root, and Spinal Mechanisms

Aδ- and C-fibers carry pain signals in peripheral nerves. At the level of the lateral spinal neural foramen, the peripheral nerve divides into ventral and dorsal roots. Motor and sensory fibers separate. The dorsal root ganglion (DRG) is located at the medial aspect of the foramen and continues as the dorsal root to the spinal cord. It emanates into several rootlets and enters at the posterolateral sulcus [23]. The proximal DRG axon passes from the ganglion to the spinal cord. It ascends spinal levels in Lissauer's tract of posterolateral white matter. Sensory fibers divide into synapses on second-order neurons in various Rexed laminae which depend on the type of sensory information transmitted [24]. The primary neurotransmitter released from first-order neurons is Glu. It induces postsynaptic excitation through receptor binding at ionotropic cation

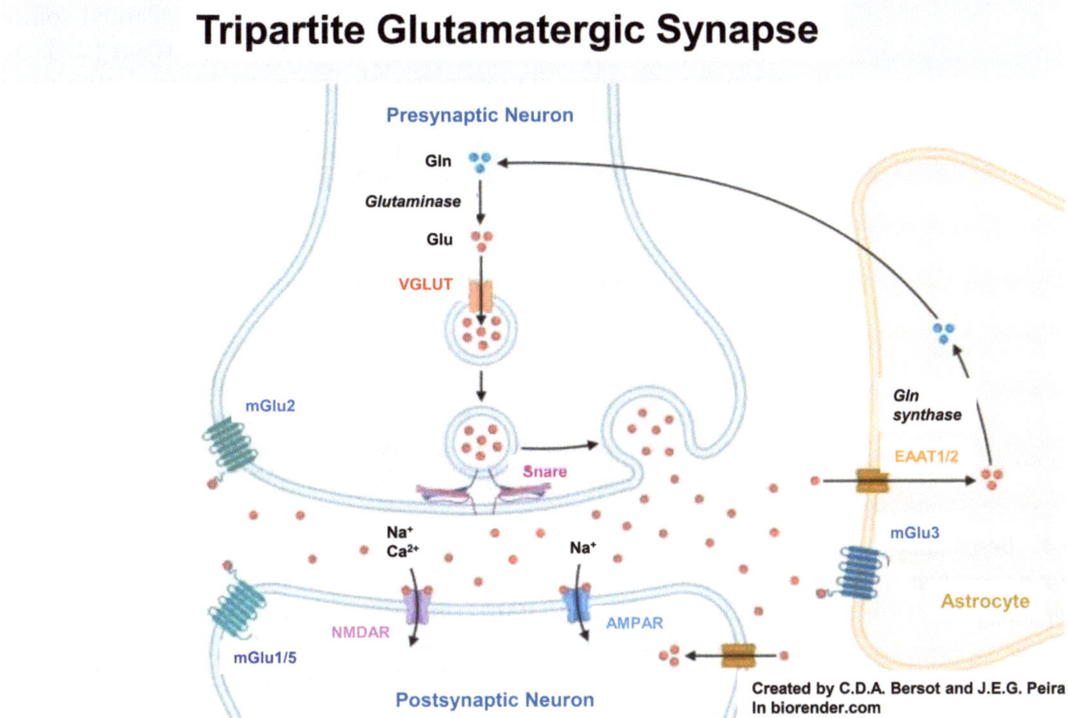

Fig. 1 The tripartite synapse concept describes the propagation of pain signals from first-order to second-order neuron at the dorsal horn. Presynaptic release of glutamate causes postsynaptic excitation through receptor binding at ionotropic cation channels. Repeated and/or prolonged activation of presynaptic fibers leads to a "windup" in response and increased signal transmission in postsynaptic fibers. This is part of long-term potentiation. In pain pathways, such algogenic mechanisms reinforce the phenomenon of central sensitization (created by C.D.A. Bersot and J.E.G. Pereira (2024) in BioRender.com/k97p918). *AMPAR* α-amino-3-hydroxy-5-methyl-4-isoxazolepropionic acid receptor, *EAAT* excitatory amino acid transporter, *Gln* glutamine, *Glu* glutamate, *mGlu* metabotropic glutamate receptor, *NMDAR* N-methyl-D-aspartate receptor, *VGLUT* vesicular glutamate transporter

channels like α-amino-3-hydroxy-5-methyl-4-isoxazolepropionic acid (AMPA), NMDA, and kainate receptors. In addition to Glu, C-fibers release peptides such as substance-P and calcitonin gene-related peptide (CGRP) [25] (Fig. 1).

In severe noxious stimulation from peripherally sensitized C-fibers and/or upregulated nociception, the interaction at postsynaptic NMDAR results in the removal of magnesium from the channel, thereby enabling a prolonged duration of opening and an increased transmission. This process is referred to as "windup" [1, 14, 26].

2.3 Spinal, Subcortical, and Cortical Mechanisms

Subsequent to the transmission's propagation to the spinal cord, second-order projection axons reach out to the brainstem and thalamus via synaptic connections. There is a direct anterolateral pathway to the thalamus, the spinothalamic tract, an indirect anterolateral pathway to the reticular formation, and collateral spinomesencephalic, spinotectal, and spinohypothalamic tracts [27]. An extensive circuitry connects the thalamus to cortical structures involved in perception, localization, and modulation of painful stimuli. It can be divided into two separate systems: one is the sensory-discriminative system, which includes the lateral thalamus and primary and secondary somatosensory cortices (SI and SII, respectively). It processes information related to nociceptive input, including the quality, intensity, and localization of pain. The other is the cognitive-affective system, which incorporates regions such as the anterior insula (AI) and anterior cingulate cortex (ACC). This system is responsible for mediating the psychological aspects of pain [28]. Central sensitization is established when excitation thresholds are reduced, spontaneous ectopic discharge is increased, and hyperexcitability is fixed [14].

2.4 Descending Antinociceptive Signaling

Descending modulation of transmission is a decisive process for endogenous pain suppression. Important players are the periaqueductal gray matter (PAG) and the rostral ventromedial medulla (RVM). The PAG sends axons to the RVM, where they synapse on raphe nuclei and gigantocellular nucleus cells. Axons from these serotonergic cells synapse on second-order DH neurons and release serotonin (Fig. 2) [29].

This inhibits pain transmission [30]. Maladaptive alterations in descending pathways are also involved in NP development and maintenance.

2.5 Cortical and Subcortical Signaling

PAG is an important center from which projections to the RVM relay inhibitory effects that travel to the DH. In addition to extensive pain modulation, reciprocal connections participate in acute to chronic pain transition. Thalamocortical dysregulation in peripheral injury affects calcium channels with the result of theta-frequency bursts of calcium spikes. After cortical projection, the aberrant oscillatory activity contributes to NP sensation [31]. NP is considered to be—at least in part—the result of neural sensitization phenomenon and maladaptive alterations in systems neurobiology. The precise mechanisms underlying the development of aberrant network communication in (nociceptive) neural signal processing and information transfer remain to be elucidated. Chronic NP persists after the initial injuries have healed. It often presents as an exaggerated response to no, subthreshold, normal, or painful stimulation which outlasts stimulus duration [1, 8].

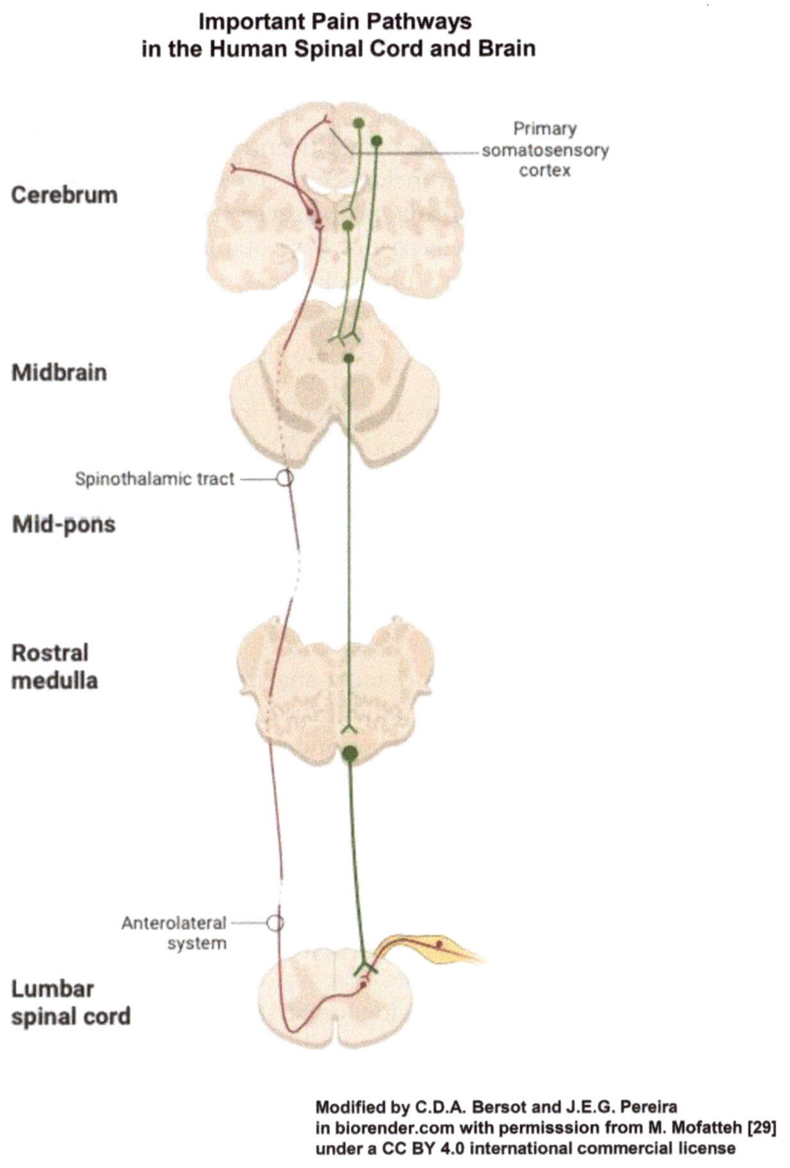

Fig. 2 The graphic illustrates neural ascending pain and descending inhibitory pathways. Red lines: The ascending pain pathway involves first-order neurons which are connected with second-order neurons by synapses at the spinal dorsal horn; pain signals are conveyed further along the spinothalamic tract to the thalamus; subsequently, following at least one additional synapse, signals reach the cerebral cortex, the hypothalamus, and the limbic system. Green lines: The descending inhibitory pathway is constituted by signals that originate in the cortex and thalamus travelling to the periaqueductal gray matter; after synaptic transmission in this important relay, signals reach the reticular formation at the rostral medulla; following another synaptic transfer, they connect downward to the dorsal horn of the spinal cord where they activate interneurons; these eventually inhibit pain signals that arrive at the ascending synapse through release of inhibitory neurotransmitters (modified by C.D.A. Bersot and J.E.G. Pereira (2024) in BioRender.com/l13q420 with permission and from Mofatteh et al. [29] under a CC BY 4.0 international commercial license; https:/ creativecommons.org/licenses/by/4.0/)

3 Perioperative Pain Prevention with Ketamine to Prevent Chronic Postoperative and Neuropathic Pain Development

In 2024, a European multicenter prospective observational trial reported a mean incidence of 10.5% of chronic postsurgical pain (CPSP) at 6 months postsurgery [32]. The prevalence of neuropathic pain characteristics was notable, occurring in 33% of sternotomy procedures, 68% of breast surgeries, 42% of total knee arthroplasties, and 41% of endometriosis cases. NP had a markedly deleterious impact on psychic and functional well-being, as well as on the quality of life of patients. In perioperative multimodal analgesia, substantial evidence indicates that subanesthetic dose ketamine improves postoperative pain relief, prevents pain and hyperalgesia, and reduces opioid needs if applied in appropriate patient-, procedure-, and setting-related schedules [33]. It may improve recovery from surgery with better sleep, mood, and rehabilitation. Inadequate perioperative pain therapy, however, increases the odds of developing complications such as CPSP and NP [34].

A recent network meta-analysis of over 20,000 patients demonstrated a reduction in the incidence of CPSP and severity following the administration of perioperative non-opioid analgesics including ketamine until 6 months after surgery [35]. Nevertheless, confidence in the results is low due to different baseline risks for postsurgical pain and possible trial imprecision. A subgroup meta-analysis of six randomized clinical trials (RCTs) investigating the development of chronic postsurgical NP found a reduction with ketamine at 3 months after surgery, but evidence was of low certainty [36].

In light of the evidence pertaining to pain pathophysiology and the dose–response relationship in acute pain, it is our contention that perioperative ketamine may prove an effective means of reducing the incidence of chronic postsurgical neuropathic pain [6, 33]. Our hypothesis is corroborated by preclinical data showing reduced NP in an animal model [37] and a recent meta-analysis reporting beneficial effects of ketamine in CPSP prevention [38]. Nevertheless, the evidence base is of low quality, and there is a paucity of knowledge regarding the most effective perioperative ketamine regimen for the prevention of NP.

As part of the NIH Helping to End Addiction Long-term initiative to address the opioid public health crisis in the USA, an ongoing RCT in oncologic mastectomy patients compares continuous perioperative ketamine infusion against one postoperative injection or placebo [39]. Because post-mastectomy syndrome is associated with a high incidence of NP and women will be evaluated until 1 year after surgery, this study will add valuable information to guide ketamine use for prevention of chronic NP.

4 Ketamine and Chronic Neuropathic Pain

The potential of ketamine as a pain-relieving agent in the treatment of chronic pain and NP is increasingly recognized. Recent systematic reviews conclude [6, 17, 40]: (1) Introducing ketamine to chronic NP therapy may result in additional benefit as compared to other clinically available NMDAR antagonists, (2) it is not yet clear how and for how long a decrease in NP persists after discontinuation of ketamine, (3) there is wide heterogeneity in settings and regimen, (4) there is a knowledge gap to inform use schedules for "optimum" efficacy, and (5) there are almost no head-to-head trials comparing ketamine with other approaches.

Ketamine is a racemic mixture of its stereoisomers esketamine and arketamine. IV esketamine has been officially approved as anesthetic, acute care analgesic, and sedative in acute and ICU therapy in a number of European and South American countries as well as China, and intranasal esketamine as antidepressant in the USA, Canada, Europe, and other countries. In addition to being twice as potent as racemic ketamine, esketamine offers the advantage of shorter neurocognitive recovery and better maintenance of logical functions [33]. It exerts less unwanted psychoactive effects and a more rapid emergence at a dose reduction of 50% of the racemate. For the purpose of this review, the term "ketamine" refers to racemic ketamine, and esketamine to its S(+)-isomer.

In NP, almost no investigation preselected ketamine responders (defined as 30–50% reduction in pain rating) versus nonresponders in neuroimaging analysis. The identification of reliable correlates or markers indicating positive effects of ketamine on abnormal network connectivity in NP [41] or other alterations of brain regions involved in NP experiences would undoubtedly be beneficial prior to ketamine administration. Relief from temporal summation of pain (TSP) reflecting sensitization in chronic NP [14, 42] and positive reaction to sympathetic blocks in CRPS [43] predicted a response to ketamine but may have been obscured by the presence of affective components or placebo effects. Functional MRI (fMRI) estimates brain connectivity and activity at rest or during tasks. Altered structural and dynamical connectivity of the default mode network (DMN: medical prefrontal cortex (mPFC), posterior cingulate cortex, and precuneus) and the salience network (SN: anterior insula, anterior cingulate cortex, and ventral striatum) has been reported in a spectrum of NP disorders, ranging from classical trigeminal neuralgia to distal NP in HIV [44–48]. fMRI showed that pretreatment, dynamic functional connectivity of the descending antinociceptive pathway may predict ketamine response and that pain relief results from neuroplastic reduction of cross-network connectivity of the DMN with sensory motor and SN [42, 47]. Diffusion-weighted imaging based on

MRI techniques allows for comprehensive assessment of white matter (WM) organization with high spatial resolution. It may predict pain after brain trauma and the transition from acute to chronic pain [48]. In ketamine responders, a recent trial reports no difference in WM structure within the DMN and the mPFC-PAG pathways as compared to nonresponders at 1 month following the initiation of therapy [49]. But pretreatment fiber density in the anterior limb of the internal capsule correlated with ketamine-induced NP relief, and pretreatment fiber density in the DMN and mPFC-PAG negatively correlated with TSP. Metrics of WM structure in pain pathways may therefore become a predictor of ketamine response and therapeutic benefits. The NP intervention protocol consisted of a targeted ketamine infusion at a dose range from 0.5 to 2 mg/kg/h for 6 h per day at 5 consecutive days. Dosages were individually titrated to achieve maximal NP relief while minimizing adverse and psychoactive effects. This was followed by oral ketamine intake at a dose of 0.5 mg/kg three times per day. NP etiology in responders and nonresponders overlapped.

Clinical evidence from study of drug combinations indicates that the administration of ketamine in conjunction with other analgesics, including paracetamol and NSAIDs, may result in enhanced NP relief. Effects may last up to 2 weeks [50] and in some cases with CRPS even 3 months [51]. The analysis of cerebrospinal fluid (CSF) metabolomics from chronic NP patients [52] and healthy volunteers [53] has revealed that paracetamol, ketamine, and metabolites can be readily detected. CSF changes featuring drug effects have been observed.

The synopsis of clinical trials shows that intravenous (IV) ketamine may induce a substantial and sustained reduction of NP with benefits that persist beyond the duration of the infusion. The duration of effect is likely longer after IV use than via any other route. In severe or refractory NP, treatment should therefore be started with an IV infusion. Recently, French pain experts highly consented that ketamine is a treatment of priority in NP settings, that IV infusions should be the initial approach for use, and that doses from 0.5 to 0.9 mg/kg/day should initially be applied for 4 days. Treatment may be repeated every 3 months in a hospital facility as in- or outpatient [54]. Our recent meta-analysis found that ketamine reduced the intensity of NP by 28–46% compared to pretreatment if infused at a dose of 0.5–1.5 mg/kg/day for maximal 10 days. This is expected to reduce NP for at least 1 month [6] (Fig. 3).

In the event of a favorable response to treatment, the dosage may be reduced on an individual basis, and/or an alternative route of administration, other than IV, may be considered.

In light of the pathophysiology of nociception, clinical experience with ketamine, and the findings of our systematic review, we believe that ketamine may offer potential advantages in the

Infobox

Proposal for Initial Ketamine Regimen to Approach Chronic Neuropathic Pain in Hospital

- ✓ Intravenous ketamine:
  0.15 – 1.5 mg/kg / day
- ✓ Duration of therapy: 4 to 10 days
- ✓ 1 Month after treatment: evaluation of response
- ✓ If success: titrate to smaller dose
  according to patient reaction and / or use other than IV route
- ✓ If necessary:
  repeat IV schedule quarterly with close monitoring for unwanted effects and steady re-evaluation of therapy
- ✓ Esketamine would have to be used at half-dose racemic ketamine

Fig. 3 The infobox illustrates an initial clinical ketamine regimen to approach chronic neuropathic pain (created by the authors in Microsoft Word, Microsoft Corp., Redmond, WA, USA)

management of NP. As described in other parts of the book, this appears also valid because ketamine exerts a myriad of effects beyond NMDAR antagonism, such as blockade of voltage-gated ion channels, inhibition of serotonin uptake, regulation of inflammatory disbalances, and many more with potential benefit in NP [1, 8]. Its neuroplastic properties may not only mitigate but reverse aberrant brain network and spinal communication and information transfer. When IV subanesthetic dose esketamine was infused in a patient with chronic peripheral arterial occlusive disease, esketamine showed enhanced CNS binding activity during periods of chronic ischemic NP as compared to periods of rest in positron emission tomography (Fig. 4).

(Es) ketamine also effectively counteracts opioid-induced respiratory depression which may be crucial in the context of high-dose opioid use in NP [55].

4.1 Specifics with Central Neuropathic Pain

Central NP is caused by diseases or lesions of the brain or spinal cord (Table 1). It is hardly possible to predict which individuals will develop central NP [56], but post-traumatic headache is one of the most prevalent complications following brain trauma, with a lifetime prevalence exceeding 80% [29, 48]. Central poststroke pain represents the most prevalent form of central NP globally, with the potential to cause disability and immobility. Additionally, indirect insults, such as the overuse of an unaffected extremity, may also

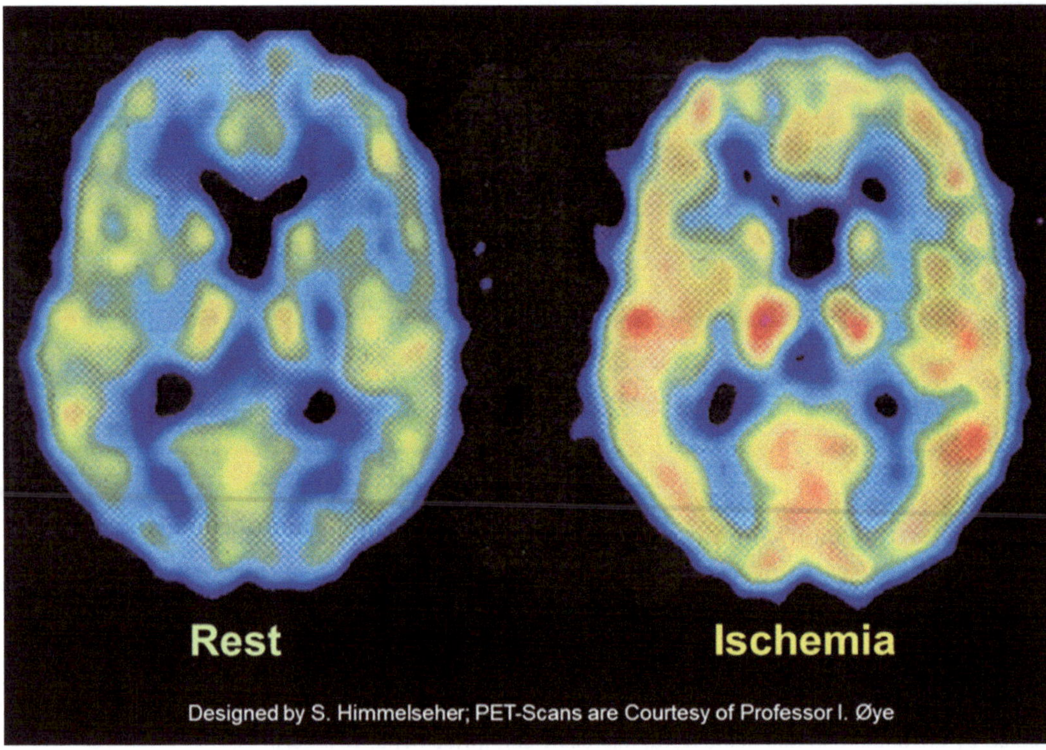

Fig. 4 Scans show the contrast between conditions at rest and chronic ischemic neuropathic pain as observed in PET images using (S) [^{11}C] ketamine as radiotracer: An increased and altered esketamine binding in ischemic regions is indicated by elevated levels of activity in areas with warmer colors, such as yellow and red (designed by S. Himmelseher in Microsoft Power Point, Microsoft Corp., Redmond, WA, USA; PET-Scans by courtesy of Professor I. Øye, Oslo University School of Medicine, Department of Pharmacology, Oslo, Norway). *PET* positron emission tomography

result in the development of this condition [57]. In debilitating poststroke pain, even oral ketamine at 30 mg increased to 50 mg three times per day reduced NP allowing for a massive opioid reduction and enormous increase in quality of life [56, 58]. After spinal cord injury, more than 80% of injured patients develop chronic NP within 1 year, but they may also be those who experience greatest relief with ketamine [59–61]: For example, when 300 mg gabapentin thrice per day was combined with one epidural injection of 0.2 mg/kg preservative-free ketamine, NP significantly decreased up to 1 month [59], or when combined with 80 mg/kg IV ketamine infused over 5 h per day for 10 days, NP was significantly reduced up to 2 weeks [61].

Taken together, the evidence for the efficacy of ketamine as a treatment for chronic central neuropathic pain (CNP) is inconclusive. This is due to the inconsistency of findings across heterogeneous trials which have employed various administration routes and schedules in varying conditions [60, 62]. The majority of patients with central NP will likely benefit from a multimodal, best individually tailored, anti-pain management strategy.

4.2 Peripheral Neuropathic Pain

Peripheral NP develops from a disorder or damage of the peripheral nervous system (PNS) and may be associated with a large disease burden [1]. It has been addressed with ketamine in various kinds of use including IV [63–65], oral [66], and topical [67] application. For example, 0.4 mg/kg IV ketamine reduced post-traumatic peripheral NP better than 2.5 mg/kg IV lidocaine [64]. In a case series of trigeminal neuralgia, IV or IM esketamine (0.4–0.5 mg/kg) always reduced NP, whereas racemic (0.5–1 mg/kg) or arketamine (0.5–2 mg/kg) did not [68] (Fig. 5).

When first titrated in hospital to an effective daily dose of 1.5–3 mg/kg IV ketamine and then converted to daily oral intake with a mean effective dose of 2 mg/kg administered over 3 months, a retrospective study reported significant chronic and NP relief in a 5-year observation period [69]. Although evidence about oral intake is still limited, promising data show a greater reduction in shooting, burning NP sensations and less allodynia with 30 mg oral esketamine as compared to 3 mg methadone given thrice per day [69]. Oral ketamine opens a window of opportunity for longer-term use as long as adverse effects are kept under control. It allows for daily administration, reducing risks of low efficacy from intermittent application with too long drug-free intervals. Because

Pt	Age Years	NP	Ketamine dose mg / kg	Without pain Duration
1	53	1	0.9 IM Rac / 0.4 IV Esket	1 / 3 Days
2	42	2	1.8 IV Rac	1 Day
3	42	3	0.4 IV Esket	1 Day
4	54	5	2 IM Arket	No Effect
5	69	11	0.5-1 IM Rac / IV Arket	No Effect
6	57	12	1 IV Arket / 0.5 IV Rac	No Effect
			0.5 IM Esket	2 Days
7	79	21	0.5 IV Rac	During infusion

Designed by S. Himmelseher with Corel Draw 6.0 according to data in [68]

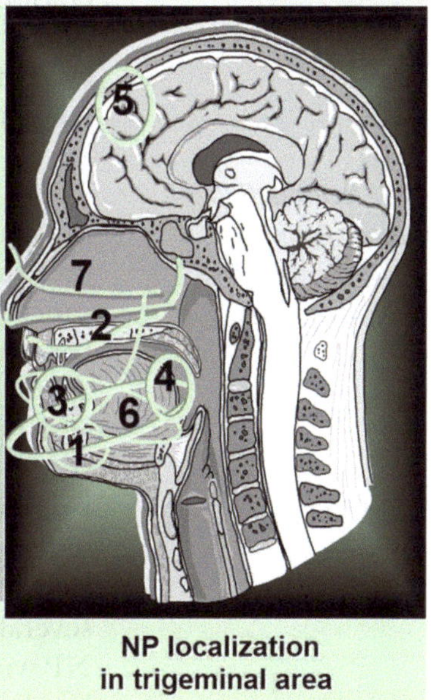

Fig. 5 Illustration indicates the cases of seven female patients with trigeminal neuropathy: Transient pain relief was achieved after a single subanesthetic dose of esketamine but not after arketamine (designed by S. Himmelseher with Microsoft Power Point, Microsoft Corp., Redmond, WA, USA, and an icon from Corel Draw Graphics 6.0, Corel Corp., Ottawa, ON, Canada, according to data in Mathisen et al. [68]). *IV* intravenous, *IM* intramuscular, *NP* neuropathic pain, *Pt* patients, *arket* arketamine, *esket* esketamine, *rac* racemic ketamine

ketamine's oral bioavailability is only 20% due to first-pass metabolism, innovative preparations are examined: In phase trials, an oral thin film with esketamine for sublingual or buccal use has demonstrated pretty predictable absorption achieving high drug plasma and metabolite concentrations with relatively long-lasting antinociceptive effects [70]. And an extended-release bonbon with racemic ketamine was effective as antidepressant with minimal cardiovascular stimulating or unwanted psychoactive effects [71]. Topical ketamine would offer the advantage of great reduction of adverse effects being basically restricted to skin irritation. In peripheral NP, 1% ketamine cream achieved only insignificant pain relief [67], but in refractory cancer pain, a 10% Lipoderm cream was more effective than nerve blocks and reduced opioid needs [72]. In addition, a novel sustained-release esketamine nanoparticle–hydrogel delivery system has been shown to provide sustained analgesia after spinal nerve ligation in a mouse model due to its ability to inhibit DH neuron overexcitability [21].

For the time being, the IV route remains the most important (primary) approach to PNP as described above. It is recommended that therapy be initiated in the hospital setting with the inconvenience for patients to stay over a few days. Sometimes such a treatment is rather limited by patient compliance and not because of unacceptable psychoactive ketamine effects or economic considerations. Nevertheless, in-hospital, effective pain relief may be achieved with IV ketamine, administered in a titrated dose under close monitoring and according to specific conditions and individual patient needs [6].

4.3 Neuropathic Cancer Pain

There is a paucity of high-quality data concerning ketamine's role in neuropathic cancer pain [73]. Approximately 40% of patients present with this NP as a direct consequence of the effects of the cancerous mass. Conditions may be significantly exacerbated (or even induced) by surgery and chemo- and radiotherapy. Management can be extremely difficult as there is always inflammation and nociceptive pain from direct local tumor effects. In a small study of patients with pain not relieved by morphine, a slow IV ketamine bolus reduced NP until 180 min after administration of 0.25–0.5 mg/kg/day [74]. For sustained pain relief, this dose, however, appears to be too low and/or should have been applied several times a day. In another small trial in patients with intractable NP with advanced cancer, oral ketamine taken at a dose of 0.5 mg/kg three times per day significantly reduced NP [75]. However, after titrating 40 mg oral ketamine to the most effective and tolerable dose up to a maximum of 400 mg per day according to individual patient response, an often-cited Scottish four-center trial did not find significantly better pain relief with ketamine compared to placebo during 16 days of administration of the titrated dose [76]. Unfortunately, there are many methodological issues in this study, such as why nonresponding patients receiving placebo were excluded as treatment failures before randomization, why 75% of patients during remission but with chemotherapy-related NP were examined, or whether other drugs simultaneously applied in multimodal analgesia could be or were reduced. To be clear, ketamine's value in cancer-related pain is without doubt. It can reduce the reliance on and the adverse effects associated with opioid analgesics while enhancing the quality of life of the patient [77, 78]. Ketamine has been recommended by professional societies for cancer pain therapy [79].

4.4 Neuropathic Breakthrough Pain

Breakthrough pain is a serious exacerbation of acute or chronic pain. It may occur even if NP is otherwise relatively controlled or stable. Its onset is unpredictable and rapid and usually more severe than the pain typically experienced. Trial evidence is heterogeneous, limited, and inconsistent with breakthrough and refractory NP pain and low-dose ketamine [65, 80]. Nevertheless, recent research and practical experience indicate some favorable outcomes.

For example, in herpes zoster (HZ) breakthrough and refractory NP, small-dose esketamine administered via patient-controlled IV analgesia induced significant relief from NP at rest and in the incidence and intensity of breakthrough pain [81]. The reduction in NP at rest was stable for 1–6 months after treatment. Esketamine was used as adjunct to usual HZ therapy with all opioids being stopped before start of esketamine. Therapy consisted of continuous delivery of 8 mg/h esketamine and possible additional demand of 10 mg/h at maximum with a 30-min lockout time. Two pumps each filled with 550 mg ketamine and 5 mg midazolam were applied. No serious adverse effects were noted.

In clinical practice, mouthwashes with ketamine in severe mucositis pain are a known approach in the treatment of children and adults with complications associated with cancer therapy and/or hematopoietic stem cell transplantation. Breakthrough toxicities with aggressive chemo- and/or radiotherapy are ubiquitous, and future studies will investigate the most efficacious schedules for the administration of ketamine in this context [82]. If epidural preservative-free ketamine was infused at 1 mg/kg at 8-h intervals for 3 weeks in case of refractory NP, pain scores decreased up to 70% which was associated with an enormous increase in quality of life until 5 weeks after epidural catheter removal [78]. Nevertheless, the off-label use of epidural and intrathecal ketamine should be restricted to palliative care settings given the concerns regarding ketamine-related neurotoxicity even with preservative-free preparations [33].

5 Psychoactive Effects with Intravenous Ketamine Therapy

In comparison to other analgesics and anesthetics, ketamine offers a significant safety advantage due to its lack of cardiorespiratory depression. Nevertheless, randomized trials in NP have reported particularly psychoactive effects with an almost fivefold increase in psychedelic events when IV ketamine was administered in awake patients in the higher dose sedative range (> 0.8 mg/kg) [6]. The question thus arises whether adverse effects may become so severe that patients experience them as "unbearable," drop out, or have to stop treatment. This must be addressed with regard to racemic and esketamine. Although esketamine allows for a 50% dose reduction as compared to the racemate, the incidence of adverse events appears to be only slightly less or comparable to that with racemic ketamine. Nevertheless, esketamine-associated psychoactive effects are less unpleasant.

The psychoactive effects of acute ketamine use are strictly dose-dependent, with a prevalence ranging from less than 5% to over 50% in chronic pain patients. They include psychedelic, psychomimetic, and dissociative events such as visual, acoustic, and proprioceptive hallucinations [33]:

- Very-low-dose ketamine (< 0.15 mg/kg/h IV) does not observably impact upon cerebral functions.
- With a single analgesic dose of 0.2–0.5 mg/kg IV ketamine (such as used for acute pain treatment), vigilance, memory function, and logical thinking may be reduced.
- At increasing doses > 0.5 mg/kg IV ketamine, mental, physical, and spiritual well-being changes. There may be blurred vision, irritability, dizziness, drowsiness, sensory delusions, delirium, "out-of-body" experiences, feelings of poisoning, perceptions of unreality and strangeness, agitation, nausea, and dysphoria. However, there may also be positive psychedelic experiences, such as euphoria, a sense of awe and liberation from anxiety, and bodily detachment with "floating."

In most cases, the mood the patients present with is intensified. In awake patients, positive suggestions such as "you will have beautiful dreams now" support positive experiences [83]. Ketamine increases hypnotizability which is sometimes used in antidepressant treatment. It is definitely decisive for success of NP therapy that caregivers guide their patients with preparation and support through the psychoactive period. Providing a peaceful, calm atmosphere is of significant benefit and should also be guaranteed during emergence and recovery from anesthesia after adjunctive intraoperative ketamine use for postsurgical pain prevention. When ketamine is administered in hospital, psychic adverse effects can effectively be reduced or prevented through intake of a benzodiazepine or α-2 adrenoceptor agonist administered at best before ketamine. Psychedelic effects after NP therapy may occur up to 1 day after ketamine but usually resolve within a few hours. Nevertheless, patients must be advised regarding impaired car driving capabilities and operating machinery on the day of ketamine use [84]. This may be very important if they receive treatment in hospital but as ambulant patients.

Because ketamine possesses abuse potential, it is essential to implement cautious prescription guidelines and rigorous control measures for treatment of NP treatment at home. Longer-term ketamine therapy may impair multiple systems such as the urinary tract, bladder, hepatobiliary functions, and more (see other chapters of this textbook). Neurocognitive decline including reduction in episodic and semantic memory has also been reported with chronic ketamine, but it must not be forgotten that its effects are completely context-dependent: There is a huge discrepancy between therapeutic effects on "diseased" and recreational impact on healthy brain. While ketamine may be beneficial in chronic NP with maladaptive transmission and aberrant neural network communication, it may be negative if applied at the same dose in an otherwise healthy brain. In addition, the high prevalence of mental comorbidities in NP patients such as depression, anxiety, and stress-related disorders is associated with worse treatment outcomes. It appears that not only does NP induce a secondary, adverse psychological impact but that mental health disorders also exert a detrimental influence on the course of NP.

In conclusion, ketamine may be an efficacious treatment for chronic NP without causing serious adverse effects if administered with skill and according to the individual patient's disposition, conditions, and circumstances. Without patient guidance and in unexperienced hands, however, especially unwanted psychoactive effects may become a problem and may lead to patient withdrawal despite NP relief.

6 Limitations

The principal limitation of our review is the low certainty of the evidence, which is a consequence of the considerable heterogeneity of the published studies and the protocols for the administration of ketamine. Presently, there is no clarity regarding the optimal design of ketamine use schedules and the optimal timing for initiating treatment. Difficulties in NP diagnosis especially relate to patients' varying clinical symptomatology, presentation, and pathophysiological background. Most systematic literature analyses had to include trials with data from patients who presented with multiple, complex, and heterogeneous clinical disorders. This reduces (precision) estimates of yielded treatment results and may blur the perspective on outcomes. It impedes the ability to conduct realistic sensitivity analyses and provide high-quality analytic reviews.

7 Conclusions

7.1 Prevention of Chronic Postoperative Pain and Neuropathic Pain Development

From a pathophysiological perspective, ketamine has the potential to reduce and/or prevent negative peripheral and central neural plastic change in NP and spinal/cortical potentiation, which contributes to pain sensitization. It may improve enhanced, maladaptive pain experiences. Perioperative IV ketamine reduces acute postoperative pain and the need for analgesics when employed in accordance with best practice. There is evidence that ketamine can prevent CPSP, although the quality of this evidence is moderate. It may also have a role in reducing the development of chronic NP, with effects that last for up to 3 months.

7.2 Therapy of Chronic Neuropathic Pain

Ketamine can safely be administered to awake patients with NP via the IV route. The increase in unwanted effects is strictly dose-dependent and occurs especially in the higher sedative dose range. Patients shall be guided through psychedelic events by experienced healthcare professionals. The administration of a GABA agonist such as midazolam or an α2-agonist prior to ketamine can effectively mitigate or avert the occurrence of adverse psychoactive effects. The adjunctive use of ketamine may result in a notable reduction in chronic NP and can improve patient quality of life. Benefits may be evident for months after IV use. Further research is required to ascertain whether alternative routes of administration, such as oral ketamine, could serve as a de-escalating strategy to maintain treatment gains. Less costs and less resources may be required if patients could be preselected by neuroimaging analysis or objective pain test results as ketamine responders or nonresponders. This should take place before initiation of in-hospital therapy. However, there may be patients who will positively respond not before several days of subanesthetic-dose IV ketamine treatment, particularly in cases involving complex constellations with chronic pain, anxiety, and depression [85]. For all settings, high-quality trials are urgently awaited.

Conflicts of Interest

We do not have any conflicts of interest.

References

1. Finnerup NB, Kuner R, Jensen TS (2021) Neuropathic pain: from mechanisms to treatment. Physiol Rev 101:259–301
2. Vieira WF, Coelho DRA, Litwiler ST et al (2024) Neuropathic pain, mood, and stress-related disorders: a literature review of comorbidity and co-pathogenesis. Neurosci Biobehav Res 161:105673
3. GBD 2017 Disease and Injury Incidence and Prevalence Collaborators (2018) Global, regional, and national incidence, prevalence, and years lived with disability for 354 diseases and injuries for 195 countries and territories, 1990–2017: a systematic analysis for the global burden of disease study 2017. Lancet 392: 1789–1858
4. Yawn BP, Wollan PC, Weingarten TN et al (2009) The prevalence of neuropathic pain: clinical evaluation compared with screening tools in a community population. Pain Med 10:586–593
5. van Hecke O, Austin SK, Khan RA et al (2014) Neuropathic pain in the general population: a systematic review of epidemiological studies. Pain 155:654–662
6. Guimarães Pereira JE, Ferreira Gomes Pereira L, Mercante Linhares L et al (2022) Efficacy and safety of ketamine in the treatment of neuropathic pain: a systematic review and meta-analysis of randomized controlled trials. J Pain Res 15:1011–1037
7. Hamdan A, Galvez R, Katati M (2024) Shedding light on neuropathic pain: current and emerging tools for diagnosis, screening, and quantification. SAGE Open Med 12: 20503121231218985
8. Cohen SP, Mao J (2014) Neuropathic pain: mechanisms and their clinical implications. BMJ 348:: f7656
9. Truini A, Aleksovska K, Anderson CC et al (2023) Joint European academy of neurology–European pain federation-neuropathic pain special interest Group of the International Association for the study of pain guidelines on neuropathic pain assessment. Eur J Neurol 30: 2177–2196
10. Colloca L et al (2017) Neuropathic pain. Nat Rev Dis Primer 3:17002
11. Cavalli E, Mammana S, Nicoletti F et al (2019) The neuropathic pain: an overview of the current treatment and future therapeutic approaches. Int Immunopathol Pharmacol 33:2058738419838383
12. Urits I, Li N, Berardino K et al (2020) The use of antineuropathic medications for the treatment of chronic pain. Best Pract Res Clin Anaesthesiol 34:493–506
13. Hylands-White N, Duarte RV, Raphael JH (2017) An overview of treatment approaches for chronic pain management. Rheumatol Int 37:29–42
14. Curatolo M (2024) Central sensitization and pain: pathophysiologic and clinical insights. Curr Neuropharmacol 22:15–22
15. Kissin I, Bright CA, Bradley EL (2000) The effect of ketamine on opioid-induced acute tolerance: can it explain reduction of opioid consumption with ketamine-opioid analgesic combinations? Anesth Analg 91:1483–1488
16. Zarate CA, Singh JB, Carlson PJ et al (2006) A randomized trial of an N-methyl-D-aspartate antagonist in treatment-resistant major depression. Arch Gen Psychiatry 63:856–864
17. Maher DP, Chen L, Mao J (2017) Intravenous ketamine infusions for neuropathic pain management: a promising therapy in need of optimization. Anesth Analg 124:661–674
18. Kamp J, van Velzen M, Olofsen E et al (2019) Pharmacokinetic and pharmacodynamic considerations for NMDA-receptor antagonist ketamine in the treatment of chronic neuropathic pain: un update of the most recent literature. Expert Opin Drug Metab Toxicol 15:1033–1041
19. Sosanya NM, Kumar R, Clifford JL et al (2020) Identifying plasma derived extracellular vesicle (EV) contained biomarkers in the development of chronic neuropathic pain. J Pain 21:82–96
20. Podvin S, Yaksh T, Hook V (2016) The emerging role of spinal dynorphin in chronic pain: a therapeutic perspective. Annu Rev Pharmacol Toxicol 56:511–533
21. Zhang H, Zhou P, Jiang Y et al (2023) Sustained-release esketamine based nanoparticle-hydrogel delivery system for neuropathic pain management. Int J Nanomedicine 18:1131–1143
22. Ren K, Torres R (2009) Role of interleukin-1beta during pain and inflammation. Brain Res Rev 60:57–64
23. Arslan M, Cömert A, Açar Hİ et al (2012) Nerve root to lumbar disc relationships at the intervertebral foramen from a surgical viewpoint: an anatomical study. Clin Anat NY N 25:218–223
24. Kirazli O, Tatarli N, Güçlü B et al (2014) Anatomy of the spinal dorsal root entry zone: its clinical significance. Acta Neurochir 156: 2351–2358

25. D'Mello R, Dickenson AH (2008) Spinal cord mechanisms of pain. Br J Anaesth 101:8–8–16
26. Farajidavar A, Saeb S, Behbehani K (2008) Incorporating synaptic time-dependent plasticity and dynamic synapse into a computational model of wind-up. Neural Netw 21:241–249
27. Patestas MA, Gartner LP (2013) A textbook of neuroanatomy, 2nd edn. Wiley, Hoboken
28. Chen C, Tassou A, Morales V, Scherrer G (2023) Graph theory analysis reveals an assortative pain network vulnerable to attacks. Sci Rep 13:21985
29. Mofatteh M (2021) Examining the association between traumatic brain injury and headache. J Integr Neurosci 20:1079–1094
30. Tracey I, Mantyh PW (2007) The cerebral signature for pain perception and its modulation. Neuron 55:377–391
31. Bentley N, Awad AJ, Patil PG (2018) Chapter 43 – physiology and pathophysiology of chronic pain. In: Krames ES, Peckham PH, Rezai AR (eds) Neuromodulation, 2nd edn. Academic Press, Cambridge, pp 565–573. https://doi.org/10.1016/B978-0-12-805353-9.00043-7
32. Martinez V, Lehman T et al (2024) Chronic postsurgical pain. A European survey. Eur J Anaesthesiol 41:351–362
33. Mion G, Himmelseher S (2024) Esketamine: less drowsiness, more analgesia. Anesth Analg 139:78
34. Ishida Y, Okada T, Kobayashi T et al (2022) Pain management of acute and chronic postoperative pain. Cureus 14:e23999
35. Doleman B, Mathiesen O, Sutton AJ et al (2023) Non-opioid analgesics for the prevention of chronic postsurgical pain: a systematic review and network meta-analysis. Br J Anaesth 130:719–728
36. Abouarab AH, Brülle R, Aboukilila MY et al (2024) Efficacy and safety of perioperative ketamine for the prevention of chronic postsurgical pain: a meta-analysis. Pain Pract 24:553–566
37. Burton AW, Lee DH, Saab C, Chung JM (1999) Preemptive intrathecal ketamine injection produces a long-lasting decrease in neuropathic pain behaviors in a rat model. Reg Anesth Pain Med 24:208–213
38. Sun W, Zhou Y, Wang J et al (2023) Effects of ketamine on chronic postsurgical pain in patients undergoing surgery: a systematic review and meta-analysis. Pain Physician 26: E111–E122
39. Wang J, Doan LV, Axelrod W et al (2024) Optimizing the use of ketamine to reduce chronic postsurgical pain in women undergoing mastectomy for oncologic indication: study protocol for the KALPAS multicenter randomized controlled trial. Trials 25:21
40. Shetty A, Delanerolle G, Cavalini H et al (2024) A systematic review and network meta-analysis of pharmaceutical interventions used to manage chronic pain. Sci Rep 14:1621
41. Fauchon C, Kim JA, El Sayed R et al (2022) Exploring sex differences in alpha activity as a potential neuromarker associated with neuropathic pain. Pain 163:1291–1302
42. Bosma AL, Cheng JC, Rogachov A et al (2018) Brain dynamics and temporal summation of pain predicts neuropathic pain relief from ketamine infusion. Anesthesiology 129:1015–1024
43. Cohen SP, Khunsriraksakul C, Yoo Y et al (2023) Sympathetic blocks as a predictor for response to ketamine infusion in patients with complex regional pain syndrome: a multicenter study. Pain Med 24:316–324
44. Motoyama Y, Oshiro Y, Takao Y et al (2019) Resting-state brain functional connectivity in patients with chronic pain who responded to subanesthetic-dose ketamine. Sci Rep 9:12912
45. Xu H, Seminowicz DA, Krimmel SR et al (2022) Altered structural and functional connectivity of salience network in patients with classical trigeminal neuralgia. J Pain 23:1389–1399
46. Hayes Weeks CC, Simmons AN, Strigo IA et al (2022) Distal neuropathic pain in HIV is associated with functional connectivity patterns in default mode and salience networks. Front Pain Res 3:1004060
47. Rogachov A, Bhatia A, Cheng JC et al (2019) Plasticity in the dynamic pain connectome associated with ketamine-induced neuropathic pain relief. Pain 160:1670–1679
48. Branco P, Bosak N, Bielefeld J et al (2023) Structural brain connectivity predicts early acute pain after mild traumatic brain injury. Pain 164:1312–1320
49. Mills EP, Bosma RL, Rogachov A et al (2024) Pre-treatment brain white matter integrity associated with neuropathic pain relief and changes in temporal summation of pain following ketamine. J Pain 25:104536. https://doi.org/10.1016/j.jpain.2024.104536
50. Song D, He A, Xu R et al (2018) Efficacy of pain relief in different postherpetic neuralgia therapies: a network meta-analysis. Pain Physician 21:19–32
51. Zhao J, Wang Y, Wang D (2018) The effect of ketamine infusion in the treatment of complex regional pain syndrome: a systemic review and meta-analysis. Curr Pain Headache Rep 22:12

52. Bäckryd E, Thordeman K, Gerdel B, Ghafouri B (2023) Cerebrospinal fluid metabolomics identified ongoing analgesic medication in neuropathic pain patients. Biomedicines 11: 2525
53. Moaddel R, Zanos P, Farmer CA et al (2022) Comparative metabolomic analysis in plasma and cerebrospinal fluid of humans and in plasma and brain of mice following antidepressant – dose ketamine administration. Transl Psychiatry 12:179
54. Voute M, Rian T, Amodeo JM et al (2022) Ketamine in chronic pain: a Delphi survey. Eur J Pain 26:873–887
55. Jonkman K, van Rijnsoever E, Olofsen E et al (2018) Esketamine counters opioid-induced respiratory depression. Br J Anaesth 120: 1117–1127
56. Gurba KN, Chaudhry R, Haroutounian S (2022) Central neuropathic pain syndromes: current and emerging pharmacological strategies. CNS Drugs 36:483–516
57. Watson JC, Sandroni P (2018) Central neuropathic pain syndromes. Mayo Clin Proc 91: 372–385
58. Angstadt R, Esperti S, Mangano A, Meyer S (2021) Palliative ketamine: the use of ketamine in central post-stroke pain syndrome – a case report. Ann Palliat Med 10:6974–6978
59. Amr YM (2011) Epidural ketamine in post spinal cord injury-related chronic pain. Anesth Essays Res 5:83–86
60. Rosner J, de Andrade DC, Davis KD et al (2023) Central neuropathic pain. Nat Rev Dis Primers 9:73
61. Amr YM (2010) Multi-day low dose ketamine infusion as adjuvant to oral gabapentin in spinal cord injury related chronic pain: a prospective, randomized, double blind trial. Pain Physician 13:245–249
62. Ling HQ, Chen ZH, He L et al (2022) Comparative efficacy and safety of 11 drugs as therapies for adults with neuropathic pain after spinal cord injury: a Bayesian network analysis based on 20 randomized controlled trials. Front Neurol 13:818522
63. Kim YH, Lee PB, Oh TK (2015) Is magnesium sulfate effective for pain in chronic postherpetic neuralgia patients comparing with ketamine infusion therapy? J Clin Anesth 27:296–300
64. Kvarnström A, Karlsten R, Quiding H et al (2003) The effectiveness of intravenous ketamine and lidocaine on peripheral neuropathic pain. Acta Anaesthesiol Scand 47:868–877
65. Pickering G, Pereira B, Morel V et al (2020) Ketamine and magnesium for refractory neuropathic pain: a randomized, double-blind, crossover trial. Anesthesiology 133: 154–164
66. Rigo FK, Trevisan G, Godoy MC et al (2017) Management of neuropathic chronic pain with methadone combined with ketamine: a randomized, double blind, active-controlled clinical trial. Pain Physician 20:207–215
67. Lynch ME, Clark AJ, Sawynok J, Sullivan MJL (2005) Topical 2% amitriptyline and 1% ketamine in neuropathic pain syndromes: a randomized, double-blind, placebo-controlled trial. Anesthesiology 103:140–146
68. Mathisen LC, Skjelbred P, Skoglund LA, Øye I (1995) Effect of ketamine, an NMDA receptor inhibitor, in acute and chronic orofacial pain. Pain 61:215–220
69. Marchetti F, Coutaux A, Bellanger A et al (2015) Efficacy and safety of oral ketamine for the relief of intractable chronic pain: a retrospective 5-year study of 51 patients. Eur J Pain 19:984–993
70. Simons P, Olofsen E, van Welzen M et al (2022) S-ketamine oral thin film – part 1: Population pharmacokinetics of S-ketamine, S-norketamine and S-hydroxynorketamine. Front Pain Res 3:946486
71. Colla M, Offenhammer B, Scherer A et al (2024) Oral prolonged-release ketamine in treatment-resistant depression - a double-blind randomized placebo-controlled multicentre trial of KET01, a novel ketamine formulation – central and safety results. J Psychiatr Res 173:124–130
72. Winegarden JA, Carr DB, Bradshaw YS (2020) Topical ketamine with other adjuvants: underutilized for refractory cancer pain? A case series and suggested revision of the World Health Organization stepladder for cancer pain. J Palliat Med 23:1167–1171
73. Shkodra M, Caraceni A (2022) Treatment of neuropathic pain directly due to cancer: an update. Cancers 14:1992
74. Mercadante S, Arcuri E, Tirelli W, Casuccio A (2000) Analgesic effect of intravenous ketamine in cancer patients on morphine therapy: a randomized, controlled, double-blind, crossover, double-dose study. J Pain Symptom Manag 20:246–252
75. Kannan T, Saxena A, Bhatnagar S, Barry A (2002) Oral ketamine as an adjuvant to oral morphine for neuropathic pain in cancer patients. J Pain Symptom Manag 23:60–65
76. Fallon MT, Wilcock A, Kelly CA et al (2018) Oral ketamine vs placebo in patients with cancer-related neuropathic pain: a randomized clinical trial. JAMA Oncol 4:870–872

77. Singh V, Gillespie TW, Lane O et al (2022) A dose-escalation clinical trial of intranasal ketamine for uncontrolled cancer-related pain. Pharmacotherapy 42:298–310
78. Lauretti GR et al (2002) Epidural ketamine versus epidural clonidine as therapeutic for refractory neuropathic chronic pain. Rev Bras Anestesiol 52:34–40
79. Aman MM, Mahmoud A, Deer T et al (2021) The American Society of Pain and Neuroscience (ASPN) best practice and guidelines for the interventional management of cancer-associated pain. J Pain Res 14:2139–2164
80. Vranken JH, Dijkgraaf MGW, Kruis MR et al (2005) Iontophoretic administration of S(+)-ketamine in patients with intractable central pain: a placebo-controlled trial. Pain 118:224–231
81. Tang J, Zhang E, Huang B et al (2023) Efficacy of patient-controlled intravenous analgesia with esketamine for herpes zoster associated with breakthrough pain. Pain Physician 26: 299–306
82. Koffler D, Eckstein J, Herman J et al (2023) Efficacy of ketamine mouthwash in the management of oral and pharyngeal toxicity associated with head and neck chemotherapy: protocol for a phase II, Simon's two-stage trials. BMJ Open 13:e064809
83. Cheong SH, Lee KM, Lim SH et al (2011) Brief report: the effect of suggestion on unpleasant dreams induced by ketamine administration. Anesth Analg 112:1082–1085
84. Edinoff AN, Sall S, Koontz CB et al (2023) Ketamine evolving clinical roles and potential effects with cognitive, motor and driving ability. Neurol Int 15:352–361
85. Marzoughi S, Ripsman D, Ong M (2023) Therapeutic effects of 7- to 14-day subanesthetic ketamine infusions for chronic pain on standardized psychiatric measures. Pain Manag 13:529–538

Chapter 5

Therapy for Chronic Pain in Children Is Challenging: Is Ketamine a "Reasonable Choice"?

Christine H. Meyer-Frießem

Abstract

The treatment of chronic pain in children is an enormous challenge. To address this problem, the potential utility of ketamine as analgesic adjuvant is discussed. However, there are yet only very limited data with low evidence that describe ketamine's analgesic efficacy in pediatric patients. Ketamine's known unwanted psychoactive and/or medical side effects often seem to be underreported, and their relevance remains unclear. Low, subanesthetic dose of adjuvant ketamine represents an off-label therapeutic option to approach selected children suffering from pain that is refractory to other treatments.

Key words Ketamine, Chronic pain, Pain, Neuropathic pain, Complex regional pain syndrome, CRPS , Non-cancer pain, Cancer, Children, Pediatric patients

1 Introduction

Pain, particularly persistent pain in children, is a complex and challenging medical condition [1]. Pain significantly impacts children's physical, emotional, and social well-being. Despite advances in pain management, some pediatric patients endure ongoing pain that does not respond to traditional pharmacological and non-pharmacological treatments. As ketamine has gained increasing attention for its analgesic properties and unique mechanism in the treatment of various acute and chronic pain conditions in adults [2, 3], it might also be considered as an option or therapeutic alternative for managing chronic pain in children.

By blocking (hyperactive) N-methyl-D-aspartate (NMDA) receptors, ketamine inhibits the amplification of pain signals. This leads to a reduction in pain perception and the transmission and modulation of pain signal processing in the peripheral and especially the central nervous system. Ketamine is therefore potentially

No conflicts of interest.

Sabine Himmelseher (ed.), *Ketamine: Pain Management, Critical Care, Depression, and Resource-Poor Settings Volume II*, Neuromethods, vol. 225, https://doi.org/10.1007/978-1-0716-4599-4_5,

effective for those types of neuropathic pain that are characterized by central sensitization. Ketamine also exerts analgesic effects through interactions with multiple other neurotransmitter systems, including opioid, monoaminergic, and cholinergic pathways. The phenomena of analgesia tolerance and (opioid) hyperalgesia, which are well-known features of pharmacotherapy in chronic pain, are also mediated by the activation of the NMDA receptor system and may beneficially be affected by ketamine as well. All these different targets within pain pathways make ketamine a potential coanalgesic for managing different pain conditions [4–6]. Finally, the antidepressant effect of ketamine is appreciated [7, 8] and has also been observed in children [9, 10]. For more details about ketamine's pharmacology, please see chapters at the beginning of this textbook.

This chapter presents and summarizes available data on the use of ketamine as an adjuvant in the management of chronic pain in children, focusing on therapeutic effects, adverse events, and administration routes and dosing.

2 Summary of Clinical Study Results

Reliable study data from pediatric patients are very limited [11]. The pharmacological treatment approach to children with chronic pain is often a trial-and-error process that is based on expert opinions and the extrapolation of findings from studies in adults. Most data lack small sample sizes and are from very heterogeneous pain conditions with different medical background. Often publications consist of individual case reports or retrospective analyses [12].

2.1 Types of Pain Disorders

Based on its characteristics, pain is classified into nonmalignant and malignant chronic pain or pain encountered in palliative care settings.

2.1.1 Non-cancer Pain

In adults, ketamine is used as a coanalgesic for different pain disorders [5], including acute (postoperative) pain [4, 6] and chronic non-cancer pain, such as migraine, fibromyalgia, complex regional pain syndrome (CRPS) type I, irritable bowel syndrome, and neuropathic pain involving ischemic limb and phantom limb pain [2, 13–15]. While substantial evidence supports the use of ketamine for acute pain [4], the level of evidence for chronic pain varies by condition and dosage [5] and lacks guidelines outside the USA and established standards [16]. In fact, mainly due to insufficient data, e.g., the German guidelines on the treatment of chronic pain in adults do not even mention ketamine as a treatment option so far [17].

In children, the lack of reliable data is even greater. For example, 12 children with different persistent pain for at least 3 months were prescribed oral ketamine at doses ranging from 0.25 to 1.5 mg/kg thrice per day for 14 days in a pilot study [18]. It turned out that study patients did not well tolerate ketamine doses exceeding 1 mg/kg. Although only five children indicated reduced pain scores and two stopped treatment due to dose-limiting adverse effects, the authors concluded that oral ketamine at a dose of 0.25–1 mg/kg thrice per day for 2 weeks can be considered as a safe analgesic option. In a single-center cohort study with 230 sick children, subanesthetic dose intravenous (IV) ketamine infusions led to significant reductions in mean pain scores [19]. Patients with cancer pain, pancreatitis, and Crohn's disease experienced greater pain relief. There were no psychotomimetic effects requiring treatment [19]. When ketamine was administered in children with pain crises in sickle cell disease, pain scores and opioid consumption decreased. Adverse events were transient [20]. In a longitudinal cohort study conducted over 15 months, 63 children and adolescents with different chronic pain states received subanesthetic ketamine infusions at a dose of 0.1–0.3 mg/kg/h over 4–8 h daily for 3 consecutive days. This was associated with a pain score reduction by an average of −1.6 points [21]. Ketamine showed the most significant pain relief in patients with CRPS, after trauma, and with postural orthostatic tachycardia syndrome. It demonstrated the slightest pain reductions in patients with chronic headaches. However, ketamine infusions could not reduce the overall morphine-equivalent analgesic drug intake [21].

In a randomized controlled trial (RCT), Sigtermans et al. showed that an S-ketamine infusion over 4 days with doses titrated up to a mean of 22 mg/h/70 kg resulted in pain relief lasting up to 3 months without improving functions in patients with CRPS [22]. Similar observations were made in the RCT by Schwartzman and colleagues who evaluated effects of a ketamine infusion with a dose of 0.35 mg/kg/h over 4 h for 10 days in CRPS settings [23]. Based on the abovementioned two RCTs [22, 23], the German guidelines conclude that IV ketamine infusions as analgesic treatment for resistant pain in adults with CRPS can only be recommended in experienced centers and under continuous monitoring. The guidelines also state that repeated uses of ketamine may increase the odds of undesirable psychoactive events, such as feelings of inebriation, hallucinations and dizziness, addiction, and organ damage [24, 25]. Recommendations for children are formulated clearly: "CRPS treatment should always be carried out by an experienced centre with paediatric expertise without invasive or potentially harmful therapy including ketamine infusions" [24].

2.1.2 Cancer Pain

In the past, no recommendations were made for the use of ketamine for pain treatment in cancer patients [26]. In a multicenter RCT analyzing 50 adults—with strong criticism of its methodology—oral ketamine was found to be equivalent to placebo for malignant neuropathic pain [27]. But the situation has since changed [15]. There is now evidence for benefits of ketamine in opioid-refractory cancer pain [28–30], because ketamine can reduce opioid consumption and tolerance in adults [31].

Ketamine has also been used in pediatric oncology for the treatment of refractory pain [32]. In two reports with small sample sizes, children without adequate pain control despite high-dose opioids received oral ketamine at a dose of 3 mg/kg thrice per day [33] or IV ketamine at a dose of 0.1–1 mg/kg/h [34]. Most children experienced pain relief, could spare opioids, and tolerated ketamine well [33, 34]. A retrospective analysis of a 10-year period in a US children's hospital evaluated all IV ketamine infusions applied for pain management in young oncology and hematology patients. IV ketamine infused at a median dose of 0.15 ± 0.11 mg/kg/h over a median time period of 3 ± 11 days was associated with significantly lower daily opioid consumption in the oncology patients, but there was no significant decrease in opioid needs in end-of-life patients, the overall study population, or in patients with acute pain [35]. These results are in agreement with Anghelescu and co-workers who summarized their retrospective analysis as follows: With monitoring, ketamine infusions at a dose range from 0.05 to 0.6 mg/kg/h can be used for treatment of persistent neuropathic pain to spare opioids in pediatric oncology patients [36]. A prospective, multicenter, observational pilot study evaluated the efficacy and safety of low-dose ketamine infused at rates below 0.05 mg/kg/h over 3 days as an adjuvant to opioid analgesics in 38 children, adolescents, and young adults with cancer-related pain. This study found that adding ketamine to opioids significantly reduced pain intensity in half of the patients 48 h after the treatment was started [37]. Although pain control tended to improve, limited statistical power weakens the results [37]. Furthermore, several case reports describe beneficial use of ketamine in desperate situations. For example, in a case report of a 12-year-old girl with a cervical spinal tumor, severe neuropathic pain was effectively managed with continuous infusions of subcutaneous morphine and IV ketamine: 36–410 mg/d over 67 days until her death was applied together with benzodiazepines to mitigate ketamine-related side effects [38]. In another report of a 5-year-old girl with cancer-related neuropathic pain unresponsive to escalating fentanyl doses, a combination of IV ketamine and lidocaine together with fentanyl effectively alleviated pain without notable side effects [39]. A retrospective case review of 14 children at their end of life showed that ketamine patient-controlled analgesia effectively relieved neuropathic pain [40]. Furthermore, ketamine's NMDA receptor blocking mechanism is hypothesized to possess immunomodulating antitumor properties [29, 41, 42].

3 Unwanted and Adverse Medical Effects

Usually, there is no cardiorespiratory depression with subanesthetic dose ketamine, and its safety margin is huge. Nevertheless, ketamine administration is not without risks. Common medical adverse events include nausea, vomiting, sedation, fatigue, drowsiness or lethargy, urinary retention, changes in blood pressure and heart rate, and myoclonic movements [12, 13, 35, 43]. These side effects are transient and can be managed with proper monitoring and supportive care. All ketamine effects are dose- and context-related, and higher dosages and more frequent infusions increase the odds of more unwanted events [5, 26]. Psychedelic, psychomimetic, and psychotropic effects with psychotic symptoms, hallucinations, feeling of inebriation, dizziness, nightmares, and dysphoric mood may develop [12, 25, 43, 44]. Long-term or repeated use of ketamine raises further concerns about abuse potential, addiction, cognitive problems [2, 24, 25, 43, 45], and development of cystitis and liver toxicity [46].

In a 9-year-old child with severe burns, a 37-day ketamine infusion provided effective analgesia as an opioid adjuvant without tolerance or psychomimetic side effects. Ketamine allowed for rapid weaning without withdrawal symptoms [47]. In contrast, in a 5-year-old child with ketamine infusions at a relatively high dose of 0.54 mg/kg/h applied over 4 months as an adjuvant to 28 mg hydromorphone per day, slight tolerance and mild withdrawal symptoms during weaning from the analgesics were observed. A clonidine patch was kept throughout the weaning process [48]. In addition, no negative effects on neurocognitive functions were reported in a prospective study conducted in 11 children with chronic pain after 2 weeks of daily oral ketamine [18, 49].

The interpretation of these data must be strictly context dependent: (1) Chronic pain, inflammation, (dis-)stress, and other potentially harmful factors lay a heavy burden of disease on a child's body and brain, including neurodegenerative impact. If pain management and distress improve with ketamine, this treatment reduces neurodegeneration and leads to better neurocognitive functions (please see chapter about the "neuroplastic narrative" of ketamine in the pharmacology section of this book). (2) When ketamine is combined with, for example, GABA-ergic or alpha-2 adrenergic agents (in sedation and analgesia), unwanted neural psychoactive effects can be reduced. (3) In ketamine-opioid combinations, ketamine reduces opioid-related cardiorespiratory depression, opioid tolerance, and hyperalgesia and often allows for opioid reduction and improves analgesic effects resulting in less opioid-induced adverse events. (4) Withdrawal symptoms from ketamine in long-term sedation and analgesia regimen can be reduced, when ketamine is stopped or its dose is early and sufficiently enough reduced before cessation of other sedatives, such as propofol. This strategy can be used to overcome dissociative and other undesirable psychoactive effects.

Nevertheless, long-term cognitive impairment and memory deficits have been reported with ketamine in countless animal studies [50, 51]. Ketamine may exert neurotoxic effects, especially in vulnerable infants and very young, seriously sick children after repeated or higher dose [52]. In such circumstances, benefits and risks of ketamine administration must be carefully weighed in individual patients, and it seems to be prudent to avoid longer and higher dose ketamine exposures.

In summary, careful patient selection, monitoring, and individualized treatment plans are crucial with ketamine treatment in pediatric pain patients. Support from psychologists should always be sought to address psychological distress and anxiety in these children. Future research should especially address the effects of drug combinations including formal evaluations of different drug dose ratio relationships in high-quality trials.

4 Administration Route and Use Protocol

Ketamine can be administered via multiple routes such as in IV, subcutaneous, oral, sublingual, buccal, rectal, topical, or intranasal (IN) application. Due to different bioavailability, equivalent ketamine doses depend on the route of administration [3, 53]. A 1:1 dose ratio for conversion from subcutaneous to oral ketamine is considered safe and effective in adult patients with cancer pain [54]. IN administration offers a rapid, convenient, and noninvasive method of drug delivery. This makes it a very suitable choice for managing breakthrough or acute on-chronic pain exacerbations. Research on IN ketamine uses in pediatric populations is presently underway, and first evidence and practical experiences show benefits. Topical ketamine administration offers the advantage of targeted delivery to an affected area, potentially minimizing systemic side effects. However, the application of not formally authorized, compounded mixtures poses risks of toxicity especially in children and should be refrained from [55]. Research on new nanoparticle-hydrogel delivery system is ongoing [56]. Of course, all types of injections are less suitable for chronic pain patients, particularly children. New extended-release oral ketamine formulations have been started to be investigated in official phase trials in healthy adult volunteers and depressed patients because bioavailability of "immediate release" oral ketamine is below $\leq$20%. However, data about oral ketamine use in children are very limited so far. According to the conclusions of the Canadian Agency for Drugs and Technologies in Health (CADTH), IV ketamine provides significant short-term pain relief for chronic non-cancer pain but lacks long-term effects after a single infusion in adults [43].

The use of ketamine in pediatric chronic pain management requires careful consideration of patient factors, dosing regimen, and monitoring protocols to ensure safe and effective treatment outcomes. The expertise of pediatric anesthesiologists and pain specialists plays a decisive role in determining the appropriate administration route and dosage based on the child's age, weight, medical history, and all other individual conditions. In general, there is no "commonly agreed standard operating procedure," and there is no—and will likely never be—a "one-size-fits-all" ketamine administration protocol. Pharmacokinetic and pharmacodynamic research has shown that children have a shorter context-sensitive half-life and lower sensitivity to ketamine than adults [57, 58]. Taken together, it is therefore highly advisable to administer ketamine at low subanesthetic dose, with titration according to effect, and under close monitoring.

5 Conclusions and Recommendation

Adjuvant, subanesthetic dose ketamine represents an off-label therapeutic option for selected children suffering from malignant and nonmalignant persistent pain that is refractory to other treatments (Fig. 1).

Ketamine's unique mechanisms and multimodal analgesic effects make it a promising substance as even low doses can reduce pain intensity and spare opioids. The relevance of ketamine's known unwanted psychoactive and/or medical side effects remains unclear. As yet, there is not much evidence on the efficacy and safety of ketamine for chronic pain management in children [5]. The lack of clinical guidelines, recommendations, and standardized use protocols highlights the need for high-quality trials with long-term follow-up [5, 43].

Experts consent that the most effective treatment for persistent pain is based on a biopsychosocial, multidimensional, interdisciplinary approach including different, pain-specific treatment modalities [55]. It is possible that ketamine's antidepressant effect is a decisive factor in all of this [59].

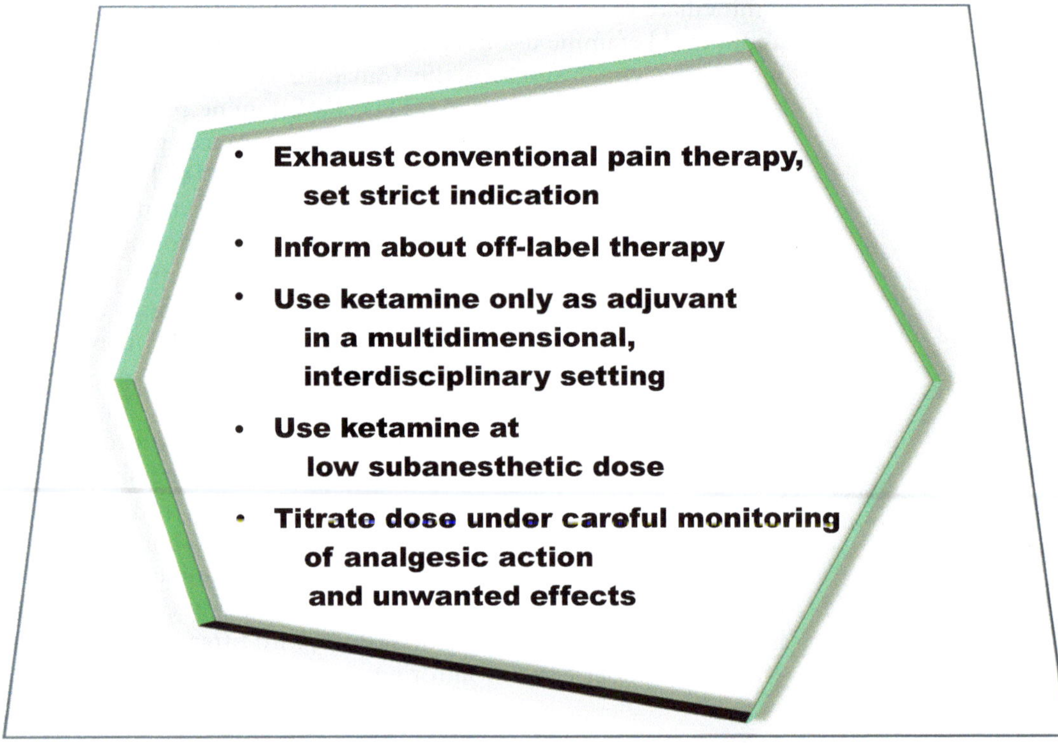

Fig. 1 Compilation of guidance for ketamine use as analgesic in selected children with chronic pain (designed by C. H. Meyer-Frießem and S. Himmelseher in Word, Microsoft Corp., Redmond, WA, USA)

References

1. Oschman A, Rao K (2024) Challenges in management of refractory pain and sedation in infants. Front Pharmacol 14:1259064. https://doi.org/10.3389/fphar.2023.1259064
2. Niesters M, Martini C, Dahan A (2014) Ketamine for chronic pain: risks and benefits. Br J Clin Pharmacol 77:357–367. https://doi.org/10.1111/bcp.12094
3. Vadivelu N, Schermer E, Kodumudi V et al (2016) Role of ketamine for analgesia in adults and children. J Anaesthesiol Clin Pharmacol 32:298–306. https://doi.org/10.4103/0970-9185.168149
4. Schwenk ES, Viscusi ER, Buvanendran A et al (2018) Consensus Guidelines on the Use of Intravenous Ketamine Infusions for Acute Pain Management From the American Society of Regional Anesthesia and Pain Medicine, the American Academy of Pain Medicine, and the American Society of Anesthesiologists. Reg Anesth Pain Med 43:456–466. https://doi.org/10.1097/AAP.0000000000000806
5. Cohen SP, Bhatia A, Buvanendran A et al (2018) Consensus guidelines on the use of Intravenous Ketamine infusions for Chronic pain from the American Society of Regional Anesthesia and Pain Medicine, the American Academy of Pain Medicine, and the American Society of Anesthesiologists. Reg Anesth Pain Med 43:521–546. https://doi.org/10.1097/AAP.0000000000000808
6. Meyer-Frießem CH, Lipke E, Weibel S et al (2022) Perioperative ketamine for postoperative pain management in patients with preoperative opioid intake: a systematic review and meta-analysis. J Clin Anesth 78:110652. https://doi.org/10.1016/j.jclinane.2022.110652
7. Johnston JN, Kadriu B, Kraus C et al (2024) Ketamine in neuropsychiatric disorders: an update. Neuropsychopharmacology 49:23–40. https://doi.org/10.1038/s41386-023-01632-1
8. Lullau APM, Haga EMW, Ronold EH, Dwyer GE (2023) Antidepressant mechanisms of ketamine: a review of actions with relevance to

treatment-resistance and neuroprogression. Front Neurosci 17:1223145. https://doi.org/10.3389/fnins.2023.1223145
9. Meshkat S, Rosenblat JD, Ho RC et al (2022) Ketamine use in pediatric depression: a systematic review. Psychiatry Res 317:114911. https://doi.org/10.1016/j.psychres.2022.114911
10. Weber G, Yao J, Binns S, Namkoong S (2018) Case report of subanesthetic intravenous ketamine infusion for the treatment of neuropathic pain and depression with suicidal features in a pediatric patient. Case Rep Anesthesiol 2018:9375910. https://doi.org/10.1155/2018/9375910
11. AlGhamdi K, Sadler K (2024) The use of ketamine for malignant and nonmalignant chronic pain in children: a review of current evidence. J Pain Palliat Care Pharmacother 38:45–55. https://doi.org/10.1080/15360288.2023.2284976
12. Bredlau AL, Thakur R, Korones DN, Dworkin RH (2013) Ketamine for pain in adults and children with cancer: a systematic review and synthesis of the literature. Pain Med Malden Mass 14:1505–1517. https://doi.org/10.1111/pme.12182
13. Bali A, Dang AK, Gonzalez DA et al (2022) clinical uses of ketamine in children: a narrative review. Cureus 14:e27065. https://doi.org/10.7759/cureus.27065
14. Voute M, Riant T, Amodéo J-M et al (2022) Ketamine in chronic pain: a Delphi survey. Eur J Pain Lond Engl 26:873–887. https://doi.org/10.1002/ejp.1914
15. Culp C, Kim HK, Abdi S (2021) Ketamine use for cancer and chronic pain management. Front Pharmacol 11:599721. https://doi.org/10.3389/fphar.2020.599721
16. Griffiths HM (2023) Low-dose ketamine infusions for chronic pain management: does this qualify as evidence-based practice? Br J Pain 17:457–467
17. Becker A, Straßner C (2023) S1-Leitlinie Chronischer nicht tumorbedingter Schmerz. AWMF Leitlinienregister. https://register.awmf.org/de/leitlinien/detail/053-036. Accessed 8 July 2024
18. Bredlau A-L, McDermott MP, Adams HR et al (2013) Oral ketamine for children with chronic pain: a pilot phase 1 study. J Pediatr 163:194–200.e1. https://doi.org/10.1016/j.jpeds.2012.12.077
19. Sheehy KA, Lippold C, Rice AL et al (2017) Subanesthetic ketamine for pain management in hospitalized children, adolescents, and young adults: a single-center cohort study. J Pain Res 10:787–795. https://doi.org/10.2147/JPR.S131156
20. Alshahrani MS, Alghamdi MA (2021) Ketamine for sickle cell vaso-occlusive crises: a systematic review. Saudi J Med Med Sci 9:3–9. https://doi.org/10.4103/sjmms.sjmms_218_20
21. Sheehy KA, Muller EA, Lippold C et al (2015) Subanesthetic ketamine infusions for the treatment of children and adolescents with chronic pain: a longitudinal study. BMC Pediatr 15:198. https://doi.org/10.1186/s12887-015-0515-4
22. Sigtermans MJ, van Hilten JJ, Bauer MCR et al (2009) Ketamine produces effective and long-term pain relief in patients with complex regional pain syndrome type 1. Pain 145:304–311. https://doi.org/10.1016/j.pain.2009.06.023
23. Schwartzman RJ, Alexander GM, Grothusen JR et al (2009) Outpatient intravenous ketamine for the treatment of complex regional pain syndrome: a double-blind placebo controlled study. Pain 147:107–115. https://doi.org/10.1016/j.pain.2009.08.015
24. Birklein F et al (2018) S1-Leitlinie: Diagnostik und Therapie komplexer regionaler Schmerzsyndrome (CRPS). In: Dtsch. Ges. Für Neurol. Hrsg Leitlinien Für Diagn. Ther. Neurol. http://www.awmf.org/leitlinien/detail/ll/030%E2%80%93116.html. Accessed 21 July 2024
25. Correll GE, Maleki J, Gracely EJ et al (2004) Subanesthetic ketamine infusion therapy: a retrospective analysis of a novel therapeutic approach to complex regional pain syndrome. Pain Med Malden Mass 5:263–275. https://doi.org/10.1111/j.1526-4637.2004.04043.x
26. Bell R, Eccleston C, Kalso E (2017) Ketamine as an adjuvant to opioids for cancer pain. Cochrane Database Syst Rev 2019:CD003351. https://doi.org/10.1002/14651858.CD003351.pub3
27. Fallon MT, Wilcock A, Kelly CA et al (2018) Oral ketamine vs placebo in patients with cancer-related neuropathic pain: a randomized clinical trial. JAMA Oncol 4:870–872. https://doi.org/10.1001/jamaoncol.2018.0131
28. Zgaia AO, Irimie A, Sandesc D et al (2015) The role of ketamine in the treatment of chronic cancer pain. Clujul Med 88:457–461. https://doi.org/10.15386/cjmed-500
29. Duan W, Hu J, Liu Y (2019) Ketamine inhibits colorectal cancer cells malignant potential via blockage of NMDA receptor. Exp Mol Pathol

107:171–178. https://doi.org/10.1016/j.yexmp.2019.02.004

30. Aman MM, Mahmoud A, Deer T et al (2021) The American Society of Pain and Neuroscience (ASPN) best practices and guidelines for the interventional management of cancer-associated pain. J Pain Res 14:2139–2164. https://doi.org/10.2147/JPR.S315585
31. Singh V, Gillespie TW, Lane O et al (2022) A dose-escalation clinical trial of intranasal ketamine for uncontrolled cancer-related pain. Pharmacotherapy 42:298–310. https://doi.org/10.1002/phar.2669
32. Prakash S, Gupta AK, Meena JP, Seth R (2021) A review of the clinical applications of ketamine in pediatric oncology. Pediatr Blood Cancer 68:e28785. https://doi.org/10.1002/pbc.28785
33. Ugur F, Gulcu N, Boyaci A (2009) Oral ketamine for pain relief in a child with abdominal malignancy. Pain Med Malden Mass 10:120–121. https://doi.org/10.1111/j.1526-4637.2008.00424.x
34. Finkel JC, Pestieau SR, Quezado ZMN (2007) Ketamine as an adjuvant for treatment of cancer pain in children and adolescents. J Pain 8:515–521. https://doi.org/10.1016/j.jpain.2007.02.429
35. Anghelescu DL, Ryan S, Wu D et al (2022) Low-dose ketamine infusions reduce opioid use in pediatric and young adult oncology patients. Pediatr Blood Cancer 69:e29693. https://doi.org/10.1002/pbc.29693
36. Anghelescu DL, Tesney JM (2019) Neuropathic pain in pediatric oncology: a clinical decision algorithm. Paediatr Drugs 21:59–70. https://doi.org/10.1007/s40272-018-00324-4
37. Courade M, Bertrand A, Guerrini-Rousseau L et al (2022) Low-dose ketamine adjuvant treatment for refractory pain in children, adolescents and young adults with cancer: a pilot study. BMJ Support Palliat Care 12:e656–e663. https://doi.org/10.1136/bmjspcare-2018-001739
38. Klepstad P, Borchgrevink P, Hval B et al (2001) Long-term treatment with ketamine in a 12-year-old girl with severe neuropathic pain caused by a cervical spinal tumor. J Pediatr Hematol Oncol 23:616
39. Kajiume T et al (2012) Continuous intravenous infusion of ketamine and lidocaine as adjuvant analgesics in a 5-year-old patient with neuropathic cancer pain. J Palliat Med 15(6):719–722
40. Taylor M, Jakacki R, May C et al (2015) Ketamine PCA for treatment of end-of-life neuropathic pain in pediatrics. Am J Hosp Palliat Care 32:841–848. https://doi.org/10.1177/1049909114543640
41. Ramirez MF, Strang A, Roland G et al (2023) Perioperative pain management and cancer outcomes: a narrative review. J Pain Res 16:4181–4189. https://doi.org/10.2147/JPR.S432444
42. Rodriguez Arango JA, Zec T, Khalife M (2024) Perioperative ketamine and cancer recurrence: a comprehensive review. J Clin Med 13:1920. https://doi.org/10.3390/jcm13071920
43. Tran K, MacDougall D (2023) Ketamine for chronic non-cancer pain: a 2023 update: CADTH Health Technology Review. Canadian Agency for Drugs and Technologies in Health, Ottawa (ON) Ketamine for Chronic Non-Cancer Pain: A 2023 Update: CADTH Health Technology Review. Accessed 21 July 2024
44. Blagrove M, Morgan CJA, Curran HV et al (2009) The incidence of unpleasant dreams after sub-anaesthetic ketamine. Psychopharmacology 203:109–120. https://doi.org/10.1007/s00213-008-1377-3
45. Kim M, Cho S, Lee J-H (2016) The effects of long-term ketamine treatment on cognitive function in complex regional pain syndrome: a preliminary study. Pain Med Malden Mass 17:1447–1451. https://doi.org/10.1093/pm/pnv112
46. Noppers IM, Niesters M, Aarts LPHJ et al (2011) Drug-induced liver injury following a repeated course of ketamine treatment for chronic pain in CRPS type 1 patients: a report of 3 cases. Pain 152:2173–2178. https://doi.org/10.1016/j.pain.2011.03.026
47. White MC, Karsli C (2007) Long-term use of an intravenous ketamine infusion in a child with significant burns. Paediatr Anaesth 17:1102–1104. https://doi.org/10.1111/j.1460-9592.2007.02329.x
48. Busse J, Phillips L, Schechter W (2015) Long-term intravenous ketamine for analgesia in a child with severe chronic intestinal graft versus host disease. Case Rep Anesthesiol 2015:834168. https://doi.org/10.1155/2015/834168
49. Bredlau AL, Harel BT, McDermott MP et al (2015) Neurocognitive changes after sustained ketamine administration in children with chronic pain. J Palliat Care Med 5:215. https://doi.org/10.4172/2165-7386.1000215

50. Yan J, Li Y, Zhang Y et al (2014) Repeated exposure to anesthetic ketamine can negatively impact neurodevelopment in infants: a prospective preliminary clinical study. J Child Neurol 29:1333–1338. https://doi.org/10.1177/0883073813517508
51. Cheung HM, Yew DTW (2019) Effects of perinatal exposure to ketamine on the developing brain. Front Neurosci 13:138. https://doi.org/10.3389/fnins.2019.00138
52. Dong C, Anand KJS (2013) Developmental neurotoxicity of ketamine in pediatric clinical use. Toxicol Lett 220:53–60. https://doi.org/10.1016/j.toxlet.2013.03.030
53. Simons P, Olofsen E, van Velzen M et al (2022) S-Ketamine oral thin film-Part 2: population pharmacodynamics of S-ketamine, S-norketamine and S-hydroxynorketamine. Front Pain Res Lausanne Switz 3:946487. https://doi.org/10.3389/fpain.2022.946487
54. Benítez-Rosario MA, Salinas-Martín A, González-Guillermo T, Feria M (2011) A strategy for conversion from subcutaneous to oral ketamine in cancer pain patients: effect of a 1:1 ratio. J Pain Symptom Manag 41:1098–1105. https://doi.org/10.1016/j.jpainsymman.2010.08.017
55. Mathew E, Kim E, Goldschneider KR (2014) Pharmacological treatment of chronic non-cancer pain in pediatric patients. Paediatr Drugs 16:457–471. https://doi.org/10.1007/s40272-014-0092-2
56. Zhang H, Zhou P, Jiang Y et al (2023) Sustained-release esketamine based nanoparticle-hydrogel delivery system for neuropathic pain management. Int J Nanomedicine 18:1131–1143. https://doi.org/10.2147/IJN.S400798
57. Grant IS, Nimmo WS, McNicol LR, Clements JA (1983) Ketamine disposition in children and adults. Br J Anaesth 55:1107–1111. https://doi.org/10.1093/bja/55.11.1107
58. Dallimore D, Anderson BJ, Short TG, Herd DW (2008) Ketamine anesthesia in children—exploring infusion regimens. Paediatr Anaesth 18:708–714. https://doi.org/10.1111/j.1460-9592.2008.02665.x
59. Yang Y, Maher DP, Cohen SP (2020) Emerging concepts on the use of ketamine for chronic pain. Expert Rev Clin Pharmacol 13:135–146. https://doi.org/10.1080/17512433.2020.1717947

Part II

Critical Care

Chapter 6

How About Esketamine for Sedation of Critically Ill COVID-19 Patients?

Sabine Himmelseher and Helmut Trimmel

Abstract

Coronavirus disease 2019 (COVID-19) affected millions of people worldwide. The pandemic had a significant impact on critical care and in particular on sedation management. Esketamine has emerged as an essential component of the analgesia and sedation regimen for both noninvasively and invasively ventilated patients. It has numerous beneficial effects, including hemodynamic stabilization, reduction of opioid requirements, neuroprotection, and anti-inflammatory effects. This article discusses the use of esketamine in the intensive care of patients with COVID-19 but also reflects on the benefits and promising perspectives for all critically ill patients. Ketamine is consistently listed as an essential drug by the World Health Organization. In a conversation with an experienced intensivist, the practical use of esketamine, its benefits, adverse effects, and differentiated use in a well-defined sedation management strategy for everyday care and perspectives for emerging new indications will be presented.

Key words Esketamine, Analgesia, Sedation, Intensive or critical care, COVID-19, Noninvasive ventilation, Opioid sparing, Neuroprotection, Delirium, Antidepressant effects

1 Introduction

At the beginning of the 2020s, coronavirus disease 2019 (COVID-19) affected millions of people worldwide [1]. A substantial proportion (10–20%) of patients required intensive care and ventilatory support and analgesia and sedation [2]. However, many of commonly used sedatives and analgesics carry peculiar risks within the pathophysiology of severe acute respiratory syndrome coronavirus-2 (SARS-CoV-2) infection.

SARS-CoV-2 binds to tissue angiotensin-converting enzyme-2 (ACE-2) receptors expressed, for example, in the lower respiratory tract, brain, small intestine enterocytes, vascular endothelium, and cardiac myocytes [3]. This results in downregulation of ACE-2, enhanced production of angiotensin II, and an increase in vascular permeability. After endocytosis or fusion with cell membranes, virus replication is considered as primary driver in early disease.

Sabine Himmelseher (ed.), *Ketamine: Pain Management, Critical Care, Depression, and Resource-Poor Settings Volume II*, Neuromethods, vol. 225, https://doi.org/10.1007/978-1-0716-4599-4_6,

Although the virus shields its RNA from host detection and antiviral interferon response, antibody-mediated virus uptake by monocytes and tissue macrophages activates inflammatory cell death [4]. This aborts virus production but is associated with an outburst of inflammatory cytokines, chemokines, and alarmins that recruit immune cells to infection sites. More tissue damage and thrombosis follow, but it remains unclear whether direct viral toxicity or the out-of-control host immune response plays the major role in patient multi-organ failure and mortality.

Critically ill COVID-19 patients often present the lungs as initial and most prominent site of damage and complication: Respiratory insufficiency and/or acute hypoxic failure results from pneumonia up to acute respiratory distress syndrome (ARDS), and supportive therapy needs to improve gas exchange, oxygen transport, and tissue oxygenation. Assisted or controlled mechanical ventilation (MV) with appropriate sedation may become necessary with the goal of preventing ventilator-induced lung injury [5]. This scenario may further be complicated by high sedation and analgesia needs, which often go along with severe hypoxia, ventilator dyssynchrony, massive elevation of inflammatory parameters, younger age, high body mass index, high metabolic rate, and use of neuromuscular paralysis [6]. All of these factors may prolong the duration of MV, predispose and precipitate delirium, and eventually increase mortality. It is therefore indispensable to provide sedation and analgesia with the lowest possible, additional "burden" for critically ill COVID-19 patients.

Ketamine appears to be an attractive option for this goal [7–9]: It offers the advantage of anti-hyperalgesic and analgesic effects at subanesthetic dose without being cardiorespiratory depressive, may spare opioids (and paralytics) in multimodal regimen, and increase sedation levels while allowing for reducing other drugs while staying within sedation targets [5, 10, 11]. Ketamine does not worsen the unique hyper- and/or hypo-inflammatory constellations seen in COVID-19 [12, 13] and may regulate inflammatory disbalances. Ketamine has recently been confirmed to prevent and treat posttraumatic stress disorder (PTSD) [14], may enhance psychic resilience [15], and is established as rapidly acting antidepressant [8, 16, 17]. This may be of great value in ICU therapy, because half of COVID-19 patients report depression during hospital stay [18], and a substantial proportion suffers from poor mental health with distress symptoms in prolonged recovery and long COVID-19 [19]. Globally used racemic ketamine consists of equal amounts of right- and left-handed enantiomers with left-handed esketamine offering the advantage of a two times higher analgesic and anesthetic potency than the racemate [11]. Esketamine is metabolized more rapidly and allows for an earlier recovery and drug emergence. Although the incidence of unwanted psychoactive effects may be similar, esketamine's brain impact is of less unpleasant quality. In addition, it induces less neurocognitive impairment and better maintains logical functions at half-dose racemic ketamine. In the following, "ketamine" refers to racemic ketamine and esketamine to its pure S(+)-isomer.

This chapter highlights aspects of postulated advantages and real-life experience with esketamine in analgesia and sedation of critically ill COVID-19 patients. **Professor Trimmel (expert), Wiener Neustadt, Austria, dedicated clinician and esketamine researcher for over three decades, will share his expertise.**

2 Sedation with Ketamine in COVID-19: Background and Expert Comments

2.1 Respiratory Failure

As advantageous, it has been reported that:

- Noninvasively ventilated, uncooperative COVID 19 patients benefit from dexmedetomidine (DEX)-ketamine combinations with faster sedation rates and better hemodynamic profiles as compared to DEX alone [20].
- A sedative regimen of subanesthetic dose esketamine (0.2–0.4 mg/kg/h) combined with inhalational sevoflurane (target end-tidal concentration 0.5–1.3 Vol%) allows for maintenance of assisted respiration even during prone positioning [21].
- In light sedation (Richmond Agitation Sedation Scale (RASS) – 2 up to +1) with propofol (PROP) or DEX, pressure support ventilation, and opioid or non-opioid analgesia, the addition of ketamine (0.6–1.2 mg/kg/h) allows for better pain management or pain control while being associated with less opioid needs (goal: Behavioral Pain Scale < 5) [5].
- Adjunctive (es-) ketamine added to a two- or three-substance, previously insufficient, sedation regimen in patients on MV mostly provides a satisfactory level of sedation and analgesia and/or deep sedation (RASS -3 to −5). This is often associated with hemodynamic stabilization while sparing opioids or even paralytics. In critically ill COVID-19 patients, a German center reported to use clonidine, midazolam, and esketamine, combined with sufentanil most often [22].

Questions

What are your experiences with esketamine as sedative/analgesic in these settings?

When do you use subanesthetic and when rather anesthetic doses?

Which drug combinations do you prefer?

(continued)

According to which criteria do you titrate esketamine (such as sedation goals, e.g., RASS values)?

Expert

Esketamine is frequently used as supplemental analgesic in our intensive care unit (ICU). We like to choose esketamine/dexmedetomidine combinations for analgesia and sedation in restless or delirious patients who require noninvasive breathing support, such as high-flow oxygen therapy or pressure support ventilation. If necessary, we increase the commonly used subanesthetic dose of 0.25–0.5 mg/kg ideal body weight (iBW) up to doses of 1.5 mg/kg iBW.

Of course, esketamine is also used in patients who are difficult to sedate, which was undoubtedly the case with critically ill COVID-19 patients. Our combination regimen consisting of propofol, remifentanil or sufentanil, and esketamine allowed for control of these patients' increased respiratory drive very effectively. Dreaded patient self-induced lung injury (P-SILI) could be prevented in almost all cases. Under such conditions, higher esketamine doses within a range from 1 to 2.5 mg/kg iBW often had to be administered. One of the special properties of esketamine is its excellent circulatory stabilizing effects, even at high dose. This allows for adequate sedation in patients undergoing multiple organ replacement therapies without the need for high exogenous vasopressor amounts. Figure 1 shows the treatment of a seriously ill COVID-19 patient necessitating multiple organ replacement procedures.

Esketamine is also particularly important in our neurotraumatological ICU, where we use the substance as adjunct in analgesia and sedation regimen on a regular basis. Esketamine additionally allows for reduction of other analgesics and sedatives including opioids, thus limiting their adverse effects. This is especially valuable in patients with severe traumatic brain injury (sTBI), who often need deep sedation for long periods of time. Ketamine is well known for its neuroprotective properties counteracting excessive NMDA-receptor over-activation in sTBI. Far beyond, esketamine is very effective in the reduction and prevention of cortical spreading depolarizations, which—untreated—may have very harmful impact on neurologic patient outcomes. Sedation requirements and therapeutic approaches to patients with sTBI differ substantially from other critically ill patients: A combination of deep sedation up to a burst-suppression electroencephalography (EEG) pattern may be necessary, and an "optimized, extremely careful handling" and nursing management concept is applied to

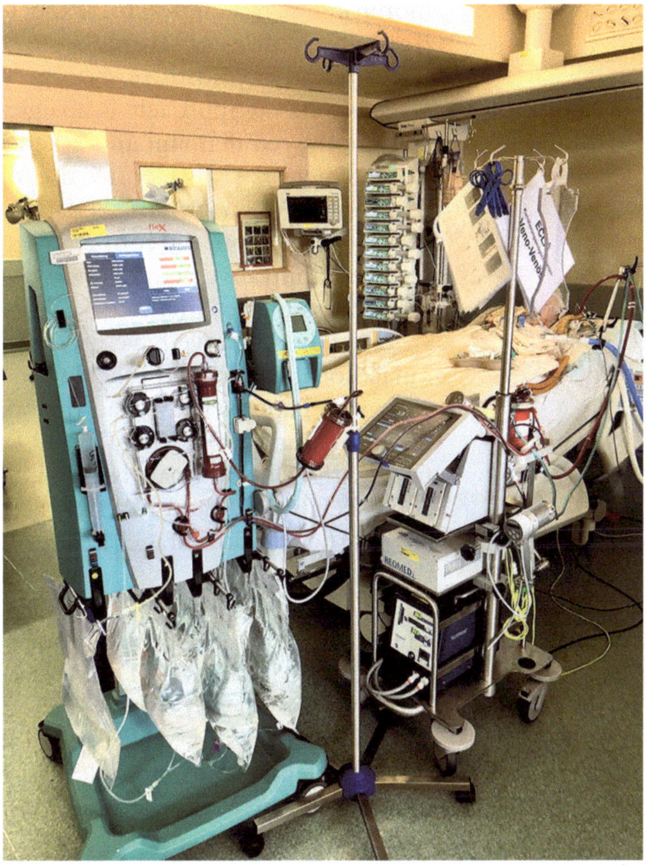

Fig. 1 A coronavirus disease 2019 (COVID-19) patient requiring multiple organ replacement procedures: extracorporeal oxygenation (ECMO), renal replacement (continuous veno-venous hemodiafiltration (CVVHDF)), and cytokine elimination (photo by H. Trimmel, Intensive Care Unit, University Hospital Wiener Neustadt, Wiener Neustadt, Austria)

all of our patients with serious intracranial pathologies such as sTBI, intracranial hemorrhage, or subarachnoid hemorrhage.

This is particularly important during most vulnerable phases with increased intracranial pressure and/or restricted cerebral blood flow. At best, this treatment strategy is combined with appropriate neuromonitoring. We use a commercially available monitoring system, which automatically analyzes raw EEG recordings based on spectral analysis. A set of parameters is created, and visually classified EEG data are provided. However, when using such a neuromonitoring device, it must be taken into account that esketamine can lead to a display of "falsely high" values. Unfortunately, depth of sedation monitoring is also still poorly validated with respect to prolonged uses in ICU settings. Nevertheless, valuable information may be gained from observation of longitudinal course of variables combined with other findings and monitored parameters. This is especially valid during early, instable treatment periods of serious acute brain pathologies.

In all other (mostly post-surgery) patients, we apply a primarily analgesia-based approach as analgesia and sedation strategy. Most commonly, remifentanil and esketamine are administered for pain management. We prefer to use our clinical judgment and the Richmond Agitation and Sedation Scale (RASS) to control the depth of sedation. In cooperative patients, including those who receive noninvasive ventilation, we also rely on the numeric rating scale for pain assessment.

Frankly open, we have little experience with a combination of inhalational agents and esketamine for analgesia and sedation. We were very happy to have had the opportunity of using inhalational sedation during the COVID-19 crisis and still use this option frequently. Our rationale is based on the fact that both sevoflurane and esketamine act on NMDA receptors so that we do not expect any particular advantage from combining those two drugs in the ICU.

2.2 Inflammatory Disbalance

Meta-analysis of the dynamic course of inflammation during development of critical COVID-19 has shown that severity and mortality go along with high levels of inflammatory biomarkers [23]. Sedatives and analgesics should therefore not additionally boost patient's immunological distress. Nonetheless, commonly used drugs may add to the momentum of inflammatory and immune overshoot and/or failure.

In consensus guidelines, pain management with opioids as mainstay of treatment is prioritized in critically ill patients [24]. However, although opioids acutely reduce pain, they also lead to tolerance, hyperalgesia, and physiological and psychological addiction. With escalating doses, for example, to facilitate lung-protective ventilation, the risk of prolonged neurologic recovery increases [10]. Far beyond, opioid treatment is associated with a high risk of immune suppression and development of inflammatory events not only caused by B and T cell-mediated immune dysfunction but also by opioid peptides and receptors expressed in tissues [25].

One of propofol's well-known adverse effects in ICU sedation is the induction of hypertriglyceridemia and even pancreatitis, which appear to be exaggerated in COVID-19 [10]. Sometimes patients present similar to those with a hemophagocytosis syndrome [12], showing an out-of-control hyper-inflammatory response with, e.g., fever, cytopenia, elevated ferritin, LDH, lever enzymes, and triglycerides.

In contrast, (es)-ketamine has repeatedly been reported to induce clinically relevant anti-inflammatory effects, for example, decreasing a cardiopulmonary bypass-related increase in pro-inflammatory IL-6 and IL-8 and increasing anti-inflammatory

IL-10 [26]. In this randomized controlled trial, esketamine effects were compared with those of sufentanil in anesthesia for coronary artery bypass graft surgery.

In retrospective observation of COVID-19's course in critically ill patients, it is described that:

- In patients on controlled MV, infusing racemic ketamine as adjunct to sedation and analgesia regimen is associated with improved sedation depth, while decreasing ferritin and lactate, but not CRP, LDH, or IL-6. No benefit in length of ICU stay or mortality occurred, but ketamine was administered "late" when other treatments were deemed "inadequate" and disease burden was high [27].
- In a comparison of a continuous infusion of propofol versus ketamine over 96 h, no difference in vasopressor needs, but higher mean arterial pressures and lower CRP values are observed with ketamine. In addition, ketamine drastically reduces opioid requirements, being only one fourth of those than with propofol [28].

Questions

Did you observe a reduction in typical COVID-19-related (inflammatory) cytokines such as C-reactive protein (CRP), interleukin (IL)-6, or other relevant (laboratory) biomarkers with esketamine?

Can you describe typical laboratory findings or any functional systems change associated with esketamine, which were likely related to less inflammation you have personally noted?

Expert

Present evidence about esketamine's influence on COVID 19-related inflammation is limited. However, there is substantial evidence of esketamine's anti-inflammatory action in other clinical settings, such as with on-pump cardiac or other surgical procedures, septic, or depressed patients: Ketamine may modulate immune responses and reduce high levels of circulating pro-inflammatory cytokines such as IL-6. This is mediated via NMDA and non-NMDA receptor-related effects and subsequent downstream signaling cascades. For example, esketamine may affect the kynurenine pathway, which is involved in the metabolism of tryptophan and linked to inflammation and neurodegeneration. Modulation of this pathway may also reduce the production of neurotoxic metabolites and inflammation. There is also some literature evidence of racemic ketamine-related inflammatory cytokine reduction in COVID-19 patients, but there is no direct trial evidence of specific anti-inflammatory action of esketamine. During the pandemic, we had the impression that esketamine might help reduce inflammatory mediators in the specific COVID-19 patient population. However, in view of our polypragmatic therapeutic approach, which included cytokine filtration and corticosteroid medication, I would not dare to claim that an anti-inflammatory effect of esketamine was of great clinical significance at that time.

2.3 Psychoactive Effects

When managing seriously ill COVID-19 patients, their mental health is of deep concern: 50% suffer from varying degrees of anxiety and depression during hospitalization [18], with some correlation to hyperinflammation [29]. Subanesthetic dose ketamine has been established as valuable treatment with rapid-acting antidepressant effects [16, 17], but its unwanted psychoactive action still causes reluctance with respect to use among intensivists [30]. Nevertheless, ketamine effects are completely dose- and context-dependent:

- The incidence of unwanted psychic action is almost negligible in awake patients in the low-subanesthetic, anti-hyperalgesic dose range (< 0.25 mg/kg IV racemic ketamine).
- In combination with a GABA- or α-2 adrenergic agonist in ICU sedation and analgesia, ketamine's psychic adverse events are enormously reduced or prevented. When ketamine is stopped or decreased in dose early enough before other sedatives will be ceased, emergence is without adverse events or of better quality than without ketamine. In addition, a recent meta-analysis on outcomes in critically ill patients reports less delirium with ketamine use. However, the number of available RCTs is still small [31].
- In scenario with heavy physic and psychic stress such as in sTBI, subarachnoid or intracranial hemorrhage, and various other brain insults, ketamine exerts neuroprotective effects with skilled use in sedation and analgesia regimen, for example, through reduction of spreading depolarization [24, 32].
- Ten to 12 months after hospitalization because of severe COVID-19, ~ 30% of patients report anxiety, ~25% depression, ~30% PTSD, and ~30% insomnia [19, 33]. In long COVID-19, a new systematic analysis of pooled trial data from >7–9 million people shows a 23% prevalence of depression and anxiety [34]. This is of immense global significance. Ketamine has the potential to alleviate such suffering: It can increase psychic resilience and prevent PTSD [15, 35, 36]. It is used to treat anxiety and PTSD and exerts antidepressant action in perioperative anesthesia care [14, 37–39].
- In addition, in a small retrospective evaluation of 12 ICU patients who were observably withdrawn from care team and family, repeated infusions of 0.5 mg/kg or 0.75 mg/kg ketamine improved mood and affect [40].

Questions

What is your impression of all of this? Do you observe less anxiety, "better mood," less depression, and less neurocognitive decline with esketamine?

Are patients "different" with regard to mental health when they leave the ICU after treatment with esketamine?

Have you ever used oral esketamine?

Expert

Esketamine has been shown to have rapid antidepressant and anxiolytic effects in diverse patient populations, including those with treatment-resistant depression. These effects could also benefit intensive care patients, who often suffer from high levels of anxiety and mood disorders due to the stress of critical illness and ICU settings. In my opinion, there is still a huge gap of data about this activity. Nevertheless, the scientific discussion on "post-ICU syndrome" has gained a lot of momentum, and we are all well aware of the fact that skillful pain management, appropriate sedation, and delirium prevention may be the key components for good functional ICU outcomes, including less pronounced post-ICU syndrome. According to my experiences, however, too much emphasis is still placed on sedation instead of implementing analgesia-based concepts across the board.

Unfortunately, it is not possible for us to follow up patients on a larger scale in everyday life after ICU and hospital discharge. I can therefore not sufficiently answer questions about our patients' "post-hospital health status." This gap of knowledge about the outcomes of our work is primarily due to limited or lack of manpower. For me personally, it clearly represents a deficit for quality improvement programs. My experience with oral administration of esketamine is very limited, so that I cannot share meaningful knowledge.

2.4 Complications with Longer-Term or Higher-Dose Use

As described in other chapters, longer-term and/or higher-dose ketamine use may especially be associated with serious adverse effects:

- Ketamine uropathy induces inflammation in the urothelium, which may manifest in all parts of the lower urinary tract, including hemorrhagic cystitis or pelvic pain, and in the upper urinary tract, including as hydronephrosis [41, 42]. Its etiopathology is uncertain, and there is no specific treatment. Cessation of ketamine is a necessity to prevent further progression and deterioration.

- For years, it has been reported that cholestasis, cholangiopathy, hepatobiliary dysfunction, and liver injury concomitantly occur with extreme ketamine exposure (> 1 mg/kg/h infusion for days) in critically ill patients. No causal relationship has been established so far [43], but COVID-19-related secondary sclerosing cholangitis and ketamine have been associated with hepatobiliary organ failure involving different underlying insults (such as biliary ischemia, direct viral damage, toxic bile, cytokine storm hyperinflammation, positive end-expiratory pressure, and possibly high-dose ketamine) [44]. Given the severe prognosis of the disease sometimes even requiring orthotopic liver transplantation, close monitoring of hepatobiliary biomarkers is especially important during and after severe COVID-19 because of virus-specific biliary tropism [45].
- Mostly in critically ill patients, all kinds of severe electrolyte disturbances have been described in case reports about ketamine use, ranging from diabetes insipidus to the syndrome of inappropriate antidiuretic hormone secretion and hyponatremia [46, 47]. Better awareness of these problems may lead to earlier identification and ketamine discontinuation.

Questions

Have you observed these complications, and how about their frequency of occurrence?

What is your treatment approach?

Have patients completely recovered after experiencing such adverse effects?

Have you noticed other serious complications, such as due to a ketamine-related increase in tracheobronchial secretions?

Expert

In the literature, ketamine-induced cystitis (KIC) is described as a manifestation of adverse ketamine effects often associated with prolonged and/or higher-dose, chronic ketamine use. KIC is characterized by chronic inflammation of the bladder, which can lead to ulcerative cystitis and, in severe cases, hemorrhagic cystitis as well as hydronephrosis. Painful urination and the presence of blood in the urine are typical symptoms of this complication. However, we do not observe KIC in our patient population, mainly because the duration of esketamine use in our ICU is rarely longer than 10–14 days.

The same is valid for cholestasis, cholangiopathy, and secondary sclerosing cholangitis. We monitor liver functions very closely. We start with enteral administration of ursodeoxycholic acid when liver parameters start to rise, particularly if alkaline phosphatase is affected. We have very good experiences with this therapy: Liver function markers usually completely return to a normal range within days or at least a few weeks. We have almost never seen long-term or lasting gallbladder or liver damage in our patients.

Electrolyte disorders are very common in ICU patients and usually result from imbalances in various organ functions. In my opinion, it is in general not permissible to attribute such disorders to a single substance. Of course, a complete reassessment of all drugs administered must be performed as part of daily rounds and visits. Whenever possible, the principle of reducing necessary and/or discontinuing unnecessary drugs must be kept.

2.5 Emergence and Outcomes

Abrupt awakening and emergence from sedation is a cause of serious disorientation and patient distress. Ketamine offers several advantages in multimodal regimen:

- Its effects allow for a graduate and adjustable return of cognitive states due to long-acting metabolites [48]. Esketamine is better controllable because of more rapid metabolism than the racemate [49–51].
- (Es-) ketamine counteracts sedation and opioid-induced respiratory depression in spontaneous, assisted, and positive pressure ventilation [8, 52]. This characteristic property is very useful in combination with remifentanil, where it may reduce dose requirements [53]. Upon cessation of remifentanil and/or propofol infusions, subanesthetic dose adjunctive ketamine (stopped early enough before propofol or reduced to very low dose) prevents confusion, agitation, and pain. It decreases remifentanil-induced hyperalgesia at small dose [54].
- Ketamine directly activates breathing and may prevent the coupling between loss of consciousness and upper airway muscle dysfunction [55, 56]. In intubated spontaneously breathing patients with propofol sedation, ketamine (0.3 or 0.6 mg/kg/h) increased inspiratory flow, reduced work of breathing, and induced an "activated" EEG pattern. All of these features may facilitate weaning.
- There is not one single long-term outcome trial after ketamine administration in critically ill patients. However, a small retrospective trial studying ketamine as sedative-analgesic in COVID-19 patients with severe ARDS reports safe use and reduced mortality with ketamine. Although complications were similar among all patients, those receiving ketamine had a longer length of hospital stay [57].

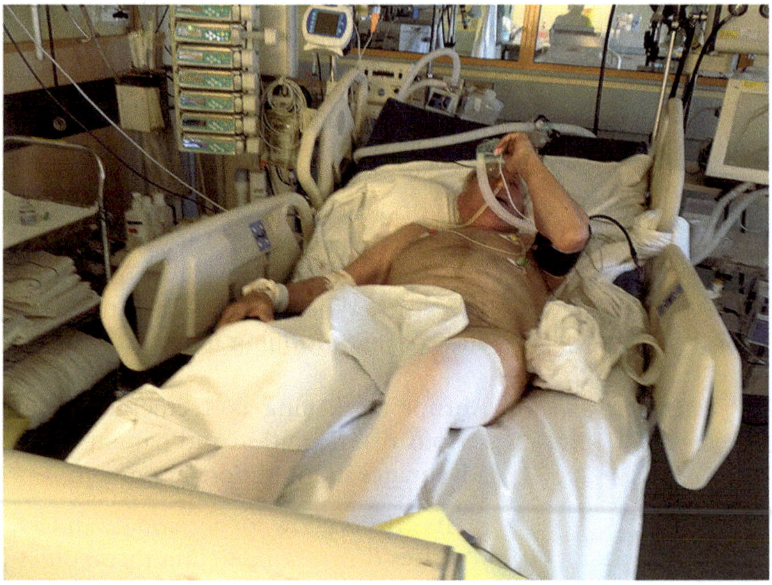

Fig. 2 Particularly critical for patient outcome is the development of a hyperactive delirium, which is characterized by restlessness, agitation, and excitement (photo by H. Trimmel, Intensive Care Unit, University Hospital Wiener Neustadt, Wiener Neustadt, Austria)

Questions

What is your perspective on esketamine in emergence and awakening from sedation?

Do you have experiences with esketamine in elderly patients and recovery from sedation and analgesia?

Do you "think" that mortality might turn out less in a controlled trial with properly addressed patient numbers after esketamine use?

Can you describe own patient cases or perhaps report own data?

Expert

Perspectives on esketamine with regard to emergence and termination of sedation in ICU patients are diverse and evolving. After short exposure time and low dosage, wake-up may be easy and occurs quickly. A smoother transition from sedation to wakefulness facilitated by esketamine may reduce the incidence of emergence delirium, which is quite common in ICU patients (as shown in Fig. 2).

Unique neuroprotective properties of esketamine as mentioned above may also contribute to better neurocognitive outcomes after sedation. Potential benefits in terms of mood and anxiety reduction could outweigh any transient dissociative esketamine effects, but available evidence and high-quality data are as yet very limited. In elderly patients, general metabolism and drug elimination, including CYP-related esketamine metabolism, are often slower than in healthy young individuals. This is of special significance in ICU settings, because likely longer-lasting drug effects require cautious drug dosing and monitoring to avoid prolonged sedation and unwanted side effects. Low-dose esketamine is generally considered safe and effective for sedation and analgesia in elderly patients. However, because of enhanced sensitivity and vulnerability to adverse effects, monitoring and individually titrated dosing are essential in this patient population. In contrast, most recent data suggest that subanesthetic dose esketamine may beneficially impact cognitive recovery after surgery and reduce delirium incidence and other central nervous system dysfunctions as outlined. It will also be very interesting to see whether such treatment gains may be achievable with oral esketamine. If so, these beneficial effects may allow for esketamine use outside of hospital settings in this extremely important field of evolving new esketamine indications.

Esketamine has a lot of promising potential not only because of lack of cardiorespiratory depression, and analgesic (and opioid-sparing) effects, but also because of its neuroprotective, anti-inflammatory, and antidepressant properties. However, large-scale, randomized, controlled trials with sufficiently powered study patient numbers are urgently needed, for example, to determine whether skillful use of esketamine in the ICU may reduce patient mortality.

3 Conclusion and Perspective

Questions

Do you include esketamine in a multimodal sedation and analgesia regimen in routine, everyday care?

Expert

Esketamine is one of the most often used medications in our ICU. It plays a central role in our multimodal analgesic and sedative regimen. Because our treatment focus is on neurotrauma patients, we attach importance to use of substances without possible neurological harm and prefer to administer agents with neuroprotective and anti-inflammatory action. Without surprise, ketamine is continuously listed on the first page of WHO's *Model List of Essential Medicines* [9].

Furthermore, there are many areas which are of great interest for future investigation, including the effects of oral esketamine use in general, or esketamine's impact on postoperative neurocognitive deficits and delirium, and symptoms associated with post-ICU syndrome.

Questions
Do you have a sedation algorithm which includes esketamine?

Expert
Our analgesia and sedation algorithm is structured according to the different patient requirements: Basis for short-term sedation of intubated (mostly postoperative) patients is inhalational therapy with sevoflurane at a target end-tidal concentration of 0.5–1.0 Vol %. As an alternative, 3–4 mg/kg/h propofol is combined with 0.1–0.3 μg/kg/min iBW remifentanil. If longer-term and/or deep sedation is necessary, 0.25–1.0 mg/kg/h iBW esketamine is added to the propofol/remifentanil combination. Clonidine is applied to allow for a reduction of other analgesic and sedative agents, while dexmedetomidine is primarily used during weaning. We try to avoid benzodiazepines whenever possible, but sometimes they have to be used, for example, in patients with sTBI. This algorithm, the S3 guideline for delirium prevention, sedation, and analgesia in intensive care ("DAS-Leitlinie") [58], and our institutional standard-operating procedures are available to all of our employees electronically at any time—even at bedside within the patient data management system (PDMS). The combination of dexmedetomidine and esketamine for noninvasive ventilation has been described above.

Questions
What do you want to say that has not been touched upon?

Expert
For me, an intensive care and emergency physician, esketamine's outstanding profile offers tremendous benefits in prehospital services. Its broad therapeutic spectrum, small respiratory depressive action, and hemodynamically stabilizing properties are of immense value in emergency care of prehospital, critically ill patients.

Questions
What are your conclusions, vistas, and recommendations *especially for esketamine*?

Expert
Esketamine has been established as indispensable component of our sedation strategy. It is an essential part of our neuroprotective approach to acute intracranial pathologies. I expect further exciting developments and emerging new indications with low-dose, subanalgesic esketamine, especially for prophylaxis and/or treatment of cognitive deficits after surgery and post-ICU syndrome. It is very important to clarify use schedules that may provide antidepressant effects in the critically ill, because depression and anxiety place a heavy burden and often lasting problem on ICU patients.

Thank you!

Conflicts of Interests

The authors declare no conflicts of interest. We do not have any (financial or other) relations to companies marketing, selling, or providing esketamine.

References

1. Grasselli G, Tonetti T, Protti A et al (2020) Pathophysiology of COVD-19-associated acute respiratory distress syndrome: a multicentre prospective observational study. Lancet Respir Med 8:1201–1208
2. Phua J, Wang L, Ling L et al (2020) Intensive care management of coronavirus disease 2019 (COVID-19): challenges and recommendations. Lancet Respir Med 8:506–517
3. Greenland JR, Michelow MD, Wang L, London MJ (2020) COVID-19 infection: implications for perioperative and critical care physicians. Anesthesiology 132:1346–1361
4. Junqueira C, Crespo A, Ranjbar S et al (2022) FcγR-mediated SARS-CoV-2 infection of monocytes activates inflammation. Nature 606:576–584
5. Payen JF, Chanques G, Futier E et al (2020) Sedation for critically ill patients with COVID-19: which specificities? Anaesth Crit Care Med 39:341–343
6. Balakrishna A, Walsh E, Hamidi A et al (2021) An examination of sedation requirements and practices for mechanically ventilated critically ill patients with COVID-19. Am J Health Syst Pharm 78:1952–1961
7. Trimmel H, Helbok R, Staudinger T et al (2018) S(+)-ketamine: current trends in emergency and intensive care medicine. Wien Klin Wochenschr 130:356–366
8. Weinbroum AA (2021) Perspectives of ketamine use in COVID-19 patients. J Korean Med Sci 36:e28
9. WHO (2023) Model list of essential medicines, 23rd list. 1.Anaesthetics.1.1 General anaesthetics. 1.1.2 Injectable medicines. Ketamine. https://iris.who.int/bitstream/handle/10665/371090/WHO-HPS-EML-2023.02-eng.pdf?sequence=1. Accessed 6 July 2024
10. Karamchandani K, Dalal R, Patel J et al (2021) Challenges in sedation management in critically ill patients with COVID-19: a brief review. Curr Anesthesiol Rep 11:107–115
11. Mion G et al (2024) Less drowsiness, more analgesia. Anesth Analg 139:78–91

12. Henderson LA, Canna SW, Schubert GS et al (2020) On the alert for cytokine storm: immune-pathology in COVID-19. Arthritis Rheumatol 72:1059–1063
13. Kim JW, Suzuki K, Kavalali ET, Monteggia KM (2024) Ketamine: mechanisms and relevance to treatment of depression. Ann Rev Med 75: 129–143
14. Almeida TM, Raianny U, da Silva L et al (2024) Effectiveness of ketamine for the treatment of post-traumatic stress disorder – a systematic review and meta-analysis. Clin Neuropsychiatry 21:22–31
15. Evers AG, Murrough JW, Charney D, Costi S (2022) Ketamine as a prophylactic resilience-enhancing agent. Front Psych 13:833259
16. Lullau APM, Haga EMW, Ronold EH, Dwyer GE (2023) Antidepressant mechanisms of ketamine: a review of actions with relevance to treatment-resistance and neuroprogression. Front Neurol 17:1223145
17. Johnston V, Kadriu B, Kraus C et al (2024) Ketamine in neuropsychiatric disorders: an update. Neuropsychopharmacology 49:23–40
18. Park HY, Jung J, Park HY et al (2020) Psychological consequences of survivors of COVID-19 pneumonia 1 month after discharge. J Korean Med Sci 35:e409
19. Sankar K, Gould MK, Prescott HC (2023) Psychological morbidity after COVID-19 critical illness. Chest 163:139–147
20. Riccardi A, Serra S, de Iaco F et al (2023) Uncovering the benefit of the ketamine-dexmedetomidine combination for procedural sedation during the Italian COVID-19 pandemic. J Clin Med 12:3124
21. Bansbach J, Wenz J, Kaufmann K et al (2022) Sevoflurane in combination with esketamine is an effective sedation regimen in COVID-19 patients enabling assisted spontaneous breathing even during prone positioning. Anaesthesiol Intensive Ther 54:23–29
22. Flinspach AN, Booke H, Zacharowski K (2021) High sedation needs of critically ill COVID-19 ARDS patients – a monocentric observational study. PLoS One 16:e0253778
23. Mahat RK, Panda S, Rathore V et al (2021) The dynamics of inflammatory markers in coronavirus disease-2019 (COVID-19) patients: a systematic review and meta-analysis. Clin Epidemiol Glob Health 11:100727
24. Peng Z, Sun T, Zhang X, Zhang H (2024) Expert consensus on the clinical utilization of ketamine and its isomers in intensive care units. Intensive Care Res Mar 4:1–8. https://doi.org/10.1007/s44231-024-00060-z
25. Malafoglia V, Ilari S, Vitiello L et al (2022) The interplay between chronic pain, opioids, and the immune system. Neuroscientist 28:613–627
26. Welters ID, Feurer MK, Preiss V et al (2011) Continuous S(+)-ketamine administration during elective coronary artery bypass graft surgery attenuates pro-inflammatory cytokine response during and after cardiopulmonary bypass. Br J Anaesth 106:172–179
27. Wyler D, Torjman MC, Leong R et al (2023) Observational study of the effects of ketamine infusions on sedation depth, inflammation, and clinical outcomes in mechanically ventilated patients with SARS-CoV-2. Anaesth Intensive Care 52:105–112
28. Phán B, Agyemang A, Klein W, Thapamagar SB (2024) Continuous infusion of ketamine in mechanically ventilated patients with SARS-CoV-2. Crit Care Res Pract 2024:7765932. https://doi.org/10.1155/2024/7765932
29. Kumari A, Itgai ABH, Rukadikar CA et al (2023) Effect of COVID-19 on stress and biomarkers: an exploratory cross-sectional study. Cureus 15:e35702
30. Sharif S, Munshi L, Burry L et al (2024) Ketamine sedation in the intensive care unit: a survey of Canadian intensivists. Can J Anaesth 71: 118–126
31. Abdildin Y, Tapinova K, Nemerenova A, Viderman D (2024) The impact of ketamine on outcomes in critically ill patients; a systematic review with meta-analysis and trial sequential analysis of randomized controlled trials. Acute Crit Care 39:34–46
32. Hartings JA, Dreier JP, Ngwenya LB et al (2023) Improving neurotrauma by depolarization inhibition with combination therapy: a phase 2 randomized feasibility trial. Neurosurgery 93:924–931
33. Klinkhammer S, Horn J, Duits AA et al (2023) Neurological and (neuro)psychological sequelae in intensive care and general ward COVID-19 survivors. Eur J Neurol 30:1880–1890
34. Seighali N, Abdollahi A, Shafiee A et al (2024) The global prevalence of depression, anxiety, and sleep disorder among patients coping with post COVID-19 syndrome (long COVID): a systematic review and meta-analysis. BMC Psychiatry 24:105
35. McGhee L, Maani CV, Garza TH et al (2008) The correlation between ketamine and post-traumatic stress disorder in burned service members. J Trauma 64(2 Suppl):S195–S198
36. Petrucci E, Cofini V, Pizzi B et al (2023) Health status perception and psychological

sequelae in buried victims: an observational study on the survivors of the earthquake in Amatrice (Italy), 3 years later. Prehosp Disaster Med 38:193–198

37. Gan SL, Long YQ, Wang QY et al (2023) Effect of esketamine on postoperative depressive symptoms in patients undergoing thoracoscopic lung cancer surgery: a randomized controlled trial. Front Psych 14:1128406
38. Hartland H, Mahdavi K, Jelen LA et al (2023) A transdiagnostic systematic review and meta-analysis of ketamine' anxiolytic effects. J Psychopharmacol 37:764–774
39. Ma S, Yuzhe D, Wang W et al (2024) Association between esketamine interventions and postpartum depression and analgesia following cesarean delivery: a systematic review and meta-analysis. Am J Obstet Gynecol MFM 6:101241
40. Giri AR, Kaur N, Yarrarapu SN et al (2022) Novel management of depression using ketamine in the intensive care unit. J Intensive Care Med 37:1654–1661
41. Sturgess G, Beckley I, Shepherd R, Downey A (2023) Ketamine uropathy: clinical experience in a high prevalence center. Neurourol Urodyn 42:1555–1562
42. Chen CL, Wu ST, Cha TL et al (2022) Molecular pathophysiology and potential therapeutic strategies of ketamine-related cystitis. Biology 11:502
43. Leonhardt S, Baumann S, Jürgensen C et al (2023) Role of intravenous ketamine in the pathogenesis of secondary sclerosing cholangitis in critically ill patients: perpetrator or innocent bystander: answers provided by forensic toxicology. Intensive Care Med 49:1549–1551
44. De Tymowski C, Depret F, Dudoignon E et al (2021) Ketamine-induced cholangiopathy in ARDS. Intensive Care Med 47:1173–1174
45. Bartoli A, Cursaro C, Seferi H, Andreone P (2023) Secondary sclerosing cholangitis after SARS-CoV2: ICU ketamine use or virus-specific biliary tropism in critically ill patients? Hepat Med 15:93–112
46. Kraus MB, Leuzinger K, Reynolds E et al (2023) Diabetes insipidus related to sedation in the intensive care unit: a review of the literature. J Crit Care 75:154233
47. Kumar K, Poonam F, Rani T et al (2022) Ketamine-induced syndrome of inappropriate antidiuretic hormone secretion and hyponatremia. Cureus 14:e25931
48. Moaddel R, Farmer CA, Yavi M et al (2023) Cerebrospinal fluid exploratory proteomics and ketamine metabolite pharmacokinetics in human volunteers after ketamine infusion. iScience 26:108527
49. Danyeli KV, Sen ZD, Colic L et al (2023) Association of the delayed changes in glutamate levels and functional connectivity with the immediate network effects of S-ketamine. Transl Psychiatry 13:60
50. Himmelseher S, Kochs EF (2021) Ready for a "breakthrough" with ketamine? A look at recent pharmacological insights. Curr Op Anaesthesiol 34:393–401
51. Ihmsen H, Geisslinger G, Schüttler J (2001) Stereoselective pharmacokinetics of ketamine: R(−)-ketamine inhibits the elimination of S (+)-ketamine. Clin Pharmacol Ther 70:431–438
52. Jonkman K, van Rijnsoever E, Olofsen E et al (2018) Esketamine counteracts opioid-induced respiratory depression. Br J Anaesth 120:1117–1127
53. Fujii T, Nishiwaki K (2022) Ketamine reduces the dose of remifentanil required during prolonged head and neck surgery: a propensity-matched analysis. Nagoya J Med Sci 84:1–6
54. Koppert W, Sittl R, Scheuber K et al (2003) Differential modulation of remifentanil-induced analgesia and postinfusion hyperalgesia by S-ketamine and clonidine in humans. Anesthesiology 99:152–159
55. Eikerman M, Grosse-Sundrup M, Zaremba S et al (2012) Ketamine activates breathing and abolishes the coupling between loss of consciousness and upper airway dilator muscle function. Anesthesiology 116:35–46
56. Suleiman A, Santer P, Munoz-Acuna R et al (2023) Effects of ketamine infusion on breathing and encephalography in spontaneously breathing ICU patients. J Intensive Care Med 38:299–306
57. Pata R, Sandeep P, Aung HM et al (2021) Ketamine infusion as a sedative-analgesic in severe ARDS (KISS). J Community Hosp Intern Med Perspect 11:619–623
58. AWMF (2024) S3-Leitlinie Analgesie, Sedierung und Delirmanagement in der Intensivmedizin [S3-Guideline Analgesia, Sedation, and Management of Delirium in Intensive Care]. Version 5. Registernumber 001–012. Valid until 30.03.2026 https://register.awmf.org/de/leitlinine/detail/001-012. Accessed 6 July 2024

Chapter 7

Adjunctive Ketamine for Sedation and Analgesia in Extracorporeal Membrane Oxygenation: Opportunities and Challenges

J. Michael Brewer, Sabine Himmelseher, and Marc O. Maybauer

Abstract

This chapter aims at exploring the use of ketamine for analgesia and sedation in patients undergoing extracorporeal membrane oxygenation (ECMO). Key considerations with the use of analgesic and sedative agents in patients on ECMO are presented, and a description of how their pharmacokinetics (PKs) may be affected is provided. Potential benefits and pitfalls with adjunctive ketamine use and a narrative review of the literature on its utilization in ECMO patients are presented; opportunities and challenges will be discussed.

Key words ECLS, Extracorporeal life support, ECMO, Extracorporeal membrane oxygenation, Sedation and analgesia, Drug pharmacokinetics, Ketamine, Esketamine, Adjunctive use

1 Introduction

In recent years, the utilization of extracorporeal membrane oxygenation (ECMO), a life support intervention for severe cardiac and respiratory failure when conventional therapies fail, has notably increased [1]. ECMO functions by extracting venous blood from the patient, circulating it through an artificial membrane lung for oxygenation and carbon dioxide removal, before returning it to the patient. This supportive intervention can augment heart and/or lung function until intrinsic organ function recovers or until the patient is bridged to other supportive or more advanced therapeutic interventions.

ECMO operates in two primary modes: venoarterial (VA) (Fig. 1a) and venovenous (VV) (Fig. 1b) [2, 3]. VV ECMO returns blood to the venous circulation after gas exchange and is primarily indicated for severe respiratory failure. VA ECMO is used in cases of cardiac failure, returning blood to the arterial circulation after gas exchange, thereby providing cardiorespiratory support by bypassing the heart and lungs.

Sabine Himmelseher (ed.), *Ketamine: Pain Management, Critical Care, Depression, and Resource-Poor Settings Volume II*, Neuromethods, vol. 225, https://doi.org/10.1007/978-1-0716-4599-4_7,

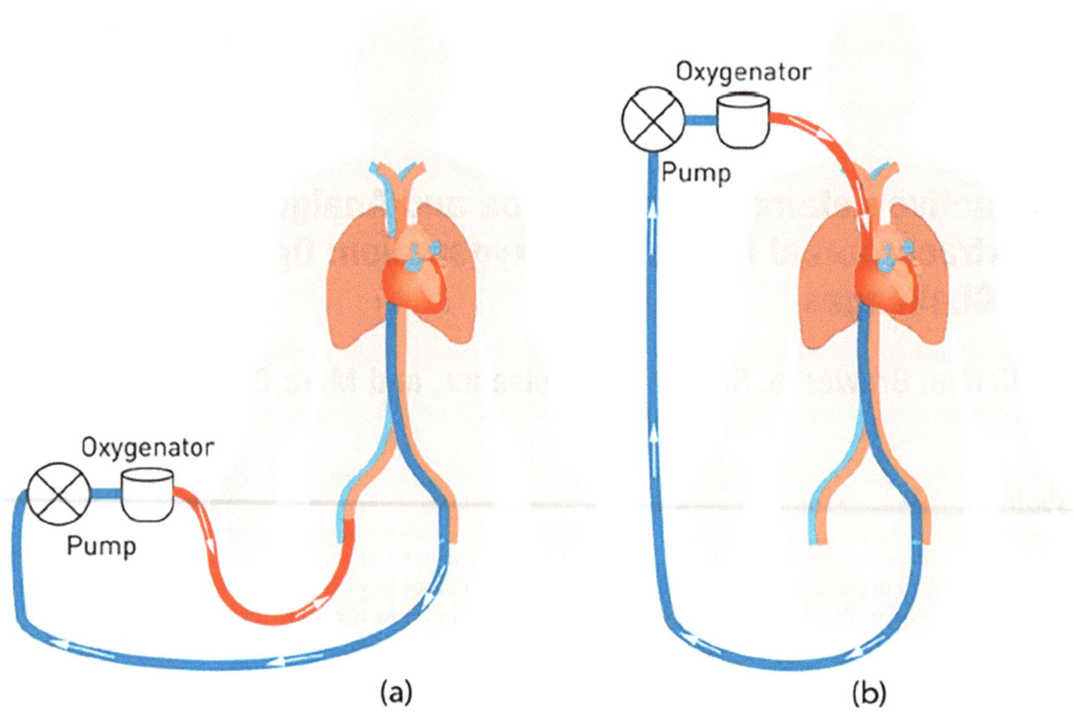

Fig. 1 Illustration of the extracorporeal membrane oxygen circuit: (**a**) Venoarterial and (**b**) venovenous setup with red color indicating oxygenated blood and blue color deoxygenated blood (from Alhomsi et al. [3] under an international commercial CC BY 4.0 license; https://creativecommons.org/licenses/by/4.0/)

In cases of severe cardiorespiratory failure, both VV and VA ECMO can be combined to create a hybrid mode—venoarteriovenous (VAV) ECMO—which can provide both heart and lung support.

The management of patients on ECMO is complex, necessitating meticulous attention to hemodynamics and gas exchange and day-to-day monitoring of coagulation. Analgesia, sedation, and patient comfort are decisive elements of therapy and impact recovery [4–7]. Ongoing efforts aim to minimize critically ill patient exposure to certain classes of narcotic and sedative medications that may contribute to adverse outcomes [5, 7–11], while new approaches to multimodal sedative regimen have emerged.

Ketamine, a noncompetitive N-methyl-D-aspartate (NMDA) receptor antagonist, possesses unique cardiocirculatory stabilizing properties that may make it particularly suitable for use in critically ill patients on ECMO [5, 12]. Traditional analgesics and sedatives may have adverse effects on hemodynamic stability, organ and immune function, and recovery from critical illness [12–15]. Adjunctive subanesthetic doses of ketamine have been reported to allow for a reduction in opioid, sedative (including propofol and midazolam), and vasopressor requirements in critically ill and postoperative patients [15–17].

This chapter aims to explore the use of ketamine for analgesia and sedation in ECMO settings. Key considerations with the use of analgesic and sedative agents in patients on ECMO are presented, and a description of how their pharmacokinetics (PKs) may be affected is provided. Potential benefits and pitfalls with ketamine use and a narrative review of the literature on its utilization in ECMO patients are presented; opportunities and challenges will be discussed.

2 Challenges and Considerations in Analgesia and Sedation During Extracorporeal Membrane Oxygenation

Analgesia and sedation management in patients receiving ECMO is challenging [4, 18]: At present, there are no *evidence-based* guidelines, and there is considerable variation in practices among centers. Only a few hospitals report an established, specific use schedule or administration protocol [4, 6, 7, 14, 19, 20]. However, especially in the early stages of therapy, ECMO-supported patients are still often deeply sedated [6, 20, 21]. Mostly, a combination of drugs that include opioids, benzodiazepines, and propofol is applied. However, oversedation of critically ill patients increases adverse effects of the agents used and has been associated with prolonged ventilator dependence, delirium, immobility, and increased hospital morbidity and mortality [5–7, 14, 20]. Therefore, oversedation should be avoided.

Patients on ECMO support are well known to require higher doses of analgesics and sedatives compared to non-ECMO patients [7, 22]. The high dose requirements in sedation and analgesia regimen may result from:

1. Patient characteristics (from genetic disposition to typical features of the respective critical illness) including high oxygen consumption or neuromuscular paralysis to prevent ventilator dyssynchrony.
2. Specific ECMO conditions and circumstances such as ECMO drainage insufficiency.
3. Particular drug PKs and/or the effect of ECMO-associated drug sequestration on plasma concentrations [5, 23, 24]. ECMO affects a drug's PKs through multiple factors, including circuit peculiarities, patient-individual aspects such as metabolism, and drug and medication interaction effects [13, 15, 25].

4. Drug adsorption and sequestration in the ECMO circuit which consists of a blood pump (centrifugal or roller pump), artificial membrane lung, cannulas for blood drainage and return, and polyvinyl chloride (PVC) tubing connecting the components [2]. Adsorption and sequestration are influenced by the physicochemical characteristics of administered drugs, such as lipophilic properties or protein binding [1, 4–7, 14, 19, 20, 23, 25–27]. Saturation at drug sequestration sites within the circuit can lead to drug toxicity if high drug dosing is maintained for a prolonged period. Moreover, the circuit may act as a reservoir, releasing sequestered drug once administration ceases, prolonging the residual analgesic or sedative effect [1, 4]. The ECMO circuit also increases a drug's volume of distribution (Vd), which may be further affected by volume resuscitation and systemic inflammation during the initiation of support [1, 5, 6, 19, 23, 26, 27]. More recent ECMO systems have been made of materials which may affect drug disposition in a different manner. They aim at decreasing drug loss compared to older systems [4, 25].

Patients on ECMO support may also experience concomitant organ system injury or failure due to a decrease in body perfusion and microcirculatory insufficiency [5]. Altered drug metabolism and clearance may result from reduced organ perfusion and function, such as with renal and liver failure [1, 5, 19, 25–27]. Additionally, patients on VA ECMO support may experience an increase in cardiac output which also alters drug clearance [1].

3 Reported and Postulated Advantages of Ketamine in Extracorporeal Membrane Oxygenation Settings

Ketamine offers several advantages in the ECMO setting, making it a compelling choice for analgesia and sedation regimen:

- Unlike many analgesics and sedatives which induce cardiorespiratory depression, ketamine stabilizes the hemodynamic situation provided that there is no decompensated volume depletion, manifested shock, or exhausted cardiogenic reserve. Ketamine exerts direct sympathomimetic stimulation and inhibits aminergic neurotransmitter reuptake. This helps to maintain or increase heart rate and blood pressure and decrease exogenous catecholamine requirement [5, 12, 14, 15, 25]. These effects are of particular value in patients who require hemodynamic support because of their critical illness, and especially in ECMO conditions.
- Ketamine preserves reflexes necessary for airway protection and spontaneous breathing [28, 29]. This may become very

advantageous in ECMO patients who will benefit from a reduction or liberation from mechanical ventilation [5, 12, 14, 25]. Patients may then require a shorter duration of controlled mechanical ventilatory support, which reduces ventilator-associated complications and ventilator-induced lung injury.

- At subanesthetic dose, ketamine is commonly used as an adjunct to opioid analgesia (such as with fentanyl or hydromorphone) and to analgesic and sedative regimens (such as with propofol and midazolam) [20]. In humans, it exhibits different affinities and weak activity on opioid receptors, and when combined with opioids at anti-hyperalgesic, low, or subanesthetic dose, ketamine acts synergistically to improve pain relief, spare opioids, and reduce opioid tolerance [5, 14, 25, 30–33]. Combining opioids with low-dose ketamine is not associated with adverse effects but may even improve outcomes in critically ill patients [6, 34].
- Ketamine is well known for its psychoactive properties. A recent survey among Canadian intensivists reported that they were reluctant to use ketamine for sedation at least partly driven by concerns of adverse psychotropism [35]. However, ketamine effects are completely dose-dependent with a small incidence of adverse events in the low, subanesthetic dose range (< 0.3 mg/kg racemic ketamine). If combined with a GABA agonist in a sedative regimen, psychiatric adverse events are mostly prevented. Moreover, a low, subanesthetic dose of ketamine has been reported to increase psychic resilience, may prevent and treat post-traumatic stress disorder, and exerts antidepressant effects in perioperative care [36–39]. All of these properties would be very valuable in critically ill patients on ECMO. In addition, ketamine exerts hypnotic and amnestic effects even at low dose, which may allow for reducing benzodiazepines during sedation [5, 14, 16, 17, 30].
- Ketamine is less lipophilic and shows lower protein binding (10–30%) than other commonly used sedatives and analgesics [13, 14, 27]. This results in less sequestration by the ECMO circuit and a Vd similar to that in critically ill, non-ECMO patients [14, 27]. Ketamine may therefore have little interaction with the ECMO circuit and may not require dramatic dose adjustments like many other agents [14, 27]. However, ketamine plasma concentrations declined rapidly in a model of a blood-primed, closed loop ex vivo ECMO circuit [22], underlining the need for in vivo PK and pharmacodynamic investigations.
- Ketamine may possess anti-inflammatory and neuroprotective effects. Wyler et al. could show that ketamine infusion improved the sedation depth in mechanically ventilated COVID-19

patients [40]. It also provided a modest anti-inflammatory benefit, but this did not result in less mortality or intensive care unit length of stay. Recent investigations by Welters and colleagues revealed the anti-inflammatory potential of the ketamine isomers such as with esketamine during cardiac anesthesia [41]. Esketamine attenuated the cardiopulmonary bypass (CPB)-induced systemic inflammatory response during coronary artery bypass grafting. ECMO, in conjunction with its lifesaving benefits, carries a significant risk of acute brain injury (ABI) and hypoxic-ischemic brain injury (HIBI), which is one of the most common types of ABI in ECMO patients. Various risk factors, such as hypertension, high lactate and low pH levels, cannulation technique, large peri-cannulation $PaCO_2$ drop, and early low pulse pressure, have been associated with the development of HIBI in ECMO patients. In the laboratory, ketamine has neuroprotective and esketamine additional neuroregenerative effects, which warrants further research in this field [42].

Despite all of these potential advantages of ketamine in critically ill patients on ECMO support, the substance is used less frequently than traditional analgesics and sedatives (e.g., opioids, propofol, and benzodiazepines [6, 19, 20, 26]).

4 The Available Evidence: Ketamine in Extracorporeal Membrane Oxygenation

4.1 Patients Without Coronavirus Disease-2019

As yet, there is a paucity of high-quality clinical studies evaluating ketamine in ECMO-supported patients. The existing literature consists of small trials, case reports, case series, and retrospective evaluation [14]. Tellor and colleagues conducted a single-center, retrospective cohort study involving 26 ECMO-supported adult patients who received ketamine infusions in addition to sedative and/or opioid infusions [15]. The authors reported that a significant reduction in vasopressor and sedative/opioid needs following ketamine initiation was possible without a change in sedation scores. Dzierba and co-authors explored the effects of ketamine in a single-center, prospective, non-blinded, randomized controlled trial with 20 adult patients receiving VV ECMO [20]. Patients who suffered from acute respiratory distress syndrome (ARDS) were randomized to treatment with ketamine plus standard sedation (as intervention) or to standard sedation alone (control) at the start of ECMO. In contrast to expectations, ketamine use did not allow for a reduction in sedative or opioid use. In fact, higher doses of fentanyl and midazolam compared to standard sedation were necessary during the study period. The unexpected increase in sedative requirements was interpreted as a result from variable sedation protocols which may have prevented effective reduction of other sedatives when combined with ketamine. Skelton and colleagues assessed sedative, analgesic, and antipsychotic ketamine usage in 53 adult patients on VA ECMO in a single-center, retrospective cohort study [43]. Their findings indicated no benzodiazepine-sparing with ketamine in four (7.5%) patients, but a reduction in post-decannulation delirium was noted. Case studies by Farrokh and co-authors [13], Lam and colleagues [27], and Floroff et al. [30] demonstrated varying effects of adjunctive ketamine use in ECMO patients. For instance, Farrokh and colleagues observed lower plasma levels of ketamine achieving targeted sedation compared to non-ECMO reference patients [13]. Lam et al. noted a similar ketamine Vd in an ECMO patient comparable to critically ill, non-ECMO patients yet a 6.5-fold increase as compared to healthy adults [27]. Floroff et al. reported a decrease in agitation and vasopressor requirements with ketamine use, coupled with a reduction in continuous benzodiazepine dose [30].

4.2 Patients with Coronavirus Disease-2019

ARDS secondary to coronavirus disease-2019 (COVID-19) expressed higher analgesia and sedation requirements in comparison to ARDS of other etiologies. In a retrospective study on the effect of esketamine in VV ECMO, Paparoupa and colleagues demonstrated that 115 patients with COVID-19 ARDS had significantly increased sedation sum scores when compared with 250 patients with ARDS from other etiologies (Fig. 2) [44].

In another retrospective evaluation, Sigala and coworkers compared 64 patients with severe acute respiratory syndrome coronavirus-2 (SARS-CoV-2) ARDS with 44 patients with non--SARS-CoV-2 ARDS on VV ECMO support: The median daily administration of dexmedetomidine was significantly higher in the SARS-CoV-2 cohort ($p = 0.03$), whereas that of propofol was significantly higher in the non-SARS-CoV-2 cohort ($p < 0.01$) [45]. No differences were detected in daily opioid, benzodiazepine, and ketamine use between groups. Nonetheless, the use of adjunct agents to facilitate weaning was significantly higher in the SARS-CoV-2 cohort ($p < 0.01$). Future research on COVID-19-induced ARDS requiring VV ECMO is unlikely. However, the pandemic with its need for higher drug doses in analgesia and sedation strategies has sparked innovative ideas for future ketamine research in ARDS.

5 Limitations

Despite reported and postulated advantages of ketamine in ECMO-supported patients, significant limitations have to be noticed which may also prevent its widespread use: The evidence in favor of ketamine is not robust, consisting mostly of small-scale, extremely heterogenous, or retrospective study data. For example,

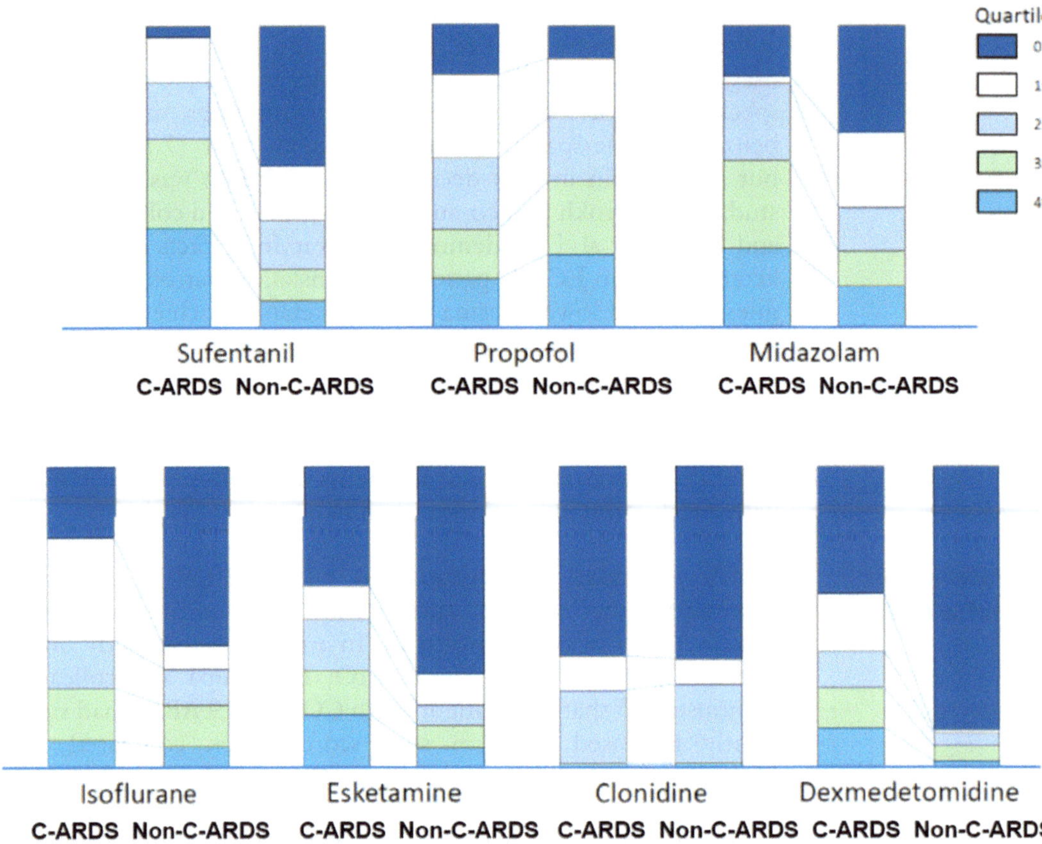

Fig. 2 Dose ranges of single substances used for analgesia and sedation in patients with COVID-19 ARDS (C-ARDS) shown in left columns as compared to patients with non-COVID-19 ARDS (non-C-ARDS) shown in right columns; ranges are indicated as quartiles. *COVID-19* corona virus disease 2019, *ARDS* acute respiratory distress syndrome (from Paparoupa et al. [44] under a CC BY 4.0 license; https:/creativecommons.org/licenses/by/4.0/)

some trials report that ketamine exerts sedative and opioid-sparing effects, whereas others do not [23]. However, multiple confounders may have played an important role. Dzierba et al.'s study, which did not find a reduction in sedative or opioid use with ketamine, underscores the importance of standardizing sedation protocols to systematically reduce other agents when ketamine is added under control of sedation depth [20]. The lack of standardization in drug dosing in sedation and analgesia regimen adds to much uncertainty and complicates implementation of new protocols and substances [1].

Most existing research on PKs and dosing relies on in vitro or ex vivo models, or pediatric studies, which may not be applicable in in vivo or adult patients [1]. The studies often involved short observation times and single drug doses, with measurements taken at various time points after use without accounting for patient-specific metabolic or elimination effects [1].

In addition, little is known about the impact of continuous ketamine infusion on drug concentration, extraction, and saturation in the circuit, and ECMO-specific PK parameters are totally under-researched [1, 4, 27].

Best possible ketamine dosing strategies remain unclear due to variable sedation targets, indication-specific dosing, and variations in concomitant analgesic and sedative drugs [5, 14, 30]. Present recommendations for analgesia and sedation in mechanically ventilated ICU patients suggest an initial bolus followed by a continuous infusion, but further trials are needed to determine optimal initial and peak doses for ECMO-supported patients [14, 30].

Lastly, potential adverse effects of ketamine, such as emergence reactions, excessive salivation, tachycardia, delirium, and increased intracranial pressure without controlled ventilation, must be considered [42]. However, all unwanted ketamine effects are dose- and context-dependent and may be prevented or reduced by competent drug use and combination with other medications. Ketamine administration schedules must always be designed according to individual patient disposition, specific conditions, and circumstances [5, 14, 30]: There is no one-size-fits-all prescription. Moreover, in ICU patients, case reports about serious electrolyte disturbances have been described, including central diabetes insipidus in ECMO settings in combination with vasopressin infusion [46]. Vasopressin may be applied frequently in ECMO patients with hypotension. In addition, especially longer-term, high-dose ketamine may also cause uropathy and cystitis and hepatobiliary dysfunction associated with sclerosing cholangitis. Patients have to be monitored for these complications, and ketamine doses need to be reduced or interrupted with occurrence.

6 Conclusion

The use of ketamine as an adjunct for analgesia and sedation in ECMO settings presents a mix of promising opportunities combined with a lack of trial evidence and inconsistent data. In addition to the medication's cardiovascular stabilizing and airway reflex preservation effects, the potential of ketamine to reduce reliance on traditional analgesics and sedatives, especially its opioid-sparing, anti-inflammatory, psychic resilience-enhancing and antidepressant properties make it a valuable option for ECMO patients. It may serve as a reasonable adjunctive agent in case of opioid and sedative failure, with sedation and analgesia protocols individualized and adjusted on patient-specific goals determined by a multidisciplinary team [23].

To integrate ketamine into ECMO protocols, high-quality research and large-scale, randomized controlled trials are essential. Studies need to focus on patient-centered outcomes, establish effective dosing strategies, and describe ketamine PKs in ECMO circuits. Standardization of protocols incorporating ketamine will be crucial to enhance the efficacy and safety of analgesia and sedation regimen. Ultimately, ketamine's adoption in ECMO care should proceed with a commitment to evidence-based practice and protocol development in this complex patient population.

Conflict of Interest

MOM serves as speaker for Abbott and is consultant for LivaNova, both unrelated to this manuscript. JMB and SH have no conflict of interest.

References

1. Romera-Ortega MA, Chamorro-Jambrina C (2023) Analgo-sedation strategies in patients with ECMO. Med Intensiva (Engl Ed) 47(3): 165–169. https://doi.org/10.1016/j.medine.2022.10.013
2. Gajkowski EF, Herrera G, Hatton L, Velia Antonini M, Vercaemst L, Cooley E (2022) ELSO guidelines for adult and pediatric extracorporeal membrane oxygenation circuits. ASAIO J 68(2):133–152. https://doi.org/10.1097/MAT.0000000000001630
3. Alhomsi Y, Alsalemi A, Noorizadeh BF, Meskin N, Hssain AA (2021) A modular approach for a patient unit for extracorporeal membrane oxygenation simulator. Membranes (Basel) 11(6):424. https://doi.org/10.3390/membranes11060424
4. Raffaeli G, Allegaert K, Koch B, Cavallaro G, Mosca F, Tibboel D et al (2018) In vitro adsorption of analgosedative drugs in new extracorporeal membrane oxygenation circuits. Pediatr Crit Care Med 19(5):e251–e258. https://doi.org/10.1097/PCC.0000000000001484
5. Maybauer MO, Koerner MM, Maybauer DM (2019) Perspectives on adjunctive use of ketamine for analgosedation during extracorporeal membrane oxygenation. Expert Opin Drug Metab Toxicol 15(5):349–351. https://doi.org/10.1080/17425255.2019.1593963
6. deBacker J, Tamberg E, Munshi L, Burry L, Fan E, Mehta S (2018) Sedation practice in extracorporeal membrane oxygenation-treated patients with acute respiratory distress syndrome: a retrospective study. ASAIO J 64(4): 544–551. https://doi.org/10.1097/MAT.0000000000000658
7. Patel M, Altshuler D, Lewis TC, Merchan C, Smith DE 3rd, Toy B et al (2020) Sedation requirements in patients on Venovenous or Venoarterial extracorporeal membrane oxygenation. Ann Pharmacother 54(2):122–130. https://doi.org/10.1177/1060028019877806
8. Devlin JW, Skrobik Y, Gelinas C, Needham DM, Slooter AJC, Pandharipande PP et al (2018) Clinical practice guidelines for the prevention and management of pain, agitation/sedation, delirium, immobility, and sleep disruption in adult patients in the ICU. Crit Care Med 46(9):e825–e873. https://doi.org/10.1097/CCM.0000000000003299
9. Extracorporeal Life Support Organization (ELSO) Extracorporeal Membrane Oxygenation (ECMO) (2024). ELSO guidelines for patient care, respiratory & cardiac support, ECMO in COVID-19. ELSO Guidelines | Extracorporeal Membrane Oxygenation (ECMO) Accessed 14 Sep 2024
10. Linden V, Palmer K, Reinhard J, Westman R, Ehren H, Granholm T et al (2000) High survival in adult patients with acute respiratory distress syndrome treated by extracorporeal membrane oxygenation, minimal sedation, and pressure supported ventilation. Intensive Care Med 26(11):1630–1637. https://doi.org/10.1007/s001340000697
11. Schmidt F, Sasse M, Boehne M, Mueller C, Bertram H, Kuehn C et al (2013) Concept of "awake venovenous extracorporeal membrane

oxygenation" in pediatric patients awaiting lung transplantation. Pediatr Transplant 17(3):224–230. https://doi.org/10.1111/petr.12001

12. Bansbach J, Wenz J, Kaufmann K, Heinrich S, Kalbhenn J (2022) Sevoflurane in combination with esketamine is an effective sedation regimen in COVID-19 patients enabling assisted spontaneous breathing even during prone positioning. Anaesthesiol Intensive Ther 54(1): 23–29. https://doi.org/10.5114/ait.2022.113950
13. Farrokh S, Kim BS, Cho SM (2024) Ketamine infusion for sedation in a patient on extracorporeal membrane oxygenation (ECMO). Perfusion 39(1):223–226. https://doi.org/10.1177/02676591221134941
14. Flanagan T, Mercer K, Johnson PN, Miller J, Yousaf FS, Fuller JA (2023) Ketamine use in adult and pediatric patients receiving extracorporeal membrane oxygenation (ECMO): a systematic review. J Pharm Pract 37: 8971900231198928. https://doi.org/10.1177/08971900231198928
15. Tellor B, Shin N, Graetz TJ, Avidan MS (2015) Ketamine infusion for patients receiving extracorporeal membrane oxygenation support: a case series. F1000Res 4:16. https://doi.org/10.12688/f1000research.6006.1
16. Garber PM, Droege CA, Carter KE, Harger NJ, Mueller EW (2019) Continuous infusion ketamine for adjunctive analgosedation in mechanically ventilated, critically ill patients. Pharmacotherapy 39(3):288–296. https://doi.org/10.1002/phar.2223
17. Umunna BP, Tekwani K, Barounis D, Kettaneh N, Kulstad E (2015) Ketamine for continuous sedation of mechanically ventilated patients. J Emerg Trauma Shock 8(1):11–15. https://doi.org/10.4103/0974-2700.145414
18. Adams CD, Altshuler J, Barlow BL, Dixit D, Droege CA, Effendi MK et al (2020) Analgesia and sedation strategies in mechanically ventilated adults with COVID-19. Pharmacotherapy 40(12):1180–1191. https://doi.org/10.1002/phar.2471
19. DeGrado JR, Hohlfelder B, Ritchie BM, Anger KE, Reardon DP, Weinhouse GL (2017) Evaluation of sedatives, analgesics, and neuromuscular blocking agents in adults receiving extracorporeal membrane oxygenation. J Crit Care 37:1–6. https://doi.org/10.1016/j.jcrc.2016.07.020
20. Dzierba AL, Brodie D, Bacchetta M, Muir J, Wasson L, Colabraro M et al (2016) Ketamine use in sedation management in patients receiving extracorporeal membrane oxygenation. Intensive Care Med 42(11):1822–1823. https://doi.org/10.1007/s00134-016-4519-9
21. Marhong JD, DeBacker J, Viau-Lapointe J, Munshi L, Del Sorbo L, Burry L et al (2017) Sedation and mobilization during venovenous extracorporeal membrane oxygenation for acute respiratory failure: an international survey. Crit Care Med 45(11):1893–1899. https://doi.org/10.1097/CCM.0000000000002702
22. Chevalier A, Green D, Hudson R, Porter Hunt J, McKnite A, Whelan A et al (2024) Ketamine extraction by extracorporeal life support circuits. Crit Care Med 52(1):S13. https://doi.org/10.1097/01.ccm.0000998492.79696.f5
23. Patel JS, Kooda K, Igneri LA (2023) A narrative review of the impact of extracorporeal membrane oxygenation on the pharmacokinetics and pharmacodynamics of critical care therapies. Ann Pharmacother 57(6):706–726. https://doi.org/10.1177/10600280221126438
24. Remmington C, McKenzie C, Camporota L, Hanks F, Barker M, Sanderson B et al (2023) Analgosedation in extracorporeal membrane oxygenation: a retrospective UK cohort study. Br J Anaesth 131(2):e50–e52. https://doi.org/10.1016/j.bja.2023.05.011
25. Dreucean D, Harris JE, Voore P, Donahue KR (2022) Approach to sedation and analgesia in COVID-19 patients on venovenous extracorporeal membrane oxygenation. Ann Pharmacother 56(1):73–82. https://doi.org/10.1177/10600280211010751
26. Franciscovich CD, Monk HM, Brodecki D, Rogers R, Rintoul NE, Hedrick HL et al (2020) Sedation practices of neonates receiving extracorporeal membrane oxygenation. ASAIO J 66(5):559–664. https://doi.org/10.1097/MAT.0000000000001043
27. Lam E, Rochani A, Kaushal G, Thoma BN, Tanjuakio J, West FM et al (2019) Pharmacokinetics of ketamine at dissociative doses in an adult patient with refractory status asthmaticus receiving extracorporeal membrane oxygenation therapy. Clin Ther 41(5):994–999. https://doi.org/10.1016/j.clinthera.2019.03.005
28. Eikermann M, Grosse-Sundrup M, Zaremba S, Henry ME, Bittner EA, Hoffmann U et al (2012) Ketamine activates breathing and abolishes the coupling between loss of consciousness and upper airway dilator muscle dysfunction. Anesthesiology 116(1):35–46. https://doi.org/10.1097/ALN.0b013e31823d010a

29. Suleiman A, Santer P, Munoz-Acuna R, Hammer M, Schaefer MS, Wachtendorf LJ et al (2023) Effects of ketamine infusion on breathing and encephalography in spontaneously breathing ICU patients. J Intensive Care Med 38(3):299–306. https://doi.org/10.1177/08850666221119716
30. Floroff CK, Hassig TB, Cochran JB, Mazur JE (2016) High-dose sedation and analgesia during extracorporeal membrane oxygenation: a focus on the adjunctive use of ketamine. J Pain Palliat Care Pharmacother 30(1):36–40. https://doi.org/10.3109/15360288.2015.1101637
31. Guillou N, Tanguy M, Seguin P, Branger B, Campion JP, Malledant Y (2003) The effects of small-dose ketamine on morphine consumption in surgical intensive care unit patients after major abdominal surgery. Anesth Analg 97(3):843–847. https://doi.org/10.1213/01.ANE.0000075837.67275.36
32. Chan K, Burry LD, Tse C, Wunsch H, De Castro C, Williamson DR (2022) Impact of ketamine on Analgosedative consumption in critically ill patients: a systematic review and meta-analysis. Ann Pharmacother 56(10):1139–1158. https://doi.org/10.1177/10600280211069617
33. Jung H, Lee J, Ahn HY, Yang JH, Suh GY, Ko RE et al (2022) Safety and feasibility of continuous ketamine infusion for analgosedation in medical and cardiac ICU patients who received mechanical ventilation support: a retrospective cohort study. PLoS One 17(9):e0274865. https://doi.org/10.1371/journal.pone.0274865
34. Zakine J, Samarcq D, Lorne E, Moubarak M, Montravers P, Beloucif S et al (2008) Postoperative ketamine administration decreases morphine consumption in major abdominal surgery: a prospective, randomized, double-blind, controlled study. Anesth Analg 106(6):1856–1861. https://doi.org/10.1213/ane.0b013e3181732776
35. Sharif S, Munshi L, Burry L, Mehta S, Gray S, Chaudhuri D et al (2024) Ketamine sedation in the intensive care unit: a survey of Canadian intensivists. Can J Anaesth 71(1):118–126. https://doi.org/10.1007/s12630-023-02608-x
36. Almeida TM, Lacerda da Silva UR, Pires JP, Borges IN, Martins CRM, Cordeiro Q et al (2024) Effectiveness of ketamine for the treatment of post-traumatic stress disorder – a systematic review and meta-analysis. Clin Neuropsychiatry 21(1):22–31. https://doi.org/10.36131/cnfioritieditore20240102
37. Evers AG, Murrough JW, Charney DS, Costi S (2022) Ketamine as a prophylactic resilience-enhancing agent. Front Psych 13:833259. https://doi.org/10.3389/fpsyt.2022.833259
38. Gan SL, Long YQ, Wang QY, Feng CD, Lai CX, Liu CT et al (2023) Effect of esketamine on postoperative depressive symptoms in patients undergoing thoracoscopic lung cancer surgery: a randomized controlled trial. Front Psych 14:1128406. https://doi.org/10.3389/fpsyt.2023.1128406
39. Ma S, Dou Y, Wang W, Wei A, Lan M, Liu J et al (2024) Association between esketamine interventions and postpartum depression and analgesia following cesarean delivery: a systematic review and meta-analysis. Am J Obstet Gynecol MFM 6(3):101241. https://doi.org/10.1016/j.ajogmf.2023.101241
40. Wyler D, Torjman MC, Leong R, Baram M, Denk W, Long SC et al (2024) Observational study of the effect of ketamine infusions on sedation depth, inflammation, and clinical outcomes in mechanically ventilated patients with SARS-CoV-2. Anaesth Intensive Care 52(2):105–112. https://doi.org/10.1177/0310057X231201184
41. Welters ID, Feurer MK, Preiss V, Muller M, Scholz S, Kwapisz M et al (2011) Continuous S-(+)-ketamine administration during elective coronary artery bypass graft surgery attenuates pro-inflammatory cytokine response during and after cardiopulmonary bypass. Br J Anaesth 106(2):172–179. https://doi.org/10.1093/bja/aeq341
42. Himmelseher S, Kochs EF (2021) Ready for a "breakthrough" with ketamine? A look at recent pharmacological insights. Curr Op Anesthesiol 34(4):393–410
43. Skelton PA, Lillyblad MP, Eckman PM, Samara MA, Williams DM, Wilson KJ et al (2020) Clinical outcomes associated with sedation and analgesia in patients supported with venoarterial extracorporeal membrane oxygenation. Int J Artif Organs 43(4):277–282. https://doi.org/10.1177/0391398819885936
44. Paparoupa M, Fischer M, Pinnschmidt HO, Grensemann J, Roedl K, Kluge S et al (2023) Impact of COVID-19 on sedation requirements during veno-venous extracorporeal membrane oxygenation for acute respiratory distress syndrome. J Clin Med 12(10):3515. https://doi.org/10.3390/jcm12103515

45. Sigala MI, Dreucean D, Harris JE, Donahue KR, Bostan F, Voore P et al (2023) Comparison of sedation and analgesia requirements in patients with SARS-CoV-2 versus non-SARS-CoV-2 acute respiratory distress syndrome on veno-venous ECMO. Ann Pharmacother 57(9):1005–1015. https://doi.org/10.1177/10600280221147695
46. Herity LB, Baker C, Kim C, Lowe DK, Cahoon WD Jr (2021) Delayed onset of central diabetes insipidus with ketamine sedation: a report of 2 cases. J Pharm Pract 34(2):314–318

Chapter 8

Ketamine in the Neuro-Intensive Care Unit

Andrew Bauerschmidt, Stuart Brill, and Jon Rosenberg

Abstract

Patients with acute brain injury have unique care challenges with regard to achieving adequate analgesia and sedation, creating wide opportunities for newer agents to improve care and expand clinical options. Of key interest in this population is the optimization of neurophysiologic parameters including intracranial pressure, cerebral blood flow, and cerebral autoregulation, with the aim of preventing secondary brain injury and allowing for brain recovery. Cortical spreading depolarizations/depression, seizures, and cerebral vasospasms are all frequent targets of monitoring and intervention in the neuro-intensive care unit. During open or endovascular neurosurgical procedures, stability of both systemic hemodynamics and cerebral perfusion is critical. The choice of sedation can make effective clinical neuromonitoring either maximally effective or nearly impossible. Ketamine has been increasingly demonstrated as a promising primary or adjunctive agent to facilitate all of these goals often while reducing the need for more established sedation agents with less favorable side effect profiles. This chapter reviews an array of specific use cases and supportive data regarding how, when, and why to consider using ketamine in the neurocritical care population.

Key words Neurocritical care, Neuroanesthesia, Pupillary reflex, Anesthesia, Intracranial pressure, Cerebral perfusion pressure, Cerebral autoregulation, Cerebral metabolism

1 Introduction

The neurocritical care population encompasses a broad array of patients, including those with acute stroke and other neurovascular brain injuries, elective neurosurgical cases, acute brain and spinal cord trauma, and others which bring specific management challenges. This chapter discusses a number of specific use cases of ketamine in this setting and highlights less well-known clinical features of importance to all practitioners managing these patients. We will present essentials of management which are valid in all patients, as well as applications specific to target subpopulations.

Sabine Himmelseher (ed.), *Ketamine: Pain Management, Critical Care, Depression, and Resource-Poor Settings Volume II*, Neuromethods, vol. 225, https://doi.org/10.1007/978-1-0716-4599-4_8,

2 Essentials of Neurocritical Care with Ketamine

2.1 Pupils and the Neurological Examination

Subanesthetic (~0.5 mg/kg body weight) and anesthetic dose (≥1 mg/kg body weight) ketamine can impact the neurological examination. Changes in pupillary size and reactivity are some of the most easily obtained and valuable exam findings in the neurocritical care population. Sudden changes suggest significant neurodeterioration from a structural process and generally indicate urgent escalation in assessment or intervention. Figure 1 shows unilateral pupil dilation of the right eye after traumatic brain injury (TBI) with increasing intracranial pressure (ICP).

Practitioners must therefore have an understanding of how medications may also inadvertently induce pupil change to avoid misattribution. For example, pentobarbital or isoflurane may cause bilaterally enlarged and nonreactive ("blown") pupils [1, 2]. At general anesthetic dose, ketamine may reduce the pupillary size and the pupillary reflex on quantitative pupillometry. In healthy beagle dogs, it also reduced the neurological pupil index (NPi) as measured by quantitative pupillometry [3]. There are only a few trials which examined the effect of ketamine and other NMDA antagonists on pupillary function in humans. In small single-center studies and case reports, ketamine infused at anesthetic dose reduced the pupillary reactivity on quantitative pupillometry [4, 5]. However, while one study found no change in pupillary size after a 1 mg/kg ketamine bolus, the second showed a sequential reduction in pupillary size with repeated ketamine boluses of 1 mg/kg after a 1.5 mg/kg induction dose in a single patient report [5]. These findings may, at least in part, be explained by ketamine's cross-reactivity with muscarinic receptors which may cause a dose-dependent receptor blockade [6, 7]. In clinical context, awareness of these effects can avoid potential confusion or alarm in patient care. For example, one case report describes a patient with normally reactive pupils that then became dilated and nonreactive after changing the sedation regimen from IV

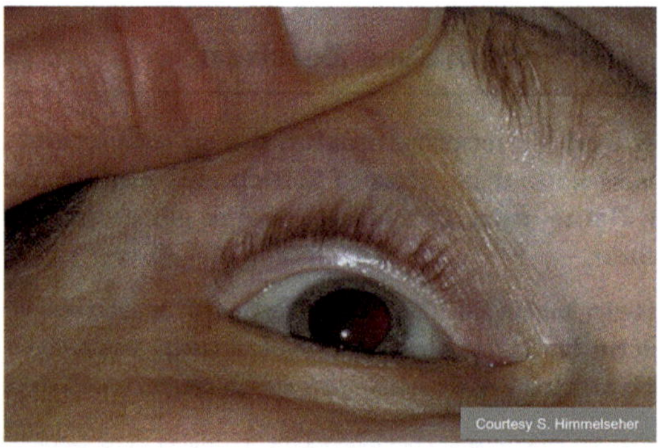

Fig. 1 Unilateral pupillary dilation of the right eye with rising intracranial pressure after traumatic brain injury (figure courtesy of S. Himmelseher)

midazolam plus esketamine to inhalational isoflurane. Following an unremarkable computed tomography (CT), the isoflurane dose was reduced, and the mydriasis disappeared [1].

At subanesthetic dose, ketamine also affects extraocular movements, specifically reducing smooth pursuit and increasing catch-up saccades during visual tracking [8]. This may be related to N-methyl-D-aspartate (NMDA) receptor inhibition of neurons involved in the frontal-thalamic-cerebellar circuitry, which connect the frontal eye fields, thalamus, and cerebellum [9].

2.2 Recovery After Surgery: Ketamine and Gut Motility

It is a generally accepted principle of intensive care that pain management is offered before patient sedation. Although pain therapy is a focus of other chapters of this book, we would like to cite a recent meta-analysis demonstrating that perioperative ketamine/esketamine enhances early subjective quality of recovery after surgery, increases pain relief, and improves mood with less anxiety and depressive symptoms without inducing more adverse events [10]. At analgesic and anesthetic dose, intravenous (IV) ketamine may also improve gut motility and reduce the frequency of postoperative ileus. A recent study examined the effects of esketamine on postoperative recovery of gastrointestinal (GI) functions. It randomized 100 patients scheduled for gynecological surgery to receive sevoflurane and remifentanil-propofol infusions with or without intraoperative subanesthetic dose esketamine (0.2 mg/kg/h) for general anesthesia. Esketamine was discontinued 30 min before the end of anesthesia. In the postoperative period, a patient-controlled analgesia (PCA) technique with 120 mL physiological saline per syringe containing 2 μg/kg sufentanil, 90 mg ketorolac, 20 mg tropisetron, and 50 mg esketamine versus placebo was offered as pain therapy. Patients receiving esketamine during surgery and in the PCA showed reduced time to first postoperative flatus despite the opioid sufentanil and reduced pain at 24 h post-surgery without any complications [11]. Similarly, an early randomized clinical trial compared patients undergoing elective lower extremity surgery with general anesthesia in a regimen of ketamine-midazolam versus fentanyl-midazolam. The study found improved postoperative GI motility in the patients receiving ketamine [12]. Moreover, a meta-analysis of 32 randomized control trials examining ketamine plus sufentanil PCA versus sufentanil PCA alone found a reduced incidence of postoperative nausea and vomiting in the ketamine-sufentanil group [13]. The exact molecular pathways by which ketamine affects the enteric nervous system or other relevant mechanisms herein remain unknown. Studies have shown that NMDA receptor inhibitors may reduce inflammation in the enteric neurons by protecting against glutamate-induced neurotoxicity and reactive oxygen species formation through inhibition of neuronal nitric oxide synthase and xanthine oxidoreductase-dependent superoxide production [14]. Ketamine also inhibits pro-inflammatory transcription factors, such as nuclear factor-kappa B (NF-κB), and activator protein-1, which may contribute to its impact on GI motility [15]. In practical application of this data, the use of esketamine over opioid-based analgesia and sedation in acute brain injury may minimize secondary GI complications. In an observational study of 24 patients with either TBI or aneurysmal subarachnoid hemorrhage (SAH), esketamine- versus fentanyl-based sedation resulted in equally adequate depth of sedation, no difference in ICP or cerebral perfusion pressure (CPP), and no difference in time to full enteral nutrition or first defecation. However, esketamine was associated with opioid-sparing effects and significantly less requirements for vasopressors to maintain target CPP [16]. In summary, depending on patient conditions, pain, and administration of other analgesics, ketamine can be used as analgesic adjunct or sole analgesic in neurocritical care patients. It may be very beneficial because of lack of negative impact on GI transit time and GI motility.

3 Key Considerations in Sedation and Analgesia with Ketamine in Neurocritical Care

At present, there is much interest in ketamine for use in sedation and analgesia regimen in the critically ill, mechanically ventilated patient population. Ketamine may even become a primary agent for optimizing analgesia and sedation, due to its relatively low impact on systemic and cerebral hemodynamics, short time to onset/offset, and potential for minimized short- and long-term neurocognitive side effects. Although high-quality data is lacking, ketamine is being actively investigated as a first-line agent, rather than playing only an adjunctive role [17–19]. Patients with acute CNS injury often have different needs with regard to analgesia and sedation than the broader critical care population [6]. In patients with significantly depressed levels of consciousness at baseline, the goal is often to avoid additional sedatives and minimize opioids while still providing analgesia and facilitating ventilator compliance. Because of its analgesic and sedative effects without inducing cardiorespiratory depression, ketamine is a useful agent in this scenario. Esketamine, its dextrorotatory isomer, is twice as potent as analgesic and anesthetic as racemic ketamine. It offers the advantage of shorter neurocognitive recovery and better maintained cognitive functions. Psychoactive events are of comparable incidence at 50% dose reduction versus the racemate but less commonly reported as unpleasant in quality. Esketamine may become a particularly preferred agent compared to the racemic formulation [20].

3.1 Ketamine and the Intracranial Pressure

A common misconception is that ketamine may problematically increase the ICP, and it is valuable to understand how that idea originated and then has been disproven over time. This idea originates from non-randomized small sample size studies as well as animal models published in the 1970s [21–23]. One notable example of this type of small study was performed by Shapiro et al. in 1972. It was meant to investigate the effect of ketamine on ICP in patients with abnormal cerebrospinal fluid (CSF) flow (e.g., patients with acute shunt obstruction or required placement of an external ventriculostomy) compared to those with normal CSF flow (e.g., patients with functional ventriculoatrial shunts) [22, 24]. The study consisted of giving IV ketamine and measuring ICP before and after administration. This was completed on 11 occasions in seven patients: With functioning ventriculoatrial shunts, the average ICP increase was 19.4 ± 6.9 mmHg, whereas with abnormal CSF flow, the average ICP increase was 41.5 ± 16.6 mmHg. The results were thought to suggest that patients with abnormal CSF flow cannot compensate for the increase in cerebral blood flow caused by ketamine as they had lost the ability to autoregulate. Subsequently, the resulting increase in ICP would then cause a reduction of CPP associated with cerebral hypoxia and ischemia [21, 24, 25]. It was therefore generally recommended to not administer ketamine in case of abnormal CSF or potential for developing abnormal CSF [25]. In the ensuing five decades, a greater body of evidence, improved study design, and more complete understanding of variables impacting brain physiology have come to light which suggest otherwise [7, 25]. Many individual studies and multiple systematic reviews regarding the relationship between ketamine and elevated ICP have found no difference in ICP with ketamine compared to controls [7, 16, 24–32]. In 1985, Pfenninger and co-workers were the first who published that ketamine does not further increase an elevated ICP under controlled ventilation versus spontaneous breathing in a hemorrhagic shock piglet model [33]. One patient study also highlighted lower ICP during tracheal suctioning of TBI patients who had received ketamine versus control [30]. This may in part be explained by ketamine's analgesic effects, which specifically appear to inhibit the unwanted cough reflex. This effect has also been noted during general anesthesia induction with fentanyl and during bronchoscopy [34, 35]. There have been several studies in present clinical settings which have found slight elevations, no change, or a decrease in ICP with ketamine compared to controls without any evidence of patient harm related to ketamine use [31, 32]. This finding has been validated in detail in the TBI population, including subpopulations of those with or without concomitant paralytic use, or with or without decompressive craniectomy. In particular, patients without craniectomy showed a more significant reduction in CPP compared to those with craniectomy [36] (Fig. 2).

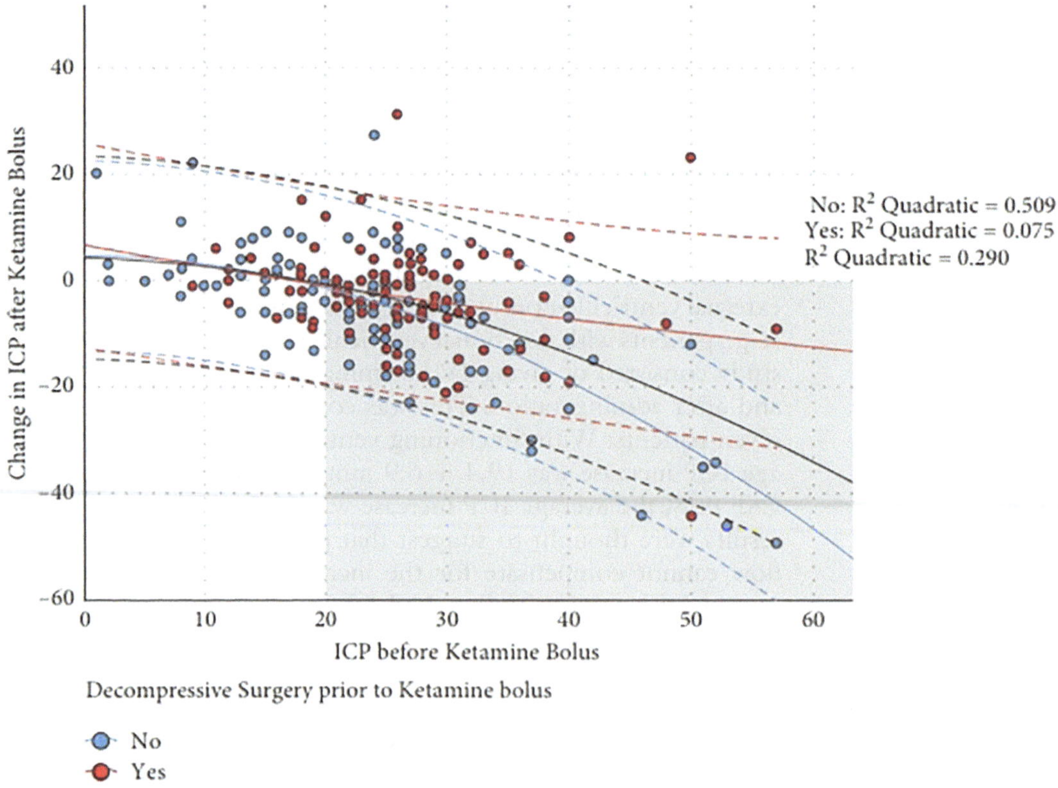

Fig. 2 Change in intracranial pressure (ICP) after a 2 mg/kg ketamine bolus injection versus baseline value. Blue circles indicate patients without (No) and red circles (Yes) with decompressive surgery; bold lines show regression lines, which are in black for whole patient population, in blue for patients without, and in red for patients with decompressive surgery; dashed lines are 95% confidence intervals around the regression lines. The gray zone represents events where ICP decreased the greatest (from Dengler et al. [36] under a CC BY 4.0 international commercial license; https:/creativecommons.org/licenses/by/4.0/)

When a sedative regimen of ketamine-midazolam was compared with fentanyl-midazolam in patients with severe TBI, similar ICP with either sedation was seen. However, with the combination of ketamine-midazolam, less exogenous vasopressors were needed, and greater CPP was reported [37]. In a comparison between a sedation regimen of ketamine-midazolam and sufentanil-midazolam, no differences in ICP or CPP between the two were observed. However, patients receiving sufentanil-midazolam required more fluids on the first day and showed a trend of greater need for vasopressors in total [28].

3.2 Ketamine and the Cerebral Pressure

Ketamine's impact on CPP has been studied to a more limited degree than its effects on ICP. One systematic review found seven studies which included CPP monitoring [31], but only one reported that ketamine increased CPP by 3.9 mmHg 2 minutes after administration in pediatric patients [27]. A retrospective, observational study published in 2017 by van der Braile and co-workers with 65 patients suffering from SAH compared a sedation regimen with ketamine in 41 patients versus controls who did not receive ketamine [38]. Ketamine decreased the ICP in 93% of patients. It also reduced vasopressor needs in 54% of those patients requiring blood pressure augmentation to maintain adequate CPP during vasospasm and delayed cerebral ischemia. Delayed cerebral ischemia associated with cerebral infarction occurred significantly less in the cohort receiving ketamine as compared to control patients without ketamine (7% infarction in patients with ketamine versus 25% in patients without ketamine, p-value 0.04). No difference in complication rates between the two cohorts was found.

Impaired cerebral autoregulation is a known feature of acute brain injury and therefore has become a specific target for both monitoring and intervention [39–41]. There is evidence that ketamine may stabilize cerebral blood flow and neurovascular coupling and that it does not negatively impact upon cerebral autoregulation. When the effects of ketamine at doses used in sedation and analgesia in the neuro-intensive care unit (NICU) were examined in a randomized, double-blind, multimodal imaging study in healthy volunteers, ketamine increased the cerebral metabolic rate of oxygen ($CMRO_2$) (measured as variable of functional cerebral metabolism) by approximately 13% in the prefrontal cortex, notably with a stable coupling between cerebral blood flow and $CMRO_2$ brain-wide [42]. Ketamine did not affect the brains' resisting state functional MRI connectivity, suggesting that metabolism and connectivity represent distinct dimensions of neural activity. However, ketamine impaired white matter-related performance and neural activity underlying working memory. In a clinical trial of 80 patients who received general anesthesia for elective surgery, participants were randomized to receive either 0.5 μg/kg sufentanil or 1.5 mg/kg esketamine as analgesic in combination with propofol for anesthesia induction. There was no significant difference in the variation of average middle cerebral artery blood flow velocity between the drugs, and esketamine better maintained hemodynamic stability during induction. The incidence of adverse effects did not increase after esketamine [43]. In patients with severe TBI and refractory ICP mostly from cerebral contusion or subdural hemorrhage, 2 mg/kg ketamine boluses had a significantly beneficial impact on the ICP and the CPP: Ketamine decreased the ICP and subsequently increased the CPP, particularly in patients with a pretreatment ICP > 20 mmHg and/or a pretreatment CPP under 80 mmHg [36] (Fig. 2). In summary, ketamine does not negatively affect cerebral autoregulatory functions and may increase the CPP in acute intracranial pathology. Future trials are warranted to assess what pharmacologic combination strategy may at best restore autoregulation in NICU patients.

3.3 Ketamine and Neuroprotection: Spreading Depolarizations

In recent years, ketamine's potential to decrease cortical spreading depolarization (CSD) has emerged. CSD is a phenomenon which occurs after injury to the cortex such as with ischemia, hypoxia, or brain trauma. It has increasingly been recognized as a major course of injury progression and is tightly linked to energetic status and metabolic impairment. Very likely, it is a component of secondary brain injury in the neurocritical care population [44]. The hallmark of CSD is an electrochemical wave of depolarizing efflux of sodium and calcium at a rate which leads to dysfunction of sodium-potassium-ATPase pumps. This involves interaction effects with extracellular glutamate and extracellular potassium [45, 46]. A reinforcing feedback loop begins when NMDA receptor activation leads to self-sustaining net inward potassium or calcium currents that result in neuronal dysregulation and cytotoxic edema [46]. Figure 3 shows typical recordings of spreading depolarizations and depressions.

A multicenter, retrospective evaluation of 43 patients with intracerebral hemorrhage or SAH found that ketamine decreased the incidence of CSD significantly as compared to midazolam

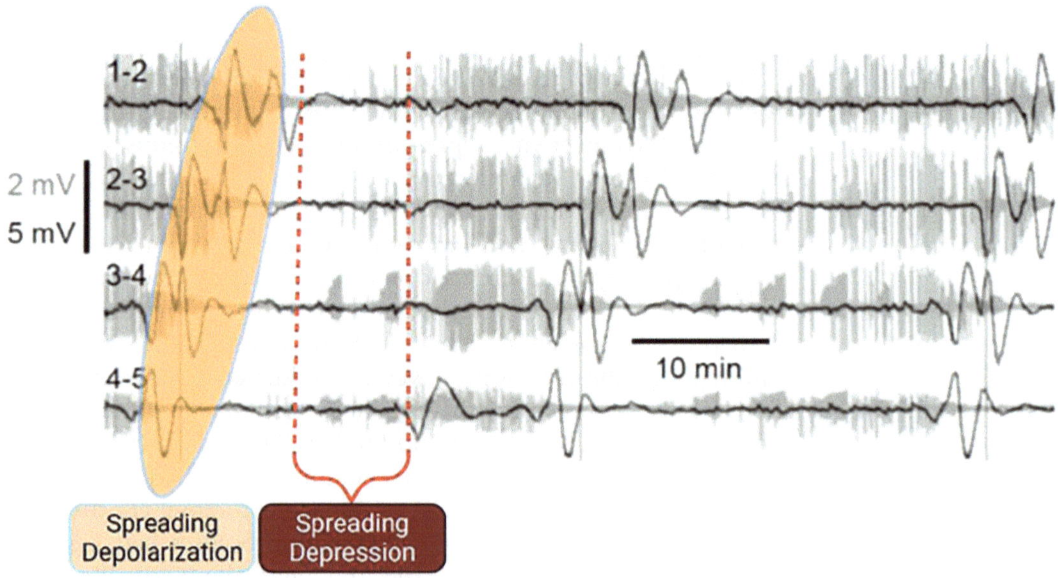

Fig. 3 Electroencephalographic recordings of cortical spreading depolarization and spreading depression: Recordings show overlaid slow 0.005–0.5 Hz activity as black lines and fast 0.5–30 Hz activity as gray lines. Slow activity indicates spreading depolarization highlighted in orange, which causes spreading depression of fast activity framed in red brackets (modified from Mehra et al. [46] under a CC BY 4.0 international commercial license; https:/creativecommons.org/licenses/by/4.0/)

which increased CSD [47]. However, these findings are limited because the investigation also found a large variability in ketamine dosing. Another retrospective cohort study with 66 patients suffering from SAH reported a significant reduction in CSD in 33 patients receiving ketamine after treatment initiation [48]. A dose-dependent correlation between higher ketamine dosage and reduction of CSD was reported. A prospective, randomized, multiple crossover pilot trial published in 2018 recruited two patients with severe TBI and eight patients with SAH [49]. The analysis revealed a strong dose-dependent effect such that hours off ketamine or hours on doses of <1.15 mg/kg/h were associated with an increased risk of CSD compared with hours on doses of >1.15 mg/kg/h or more (odds ratio (OR) = 13.838, 95% confidence interval (CI) = 1.99–1000) [49]. The authors conclude that ketamine effectively inhibited CSD over a wide range of doses commonly used for sedation, even in non-intubated patients. A prospective phase 2 trial is presently underway to assess the impact of a three-tiered intervention strategy to reduce CSD in patients with severe TBI, with ketamine infusion included in the top two tiers [50]. In summary, treatment of CSD will likely become an important component of prevention of secondary brain injury in the neurocritical care population, with ketamine playing a major role.

Ketamine shows therapeutic utility in treatment of refractory (RSE) and superrefractory status epilepticus (SRSE) [51]. In a retrospective study published in 2020, Alkhachroum and co-workers investigated a consecutive series of 68 patients admitted with SRSE [52]. Within 24 h of ketamine initiation plus midazolam infusion, 81% of patients had at least a 50% decrease in seizure burden, and 63% of patients experienced seizure cessation. Average dose of ketamine infusion was 2.2 ± 1.8 mg/kg/h, with a median duration of 2 days. As a mechanistic explanatory approach, the authors attributed the efficacy of high-dose ketamine to result from pathophysiological changes underlying SRSE. This includes the internalization of gamma-aminobutyric acid A ($GABA_A$) receptors and upregulation of NMDA receptors on neuronal cell membranes. Ketamine may thus be very beneficial for the treatment of RSE and SRSE by antagonizing NMDA receptor-related signaling and inhibiting excessive glutamatergic transmission [52]. Because its anti-epileptic effects function predominantly via these distinct neurotransmitter systems, there is a clear rationale for utilizing them earlier in the course of therapy of RSE and SRSE. Ketamine may also be used in parallel with established second-line agents for status epilepticus, before anesthetic drugs such as midazolam or propofol are administered [51].

3.4 Ketamine Sedation and Neurocognition

Sedative drugs can have a major impact on long-term neurocognitive outcomes in the general critical care population. This may be particularly true in patients being treated for primary acute brain insult and injury where sedatives and analgesics may potentiate neurologic injury. In a retrospective study of 91 pediatric patients with acute acquired brain injury, any exposure or higher cumulative doses of ketamine and benzodiazepines were associated with worse neurocognitive outcomes at 1–3 months after discharge, but benzodiazepines were independently associated with very poor outcomes [53].

Delirium is the fluctuating yet ultimately reversible deterioration in neurocognitive function and attention which is frequently encountered in ICU patients. In patients with delirium superimposed on acute brain injury, detection of delirium via established means, such as the Confusion Assessment Method for the ICU (CAM-ICU), may be particularly challenging [6]. Pharmacologic therapy for sedation and analgesia in these patients adds an extra layer of challenges and burden on patients. Practitioners must have a clear rationale to utilize agents with as minimal negative short- and long-term cognitive impact as possible.

The mental health of NICU patients often is of serious concern, and many patients suffer from acute depressive symptoms and anxiety. Short-term, repeated infusions of subanesthetic dose ketamine could be very valuable in all ICU settings as antidepressive therapy. When a controlled trial of patients undergoing brain tumor resection surgery compared the impact of a single intraoperative subanesthetic ketamine dose (0.5 mg/kg infused over 40 min at dural opening) to placebo, the ketamine intervention resulted in a significant reduction in depressive symptoms on 3 postoperative days [54]. A meta-analysis of nearly 1700 patients found a significant improvement in postoperative depression in the broader surgical population with perioperative ketamine use until 7 days after surgery compared to placebo [55]. However, this benefit was associated with an increased risk of nausea and vomiting, headache, hallucinations, and dizziness. All of these unwanted ketamine effects typically occur when the substance is not stopped early enough before the surgical procedure and anesthesia finish. Ketamine always exerts dose- and context-dependent effects and must be used according to individual patient disposition, conditions, and circumstances. In this context, a small retrospective evaluation was made of 12 ICU patients who were visibly withdrawn as per the care team and family. They received repeated infusions of 0.5 mg/kg or 0.75 mg/kg ketamine specifically for depression with the result of improved mood and affect [56].

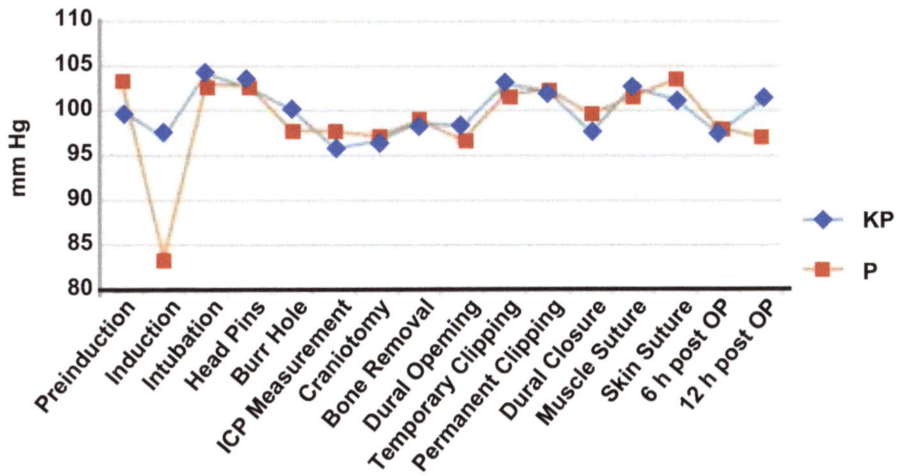

Fig. 4 Mean arterial pressure (mm of Hg) decrease after anesthesia induction with propofol alone versus a combined regimen of ketamine-propofol in aneurysmal clipping surgery. Red dots and lines indicate effects of propofol (P) alone, and blue show combined effects of ketamine-propofol (KP). *ICP* intracranial pressure, *Op* operation (modified from Bhardwaj et al. [57] with permission from Thieme conveyed through Copyright Clearance Center, Inc.)

3.5 Ketamine and Hemodynamic Stabilization

Ketamine is a useful component for procedural sedation in the acute stroke population and other neurovascular patients where cerebrovascular injury makes hemodynamic stability of utmost importance.

In the setting of surgical aneurysm clipping, a prospective randomized trial of 40 patients compared the use of propofol-ketamine ("ketofol") with propofol alone. In comparison to propofol, administration of ketofol especially improved the hemodynamic stability after anesthesia induction and throughout the clipping procedure [57] (Fig. 4).

The effects of ketofol were also compared with propofol for anesthesia induction and maintenance during decompressive surgery in patients with moderate to severe TBI. The combination regimen was associated with improved hemodynamic stability, lower vasopressor requirement, and similar brain relaxation [58]. In acute ischemic stroke patients undergoing mechanical thrombectomy, an approach for sedation with ketamine as first-line agent was assessed in a sub-study of the prospective Sedation versus Intubation for Endovascular Stroke TreAtment (SIESTA) trial. In a protocol in which lighter, opioid-sparing procedural sedation for thrombectomy was utilized, esketamine and propofol were compared to propensity score matched patients receiving conscious sedation with propofol and remifentanil in the SIESTA trial. The esketamine-propofol sedative regimen resulted in significantly reduced procedure-related treatment delays [59]. Taken together, ketamine's cardiocirculatory stimulating effects can safely and effectively balance cardiocirculatory depressant actions of other sedatives and anesthetics such as propofol. In skillful combinations, spontaneous ventilation may be maintained with subanesthetic dose ketamine.

3.6 Ketamine and Ventilatory Stimulation

In patients with depressed level of consciousness or otherwise impaired ventilatory drive, ketamine is valued for its lack of significant respiratory suppression. There is evidence that it may even increase respiratory drive, which can prove useful in patients with brainstem injury, other causes of central apnea, or during ventilator weaning. A prospective observational study of 12 intubated patients undergoing a spontaneous breathing trial showed a significant increase in inspiratory flow, decreased work of breathing, and a concomitant increase in EEG power in the beta through gamma ranges with ketamine. This was interpreted as ketamine having an increasing effect on central respiratory drive, in conjunction with direct effects on pulmonary compliance and airway resistance [60]. Similarly, a randomized trial of 54 patients receiving adjunctive ketamine during operative procedures in deep sedation with propofol or midazolam suggested that ketamine increases central respiratory drive and can prevent severe sedation-associated hypoventilation [61]. A combination regimen of ketamine and propofol or midazolam may therefore be very valuable for short procedures in neurocritical care patients. (Es-) ketamine also counteracts opioid-induced respiratory depression in spontaneous, assisted, and positive pressure ventilation [62]. This is useful in combination with remifentanil, where it may reduce dose requirements [63]. Ketamine is therefore an excellent agent to facilitate ventilator weaning, due to its support of the ventilatory drive, inhibition of cough reflexes, and synergistic opioid-dose-sparing effect. Similarly, in coronavirus disease 2019 (COVID-19) patients, subanesthetic dose esketamine infused at a rate 0.2–0.4 mg/kg/h combined with inhalational sevoflurane (target end-tidal concentration 0.5–1.3 Vol%) as sedative regimen allowed for maintenance of assisted respiration even during prone positioning [64].

4 Spinal Cord Injury, Pain Management, and Ketamine

Patients with traumatic spinal cord injury (SCI) often pose unique challenges with regard to pain control. SCI patients may suffer from significant pain and yet are at particular risk of constipation and spinal neuropathic pain syndromes. The use of opioids should therefore be minimized. Multiday infusions of ketamine have been studied with good results in chronic pain SCI patients. In the acute phase, a series of 13 patients received continuous ketamine infusions in conjunction with oral agents. They experienced significant pain relief with pain scores being less than 30% of premedication [65]. In a pediatric acute SCI patient, refractory neuropathic pain was successfully treated with continuously infused ketamine [66]. In the chronic SCI patient population, a prospective randomized trial of 40 patients could also show significant improvement in pain with once-weekly infusions of ketamine combined with oral gabapentin [67].

5 Limitations

There is an increasingly robust array of data supporting the safety, utility, and benefits of using ketamine in the neurocritical care population. However, this data is often limited to single-center cohort studies. Advanced multimodal neuromonitoring of complex neurophysiology data, such as ICP, cerebral blood flow, brain tissue oxygenation, micro-dialysis, evoked potentials, or quantitative electroencephalography, is generally obtained in a rather limited number of patients treated in highly specialized centers. As can be seen in trial registries, investigation of ketamine effects on NICU patients has only recently been reinitiated in such centers. The complex relationships between variables of interest in acute intracranial pathologies and NICU patients' outcomes are still not well understood, so that much hope is placed on future study findings.

6 Conclusion

In summary, ketamine has specific applications and considerations in the neurocritical care population. In short-term study, ketamine is safe and often beneficial with regard to cerebrovascular dynamics, ICP, and minimization of secondary brain injury. For optimizing sedation and analgesia, especially formal evaluation of dose relationships within different drug combination regimens appears to be of priority. Far beyond, long-term outcome data after ketamine use in the NICU are still lacking.

Conflicts of Interest

No potential conflicts of interest or specific funding sources to report.

References

1. Korf EM, Tronnier VM, Gliemroth J, Kuchler JN (2017) Isoflurane-associated mydriasis mimicking blown pupils in a patient treated in a neurointensive care unit. J Neurol Surg A Cent Eur Neurosurg 78:403–406. https://doi.org/10.1055/s-0036-1594237
2. Mullaguri N et al (2017) Pitfall in pupillometry: exaggerated ciliospinal reflex in a patient in barbiturate coma mimicking a nonreactive pupil. Cureus 9:e2004. https://doi.org/10.7759/cureus.2004
3. Kim J, Heo J, Ji D, Kim MS (2015) Quantitative assessment of pupillary light reflex in normal and anesthetized dogs: a preliminary study. J Vet Med Sci 77:475–478. https://doi.org/10.1292/jvms.14-0387
4. Eilers H, Larson MD (2010) The effect of ketamine and nitrous oxide on the human pupillary light reflex during general anesthesia. Auton Neurosci 152:108–114. https://doi.org/10.1016/j.autneu.2009.10.004
5. Vide S, Costa CM, Gambus PL, Amorim PP (2018) Effects of ketamine on pupillary reflex dilation: a case report. A&A Pract 10:39–41. https://doi.org/10.1213/XAA.0000000000000633
6. Bauerschmidt A, Al-Bermani T, Ali S, Bass B, Dorilio J, Rosenberg J, Al-Mufti F (2023) Modern sedation and analgesia strategies in neurocritical care. Curr Neurol Neurosci Rep 23:149–158. https://doi.org/10.1007/s11910-023-01261-7
7. Rueda Carrillo L, Garcia KA, Yalcin N, Shah M (2022) Ketamine and its emergence in the field of neurology. Cureus 14:e27389. https://doi.org/10.7759/cureus.27389
8. Weiler MA, Thaker GK, Lahti AC, Tamminga CA (2000) Ketamine effects on eye movements. Neuropsychopharmacology 23:645–653. https://doi.org/10.1016/S0893-133X(00)00156-1
9. Lencer R, Trillenberg P (2008) Neurophysiology and neuroanatomy of smooth pursuit in humans. Brain Cogn 68:219–228. https://doi.org/10.1016/j.bandc.2008.08.013
10. Hung KC et al (2024) The impact of perioperative ketamine or esketamine on the subjective quality of recovery after surgery: a meta-analysis of randomised controlled trials. Br J Anaesth 132:1293–1303. https://doi.org/10.1016/j.bja.2024.03.012
11. Zhang T et al (2023) S-ketamine promotes postoperative recovery of gastrointestinal function and reduces postoperative pain in gynecological abdominal surgery patients: a randomized controlled trial. BMC Surg 23:74. https://doi.org/10.1186/s12893-023-01973-0
12. Freye E, Knufermann V (1994) No inhibition of intestinal motility following ketamine-midazolam anesthesia. A comparison of anesthesia with enflurane and fentanyl/midazolam. Anaesthesist 43:87–91. https://doi.org/10.1007/s001010050036
13. Yao M, Fang B, Yang J, Chen P, Chen F (2024) Esketamine combined with sufentanil versus sufentanil in patient-controlled intravenous analgesia: a meta-analysis. Front Pharmacol 15:1247646. https://doi.org/10.3389/fphar.2024.1247646
14. Kaszaki J et al (2012) Kynurenines and intestinal neurotransmission: the role of N-methyl-D-aspartate receptors. J Neural Transm (Vienna) 119:211–223. https://doi.org/10.1007/s00702-011-0658-x
15. Ostovic H, Simac B, Prazetina M, Bradic N, Persec J (2023) The effect of intravenous lidocaine, ketamine, and lidocaine-ketamine combination in colorectal cancer surgery: a randomized controlled trial. Anesth Analg 140:67. https://doi.org/10.1213/ANE.0000000000006555
16. Schmittner MD et al (2007) Effects of fentanyl and S(+)-ketamine on cerebral hemodynamics, gastrointestinal motility, and need of vasopressors in patients with intracranial pathologies: a pilot study. J Neurosurg Anesthesiol 19:257–262. https://doi.org/10.1097/ANA.0b013e31811f3feb
17. Richards ND, Weatherhead W, Howell S, Bellamy M, Mujica-Mota R (2024) Continuous infusion ketamine for sedation of mechanically ventilated adults in the intensive care unit: a scoping review. J Intensive Care Society 25:59–77. https://doi.org/10.1177/17511437231182507
18. Yagnala N, Brown CS, Peters B et al (2024) Continuous-infusion ketamine versus opioids for primary analgosedation in ventilated STICU patients. Crit Care Med 52:S484. https://doi.org/10.1097/01.ccm.0001002260.67644.3a
19. Peters BJ et al (2024) Ketamine for primary analgosedation in critically ill surgery and trauma patients requiring mechanical ventilation. Crit Care Explor 6:e1041. https://doi.org/10.1097/CCE.0000000000001041
20. Mion G, Himmelseher S (2024) Esketamine: less drowsiness, more analgesia. Anesth Analg

139:78–91. https://doi.org/10.1213/ANE.0000000000006851

21. Dawson B, Michenfelder JD, Theye RA (1971) Effects of ketamine on canine cerebral blood flow and metabolism: modification by prior administration of thiopental. Anesth Analg 50:443–447
22. Shaprio HM, Wyte SR, Harris AB (1972) Ketamine anaesthesia in patients with intracranial pathology. Br J Anaesth 44:1200–1204. https://doi.org/10.1093/bja/44.11.1200
23. Takeshita H, Okuda Y, Sari A (1972) The effects of ketamine on cerebral circulation and metabolism in man. Anesthesiology 36:69–75. https://doi.org/10.1097/00000542-197201000-00013
24. Reede K, Bartholomew R, Nielsen D, Ahmeti M, Zreik K (2023) Ketamine in trauma: a literature review and administration guidelines. Cureus 15:e48099. https://doi.org/10.7759/cureus.48099
25. Himmelseher S, Durieux ME (2005) Revising a dogma: ketamine for patients with neurological injury? Anesth Analg 101:524–534. https://doi.org/10.1213/01.Ane.0000160585.43587.5b
26. Albanèse J et al (1997) Ketamine decreases intracranial pressure and electroencephalographic activity in traumatic brain injury patients during propofol sedation. Anesthesiology 87:1328–1334. https://doi.org/10.1097/00000542-199712000-00011
27. Bar-Joseph G, Guilburd Y, Tamir A, Guilburd JN (2009) Effectiveness of ketamine in decreasing intracranial pressure in children with intracranial hypertension: clinical article. J Neurosurgery: Pediatrics PED 4:40–46. https://doi.org/10.3171/2009.1.PEDS08319
28. Bourgoin A et al (2005) Effects of sufentanil or ketamine administered in target-controlled infusion on the cerebral hemodynamics of severely brain-injured patients. Crit Care Med 33:1109–1113. https://doi.org/10.1097/01.ccm.0000162491.26292.98
29. Bourgoin A et al (2003) Safety of sedation with ketamine in severe head injury patients: comparison with sufentanil. Crit Care Med 31: 711–717. https://doi.org/10.1097/01.Ccm.0000044505.24727.16
30. Caricato A et al (2013) Racemic ketamine in adult head injury patients: use in endotracheal suctioning. Crit Care 17:R267. https://doi.org/10.1186/cc13097
31. Gregers MCT, Mikkelsen S, Lindvig KP, Brøchner AC (2020) Ketamine as an anesthetic for patients with acute brain injury: a systematic review. Neurocrit Care 33:273–282. https://doi.org/10.1007/s12028-020-00975-7
32. Patanwala AE, Martin JR, Erstad BL (2015) Ketamine for analgosedation in the intensive care unit: a systematic review. J Intens Care Med 32:387–395. https://doi.org/10.1177/0885066615620592
33. Pfenninger E, Grünert A, Bowdler I, Kilian J (1985) The effect of ketamine on intracranial pressure during haemorrhagic shock under the conditions of both spontaneous breathing and controlled ventilation. Acta Neurochir 78: 113–118. https://doi.org/10.1007/bf01808689
34. Guler G, Aksu R, Bicer C, Tosun Z, Boyaci A (2010) Comparison of the effects of ketamine or lidocaine on fentanyl-induced cough in patients undergoing surgery: a prospective, double blind, randomized, placebo-controlled study. Curr Ther Res Clin Exp 71:289–297. https://doi.org/10.1016/j.curtheres.2010.10.004
35. Shafa A, Shahhosseini S, Rajaee E (2021) Comparison of two dosages of ketamine in preventing fentanyl-induced coughs in children. Adv Biomed Res 10:20. https://doi.org/10.4103/abr.abr_218_20
36. Dengler BA et al (2022) Ketamine boluses are associated with a reduction in intracranial pressure and an increase in cerebral perfusion pressure: a retrospective observational study of patients with severe traumatic brain injury. Crit Care Res Pract 2022:3834165. https://doi.org/10.1155/2022/3834165
37. Kolenda H, Gremmelt A, Rading S, Braun U, Markakis E (1996) Ketamine for analgosedative therapy in intensive care treatment of head-injured patients. Acta Neurochir 138:1193–1199. https://doi.org/10.1007/bf01809750
38. Von der Brelie C et al (2017) Sedation of patients with acute aneurysmal subarachnoid hemorrhage with ketamine is safe and might influence the occurrence of cerebral infarctions associated with delayed cerebral ischemia. World Neurosurg 97:374–382. https://doi.org/10.1016/j.wneu.2016.09.121
39. Czosnyka M, Smielewski P, Kirkpatrick P, Menon DK, Pickard JD (1996) Monitoring of cerebral autoregulation in head-injured patients. Stroke 27:1829–1834. https://doi.org/10.1161/01.str.27.10.1829
40. Svedung Wettervik T, Fahlström M, Enblad P, Lewén A (2022) Cerebral pressure autoregulation in brain injury and disorders-a review on monitoring, management, and future directions. World Neurosurg 158:118–131. https://doi.org/10.1016/j.wneu.2021.11.027

41. Park S, Beqiri E, Smielewski P, Aries M (2024) Inaugural state of the union: continuous cerebral autoregulation monitoring in the clinical practice of neurocritical care and anesthesia. Neurocrit Care 40:855–864. https://doi.org/10.1007/s12028-023-01860-9
42. Driesen NR et al (2023) Ketamine effects on energy metabolism, functional connectivity and working memory in healthy humans. https://doi.org/10.1101/2023.02.21.529425
43. Yan S, Li Q, He K (2024) The effect of esketamine combined with propofol-induced general anesthesia on cerebral blood flow velocity: a randomized clinical trial. BMC Anesthesiol 24:66. https://doi.org/10.1186/s12871-024-02446-4
44. Sakowitz OW et al (2009) Preliminary evidence that ketamine inhibits spreading depolarizations in acute human brain injury. Stroke 40:e519–e522. https://doi.org/10.1161/strokeaha.109.549303
45. Lindquist BE (2024) Spreading depolarizations pose critical energy challenges in acute brain injury. J Neurochem 168:868–887. https://doi.org/10.1111/jnc.15966
46. Mehra A, Gomez F, Bischof H, Diedrich D, Laudanski K (2023) Cortical spreading depolarization and delayed cerebral ischemia; rethinking secondary neurological injury in subarachnoid hemorrhage. Int J Mol Sci 24:9883
47. Hertle DN et al (2016) Changes in electrocorticographic beta frequency components precede spreading depolarization in patients with acute brain injury. Clin Neurophysiol 127:2661–2667
48. Santos E et al (2019) Lasting s-ketamine block of spreading depolarizations in subarachnoid hemorrhage: a retrospective cohort study. Crit Care 23:1–14
49. Carlson AP, Abbas M, Alunday RL, Qeadan F, Shuttleworth CW (2018) Spreading depolarization in acute brain injury inhibited by ketamine: a prospective, randomized, multiple crossover trial. J Neurosurg 130:1513–1519
50. Hartings JA et al (2023) Improving neurotrauma by depolarization inhibition with combination therapy: a phase 2 randomized feasibility trial. Neurosurgery 93:924–931. https://doi.org/10.1227/neu.0000000000002509
51. Buratti S et al (2023) (2023) ketamine as advanced second-line treatment in benzodiazepine-refractory convulsive status epilepticus in children. Epilepsia 64:797–810. https://doi.org/10.1111/epi.17550
52. Alkhachroum A et al (2020) Ketamine to treat super-refractory status epilepticus. Neurology 95:e2286–e2294. https://doi.org/10.1212/wnl.0000000000010611
53. Drury KM et al (2024) Exposure to sedation and analgesia medications: short-term cognitive outcomes in pediatric critical care survivors with acquired brain injury. J Intensive Care Med 39:374–386. https://doi.org/10.1177/08850666231210261
54. Zhou Y et al (2021) Ketamine alleviates depressive symptoms in patients undergoing intracranial tumor resection: a randomized controlled trial. Anesth Analg 133:1588–1597. https://doi.org/10.1213/ane.0000000000005752
55. Guo J et al (2023) Efficacy and safety of perioperative application of ketamine on postoperative depression: a meta-analysis of randomized controlled studies. Mol Psychiatry 28:2266–2276. https://doi.org/10.1038/s41380-023-01945-z
56. Giri AR et al (2022) Novel management of depression using ketamine in the intensive care unit. J Intensive Care Med 37:1654–1661. https://doi.org/10.1177/08850666221088220
57. Bhardwaj A et al (2020) Comparison of ketofol (combination of ketamine and propofol) and propofol anesthesia in aneurysmal clipping surgery: a prospective randomized control trial. Asian J Neurosurg 15:608–613. https://doi.org/10.4103/ajns.AJNS_346_19
58. Maheswari N et al (2023) Ketofol as an anesthetic agent in patients with isolated moderate to severe traumatic brain injury: a prospective, randomized double-blind controlled trial. J Neurosurg Anesthesiol 35:49–55. https://doi.org/10.1097/ANA.0000000000000774
59. Schönenberger S et al (2019) The keep simplest study: improving in-house delays and periinterventional management in stroke thrombectomy – a matched pair analysis. Neurocrit Care 31:46–55. https://doi.org/10.1007/s12028-018-00667-3
60. Suleiman A et al (2023) Effects of ketamine infusion on breathing and encephalography in spontaneously breathing ICU patients. J Intensive Care Med 38:299–306. https://doi.org/10.1177/08850666221119716
61. De Oliveira GS et al (2014) The effect of ketamine on hypoventilation during deep sedation with midazolam and propofol: a randomised, double-blind, placebo-controlled trial. Eur J Anaesthesiol 31:654–662. https://doi.org/10.1097/eja.0000000000000025
62. Jonkman K et al (2018) Esketamine counters opioid-induced respiratory depression. Br J

Anaesth 120:1117–1127. https://doi.org/10.1016/j.bja.2018.02.021

63. Fujii T, Nishiwaki K (2022) Ketamine reduces the dose of remifentanil required during prolonged head and neck surgery: a propensity-matched analysis. Nagoya J Med Sci 84:1–6. https://doi.org/10.18999/nagjms.84.1.1
64. Bansbach J, Wenz J, Kaufmann K, Heinrich S, Kalbhenn J (2022) Sevoflurane in combination with esketamine is an effective sedation regimen in COVID-19 patients enabling assisted spontaneous breathing even during prone positioning. Anaesthesiol Intensive Ther 54:23–29. https://doi.org/10.5114/ait.2022.113950
65. Kim K et al (2013) Ketamine for acute neuropathic pain in patients with spinal cord injury. J Clin Neurosci 20:804–807. https://doi.org/10.1016/j.jocn.2012.07.009
66. Shiwlochan DG, Shah M, Baldev K, Thomas DA, Debrosse M (2020) The effectiveness of ketamine in pediatric acute deafferentation pain after spinal cord injury. Case Rep Anesthesiol 2020:8835292. https://doi.org/10.1155/2020/8835292
67. Amr YM (2010) Multi-day low dose ketamine infusion as adjuvant to oral gabapentin in spinal cord injury related chronic pain: a prospective, randomized, double blind trial. Pain Physician 13:245–249

Chapter 9

Racemic Ketamine and Esketamine for Treatment of Refractory and Super-Refractory Status Epilepticus

Mohamed ElSayed, Kaveh Latifi, Lobna Abdelwahab, Marina Ramzy Mourid, Hassan Sayed Kandil, Dania Imtiyaz Khan, and Islam Mohammad Shehata

Abstract

This narrative review examines the potential role of ketamine and its enantiomers in the treatment of refractory and super-refractory status epilepticus (RSE and SRSE, respectively). While early studies and a few recent case reports have argued that ketamine has a pro-epileptic profile, recent studies (including one systematic review and meta-analysis) have proposed a promising role for using ketamine/esketamine in the treatment of RSE/SRSE, albeit the current evidence for this role remains of a low quality. There is still no general consensus on when to start ketamine after the seizure onset or what dose the bolus and/or infusion should be. The authors propose a titrating, individualized, and patient-centered approach when considering ketamine/esketamine for treatment-refractory seizures.

Key words Ketamine, Esketamine, Status epilepticus, Treatment-refractory

1 Introduction

Refractory and super-refractory status epilepticus (RSE and SRSE) are life-threatening emergencies that require prompt management to prevent severe morbidity and mortality. In both conditions, irreversible harm at the molecular and cellular neuronal level may occur if epileptic pathways and vicious circles are not interrupted. A timely therapeutic intervention is of utmost importance. Currently, there are no specific guidelines supporting the choice of a specific medication in RSE/SRSE, and the available recommendations favor personalized treatment [1].

Ketamine, a dissociative anesthetic and a potent antagonist of the N-methyl-D-aspartate receptor (NMDAR), has shown effectiveness in treating RSE/SRSE. Since its approval for clinical anesthesia in 1964 [2], ketamine has also gained popularity because of its neuroprotective potential. Several studies have been conducted

Sabine Himmelseher (ed.), *Ketamine: Pain Management, Critical Care, Depression, and Resource-Poor Settings Volume II*, Neuromethods, vol. 225, https://doi.org/10.1007/978-1-0716-4599-4_9,

to investigate the role of ketamine in aborting RSE/SRSE. Herein, we aim to discuss the potential role of racemic and esketamine as a therapeutic approach to RSE/SRSE [3].

2 Overview of Use of Ketamine in Epilepsy

2.1 Background Knowledge I: Ketamine and Isomers

Ketamine is a chiral drug with two enantiomers, S(+)-ketamine and R(−)-ketamine, and is worldwide used as racemic formulation of an equimolar mixture of both [4]. The structure of racemic ketamine is a cyclohexanone ring attached to a phenyl ring, with an amino group and two methyl groups [5–7] (Fig. 1).

Ketamine first came into existence in 1962 and was introduced for medical use as an anesthetic agent in 1970 [8]. Esketamine (S (+)-ketamine) is the more potent enantiomer of racemic ketamine. With intravenous (IV) use, esketamine shows an approximately double analgesic, antidepressant, and anesthetic potency as IV racemic ketamine with less unwanted psychoactive effects and more rapid drug recovery. In addition, esketamine has been found to induce less neurocognitive impairment than racemic and arketamine. IV esketamine gained clinical approval in 1997 in Germany and then in other European and South American countries and was recently authorized for clinical use in France and China [7]. In the USA and Canada, only the intranasal spray formulation of esketamine has been approved in 2019 so far. Some expert authors consider the effects of esketamine to be mostly responsible for its antiepileptic impact of the racemic ketamine mixture [9] (for more detailed information about racemic ketamine, its enantiomers, and esketamine, please see previous chapters in this textbook). Arketamine (R(−)-ketamine) has a different orientation of atoms compared to S(+)-ketamine which results in distinct pharmacologic

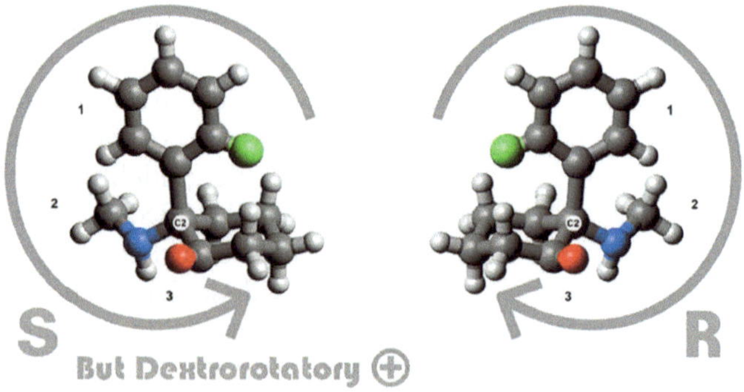

Fig. 1 Chemical structure of the two stereoisomers of ketamine, S(+)-ketamine and R(−)-ketamine; racemic ketamine is a mixture of equal amounts of both (from Mion and Himmelseher [7] with permission from Wolters Kluwer Health, Inc. via Copyright Clearance Center, Inc.)

properties. However, as yet, there are only very few human studies with arketamine, and there is no clinical approval for arketamine. The substance has not been examined in the context of epilepsy. For the purpose of this review, the expression "ketamine" refers to racemic ketamine and "esketamine" to the clinical preparation of pure S(+)-ketamine.

2.2 Background Knowledge II: Pharmacokinetics

Ketamine is commonly used as IV administration, demonstrating rapid onset of action due to high lipophilicity and fast distribution. The initial distribution phase has a half-life of approximately 10–15 min, corresponding to its anesthetic effects [10]. Oral ketamine intake is a comfortable way of ketamine administration, but the substance undergoes extensive first-pass metabolism with the result of a systemic bioavailability of less than 20%. This low bioavailability limits its clinical use for anti-convulsive therapy, although enteral ketamine has been used in settings of RSE/SRSE.

Intranasal ketamine is used in various scenarios, such as for acute and chronic pain therapy, premedication in children, or as antidepressant. In the USA and Canada, only intranasal esketamine has gained clinical approval so far [11]; its absolute bioavailability is approximately 48%. The fraction absorbed through the nasal cavity is about 54%. From the remaining, a part gets uncontrollably swallowed and undergoes first-pass metabolism [11]. The time to reach maximum plasma concentration is 20–40 min post-intranasal administration [11, 12]. IV esketamine is clinically used in those countries where it is available. Esketamine has a larger volume of distribution at steady state and a greater clearance than racemic ketamine.

Ketamine and its isomers have a low plasma protein binding of 10–30% and an extensive distribution owing to their high liposolubility. Ketamine has a large central compartment volume of ~70 L and a steady-state distribution volume of approximately 200 L (or 2.3 L/kg) [13].

2.3 Background Knowledge III: Metabolism and Excretion

Ketamine is primarily metabolized in the liver through a series of enzymatic reactions. The initial and major metabolic pathway involves N-demethylation to form norketamine, an active metabolite. This reaction is predominantly catalyzed by cytochrome P450 enzymes, specifically CYP2B6 and CYP3A4, with minor contributions from CYP2C9 [14, 15]. Following N-demethylation, norketamine undergoes further biotransformation. The cyclohexanone ring of norketamine is hydroxylated to form hydroxynorketamine (HNK) compounds via CYP-dependent pathways. These hydroxylated metabolites are then conjugated with glucuronic acid, leading to the formation of glucuronides. Hydroxynorketamine can undergo dehydration to form dehydroxynorketamine [14, 15]. Ketamine is excreted via bile and urine following glucuronide conjugation [16]. Other minor metabolites are also formed.

Ketamine's elimination clearance is high (1000–1600 mL/min or 12–20 mL/min/kg) and largely depends on liver blood flow [16]. Its elimination half-life is 2–3 hours because of its rapid metabolism [17]. Ketamine metabolism differs by gender, with higher clearance rates observed in women than men [18]. Hence, a sex-specific dosing strategy may be required to optimize therapeutic outcomes.

2.4 Background Knowledge IV: Use in Different Patient Populations

Preclinical studies suggest a minuscule but noticeable change in ketamine plasma concentrations related to variations in the estrous cycle. However, this appears to be more important in ketamine's antidepressant effects at subanesthetic dose [19, 20] as compared to its anticonvulsant actions in RSE and SRSE within the higher dose range.

In children, ketamine is absorbed faster when applied via the intramuscular (IM) route than in adults, possibly because of the differences in muscle composition and regional blood flow. Chil dren have a lower distribution volume (1.9 L/kg), higher plasma clearance rates (16.8 mL/kg/min), and a shorter ketamine elimination half-life (100 min) than adults [19]. Newborns and infants also have an increased elimination half-life secondary to shorter ketamine plasma clearance. Dosing and frequency of ketamine administration should be adjusted accordingly [21].

Patients with kidney dysfunction show higher ketamine concentrations after the same dose compared to those with normal kidney function. This may lead to a moderate decrease in clearance and may necessitate dose adjustments and/or close patient monitoring to avoid potential adverse effects associated with higher drug doses used in RSE/SRSE.

2.5 Background Knowledge V: Pharmacodynamics

Ketamine is a noncompetitive antagonist of glutamatergic NMDA receptors and has demonstrated anticonvulsant properties in clinical study [22]. Several trial findings point to ketamine's effectiveness in therapeutic approaches to RSE when conventional antiepileptic drugs (AED) targeting the gamma amino butyric (GABA)- ergic system have often lost efficacy. During prolonged seizures, GABA receptors may not be accessible to AED [23–27]. The mechanism of ketamine's action in RSE/SRSE will be discussed below.

2.6 Background Knowledge VI: Ketamine Effects Are Dose- and Context-Dependent

The pharmacodynamic profile of ketamine is highly dose-dependent: *With small, subanesthetic doses of 0.1 to 0.5 mg/kg IV* ketamine, the substance demonstrates rapid antidepressant and analgesic effects, and there is mounting research about its antidepressant role in Alzheimer's disease due to its neuroprotective action [28]. *Anesthesia induction doses consist of approximately 1–2 mg/kg* IV ketamine and are associated with psychosis-like symptoms [29]. In multimodal perioperative care, analgesic dose use has, for example, been proposed for patients undergoing major complex, painful spinal surgery: a use schedule with an intraoperative 0.1–1 mg/kg IV bolus after anesthesia induction followed by an intraoperative infusion of 0.1–0.25 mg/kg/h possibly continued during postoperative care has been recommended in the clinical practice guidelines from the Society for Neuroscience in Anesthesiology and Critical Care [30]. In addition, a wide range of doses has been reported for recreational use, including high doses of 3 mg/kg and more, which were reported to increase locomotor activity and brain temperature.

3 Refractory and Super-Refractory Status Epilepticus

3.1 Definitions

The International League Against Epilepsy (ILAE) defines a refractory status epilepticus (RSE) as "... a status epilepticus that continues despite the use of two or more antiseizure medications, administered in appropriate and adequate doses, usually including IV benzodiazepines as first-line drugs ..." [31]. Far beyond, the ILAE defines a super-refractory status epilepticus (SRSE) as "... a persisting or recurring SE after 24 h or longer treatment with highly sedating antiepileptic drugs (AED) such as midazolam, phenobarbital, or propofol infusion ...". If seizures recur after tapering anti-seizure medications within 24 h of use, this is also defined as SRSE [32].

3.2 Pathophysiology of Refractory Status Epilepticus

RSE and SRSE are complex medical conditions characterized by various pathophysiological changes in different neurotransmitter systems. To provide a basis for understanding why classic AEDs often fail in RSE and SRSE and the mechanisms underlying ketamine's effectiveness, we will now focus on distinct neurobiological alterations described for RSE/SRSE settings.

RSE and SRSE are usually resistant to the effects of benzodiazepines (BZD). This is likely secondary to the internalization of GABA-A receptors during prolonged seizures. This, in turn, leads to a decrease in GABA receptor accessibility and therapeutic options for seizure inhibition [32]. As soon as GABA-A receptors are internalized, they undergo structural changes that affect their function and transmembrane gradient to chloride. GABA-A receptors can also undergo relocation to extra-synaptic sites [33]. Excessive glutamate release during RSE/SRSE activates postsynaptic NMDA receptors and further upregulates NMDA receptor transmission and expression. This induces neuronal death through receptor-mediated calcium influx, a process referred to as excitotoxicity [34]. Ketamine exerts its neuropharmacological effects primarily by antagonizing glutamate receptors, notably of the NMDA type [35].

3.3 Potential Mechanisms of Ketamine

Previous animal models have demonstrated that prolonged seizures are associated with a different receptor expression than non-prolonged seizures [9, 36–38]. In prolonged seizures, the main difference lies in NMDA receptor upregulation and GABAergic receptor downregulation [39–41]. On one hand, first- and second-line antiepileptic medications, which act by facilitating inhibitory GABA-A receptor function, are therefore less effective. On the other hand, ketamine, with its NMDA receptor antagonizing action, may thus be more effective in persistent or refractory seizures [42]. NMDA receptor antagonism leads to a reduction in the excitotoxic effects of glutamate, as demonstrated, for example, in models of global forebrain ischemia [43]. In addition, ketamine modulates the GABA-ergic system by facilitating the release of GABA and binding to GABA-A receptors. This decreases neuronal over-excitability, which leads to an overall decrease in seizure sensitivity [17, 44, 45]. Ketamine also interacts with numerous other receptors and neurobiological systems, such as μ, δ, κ opioid, monoaminergic, cholinergic, nicotinic, and muscarinic mechanism, but it is not yet entirely understood how this action affects its antiepileptic efficacy in RSE and SRSE [21].

Ketamine also reduces neuroinflammation by suppressing pro-inflammatory cytokines, including interleukin-1β (IL-1β), tumor necrosis factor-alpha (TNF-α), and interleukin-6 (IL-6), and enhances antioxidant defenses [46]. Ketamine regulates neurotrophic factors, especially from the mammalian target of rapamycin (mTOR) signaling pathway, which leads to reduced expression of autophagic proteins and a decrease in epileptic cascades [42, 47]. Esketamine was also reported to significantly reduce cortical neuronal loss [43] (Fig. 2).

Additionally, ketamine improves neuronal recovery following spreading depolarizations in peri-infarct tissues, reducing intracellular calcium accumulation and preserving functional recovery [48].

In summary, ketamine offers neuroprotective effects in various animal models of brain insults and in clinical trials, which point to its potential for also mitigating neuronal damage caused by RSE and SRSE.

4 The Clinical Role of Racemic and Esketamine as Therapeutic Approach to Refractory Status Epilepticus

Under clinical conditions of RSE and SRSE, ketamine is preferable to other sedatives and anesthetics because of its unique mechanisms of action and lack of cardiorespiratory depression [49]. With skilled use, such as slow administration, ketamine exerts minimal inhibition on the central respiratory drive. It is, therefore, a great choice in non-intubated patients and may sometimes even prevent

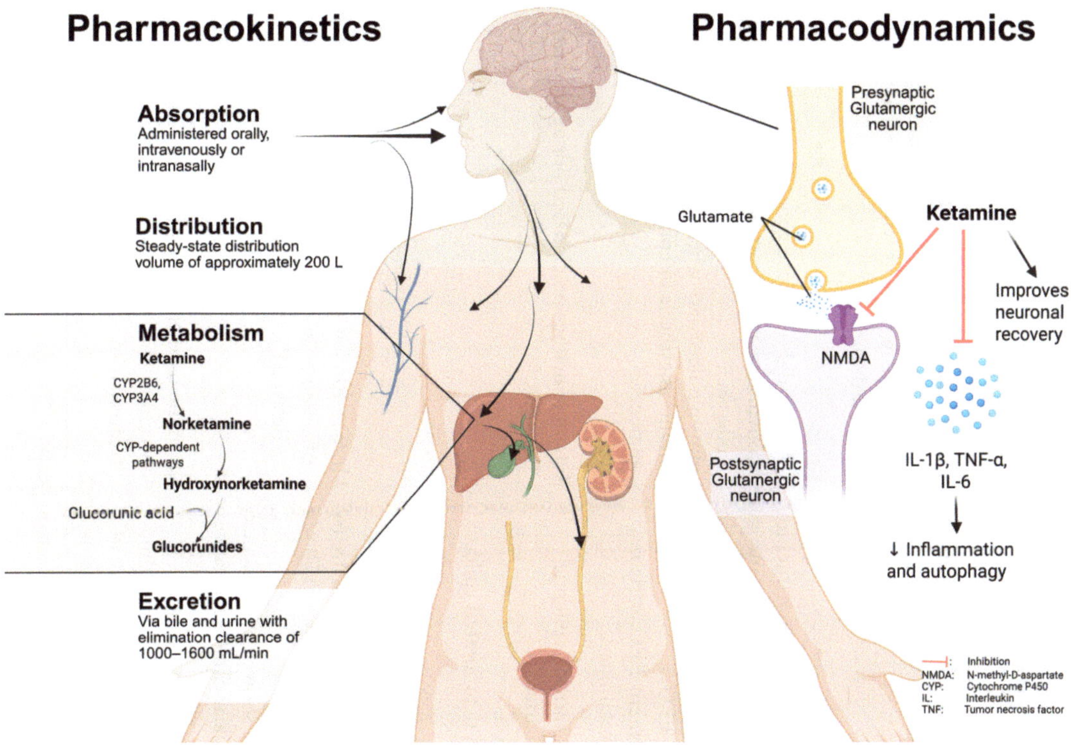

Fig. 2 Symbolized diagram of ketamine pharmacokinetics and pharmacodynamics relevant in treatment approaches to status epilepticus. Some modes of administration, ketamine distribution, liver metabolism, and excretion, as well as assumed mechanisms of action are shown (designed by ElSayed M, all authors, and Ro'a Azzam Hassouna Mohammad, Faculty of Medicine, MTI University, Cairo, Egypt, (2024) in BioRender.com/c20n933)

intubation [50–52]. Most reports describe the effects of IV racemic ketamine as the main preparation used in RSE and SRSE [24], but there are trial data indicating that IV esketamine is associated with promising results in the treatment of RSE/SRSE. In one retrospective study of 42 patients with RSE and SRSE, infusion of esketamine was found to be effective [53]: seven patients received a median dose of 200 mg as initial bolus, and all patients were treated with a maximum median dose infused at 2.55 mg/kg/h. Other uses than the IV administration route, such as oral or rectal intake, have so far been reported in only a few reports. Nonetheless, ketamine may be used successfully for termination of RSE and SRSE with oral use, for example, in elderly patients.

4.1 The Pro-epileptic Claim

In clinical practice, a common, albeit relative concern with respect to ketamine use is and has been the notion of a pro-epileptic potential. This concern is rooted in studies from 50 years ago, and a few cases reports that describe seizure activity after ketamine administration (Table 1).

What is remarkable in all landmark studies and reported cases is the fact that ketamine has been used as mono-anesthetic or mono-

Table 1
Early trial results and case reports of ketamine showing pro-epileptic activity

Article	Study participants	Results and findings	Conclusions
Ferrer-Allado et al. [65]	9 patients with seizure disorders and cortical, limbic, and thalamic electrode implants. Effects of thiopental versus N_2O versus mono-anesthesia with IV ketamine at a dose of 0.5, 1, 2, and 4 mg/kg were compared.	Seizures were noted in 1 patient who received 0.5 mg/kg ketamine and in 6 patients after 2 or 4 mg/kg ketamine.	Authors recommended not to use or very cautiously use ketamine as a mono-anesthetic in patients who have seizure-related disorders.
Corssen et al. [66]	This study compared the impact of ketamine on parameters of the EEG in healthy human volunteers without a history of epilepsy versus volunteers with a history of active epilepsy.	Ketamine did not trigger seizure discharges in individuals without epilepsy or cause any EEG changes that indicate seizure activity in patients with a history of epilepsy and a normal pre-treatment EEG. With abnormal pre-treatment EEG, variables rarely worsened after ketamine.	Ketamine did not cause generalized convulsions in patients with a history of epilepsy. Caution is necessary when ketamine is used as a mono-anesthetic in epilepsy-susceptible patients.
Celesia et al. [67]	IV ketamine was administered to 26 patients with epilepsy to examine EEG effects.	17 (65%) awake patients showed epileptic discharges in the EEG after ketamine administration. 8 (31%) patients showed deteriorated epileptic discharges. No clinical seizures were observed during ketamine anesthesia.	Ketamine neither precipitated nor aggravated seizures.
Meaden et al. [68]	A case report of a 15 years old female with autism spectrum disorder who received IM ketamine as a sole analgesic/sedative for digit laceration wound care and stitching in ER. No use of GABA-ergic drug or local anesthetic.	The patient developed a tonic-clonic seizure for 1 min at approximately 10 min after 150 mg IM ketamine. The following cranial CT showed a ventricular colpocephaly with unclear clinical relevance.	More research is needed to understand the role of ketamine as a mono-analgesic and sedative in delayed therapy of acute injury in children.
Kim et al. [69]	A case report of a 7 years old healthy male who received IM ketamine as a sole analgesic/sedative for nose laceration surgery in ER. No use of GABA-ergic drug or local anesthetic.	Patient developed twitches and stiffness followed by bilateral symmetrical rhythmic tonic-clonic movements with up-rolling of eyeballs for 90 s after 4 mg/kg IM ketamine	Ketamine used alone may induce seizures in previously healthy children in delayed treatment of acute injury.

CT computed tomography, *EEG* electroencephalogram, *ER* emergency room, *GABA* gamma amino butyric acid, *IM* intramuscular, *IV* intravenous, ***min*** minute(s), *s* seconds analgesic. For decades, ketamine has not been used as a mono-anesthetic in clinical anesthesia care anymore because of its unwanted psychoactive effects, including induction of dissociation, especially in the anesthetic dose range. Ketamine is always combined with a GABA agonist or an alpha-adrenergic agent, which is also valid with use at subanesthetic doses in acute and chronic pain. In prehospital emergency care, disaster situations, and poor resource settings, ketamine is also administered together with a GABA agonist if the cardiocirculatory situation is stable enough. The use of a GABA agonist in a patient with a history of epilepsy will, of course, reduce ketamine's pro-epileptic potential in vulnerable scenarios. However, there may be situations in which patients with a history of epilepsy where it appears to be prudent to refrain from ketamine use. These include, for example, asthma patients who take seizure-threshold-lowering drugs, such as theophylline.

In Table 1, two recent case reports of children describe ketamine use for analgesia and sedation without the administration of a GABA agonist. This is not really "understandable," especially in view of the relatively high IM dose applied. Nonetheless, the insight provided from both cases is that in a scenario of approximately 24 h delayed care after acute injury, a relatively high dose of mono-ketamine could potentially induce a seizure in children. As part of an explanatory approach, it needs to be remembered that seizure thresholds in children are lower than their adult counterparts and that the stressful situation of not being able to control the injury by the parents has likely not allowed for a good sleep the night before presenting in the emergency room. We can, therefore, only underscore the importance of combining ketamine with a GABA agonist, administered at best before ketamine, in such situations, even if this may prolong the stay in the hospital.

4.2 Anticonvulsive Effects in Selected Studies and Case Reports in (Super) Refractory Status Epilepticus

As outlined, the pathophysiological conditions in RSE and SRSE are substantially different from those in patients with a history of epilepsy but without a status epilepticus. The results from systematic analysis and case series argue that ketamine possesses clinically valuable anticonvulsant effects in RSE and SRSE (Table 2), although the evidence is of rather low quality.

We, therefore, recommend the use of ketamine as a therapeutic approach to refractory seizures as a sole agent or better in combination with other anticonvulsant agents like midazolam. This enhances ketamine's safety by reducing adverse psychiatric and medical events [42].

4.3 Timing and Dose of Ketamine Administration in (Super) Refractory Status Epilepticus

At present, there is no consensus about the best possible time point for the initiation of ketamine administration or about the" optimum" duration of ketamine treatment in RSE and SRSE. In general, most studies have shown that ketamine administration is most effective if used early after an RSE or SRSE has begun. In a systematic review, ketamine was shown to be effective in 64% of patients if

Table 2
Trial results of ketamine used as antiepileptic medication in RSE and SRSE

Article	Study participants	Treatment	Results and findings	Conclusions
Systematic review and meta-analysis				
Rosati et al. [54]	27 case reports with 30 cases, 14 case series, 6 including children; overall, 268 patients with a median age of 43.5 years	Ketamine and conventional antiepileptic drugs	In adults, ketamine controlled seizures in 156/222 patients (70.3%). In children, ketamine controlled 28/46 RSE episodes (61%). Regardless of the status epilepticus type, ketamine was twice as effective if used early.	The available evidence of therapeutic effects of ketamine in RSE/SRSE is of low quality.
Observational studies				
Dericioglu et al. [70]	7 patients with nonconvulsive RSE/SRSE; age: 44–86 years	Ketamine and midazolam	Ketamine was effective in 4/7 patients, and possibly effective in 1/7; Earlier initiation correlated with higher efficacy ($p = 0.47$)	Ketamine was effective in controlling nonconvulsive RSE/SRSE.
Alkhachroum et al. [24]	261 adult patients with SRSE 68 patients received ketamine	Ketamine was infused at a dose of 2.2 ± 1.8 mg/kg/h over a median duration of 2 (1–4) days together with midazolam	55 patients had at least a 50% reduced seizure burden within 24 h of ketamine start, – 43 patients had seizure cessation with ketamine. After ketamine was stopped, 54 patients achieved at least a 50% decrease in seizure burden, and 44 patients had complete seizure cessation.	Ketamine infusion effectively treated SRSE. High-dose ketamine was effective as an anticonvulsant, improved hemodynamics with less vasopressor needs, and did not negatively affect the intracranial pressure or cerebral blood flow.

Case series				
Santoro et al. [71]	3 patients with anti-NMDA receptor antibody encephalitis and RSE	Ketamine loading dose followed by infusion at a rate of 3 mg/kg/h after multiple other anticonvulsants had failed.	Seizures responded to the addition of IV ketamine. One patient expired 14 days after starting ketamine.	Ketamine could be an effective treatment in RSE associated with anti-NMDA receptor antibody encephalitis.
Souidan et al. [72]	25 patients with status epilepticus	4 patients received ketamine and perampanel. 21 patients received ketamine and other antiepileptics.	Burst suppression pattern was achieved in 75% of patients with ketamine and perampanel. But, burst suppression was reached in only 23.5% of patients with ketamine and other antiepileptics.	A dual anti-glutamatergic regimen could be considered a favorable approach to status epilepticus
Devine et al. [73]	3 young infants with RSE/SRSE after abusive head trauma, ischemic stroke, ischemia after cardiac arrest	Ketamine use in SE refractory to at least 6 other antiepileptic drugs	1 mg/kg ketamine bolus (used 1–2 times) followed by an infusion of 1–6 mg/kg/h was effective to control the electro-clinical seizure presentation in 3/3 of infants.	Continuous ketamine infusion may be considered for infantile RSE and SRSE after failure of first and second-line therapy modalities.
Case reports				
Brisca et al. [74]	A 4-month-old girl with hemimegalencephaly and intractable seizures resistant to multiple antiepileptics	Ketamine and 0.1 mg/kg/h midazolam	1 mg/kg ketamine bolus followed by infusion of up to 1.7 mg/kg/h reduced seizure burden to more than 80% within 24 h.	Ketamine could be effective in controlling hemimegaloencephaly-associated refractory seizures.
Mutkule et al. [75]	An 18-year-old man with cannabis-use-induced SRSE	Ketamine use in SRSE resistant to multiple other conventional anticonvulsants such as thiopentone	1 mg/kg ketamine bolus followed by 2 mg/kg/h infused over 48 h achieved burst suppression pattern; then tapering off over 5 days.	Ketamine is useful in the treatment of cannabis-induced SRSE.

(continued)

Table 2
(continued)

Article	Study participants	Treatment	Results and findings	Conclusions
Huntsman et al. [76]	A neonate with RSE	Ketamine was used in RSE with resistant to many other antiepileptics.	At first, 0.5 mg/kg/h was used, then the dose was increased to 1 mg/kg/h to achieve a burst suppression pattern.	Ketamine could be considered for neonatal refractory seizures, but there is a fear of neurodevelopmental toxicity.
Samanta et al. [77]	2 children, 6 and 8 years old, with alternating hemiplegia of childhood and SRSE	Ketamine use after other antiepileptic drugs failed.	Seizures were controlled in both patients.	Early ketamine use may be considered for the treatment of alternating hemiplegia of childhood-associated SRSE.

NMDA N-mehtyl-D-aspartate, *RSE* refractory status epilepticus, *SRSE* super- refractory status epilepticus administered within the first 3 days of an RSE/SRSE, but this effectiveness decreased to 32% when the mean duration of RSE/SRSE was already 26 days [54]. Another study argued that ketamine is most effective if administered within the first week of SRSE onset [26]. A similar conclusion was drawn in a retrospective study published in 2024, which evaluated 58 patients, of whom 26 (44.8%) had RSE and 32 (55.2%) SRSE treated in hospital between 2017 and 2023. The authors of this study found out that the time elapsed between the onset of RSE/SRSE and ketamine use and the presence of seizures at the initiation of ketamine were decisive factors for a response or a lack of response to ketamine [55]. Mechanical ventilation was not associated with a response to ketamine or seizure recurrence.

Similar to the variable time use schedules, the ketamine doses administered vary among different studies with no general consensus. IV ketamine administration reaches maximum plasma concentrations within 1–5 min [13], and dose calculations were based on pharmacokinetic principles in the following studies: Kremer et al. proposed to administer IV ketamine as a bolus of 50 mg followed by infusion at a rate of 0.6 mg/kg/h. This protocol reduced seizure frequency and duration and was associated with the termination of RSE [56]. Another study concluded that the onset of ketamine effect was 90 min after injection, which suggested that ketamine best be used within the first 24 h of RSE [57]. A different study proposed the use of an average bolus dose of 1.5 mg/kg ketamine and an average infusion rate of 2.75 mg/kg/h for a median of 4 days (range 0–24 days) [26]. The same expert research group stated that the maximum loading dose of ketamine is 5 mg/kg, and the maximum infusion rate is 10 mg/kg/h [26]. Another study found a loading dose of 1–2 mg/kg ketamine followed by an infusion at a dose of 1.3 mg/kg/h (range 0.45–2.1 mg/kg/h) for a median of 9.8 days (range 4–28 days) was successful in achieving control of RSE in 100% of adult patients [58]. Table 3 summarizes the ketamine doses used and proposes a dose regimen to approach RSE/SRSE.

Unlike previous studies [26], Sabharwal et al. [59] investigated the use of ketamine with and without other AEDs. In one of the largest single-center case series studies, the efficacy of a combination therapy of ketamine and propofol for the treatment of SRSE was investigated. Six patients (9%) from the 67 patients studied received ketamine infusion at a dose of 1.5 mg/kg/h in the first 24–48 h for an average period of 6 days, with a maximum ketamine dose of 10.5 mg/kg/h. In the other 61 patients (91%), the treatment was started with a propofol infusion at a dose of 8.4–10.5 mg/kg/h for an average of 6.5 days. The mean time period during which ketamine and propofol were used was 3.6 days. Short-term or long-term infusions of ketamine, with or without propofol, were successful in controlling SRSE. These

Table 3
Proposals for dose regimen and doses used with ketamine in refractory and super-refractory status epilepticus

Study	Proposed/used regimen
Kramer [56]	50 mg IV ketamine bolus followed by infusion at a rate of 0.6 mg/kg/h.
Sheth and Gidal [57]	0.002 mg/kg IV ketamine bolus that could be repeated, followed by continuous infusion at a maximum rate of 0.0075 mg/kg/h.
Synowiec et al. [58]	1–2 mg/kg (max. 5 mg/kg) average IV ketamine bolus followed by infusion at a dose range of 0.45–2.1 mg/kg/h.
Espinosa et al. [9]	70% of studies use a 0.5–5 mg/kg loading dose as a bolus followed by continuous infusion at a dose of 0.05–10.5 mg/kg/h. With enteral administration, doses ranged from 50 to 250 mg twice a day. Ketamine was administered within 24 h after seizure onset up to 140 days after diagnosis of RSE. Treatment duration ranged from 2 h to 29 days. The electroencephalographic response—determined by burst-suppression pattern—was as rapid as 2 h or as delayed as 28 days.

findings propose that such a combination regimen can be a viable strategy to treat SRSE. The authors are of the opinion that a combination of propofol and ketamine may reach a resolution rate of 91% in SRSE.

4.4 Seizure Score to Determine Ketamine Use in (Super) Refractory Status Epilepticus and Prognostic Factors

The status epilepticus severity score (STESS) was designed as a prognostic tool for seizures and plays an essential role in determining the severity of seizures to guide decisions for treatment. STESS works by assessing four simple clinical variables: age of the patient, history of epilepsy, level of consciousness, and seizure type (Table 4).

The higher the STESS score, the higher the mortality from RSE/SRSE. Factors associated with a worse seizure outcome include older age (> 65 years), no previous seizure history, level of consciousness (stupor and coma), nonconvulsive status epilepticus (NCSE), and if the patient was in coma at seizure onset [60].

Overall, the underlying etiology of RSE/SRSE can be decisive in determining the prognosis and response to ketamine treatment. In one study, the presence of febrile infection-related epilepsy syndrome (FIRES), new-onset refractory status epilepticus (NORSE), and infection-related etiologies increased mortality and morbidity [61]. An important retrospective study evaluated the predictors of outcome and mortality in new-onset RSE between 2005 and 2013. Of 114 patients with SE, 52 progressed to RSE. The decisive factor toward the development of refractory status epilepticus was a cryptogenic etiology of the SE, while acidosis was linked to mortality in the entire cohort. The mortality rate was higher in new-onset RSE patients [62].

Table 4
Status Epilepticus Severity Score (STESS) [60]

Clinical feature	Findings	Score value
Level of consciousness at onset	Alert or somnolent; confused Stuporous or comatose	0 1
Seizure type at onset	Simple partial, complex, or absent Generalized convulsive Nonconvulsive	0 1 2
Age at onset	$\leq$ 65 years > 65 years	0 2
History of seizures at onset	Prior seizures No prior seizures	0 1

5 Conclusion

To guide decisions on whether to administer ketamine in the treatment of RSE/SRSE or not, we recommend the consideration of key factors, such as patient age, previous history of seizures, and the underlying cause of the seizures. It is advisable to be careful in the practical use of ketamine for RSE/SRSE, including adapting a titrating, patient-individualized approach. Despite the low quality of evidence in favor of ketamine use, there is a current trend to apply ketamine earlier than as a third-line therapy to increase response and therapeutic success. Ketamine can be used together with or without other antiepileptic drugs. However, we recommend that ketamine be combined with midazolam to prevent unwanted psychiatric and medical effects [25, 63, 64]. Nevertheless, it must also not be forgotten that these are often desperate situations when ketamine is added to a multiple anticonvulsant therapy that has failed to control seizure activity, sometimes over the course of days.

Acknowledgments

The authors thank Ro'a Azzam (Ro'a Azzam Hassouna Mohammad; Faculty of Medicine, MTI University, Cairo, Egypt; roaaazzam14@gmail.com) for graphic design support with Fig. 2.

Conflict of Interest

The authors have no conflict of interest.

References

1. Burman RJ, Rosch RE, Wilmshurst JM, Sen A, Ramantani G, Akerman CJ et al (2022) Why won't it stop? The dynamics of benzodiazepine resistance in status epilepticus. Nat Rev Neurol 18(7):428–441
2. Hirota K, Lambert DG (2022) Ketamine; history and role in anesthetic pharmacology. Neuropharmacology 216(216):109171
3. Jelen LA, Young AH, Stone JM (2021) Ketamine: A tale of two enantiomers. J Psychopharmacol Oxf Engl 35(2):109–123
4. Muller J, Pentyala S, Dilger J, Pentyala S (2016) Ketamine enantiomers in the rapid and sustained antidepressant effects. Ther Adv Psychopharmacol 6(3):185–192
5. Südhof TC, Starke K (eds) (2008) Pharmacology of neurotransmitter release. Handbook of experimental pharmacology. Springer, Berlin, p 184
6. Chang L, Wei Y, Hashimoto K (2020) Antidepressant actions of ketamine and its two enantiomers. In: Hashimoto K, Ide S, Ikeda K (eds) Ketamine. Springer, Singapore, pp 105–125
7. Mion G, Himmelseher S (2024) Esketamine: less drowsiness, more analgesia. Anesth Analg 139(1):78–91
8. Tyler MW, Yourish HB, Ionescu DF, Haggarty SJ (2017) Classics in chemical neuroscience: ketamine. ACS Chem Neurosci 8(6): 1122–1134
9. Espinosa L, Gomez M, Zamora A, Molano-Franco D (2023) Refractory and super-refractory status epilepticus and evidence for the use of ketamine: a scope review. J Neurocritical Care 16(1):1–9
10. Keiser M, Hasan M, Oswald S (2018) Affinity of ketamine to clinically relevant transporters. Mol Pharm 15(1):326–331
11. Nguyen TML, Defaix C, Mendez-David I, Tritschler L, Etting I, Alvarez JC et al (2023) Intranasal (R, S)-ketamine delivery induces sustained antidepressant effects associated with changes in cortical balance of excitatory/inhibitory synaptic activity. Neuropharmacology 225:109357
12. Perez-Ruixo C, Roossenu S, Zannikos P et al (2021) Population pharmacokinetics of esketamine nasal spray and its metabolite noresketamine in healthy subjects and patients with treatment-resistant depression. Clin Pharmacokinet 60:501–516
13. Mion G, Villevieille T (2013) Ketamine pharmacology: an update (pharmacodynamics and molecular aspects, recent findings). CNS Neurosci Ther 19(6):370–380
14. Hijazi Y, Boulieu R (2002) Contribution of CYP3A4, CYP2B6, and CYP2C9 isoforms to *N*- demethylation of ketamine in human liver microsomes. Drug Metab Dispos 30(7): 853–858
15. Schep LJ, Slaughter RJ, Watts M, Mackenzie E, Gee P (2023) The clinical toxicology of ketamine. Clin Toxicol 61(6):415–428
16. Dinis-Oliveira RJ (2017) Metabolism and metabolomics of ketamine: a toxicological approach. Forensic Sci Res 2(1):2–10
17. Zanos P, Moaddel R, Morris PJ, Riggs LM, Highland JN, Georgiou P et al (2018) Ketamine and ketamine metabolite pharmacology: insights into therapeutic mechanisms. Pharmacol Rev 70(3):621–660
18. Highland JN, Farmer CA, Zanos P, Lovett J, Zarate CA, Moaddel R et al (2022) Sex-dependent metabolism of ketamine and (2R,6R)-hydroxynorketamine in mice and humans. J Psychopharmacol Oxf Engl. 36(2): 170–182
19. Saland SK, Kabbaj M (2018) Sex differences in the pharmacokinetics of low-dose ketamine in plasma and brain of male and female rats. J Pharmacol Exp Ther 367(3):393–404
20. Dossat AM, Wright KN, Strong CE, Kabbaj M (2018) Behavioral and biochemical sensitivity to low doses of ketamine: influence of estrous cycle in C57BL/6 mice. Neuropharmacology 130(130):30–41
21. Flint RB, Brouwer CNM, Kränzlin ASC, Lie-A-Huen L, Bos AP, Mathôt RAA (2017) Pharmacokinetics of S-ketamine during prolonged sedation at the pediatric intensive care unit. Paediatr Anaesth 27(11):1098–1107
22. Fang Y, Wang X (2015) Ketamine for the treatment of refractory status epilepticus. Seizure 30:14–20
23. Kirmani BF, Au K, Ayari L, John M, Shetty P, Delorenzo RJ (2021) Super-refractory status epilepticus: prognosis and recent advances in management. Aging Dis 12(4):1097–1119
24. Alkhachroum A, Der-Nigoghossian CA, Mathews E, Massad N, Letchinger R, Doyle K et al (2020) Ketamine to treat super-refractory status epilepticus. Neurology 95(16):e2286–e2294
25. Wu YW, Shek DW, Garcia PA, Zhao S, Johnston SC (2002) Incidence and mortality of generalized convulsive status epilepticus in California. Neurology 58(7):1070–1076

26. Gaspard N, Foreman B, Judd LM, Brenton JN, Nathan BR, McCoy BM et al (2013) Intravenous ketamine for the treatment of refractory status epilepticus: a retrospective multicenter study. Epilepsia 54(8):1498–1503
27. Höfler J, Trinka E (2018) Intravenous ketamine in status epilepticus. Epilepsia 59(S2): 198–206
28. Mohammad Shehata I, Masood W, Nemr N, Anderson A, Bhusal K, Edinoff AN et al (2022) The possible application of ketamine in the treatment of depression in Alzheimer's disease. Neurol Int 14(2):310–321
29. Andrade C (2017) Ketamine for depression, 4: in what dose, at what rate, by what route, for how long, and at what frequency? J Clin Psychiatry 78:e852–e857
30. Blacker SN, Vincent A, Burbridge M, Bustillo M, Hazard SW, Heller BJ et al (2022) Perioperative Care of Patients Undergoing Major Complex Spinal Instrumentation Surgery: clinical practice guidelines from the Society for Neuroscience in anesthesiology and critical care. J Neurosurg Anesthesiol 34(3): 257–276
31. Novy J, Logroscino G, Rossetti AO (2010) Refractory status epilepticus: a prospective observational study. Epilepsia 51(2):251–256
32. Shorvon S, Ferlisi M (2011) The treatment of super-refractory status epilepticus: a critical review of available therapies and a clinical treatment protocol. Brain J Neurol 134(Pt 10): 2802–2818
33. French JA, Krauss GL, Wechsler RT, Wang XF, DiVentura B, Brandt C et al (2015) Perampanel for tonic-clonic seizures in idiopathic generalized epilepsy. Neurology 85(11): 950–957
34. Fujikawa DG (2005) Prolonged seizures and cellular injury: understanding the connection. Epilepsy Behav 7(Suppl 3):S3–S11
35. Herd D, Anderson BJ (2007) Ketamine disposition in children presenting for procedural sedation and analgesia in a children's emergency department. Paediatr Anaesth 17(7): 622–629
36. Naylor DE, Liu H, Niquet J, Wasterlain CG (2013) Rapid surface accumulation of NMDA receptors increases glutamatergic excitation during status epilepticus. Neurobiol Dis 54: 225–238
37. Goodkin HP (2005) Status epilepticus increases the intracellular accumulation of GABAA receptors. J Neurosci 25(23): 5511–5520
38. Niquet J, Baldwin R, Suchomelova L, Lumley L, Eavey R, Wasterlain CG (2017) Treatment of experimental status epilepticus with synergistic drug combinations. Epilepsia 58(4):13695
39. Deeb TZ, Maguire J, Moss SJ (2012) Possible alterations in GABAA receptor signaling that underlie benzodiazepine-resistant seizures. Epilepsia 53(Suppl 9):79–88
40. Feng HJ, Mathews GC, Kao C, Macdonald RL (2008) Alterations of GABA A-receptor function and allosteric modulation during development of status epilepticus. J Neurophysiol 99(3):1285–1293
41. Traynelis SF, Wollmuth LP, McBain CJ, Menniti FS, Vance KM, Ogden KK et al (2010) Glutamate receptor ion channels: structure, regulation, and function. Pharmacol Rev 62(3):405–496
42. Shehata IM, Kohaf NA, ElSayed MW, Latifi K, Aboutaleb AM, Kaye AD (2024) Ketamine: pro or antiepileptic agent? A systematic review. Heliyon 10(2):e24433
43. Proescholdt M, Heimann A, Kempski O (2001) Neuroprotection of S(+) ketamine isomer in global forebrain ischemia. Brain Res 904(2):245–251
44. Zhang JC, Li SX, Hashimoto K (2014) R (−)-ketamine shows greater potency and longer lasting antidepressant effects than S (+)-ketamine. Pharmacol Biochem Behav 116: 137–141
45. Ghosal S, Duman CH, Liu RJ, Wu M, Terwilliger R, Girgenti MJ et al (2020) Ketamine rapidly reverses stress-induced impairments in GABAergic transmission in the prefrontal cortex in male rodents. Neurobiol Dis 134:104669
46. Wang R, Zhang Z, Kumar M, Xu G, Zhang M (2019) Neuroprotective potential of ketamine prevents developing brain structure impairment and alteration of neurocognitive function induced via isoflurane through the PI3K/AKT/GSK-3β pathway. Drug Des Devel Ther 13:501–512
47. Wang CQ, Ye Y, Chen F, Han WC, Sun JM, Lu X et al (2017) Posttraumatic administration of a sub-anesthetic dose of ketamine exerts neuroprotection via attenuating inflammation and autophagy. Neuroscience 343:30–38
48. Reinhart KM, Morton RA, Brennan KC, Carlson AP, Shuttleworth CW (2024) Ketamine improves neuronal recovery following spreading depolarization in peri-infarct tissues. J Neurochem 168(5):855–867
49. Elsaeidy AS, Ahmad AHM, Kohaf NA, Aboutaleb A, Kumar D, Elsaeidy KS et al (2024) Efficacy and safety of ketamine-dexmedetomidine versus ketamine-propofol

combination for Periprocedural sedation: a systematic review and meta-analysis. Curr Pain Headache Rep 28(4):211–227

50. Ilvento L, Rosati A, Marini C, L'Erario M, Mirabile L, Guerrini R (2015) Ketamine in refractory convulsive status epilepticus in children avoids endotracheal intubation. Epilepsy Behav 49:343–346
51. Sutter R, Kaplan PW, Rüegg S (2013) Outcome predictors for status epilepticus – what really counts. Nat Rev Neurol 9(9):525–534
52. Yuan F, Gao Q, Jiang W (2018) Prognostic scores in status epilepticus-a critical appraisal. Epilepsia 59(Suppl 2):170–175
53. Höfler J, Rohracher A, Kalss G, Zimmermann G, Dobesberger J, Pilz G et al (2016) (S)-ketamine in refractory and super-refractory status epilepticus: a retrospective study. CNS Drugs 30(9):869–876
54. Rosati A, De Masi S, Guerrini R (2018) Ketamine for refractory status epilepticus: a systematic review. CNS Drugs 32(11):997–1009
55. García-Ruiz M, Rodríguez PM, Palliotti L, Lastras C, Romeral-Jiménez M, Morales IG et al (2024) Ketamine in the treatment of refractory and super-refractory status epilepticus: experience from two centres. Seizure 117: 13–19
56. Kramer AH (2012) Early ketamine to treat refractory status epilepticus. Neurocrit Care 16(2):299–305
57. Sheth RD, Gidal BE (1998) Refractory status epilepticus: response to ketamine. Neurology 51(6):1765–1766
58. Synowiec AS, Singh DS, Yenugadhati V, Valeriano JP, Schramke CJ, Kelly KM (2013) Ketamine use in the treatment of refractory status epilepticus. Epilepsy Res 105(1–2):183–188
59. Sabharwal V, Ramsay E, Martinez R, Shumate R, Khan F, Dave H et al (2015) Propofol-ketamine combination therapy for effective control of super-refractory status epilepticus. Epilepsy Behav 52(Pt A):264–266
60. Millán Sandoval JP, Escobar Del Rio LM, Gómez EA, Ladino LD, Ospina LML, Díaz DM et al (2020) Validation of the status epilepticus severity score (STESS) at high-complexity hospitals in Medellín, Colombia. Seizure 81:287–291
61. Caranzano L, Novy J, Rossetti AO (2022) Ketamine in adult super-refractory status epilepticus: efficacy analysis on a prospective registry. Acta Neurol Scand 145(6):737–742
62. Jayalakshmi S, Vooturi S, Sahu S, Yada PK, Mohandas S (2016) Causes and outcomes of new onset status epilepticus and predictors of refractoriness to therapy. J Clin Neurosci 26: 89–94
63. McGinn KA, Bishop L, Sarwal A (2016) Use of ketamine in barbiturate coma for status epilepticus. Clin Neuropharmacol 39(1):62–65
64. Walker MC, Howard RS, Smith SJ, Miller DH, Shorvon SD, Hirsch NP (1996) Diagnosis and treatment of status epilepticus on a neurological intensive care unit. QJM: Int J Med 89(12): 913–920
65. Ferrer-Allado T, Brechner VL, Dymond A, Cozen H, Crandall P (1973) Ketamine-induced electroconvulsive phenomena in the human limbic and thalamic regions. Anesthesiology 38(4):333–344
66. Corssen G, Little SC, Tavakoli M (1974) Ketamine and epilepsy. Anesth Analg 53(2): 319–335
67. Celesia GG, Chen RC, Bamforth BJ (1975) Effects of ketamine in epilepsy. Neurology 25(2):169–172
68. Meaden CW, Barnes S (2019) Ketamine implicated in new onset seizure. Clin Pract Cases Emerg Med 3(4):401–404
69. Kim JH, Lee CK, Yu SH, Min BD, Chung CE, Kim DC (2021) Ketamine-induced generalized convulsive seizure during procedural sedation. Arch Craniofacial Surg 22(2): 119–121
70. Dericioglu N, Arslan D, Arsava EM, Topcuoglu MA (2021) Efficacy and safety of ketamine in refractory/super-refractory nonconvulsive status epilepticus: single-center experience. Clin EEG Neurosci 52(5):345–350
71. Santoro JD, Filippakis A, Chitnis T (2019) Ketamine use in refractory status epilepticus associated with anti-NMDA receptor antibody encephalitis. Epilepsy Behav Rep 12:100326
72. Souidan H, Machado RA, Lagnf A, Elsayed M (2022) Double antiglutamatergic treatment in patients with status epilepticus: a case series. Epilepsy Behav 137(Pt A):108954
73. DeVine MN, Gordon SE, Press CA (2023) Use of continuous ketamine infusion as an adjunctive agent in young infants with refractory and super refractory status epilepticus: a case series. J Pediatr Pharmacol Ther 28(2): 161–166
74. Brisca G, Pirlo D, Prato G, Tortora D, Siri L, Nobili L et al (2022) Efficacy of ketamine use

in refractory status epilepticus associated with hemimegalencephaly. Pediatr Emerg Care 38(3):e1090
75. Mutkule DP, Rao SM, Chaudhuri JR, Rajasri K (2018 Jan) Successful use of ketamine for burst suppression in super refractory status epilepticus following substance abuse. Indian J Crit Care Med 22(1):49–50
76. Huntsman RJ, Strueby L, Bingham W (2020 Feb) Are ketamine infusions a viable therapeutic option for refractory neonatal seizures? Pediatr Neurol 103:8–11
77. Samanta D (2020) Ketamine infusion for super refractory status epilepticus in alternating hemiplegia of childhood. Neuropediatrics 51(3):225–228

Chapter 10

Racemic or Esketamine for Treatment of Septic Encephalopathy: First Preclinical and Clinical Research Data and Registered Clinical Trials

Yuming Wu, Shiying Yuan, and Jiancheng Zhang

Abstract

Septic-associated encephalopathy (SAE) is a frequent neurological complication of sepsis, posing significant challenges in clinical management due to its complex pathophysiology and limited therapeutic options. Recent research has explored the potential of both racemic ketamine and its S-enantiomer, esketamine, in mitigating SAE symptoms by modulating inflammation, protecting the blood-brain barrier, and preserving cognitive function. This chapter summarizes preclinical and clinical research findings, as well as ongoing clinical trials, regarding the efficacy and safety of racemic and esketamine in SAE treatment.

Key words Septic-associated encephalopathy, Racemic ketamine, Esketamine, Neuroinflammation

1 Introduction

Sepsis-associated encephalopathy (SAE) is a severe complication of sepsis, presenting with a spectrum of neurological manifestations, ranging from mild confusion and delirium to deep coma and long-term cognitive impairment. SAE is a multifactorial syndrome, characterized by systemic inflammation, oxidative stress, blood-brain barrier (BBB) disruption, and neurotransmitter imbalance [1, 2]. It affects up to 70% of sepsis patients and is associated with adverse outcomes such as increased mortality and potential long-term neurological damage. Symptoms can range from mild confusion to coma, often preceding the clinical presentation of sepsis itself. While death is commonly caused by multiple organ failure, survivors frequently suffer from cognitive and psychological deficits [3].

Therapeutic options for SAE remain limited. In the intensive care unit, sedative treatment is often required, but conventional

The authors declare no competing interests.

Sabine Himmelseher (ed.), *Ketamine: Pain Management, Critical Care, Depression, and Resource-Poor Settings Volume II*, Neuromethods, vol. 225, https://doi.org/10.1007/978-1-0716-4599-4_10,

sedatives and analgesics like benzodiazepines and opioids can exacerbate SAE by impairing cerebral perfusion and disrupting intracranial pressure regulation [4, 5]. These medications may cause hemodynamic instability, complicating the management of SAE [6]. Although electroencephalography (EEG) can detect functional abnormalities, its specificity is limited. Magnetic resonance imaging (MRI) is more sensitive for detecting brain abnormalities such as vasogenic edema. Biomarkers such as neurofilament light chain (NfL) and S100B show promise in quantifying neuronal injury, while cytokines like interleukin (IL)-6, IL-8, IL-10, and tumor necrosis factor-alpha (TNF-α) can predict encephalopathy and prognosis. Neuron-specific enolase (NSE) levels are affected by hemolysis, underscoring the need for non-hemolytic samples in estimating SAE outcomes [7]. The lack of specific treatments has fueled growing interest in alternative therapies targeting the inflammatory mechanisms underlying SAE.

Ketamine, a dissociative anesthetic, has garnered attention for its potential to modulate inflammatory responses, protect the BBB, and reduce neuronal damage, particularly in sepsis-like states induced by lipopolysaccharide (LPS) [8]. Its neuroprotective effects are primarily mediated through antagonism of N-methyl-D-aspartate (NMDA) receptors, which play a critical role in excitotoxicity, a key contributor to neuronal death [9]. Esketamine, the S-enantiomer of ketamine, exhibits higher NMDA receptor affinity and may provide additional neuroprotective benefits while reducing some of the psychotropic side effects associated with racemic ketamine [10].

2 Preclinical Research Data

2.1 Animal Models of Sepsis-Associated Encephalopathy

Preclinical models are crucial for investigating the complex mechanisms underlying sepsis-associated encephalopathy (SAE) and evaluating the therapeutic potential of drugs such as ketamine and esketamine. Animal models, including cecal ligation and puncture (CLP) and lipopolysaccharide (LPS) injection, effectively replicate the pathophysiological features of human sepsis, including systemic inflammation, microcirculatory dysfunction, and blood-brain barrier (BBB) disruption [11].

The CLP model, which induces polymicrobial sepsis, generates a robust inflammatory response that closely mirrors the human condition. Studies using this model have demonstrated significant neuroinflammation, characterized by the activation of microglia and astrocytes, and elevated levels of pro-inflammatory cytokines such as TNF-α, IL-1β, and IL-6 [12]. The activation of glial cells and Toll-like receptor signaling disrupts immune homeostasis and promotes cerebrovascular injury. Increased BBB permeability

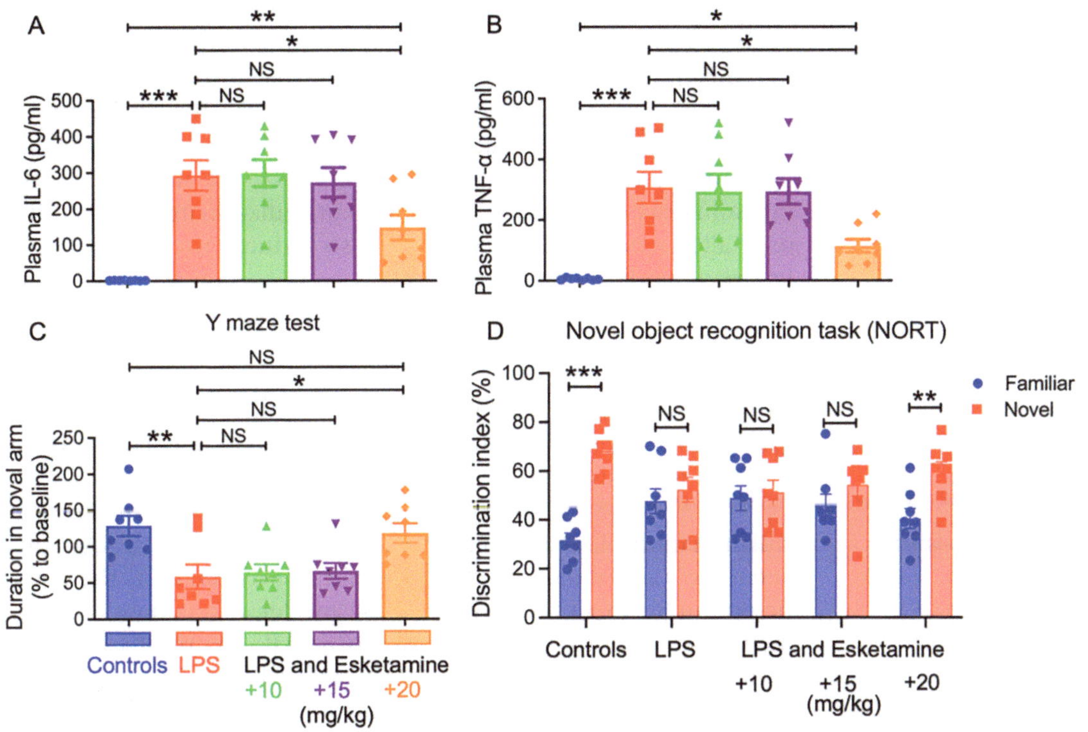

Fig. 1 Esketamine reduced a second lipopolysaccharide (LPS) exposure-induced increase in plasma proinflammatory cytokines (**a**) interleukin (IL-) 6 and (**b**) tumor-necrosis factor (TNF)-α and (**c**, **d**) better preserved cognitive functions. In this simulation of septic conditions, esketamine was therapeutically administered after a first intraperitoneal (IP) injection of LPS which was followed by another LPS injection as a second infectious injury hit model 10 days later. Bars indicate mean and SEM from measurements and behavioral tests performed on study day 11. Especially the highest subanesthetic dose of 30 mg/kg IP esketamine examined (orange bars) was effective in reducing cytokines and improving performance in the Y maze test (YMT) and in the novel object recognition task (NORT). The YMT assesses spatial working memory in rodents based on their natural behavior to explore new environments. The NORT analyzes recognition memory and can be used to test the efficacy of memory enhancing agents. (**c**) shows the duration of stay in the novel arm in the YMT and (**d**) the discrimination index in familiar and novel object recognition (from Wu et al. [16] with permission from Elsevier conveyed via Copyright Clearance Center, Inc.). $^{*}p < 0.05$, $^{**}p < 0.01$, $^{***}p < 0.001$, *NS* not significant

allows inflammatory and toxic mediators to enter the brain, triggering a cascade of events that lead to neuronal damage [13–15].

The LPS model, commonly used to induce endotoxemia, triggers a rapid systemic inflammatory response through the introduction of bacterial endotoxins. Research utilizing the LPS model has demonstrated that esketamine can significantly reduce inflammatory mediator levels and preserve cognitive function [16] (Fig. 1).

Advanced imaging techniques, including magnetic resonance imaging (MRI) and functional imaging, have been employed to assess brain structure and function in animal models of SAE, revealing alterations in metabolic activity and neuronal connectivity [17].

2.2 General Effects of Ketamine in Sepsis

2.2.1 Survival

Preclinical studies have shown that ketamine improves survival in various sepsis models by modulating immune responses and reducing inflammation [18, 19]. Esketamine has also been found to increase the 14-day survival rate in septic mice [20]. For instance, ketamine reduces mortality by decreasing the production of key cytokines, such as TNF-α and IL-6, which are involved in the inflammatory response [21]. This anti-inflammatory effect not only alleviates systemic sepsis but also reduces the risk of organ failure, including brain damage [22].

Additionally, Gurfinkel et al. (2006) demonstrated that ketamine improved survival in a rat model of burn injury followed by sepsis, further supporting its benefits in severe systemic inflammatory response syndrome (SIRS). In this model, ketamine reduced cytokine levels and prevented multiorgan failure, a common cause of death in sepsis [19].

2.2.2 Inflammation

Inflammation plays a critical role in the development of SAE, and ketamine's anti-inflammatory properties are central to its therapeutic potential. In an acute SAE mouse model, esketamine pretreatment significantly reduced hippocampal neuroinflammation, oxidative stress, and neuronal apoptosis [23] (Fig. 2).

Additionally, low-dose ketamine inhibits neuronal apoptosis and neuroinflammation in PC12 cells via the α7nAChR-mediated TLR4/MAPK/NF-κB signaling pathway, highlighting its anti-inflammatory effects [24].

Ketamine suppresses the release of pro-inflammatory cytokines, including TNF-α, IL-1β, and HMGB1, all of which are crucial in SAE progression. In vitro and in vivo studies demonstrated that ketamine significantly inhibited HMGB1 release, reducing inflammation and improving survival in septic animals [25].

Moreover, ketamine mitigates inflammatory responses by inhibiting TNF-α and IL-6 secretion, likely through NMDA receptor antagonism [26, 27]. Ketamine also downregulates HMGB1 expression, a late mediator of sepsis-related inflammation, further reducing systemic inflammation [28, 29]. These findings suggest that ketamine not only provides hemodynamic support but also exerts a protective anti-inflammatory effect, making it a promising candidate for sepsis management. However, further clinical studies are needed to explore its broader therapeutic implications in human sepsis.

Additionally, (R)-ketamine has demonstrated the ability to modulate immune responses in a CLP-induced sepsis model, reducing inflammatory cytokine levels and preventing multiorgan injury [30]. This highlights its systemic effects and potential to improve outcomes in sepsis patients.

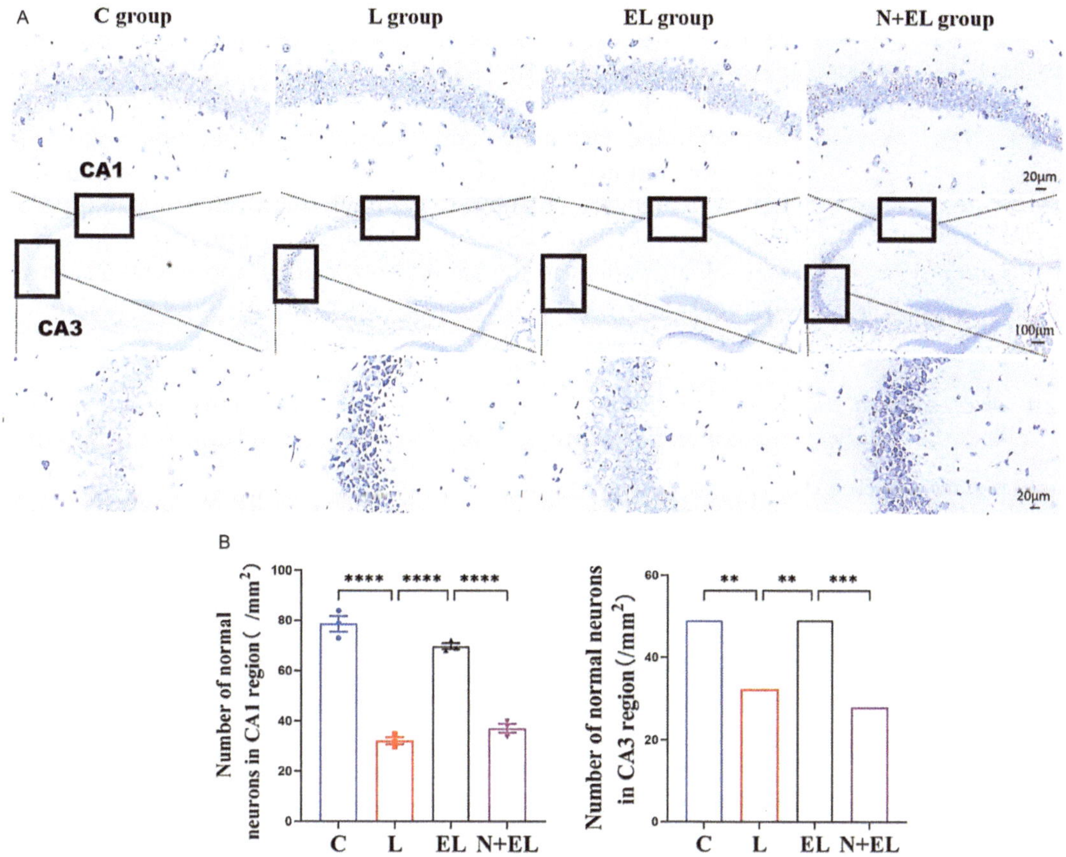

Fig. 2 (**a**) Cresyl violet (Nissl) staining of hippocampal histopathological changes after esketamine pretreatment in a lipopolysaccharide model of sepsis-associated encephalopathy: upper images and lying rectangles show CA1 and lower images and upright rectangles CA3 regions in saline-treated controls (C group), intraperitoneal lipopolysaccharide (LPS) injected animals (L group), esketamine pre-treated and LPS injected animals (EL group), and animals injected with a nuclear factor erythroid 2–related factor 2 (Nrf2) inhibitor, esketamine and LPS (N + EL group). (**b**) The diagram below shows numbers of normal neuron counts in CA1 and right in CA3 regions. Bars indicate mean and standard error of the mean. Repeated esketamine pretreatment protected neurons in LPS-induced acute sepsis, mitigating sepsis-associated encephalopathy. Its neuroprotective effects were almost repealed by an Nrf2 inhibitor. Nrf2 has emerged as an important regulator of cellular resistance to oxidant exposure. It is a key upstream regulator of heme oxygenase-1, which is upregulated in sepsis models exhibiting anti-inflammatory and cytoprotective effects (from Wang et al. [23] with permission from Elsevier conveyed via Copyright Clearance Center, Inc.)

2.2.3 Heart and Microcirculation

Sepsis severely affects the heart and microcirculatory systems, often leading to septic cardiomyopathy. Studies in isolated septic rat heart models have shown that esketamine provides superior cardiovascular stability compared to other induction agents. It exerts minimal negative effects on left ventricular pressure, contractility, and lusitropy across a wide range of concentrations, making it an advantageous choice for managing septic patients with compromised cardiac function [31, 32]. In contrast, agents like propofol significantly reduce cardiac work and increase coronary flow, indicating detrimental vasodilation [32].

Research on microcirculation during endotoxemia indicates that ketamine, even when combined with pentobarbital anesthesia, does not significantly affect intestinal microcirculatory function or leukocyte behavior, helping to maintain microvascular stability during sepsis [31]. These findings emphasize the potential of ketamine as a valuable agent in managing both cardiac and microcirculatory dysfunction in sepsis, offering cardiovascular stability with minimal microvascular interference.

2.2.4 Airway and Lung

Ketamine exerts protective effects on lung function, particularly in the context of acute lung injury (ALI) and inflammation. In septic models, ketamine has been shown to promote autophagy and reduce apoptosis in alveolar epithelial cells by regulating the AMPK/mTOR pathway. This reduces the expression of pro-inflammatory cytokines such as IL-1β, IL-6, and TNF-α, while enhancing the survival of alveolar type II epithelial cells [33].

Ketamine also affects alveolar fluid clearance (AFC), a key mechanism in resolving pulmonary edema. While alveolar administration of esketamine impairs sodium transport and AFC, intravenous esketamine does not impair AFC at clinically relevant doses, even in the presence of endotoxemia [34]. This indicates a safer profile for intravenous use in managing sepsis-induced lung injury. Furthermore, studies on intravenous esketamine in septic rat models confirm that it does not inhibit AFC, reinforcing its potential role in managing lung complications without compromising fluid balance [35].

3 Clinical Research Data

3.1 Anesthesia and General Effects

Ketamine has been extensively studied in clinical settings for its benefits in managing sepsis and septic shock. In a case report of a septic shock patient undergoing anesthesia with ketamine, it was found that ketamine stabilized hemodynamics by increasing arterial pressure and supporting cardiac output, reducing the need for other inotropic agents [36] (Fig. 3).

This highlights ketamine's ability to improve cardiovascular function in critically ill patients. Additionally, ketamine modulates immune responses by inhibiting neutrophil adhesion molecule expression and reducing the production of reactive oxygen species (ROS) and pro-inflammatory cytokines, such as IL-6, thereby mitigating excessive inflammatory responses in sepsis [37].

From an anesthetic management perspective, choosing agents that do not exacerbate hemodynamic instability in septic patients is crucial. Ketamine, along with etomidate, is considered a preferred

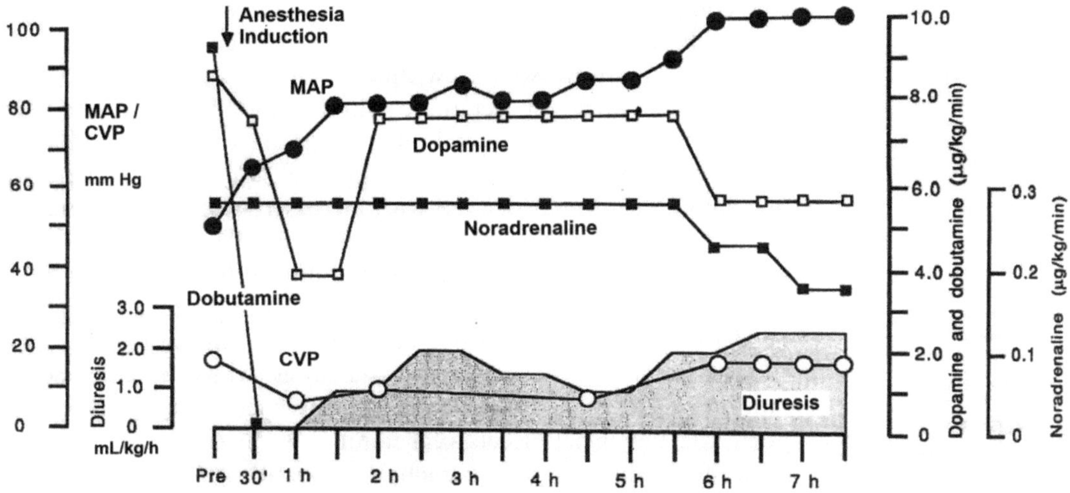

Fig. 3 Typical hemodynamic changes, diuresis, and needs for ionotropic support in a 71-year-old patient with septic shock after anesthesia induction with an adjunctive 40 mg intravenous racemic ketamine bolus and anesthesia maintenance with 0.38–1.0 mg/kg/h continuous infusion combined with diazepam and fentanyl in an early report. After the ketamine bolus, dobutamine could be stopped. During surgery, dopamine and noradrenaline could be reduced while the mean arterial pressure increased. The central venous pressure decreased post-induction but re-increased in the 5 h lasting procedure, at least in part indicating the volume load the patient received during surgery. Excretion of urine started 80 min after induction. In the intensive care unit, ketamine was continued at a dose of 0.5 mg/kg/h until the first postoperative day. The authors report that the patient left hospital in a "satisfactory" condition (from Yli-Hankala et al. [36] with permission from John Wiley and Sons conveyed via Copyright Clearance Center, Inc.)

anesthetic agent due to its limited hemodynamic suppression, unlike propofol, which can cause significant vasodilation and myocardial depression. Ketamine's ability to maintain cardiovascular stability makes it an advantageous choice for the perioperative management of septic patients [38]. Additionally, ketamine anesthesia has been shown to prevent further neurological deterioration in septic patients during surgery, emphasizing its neuroprotective and hemodynamic benefits in critical care and surgical environments [39].

3.2 Anesthesia Induction

The choice of induction agent in septic patients undergoing rapid-sequence intubation is critical due to the risk of hemodynamic instability. Ketamine has demonstrated superiority in maintaining cardiovascular stability compared to agents such as fentanyl. In a randomized controlled trial, patients receiving ketamine had a significantly lower incidence of postintubation hypotension (47.8% vs. 84.2%) compared to those receiving fentanyl, attributed to ketamine's sympathomimetic properties [40].

Etomidate, another commonly used induction agent, is favored for its minimal cardiovascular depression. However, a randomized trial comparing etomidate and ketamine in septic patients showed that a higher proportion of patients in the etomidate group required vasopressors within 24 h postintubation (43.9% vs. 17.7% for ketamine), indicating that ketamine may offer better hemodynamic stability during the acute postintubation phase [41]. While postintubation hypotension (PIH) is a recognized complication associated with increased mortality, ketamine's hemodynamic benefits may play a protective role in preventing severe hypotension [42]. The combination of lidocaine with ketamine has further been shown to reduce severe hypotension and the need for vasopressors during rapid-sequence intubation in septic shock patients [43]. A recent meta-analysis, which included seven randomized trials and one propensity-matched study, further supports the beneficial effects of ketamine as an induction agent in critically ill adults requiring tracheal intubation. The analysis found a moderate probability that induction with ketamine, compared to etomidate, was associated with a reduced risk of mortality. However, more research is required to confirm ketamine's potential for improving clinically relevant outcomes in this context. This finding adds further weight to the use of ketamine, particularly in septic patients, where hemodynamic stability and mortality are critical considerations [44].

3.3 Sedation

In septic shock, effective sedation is critical yet challenging due to associated hemodynamic instability. Ketamine, as NMDA receptor antagonist, provides sedation while preserving respiratory drive and offering analgesia without significant cardiovascular depression. A pilot study evaluating ketamine for sedation in septic shock patients demonstrated multiple benefits, including reduced vasopressor requirements and decreased reliance on adjunctive sedatives like benzodiazepines and opioids [45].

Recent studies have further supported these findings, showing that ketamine improves hemodynamic stability and reduces the need for additional sedative and analgesic medications in critical care settings [46, 47]. Ketamine's favorable safety profile, particularly in mechanically and noninvasive ventilated septic patients, results in fewer respiratory complications compared to traditional sedatives like benzodiazepines [48]. Although some studies suggest a trend toward reduced vasopressor use in ketamine-treated patients, larger randomized controlled trials are needed to confirm these results [45]. Overall, ketamine's combination of sedation, analgesia, hemodynamic stability, and anti-inflammatory effects positions it as a valuable agent for sedation in septic shock [49].

3.4 Specific Brain Effects

Ketamine and especially its S-enantiomer have shown significant neuroprotective effects, particularly in models of neuronal injury. For instance, in studies using rat hippocampal neurons, esketamine enhanced survival and promoted axonal regeneration following glutamate-induced excitotoxicity and axonal injury [50]. This effect was more pronounced with esketamine compared to racemic or R(−)-ketamine, likely due to the S-enantiomer's stronger NMDA receptor antagonism. This may be crucial in mitigating secondary excitotoxic damage observed in conditions such as traumatic brain injury and stroke [51]. Combining esketamine with propofol has been reported to maintain the dynamic cerebrovascular autoregulation [52], which may be important in SAE-related alterations in brain perfusion and impaired cerebrovascular autoregulation [53].

Ketamine's neuroprotective effects extend beyond acute brain injuries to conditions like treatment-resistant depression (TRD). Elevated levels of neurofilament light chain (NFL), a marker of neuronal injury, were reduced in patients with TRD following ketamine treatment, suggesting a link between ketamine's antidepressant effects and its ability to stabilize neuronal integrity [54].

Moreover, ketamine has been shown to prevent neuronal necrosis [55] and attenuate postoperative cognitive dysfunction in patients undergoing major surgeries, including those involving cardiopulmonary bypass [56–59]. Overall, ketamine's neuroprotective, anti-inflammatory, and regenerative properties make it a promising therapeutic agent for both acute brain injuries and chronic neurodegenerative conditions. However, further clinical studies are needed to fully explore its long-term efficacy and safety in human patients.

3.5 Immuno-inflammatory Regulation of Ketamine in Septic Patients

Ketamine has demonstrated significant immunomodulatory effects, particularly in septic patients. In severe sepsis, hyperactivation of the immune system leads to excessive release of pro-inflammatory cytokines such as IL-6, TNF-α, and IL-1β, resulting in systemic inflammation and multiple organ failure. Ketamine has been shown to attenuate this cytokine storm, positioning it as a useful agent in modulating the immune response in sepsis.

A study on mechanically ventilated COVID-19 patients highlighted ketamine's ability to reduce inflammatory markers, especially IL-6, which is closely associated with poor outcomes in both COVID-19 and sepsis [60]. This multicenter study revealed that patients treated with ketamine experienced shorter hospital stays and faster normalization of lactic acid levels, a marker of systemic inflammation and tissue hypoperfusion. These findings suggest that ketamine's anti-inflammatory properties, alongside its sedative and analgesic effects, can significantly improve clinical outcomes in septic patients.

Other studies focused on ketamine use in severe acute respiratory distress syndrome (ARDS) in septic patients. One trial found that ketamine improved sedation depth and provided anti-inflammatory effects without reducing mortality rates [61], but the other one reported reduced mortality [62]. Ketamine's ability to maintain blood pressure and reduce inflammation without increasing the risk of complications, such as acute kidney injury or ventilator-associated pneumonia, positions it as a favorable sedation option for critically ill septic patients.

Overall, ketamine's dual role as a sedative and immune modulator makes it a promising agent for managing sepsis. Its ability to suppress pro-inflammatory cytokines while preserving hemodynamic and respiratory stability offers a comprehensive approach to addressing both the symptoms and underlying immune dysregulation in sepsis.

4 Conclusion

Sepsis-associated encephalopathy (SAE) remains a significant challenge in critical care, with limited therapeutic options and poor patient outcomes (Fig. 4).

Both racemic ketamine and esketamine offer promising therapeutic potential for SAE due to their neuroprotective, anti-inflammatory, and hemodynamic-stabilizing properties. Preclinical and clinical studies have demonstrated that ketamine can reduce neuroinflammation, protect the blood-brain barrier (BBB), and preserve cognitive function, positioning it as a valuable option for SAE treatment.

However, further research is needed to determine the optimal dosing, timing, and administration routes for ketamine and esketamine in SAE patients. Ongoing clinical trials will provide critical data on esketamine's efficacy in reducing inflammation and improving neurological outcomes, potentially paving the way for new therapeutic strategies in SAE management (NCT04843982, NCT04873479, NCT05551910, NCT05839366).

By targeting the underlying mechanisms of SAE, ketamine and esketamine have the potential to improve outcomes for septic patients, reducing the long-term cognitive and neurological sequelae associated with this condition. Continued collaboration between researchers, clinicians, and regulatory bodies will be essential to bring these promising therapies into widespread clinical use.

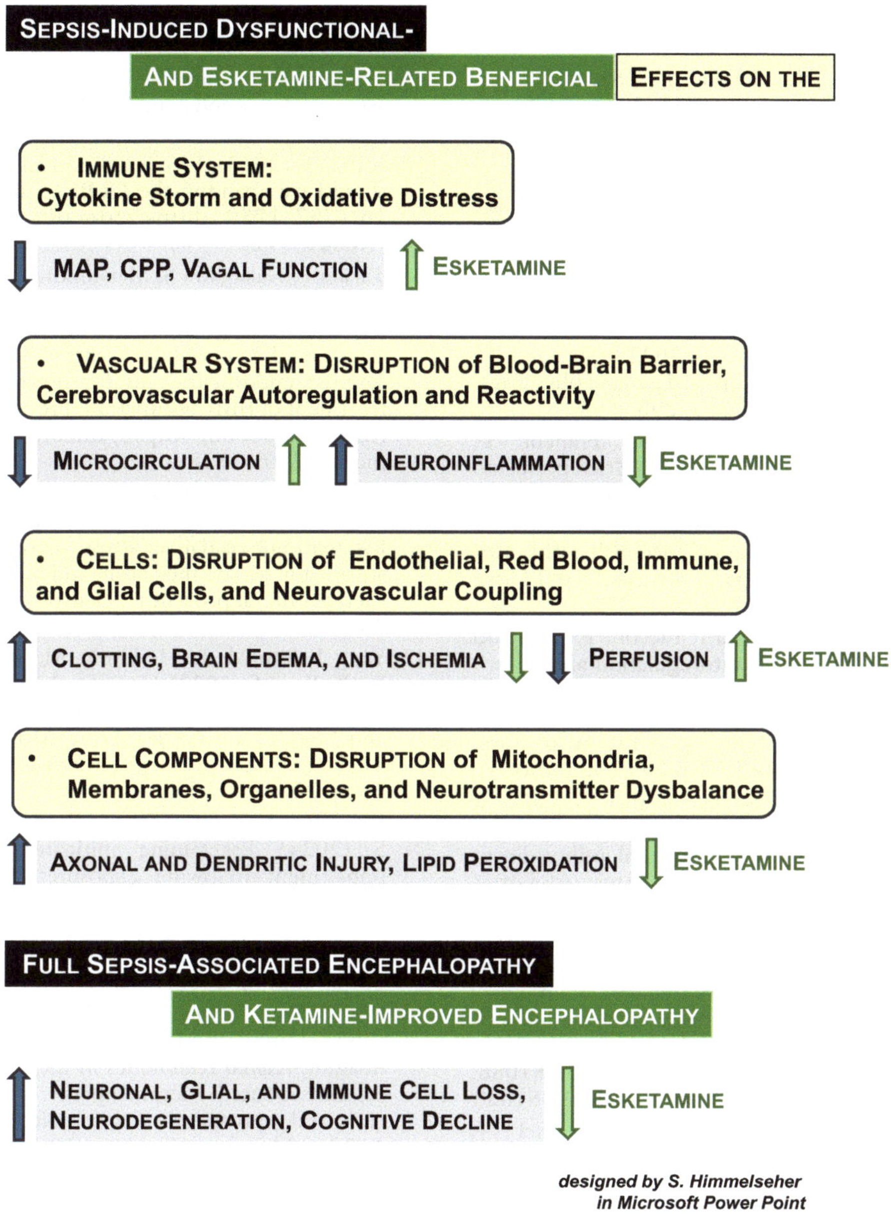

Fig. 4 Synthesis of pathophysiological events and mechanisms underlying the development of sepsis-associated encephalopathy. Often these processes occur in a self-perpetuating, vicious cycle. All contribute to acute sepsis-related brain dysfunction and can enhance neuroinflammation and neurotoxic overexpression of glutamate, finally leading to cell death. In the long-term, they may cause brain atrophy. As outlined in this review, (es-)ketamine can exert beneficial effects at all levels shown in this visualization (created by S. Himmelseher in Power Point, Microsoft Corp., Redmond, WA, USA)

References

1. Krzyzaniak K, Krion R, Szymczyk A, Stepniewska E, Sieminski M (2023) Exploring neuroprotective agents for sepsis-associated encephalopathy: a comprehensive review. Int J Mol Sci 24:10780
2. Dumbuya JS, Li S, Liang L, Zeng Q (2023) Paediatric sepsis-associated encephalopathy (SAE): a comprehensive review. Mol Med 29: 27. https://doi.org/10.1186/s10020-10023-00621-w
3. Carter BL, Underwood J (2022) Sepsis and the brain: a review for acute and general physicians. Clin Med (Lond) 22:392–395
4. Taccone FS, Scolletta S, Franchi F, Donadello K, Oddo M (2013) Brain perfusion in sepsis. Curr Vasc Pharmacol 11:170–186
5. Rivera-Lara L (2019) The role of impaired brain perfusion in septic encephalopathy. Crit Care 23:54. https://doi.org/10.1186/s13054-13018-12299-z
6. Ferlini L, Su F, Creteur J, Taccone FS, Gaspard N (2020) Cerebral autoregulation and neurovascular coupling are progressively impaired during septic shock: an experimental study. Intensive Care Med Exp 8:44. https://doi.org/10.1186/s40635-40020-00332-40630
7. Schütze S, Drevets DA, Tauber SC, Nau R (2023) Septic encephalopathy in the elderly - biomarkers of potential clinical utility. Front Cell Neurosci 17:1238149. https://doi.org/10.3389/fncel.2023.1238149
8. Taheri G, Sardari M, Hermann DM, Sepehri H (2022) N-methyl-D-aspartate receptors antagonist prevents secondary ischemic brain injury associated with lipopolysaccharide-induced sepsis-like state presumably via immunomodulatory actions. Front Cell Neurosci 16:881088. https://doi.org/10.3389/fncel.2022.881088
9. Villéga F, Fernandes A, Jézéquel J, Uyttersprot F, Benac N, Zenagui S, Bastardo L, Gréa H, Bouchet D, Villetelle L et al (2024) Ketamine alleviates NMDA receptor hypofunction through synaptic trapping. Neuron 112:3311–3328.e9. https://doi.org/10.1016/j.neuron.2024.06.028
10. Bieber M, Schwerin S, Kreuzer M, Klug C, Henzler M, Schneider G, Haseneder R, Kratzer S (2022) S-ketamine enhances thalamocortical and corticocortical synaptic transmission in acute murine brain slices via increased AMPA-receptor-mediated pathways. Front Syst Neurosci 16:1044536. https://doi.org/10.3389/fnsys.2022.1044536
11. Qin M, Gao Y, Guo S, Lu X, Zhao Q, Ge Z, Zhu H, Li Y (2023) Establishment and evaluation of animal models of sepsis-associated encephalopathy. World J Emerg Med 14:349–353
12. Joffre J (2023) Preclinical model in sepsis: should we abandon the CLP? J Inflamm Res 16:1757–1759. https://doi.org/10.2147/JIR.S415972
13. Takata F, Nakagawa S, Matsumoto J, Dohgu S (2021) Blood-brain barrier dysfunction amplifies the development of neuroinflammation: understanding of cellular events in brain microvascular endothelial cells for prevention and treatment of BBB dysfunction. Front Cell Neurosci 15:661838. https://doi.org/10.3389/fncel.2021.661838
14. Gu M, Mei XL, Zhao YN (2021) Sepsis and cerebral dysfunction: BBB damage, neuroinflammation, oxidative stress, apoptosis and autophagy as key mediators and the potential therapeutic approaches. Neurotox Res 39: 489–503
15. Yang J, Ran M, Li H, Lin Y, Ma K, Yang Y, Fu X, Yang S (2022) New insight into neurological degeneration: inflammatory cytokines and blood-brain barrier. Front Mol Neurosci 15:1013933. https://doi.org/10.3389/fnmol.2022.1013933
16. Wu Y, Zhang Y, Xie B, Zhang X, Wang G, Yuan S (2024) Esketamine mitigates cognitive impairment following exposure to LPS by modulating the intestinal flora/subdiaphragmatic vagus nerve/spleen axis. Int Immunopharmacol 126:111284. https://doi.org/10.1016/j.intimp.2023.111284
17. Wu S, Wang Y, Song Y, Hu H, Jing L, Zhu W (2023) Application of magnetic resonance imaging-related techniques in the diagnosis of sepsis-associated encephalopathy: present status and prospect. Front Neurosci 17: 1152630. https://doi.org/10.3389/fnins.2023.1152630
18. Taniguchi T, Takemoto Y, Kanakura H, Kidani Y, Yamamoto K (2003) The dose-related effects of ketamine on mortality and cytokine responses to endotoxin-induced shock in rats. Anesth Analg 97:1769–1772
19. Gurfinkel R, Czeiger D, Douvdevani A, Shapira Y, Artru AA, Sufaro Y, Mazar J, Shaked G (2006) Ketamine improves survival in burn injury followed by sepsis in rats. Anesth Analg 103:396–402. https://doi.org/10.1007/s44231-024-00060-z
20. Xie B, Wang M, Zhang X, Zhang Y, Qi H, Liu H, Wu Y, Wen X, Chen X, Han M et al (2024) Gut-derived memory γδ T17 cells

exacerbate sepsis-induced acute lung injury in mice. Nat Commun 15:6737

21. Deng B, Yang D, Wu H, Wang L, Wu R, Zhu H, Huang A, Song J, Cai T, Liu S et al (2022) Ketamine inhibits TNF-α-induced cecal damage by enhancing RIP1 ubiquitination to attenuate lethal SIRS. Cell Death Discov 8:72. https://doi.org/10.1038/s41420-41022-00869-x
22. Zhang J, Ma L, Wan X, Shan J, Qu Y, Hashimoto K (2021) (R)-ketamine attenuates LPS-induced endotoxin-derived delirium through inhibition of neuroinflammation. Psychopharmacology 238:2743–2753
23. Wang CM, Zhang Y, Yang YS, Lin S, He HF (2024) Effect of esketamine pretreatment on acute sepsis-associated encephalopathy. Exp Neurol 372:114646. https://doi.org/10.1016/j.expneurol.2023.114646. Epub 112023 Dec 114647
24. Zhao J, Zhang R, Wang W, Jiang S, Liang H, Guo C, Qi J, Zeng H, Song H (2023) Low-dose ketamine inhibits neuronal apoptosis and neuroinflammation in PC12 cells via α7nAChR mediated TLR4/MAPK/NF-κB signaling pathway. Int Immunopharmacol 117:109880. https://doi.org/10.1016/j.intimp.2023.109880
25. Zhang Z, Zhang L, Zhou C, Wu H (2014) Ketamine inhibits LPS-induced HGMB1 release in vitro and in vivo. Int Immunopharmacol 23:14–26. https://doi.org/10.1016/j.intimp.2014.1008.1003
26. Besnier E, Clavier T, Tonon MC, Selim J, Lefevre-Scelles A, Morin F, Tamion F, Dureuil B, Castel H, Compere V (2017) Ketamine and etomidate down-regulate the hypothalamic-pituitary-adrenal axis in an endotoxemic mouse model. Anesthesiology 127:47–354
27. Spentzas T, Shapley RK, Aguirre CA, Meals E, Lazar L, Rayburn MS, Walker BS, English BK (2011) Ketamine inhibits tumor necrosis factor secretion by RAW264.7 murine macrophages stimulated with antibiotic-exposed strains of community-associated, methicillin-resistant Staphylococcus aureus. BMC Immunol 12:11. https://doi.org/10.1186/1471-2172-1112-1111
28. DeWulf B, Minsart L, Verdonk F, Kruys V, Piagnerelli M, Maze M, Saxena S (2023) High mobility group box 1 (HMGB1): potential target in sepsis-associated encephalopathy. Cells 12:1088
29. Li K, Yang J, Han X (2016) Ketamine attenuates sepsis-induced acute lung injury via regulation of HMGB1-RAGE pathways. Int Immunopharmacol 34:114–128. https://doi.org/10.1016/j.intimp.2016.1001.1021
30. Zhang J, Ma L, Hashimoto Y, Wan X, Shan J, Qu Y, Hashimoto K (2021) (R)-ketamine ameliorates lethal inflammatory responses and multi-organ injury in mice induced by cecum ligation and puncture. Life Sci 284:119882. https://doi.org/10.1016/j.lfs.2021.119882
31. Lehmann C, Feyerherd F, Feyerherd T, Fogliata M, Gründling M, Usichenko TI, Meissner K, Wendt M, Pavlovic D (2007) Ketamine does not affect intestinal microcirculation in pentobarbital-anaesthetized rats during experimental endotoxaemia. Lab Anim 41:55–62. https://doi.org/10.1258/002367707779399491
32. Zausig YA, Busse H, Lunz D, Sinner B, Zink W, Graf BM (2009) Cardiac effects of induction agents in the septic rat heart. Crit Care 13:R144
33. Cao J, Mai H, Chen Y, Yuan R, Fang E, Liu M, Fu C (2022) Ketamine promotes LPS-induced pulmonary autophagy and reduces apoptosis through the AMPK/mTOR pathway. Contrast Media Mol Imaging 2022:8713701. https://doi.org/10.1155/2022/8713701
34. Berger MM, Pitzer B, Zügel S, Wieland CW, Vlaar AP, Schultz MJ, Dahan A, Bärtsch P, Hollmann MW, Mairbäurl H (2010) Alveolar but not intravenous S-ketamine inhibits alveolar sodium transport and lung fluid clearance in rats. Anesth Analg 111:164–170
35. Fastner C, Mairbäurl H, Weber NC, van der Sluijs K, Hackl F, Hotz L, Dahan A, Hollmann MW, Berger MM (2014) Intravenous S-ketamine does not inhibit alveolar fluid clearance in a septic rat model. PLoS One 9:e112622
36. Yli-Hankala A, Kirvelä M, Randell T, Lindgren L (1992) Ketamine anaesthesia in a patient with septic shock. Acta Anaesthesiol Scand 36:483–485
37. Weigand MA, Schmidt H, Zhao Q, Plaschke K, Martin E, Bardenheuer HJ (2000) Ketamine modulates the stimulated adhesion molecule expression on human neutrophils in vitro. Anesth Analg 90:206–212
38. Carsetti A, Vitali E, Pesaresi L, Antolini R, Casarotta E, Damiani E, Adrario E, Donati A (2023) Anesthetic management of patients with sepsis/septic shock. Front Med (Lausanne) 10:1150124. https://doi.org/10.3389/fmed.2023.1150124
39. Sokołowska EM, Wityk P, Szypenbejl J, Petrosjan R, Raczak-Gutknecht J, Waszczuk-Jankowska M, Dudzik D, Markuszewski M, Siemiński M (2024) Clinical image of sepsis-

associated encephalopathy midst E. coli urosepsis: emergency department database study. Heliyon 10:e29530

40. Ali H, Abdelhamid BM, Hasanin AM, Amer AA, Rady A (2021) Ketamine-based versus fentanyl-based regimen for rapid-sequence endotracheal intubation in patients with septic shock: a randomised controlled trial. Rom J Anaesth Intensive Care 28:98–104
41. Srivilaithon W, Bumrungphanithaworn A, Daorattanachai K, Limjindaporn C, Amnuaypattanapon K, Imsuwan I, Diskumpon N, Dasanadeba I, Siripakarn Y, Ueamsaranworakul T et al (2023) Clinical outcomes after a single induction dose of etomidate versus ketamine for emergency department sepsis intubation: a randomized controlled trial. Sci Rep 13:6362
42. Tangkulpanich P, Angkoontassaneeyarat C, Trainarongsakul T, Jenpanitpong C (2023) Factors associated with postintubation hypotension among patients with suspected sepsis in emergency department. Open Access Emerg Med 15:427–436. https://doi.org/10.2147/OAEM.S426822
43. Fathy S, Hasanin A, Mostafa M et al (2021) The benefit of adding lidocaine to ketamine during rapid sequence endotracheal intubation in patients with septic shock: a randomised controlled trial. Anaesth Crit Care Pain 40:100731. https://doi.org/10.1016/j.accpm.2020.06.017
44. Koroki T, Kotani Y, Yaguchi T, Shibata T, Fujii M, Fresilli S, Tonai M, Karumai T, Lee TC, Landoni G, Hayashi Y (2024) Ketamine versus etomidate as an induction agent for tracheal intubation in critically ill adults: a Bayesian meta-analysis. Crit Care 28:48
45. Reese JM, Sullivan VF, Boyer NL, Mount CA (2018) A non-comparative prospective pilot study of ketamine for sedation in adult septic shock. Mil Med 183:e409–e413
46. Phan B, Agyemang A, Klein W, Thapamagar SB (2024) Continuous infusion of ketamine in mechanically ventilated patients with SARS-CoV-2. Crit Care Res Pract 2024:7765932
47. Chan K, Burry LD, Tse C et al (2022) Impact of ketamine on analgosedative consumption in critically ill patients: a systematic review and meta-analysis. Ann Pharmacother 56:1139–1158. https://doi.org/10.1177/10600280211069617
48. Riccardi A, Serra S, De Iaco F, Fabbri A, Shiffer D, Voza A (2023) Uncovering the benefits of the ketamine-dexmedetomidine combination for procedural sedation during the Italian COVID-19 pandemic. J Clin Med 12:3124
49. Peng Z, Sun T, Zhang X, Zhang H (2024) Expert consensus on the clinical utilization of ketamine and its isomers in intensive care units. Intensive Care Res 4:1–8
50. Himmelseher S, Pfenninger E, Georgieff M (1996) The effects of ketamine-isomers on neuronal injury and regeneration in rat hippocampal neurons. Anesth Analg 83:505–512
51. Sánchez-Porras R, Kentar M, Zerelles R (2022) Eighteen-hour inhibitory effect of s-ketamine on potassium- and ischemia-induced spreading depolarizations in the gyrencephalic swine brain. Neuropharmacology 216:109176. https://doi.org/10.1016/j.neuropharm.2022.109176
52. Engelhard K, Werner C, Möllenberg O, Kochs E (2001) S(+)-ketamine/propofol maintain dynamic cerebrovascular autoregulation in humans. Can J Anaesth 48:1034–1039
53. Crippa IA, Salvagno M, Calabro L, Taccone F (2024) Brain hemodynamics in septic patients. Minerva Anestesiol 90:814–829
54. Lin WC, Su TP, Li CT, Wu HJ, Bai YM, Liu YL, Tu PC, Chen MH (2023) Association of neurofilament light chain with the antidepressant effects of low-dose ketamine infusion among patients with treatment-resistant depression. Int J Neuropsychopharmacol 26:649–653
55. Giuliano K, Etchill E, Velez AK, Wilson MA, Blue ME, Troncoso JC, Baumgartner WA, Lawton JS (2023) Ketamine mitigates neurobehavioral deficits in a canine model of hypothermic circulatory arrest. Semin Thorac Cardiovasc Surg 35:251–258
56. Han C, Haiyan J, Guo Y et al (2023) Effect of subanesthetic dose of esketamine on perioperative neurocognitive disorders in elderly undergoing gastrointestinal surgery: a randomized controlled trial. Drug Des Develop Ther 17:863–873
57. Wang Y, Ma B, Wang C et al (2024) The influence of low-dose s-ketamine on postoperative delirium and cognitive function in older adults undergoing thoracic surgery. J Cardiothorac Surg 19:324. https://doi.org/10.1186/s13019-024-02811-x
58. Welters ID, Feurer MK, Preiss V et al (2011) Continuous S-(+)-ketamine administration during elective coronary artery bypass graft surgery attenuates pro-inflammatory cytokine response during and after cardiopulmonary bypass. Br J Anaesth 106:172–179. https://doi.org/10.1093/bja/aeq341
59. Xiong X, Shao Y, Chen D et al (2024) Effect of Esketamine on postoperative delirium in patients undergoing cardiac valve replacement

with cardiopulmonary bypass: a randomized controlled trial. Anesth Analg 139:743–753

60. Aljuhani O, Al Sulaiman K, Korayem GB, Altebainawi AF, Alshaya A, Nahari M, Alsamnan K, Alkathiri MA, Al-Dosari BS, Alenazi AA et al (2024) Ketamine-based sedation use in mechanically ventilated critically ill patients with COVID-19: a multicenter cohort study. Saudi Pharm J 32:102061

61. Wyler D, Torjman MC, Leong R et al (2023) Observational study of the effects of ketamine infusions on sedation depth, inflammation, and clinical outcomes in mechanically ventilated patients with SARS-CoV-2. Anaesth Intensive Care 52:105–112. https://doi.org/10.1177/0310057X231201184

62. Pata R, Sandeep P, Aung HM, Patel MJ, Dolkar T, Nway N, Bhanu K, Ahmady A, Kiani R, Swaroop R (2021) Ketamine infusion as a sedative-analgesic in severe ARDS (KISS). J Community Hosp Intern Med Perspect 11: 619–623

Part III

Difficult-to-Treat Depression and Other Severe Mental Health Disorders

Chapter 11

The Global Prevalence of Depression Calls for Action: A Look to the Situation and Specifics with Ketamine

Sabine Himmelseher, Chimaobi Tim Nnaji, and Aletta M. E. Millen

Abstract

The Global Burden of Disease, Injuries, and Risk Factors Study published in the Lancet in 2024 for the period 2010 to 2021 identified anxiety and depressive disorders as the most prevalent and among the leading 25 causes of morbidity according to age-standardized, disability-adjusted life-years. Mental illness and distress are associated with premature mortality and an increased risk of suicide and likelihood of physical comorbidities. Despite the availability of antidepressant treatments for decades, remission rates of depression remain low and relapse rates high. Only approximately 15% of patients receive minimally adequate treatment. This introductory chapter will provide an overview of the global situation and specifics with ketamine treatment for severe depressive disorders. Obstacles to implement ketamine therapy more widely will be described and steps that should be prioritized for implementation will be presented.

Key words Depression, Global prevalence, Ketamine therapy, Implementation, Regulatory issues

1 Introduction

The most recent Global Burden of Disease, Injuries, and Risk Factors Study published in the Lancet in April 2024 for the period 2010 to 2021 listed anxiety (17%) and depressive disorders (16%) as the most increased and among the leading 25 causes of morbidity according to age-standardized, disability-adjusted life-years [1]. Despite the availability of antidepressant treatments for decades, remission rates are low and relapse rates high, and only approximately 15% of patients may receive minimally adequate treatment [1–3]. The widespread lockdown and impact of social isolation during the coronavirus disease 2019 (COVID-19) pandemic contributed to an overall increase in the prevalence of depression. This included even postpartum depression, which was found to be as high as 31% regardless of the country's developmental index [4]. In general, mental illness and distress are associated

Sabine Himmelseher (ed.), *Ketamine: Pain Management, Critical Care, Depression, and Resource-Poor Settings Volume II*, Neuromethods, vol. 225, https://doi.org/10.1007/978-1-0716-4599-4_11,

with an increased likelihood of premature mortality, an elevated risk of suicide, and a greater prevalence of physical comorbidities [5].

Depression and mental health disorders are indiscriminate of age. In most recent publications, the prevalence of depression among older adults residing in care facilities in China was reported to be as high as 37% [6], while in young Australians aged between 10 and 24 years, it was as high as 25% [7]. Moreover, low- and middle-income countries are estimated to bear the greatest burden of mental health disorders and suicide, in part due to under-resourced public healthcare systems. In a rural Brazilian population, the prevalence of depressive symptoms was approximately 17%, with pesticide poisoning identified as a significant risk factor [8]. Although there is a dearth of data on the epidemiology of depression in Africa, its prevalence is likely grossly underestimated [9]. Taken together, the time to act is now, and the need for improvements to therapy feels urgent.

The precise mechanisms underlying depression remain incompletely elucidated. However, its phenomenological presentation can be understood as the result of a complex interaction between multiple risk, genetic, molecular, environmental, and developmental factors, which are unique to each individual. Figure 1 illustrates the brain regions that are involved in mediating the features of depression [2].

In view of the heterogeneous mechanisms of depression, it is not surprising that traditional antidepressants have a high failure rate. After the therapeutic "breakthrough"of demonstrating ketamine's rapid antidepressant effects in treatment-resistant depression (TRD) and major depressive disorder, the results of clinical studies and real-world observations from the last two decades have raised hopes to turn a corner in the management of severely depressed patients with the substance [10, 11]. Despite the ongoing debate regarding the precise definition of TRD, it is typically conceptualized as a lack of response or achievement of remission from depression following the utilization of two or more antidepressant agents with disparate mechanisms of action [12]. The administration of ketamine frequently elicits a response exhibiting a greater than 50% improvement in depressive symptoms in patients who have not responded to conventional pharmaceutical combinations and other augmentation therapies. The beneficial effects of ketamine may even be extended to other severe neuropsychiatric conditions [13] and emerging medical indications [3]. Figure 2 presents a synthesis of the rapid molecular antidepressant effects of ketamine, which have been described in successful applications [3] and are discussed in detail in the various chapters of this textbook.

Nonetheless, the global interest and necessity to implement novel antidepressant strategies will have to overcome considerable obstacles with ketamine in the transition from research and

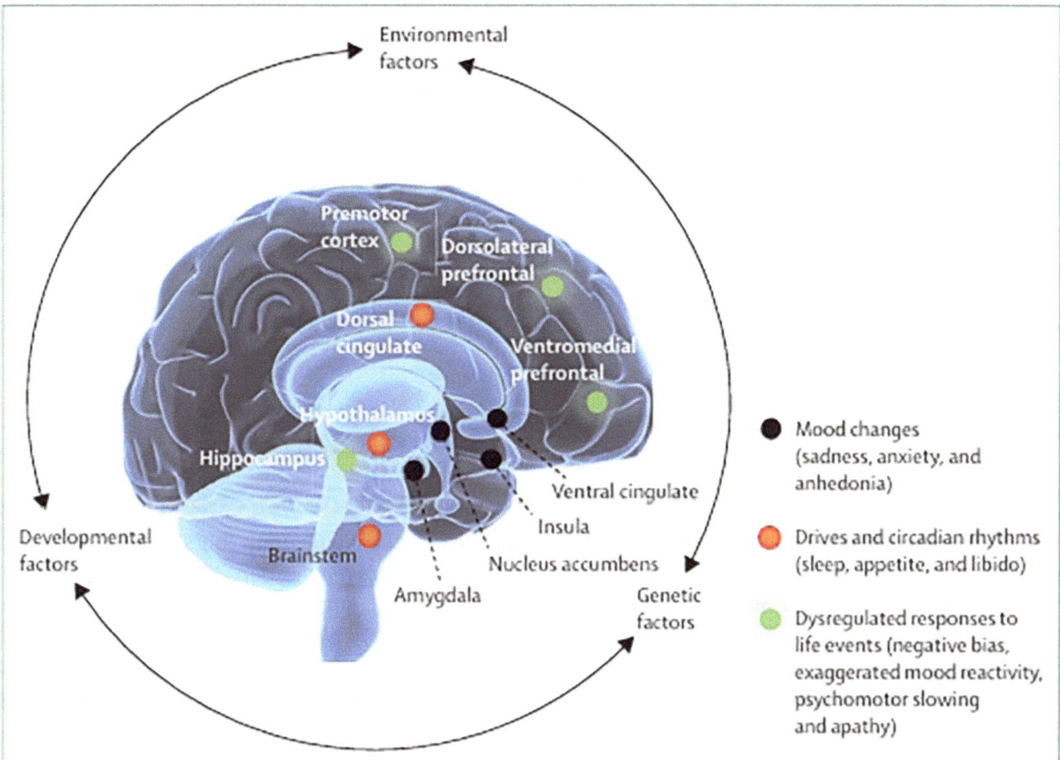

Fig. 1 Depression results from multiple interacting factors which are unique for each individual. Eventually, pathways converge and get expressed in brain regions that mediate depressive features. Patterns are highly variable and may reflect the underlying etiology, different illness severities, and disorder persistence. Likely, they indicate the heterogeneous expression of mood, motor, cognitive, and vegetative depressive symptomatology (from Herrman et al. [2] with permission from Elsevier conveyed through Copyright Clearance Center, Inc.)

specialist centers to global clinical practice in diverse settings: **(1)** Intravenous (IV) worldwide-used racemic ketamine has not officially been approved for clinical use in depression and other neuropsychiatric disorders. It is currently applied as an off-label treatment. Before treating a patient with racemic ketamine, it is therefore important to clarify the coverage, payment, and conditions in case of legal issues provided by the patient's health insurance. On the other hand, S(+)-ketamine (esketamine), the approximately two times more potent antidepressant, analgesic, and anesthetic stereoisomer of racemic ketamine, has been granted authorization in over 70 countries, including the USA, Canada, Europe, Australia, New Zealand, South America, and Asia for the therapy of TRD, major depression, and depression with suicidal ideation as a patented intranasal (IN) preparation [10, 11, 14]. Despite the higher costs, IN esketamine is typically covered on a patient-by-patient basis by most insurances, for example, in the

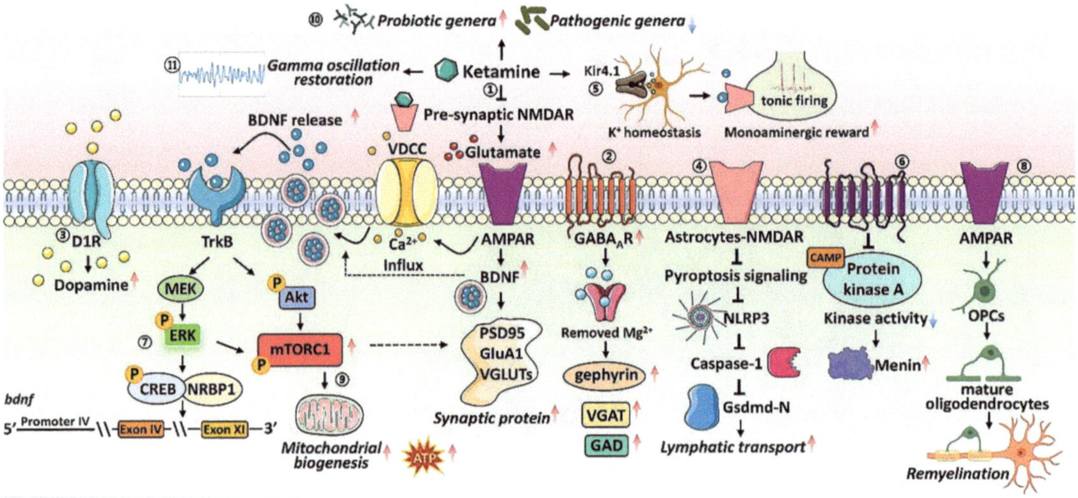

Fig. 2 When used as rapid acting antidepressant (**1**) ketamine induces a glutamate "burst" increasing brain-derived neurotrophic factor (BDNF) and synaptic proteins (e.g., postsynaptic density protein 95 (PSD95), glutamate (Glu) A1, vesicular glutamate transporters (VGLUTs)) and (**2**) amplifies gamma-amino-butyric acid (GABA) function augmenting levels of vesicular GABA transporter (VGAT), glutamic acid decarboxylase (GAD), and gephyrin; (**3**) ketamine causes increased firing of dopaminergic neurons and dopamine release; (**4**) it reverses lymphatic dysfunction by decreasing the expression of pyroptosis-related proteins NLRP3/Caspase-1/Gsdmd-N and (**5**) modulates the potassium (K^+) homeostasis through reduction of astrocyte-Kir4.1 (astrocyte channel regulating synaptic K^+ buffering) with the result of relieved inhibition of downstream monoaminergic reward centers. (**6**) Ketamine also stabilizes menin levels by inhibiting protein kinase A activity thus maintaining its antidepressant effect (menin is primarily a nuclear scaffold protein which regulates transcription by coordinating chromatin remodeling). (**7**) Ketamine's sustained antidepressant responses are also related to microglial ERK-NRBP1-CREB-BDNF signaling; (**8**) in addition, ketamine promotes the differentiation of oligodendrocyte precursor cells (OPCs) into mature oligodendrocytes through activation of α-amino-3-hydroxy-5-methyl-4-isoxazolepropionic acid receptor (AMPAR) signaling, thus facilitating remyelination; (**9**) Ketamine activates mammalian target of rapamycin complex 1 (mTORC1) to promote anabolic processes such as protein synthesis and mitochondrial biogenesis; (**10**) with regard to the brain-gut-microbiota axis, ketamine may increase probiotic genera and decrease pathogenic genera; (**11**) it also ameliorates depression-like behaviors in animal models though restoration of gamma oscillations (from Deng et al. [3] under a CC BY 4.0 international commercial license; https:/creativecommons.org/licenses/by/4.0/). *Akt* protein kinase B, *CAMP* cyclic adenosine monophosphate, *CREB* CAMP-responsive element binding protein, *Ca2+* calcium cation, *D1R* dopamine 1 receptor, *ERK* extracellular signal-regulated kinase *Gsdmd-N* gasdermin-N pore forming protein = executioner of lytic pyroptotic cell death, *MEK* mitogen-activated protein kinase kinase, *Mg2+* magnesium cation, *NMDAR* N-methyl-D-aspartate receptor, *NLRP3* inflammasome pyrin-domain containing multiprotein 3 complex, *NRBP1* nuclear receptor binding protein 1, *TrkB* tropomyosin receptor kinase B, *VDCC* voltage-gated calcium channel

USA. Until recently, the prerequisites for treatment in the USA necessitated that patients receive an oral antidepressant concomitantly, but in January 2025, the FDA expanded its approval for IN esketamine as standalone treatment in major depression. Anyhow, in order to achieve the maximum prescribed dose of 84 mg per session, up to three activations of the nasal spray device are required, with a 2-h observation period following administration. Nevertheless, the number of prescriptions for esketamine increased by approximately 2.7 times between 2020 and 2022 solely within the USA [14]. One advantage of the IN application route is that it obviates the need for insertion of an IV line. Furthermore, slow-release oral racemic ketamine preparations are already undergoing phase 2 trials [15]. Head-to-head comparisons of the efficacy of antidepressant doses of IV racemic ketamine versus IN esketamine, are, however, almost impossible. This is true not only because the bioavailability of these drugs differs depending on the route of administration but also because the formation of metabolites varies due to route-dependent differences, and numerous additional factors (*please see* chapter **13** by Singh et al. or chapter **17** by Zaremba et al). In the meantime, the data from recent real-world observations have corroborated the antidepressant efficacy of esketamine, as previously reported in clinical trials, without the emergence of any unexpected safety signals [14, 16]. The sustained improvement in depressive symptoms observed with flexible IN esketamine dosing has now been reported for a period of up to 6.5 years. This has been associated with a reduction in mortality and hospitalization due to psychiatric causes when compared to previous investigations of standard care for TRD [16]. However, the interpretation of the low relapse rates observed in long-term safety studies is significantly limited by open-label trial designs and adjustments to oral medications employed. Furthermore, discontinuations due to tolerability or loss of efficacy must be considered, although lack of efficacy was seldomly reported (usually <5%). Adverse effects were perceived to be more a "transient annoyance" than a significant medical or psychological issue. Nevertheless, it seems unlikely that IN patented esketamine devices will be accessible or affordable in low- and middle-income countries in the near future. **(2)** Racemic ketamine has been illicitly employed as a "recreational substance" and is a scheduled drug in numerous countries, with discrepancies in dispensing regulations even across European Medicine Agency (EMA) member states. The psychiatric literature has issued cautionary advice regarding the addictive potential of ketamine, particularly in less supervised use. Nevertheless, it must be acknowledged that the risks associated with misuse are not different from those of other established, commonly prescribed agents such as benzodiazepines [17]. In professionally controlled settings, limited evidence from single or repeated ketamine therapy indicates that dependence, diversion, and gateway activity have not been observed in patients with TRD [18]. Nevertheless, to date, there is a paucity of well-structured studies evaluating clinical adverse effects of ketamine, particularly in patients treated with the racemate. The implementation of an oversight system across healthcare facilities offering ketamine therapies to achieve surveillance therefore appears to be necessary, particularly for private treatment settings without official control [19]. **(3)** The psychoactive effects of ketamine represent a significant concern for many practitioners, particularly those lacking personal experience with its administration to awake patients. Nevertheless, ketamine is employed at a low—subanesthetic dose for antidepressant purposes, and all effects of ketamine are dose- and context-dependent. Currently, the most evidence-based protocol involves a 40-min infusion of 0.5 mg/kg IV racemic ketamine, during which many patients experience perceptual disturbances and minor dissociative phenomena. The adverse effects typically dissipate within 1 h following infusion. It is advised that patients be monitored for a minimum of that duration. However, the dosage and frequency of ketamine infusions are currently unregulated, and uses of higher doses have been documented. Esketamine is currently accessible solely through a restricted distribution system under a "Risk and Mitigation Strategy" delineating safety standards in the USA. A post-treatment observation period of 2 h following IN esketamine is therefore mandated. Nevertheless, with the implementation of pre-treatment patient preparation, guidance, and support during psychoactive events, as well as the incorporation of potential adverse effects into the therapy by ketamine-experienced healthcare professionals, no additional intervention is typically required. This is valid for IV racemic and IN esketamine. Interestingly, a recent secondary safety and tolerability analysis of the phase III trial comparing esketamine nasal spray with the atypical neuroleptic quetiapine in patients with TRD revealed a higher incidence of adverse effects with esketamine [20]. However, these were typically mild to moderate, transient, and more likely to resolve on the day of therapy as compared to quetiapine. In light of the greater proportion of days spent with adverse effects associated with quetiapine and the more favorable tolerability profile and superior treatment efficacy of esketamine, the authors concluded that esketamine should be the preferred therapeutic approach for TRD.

A number of additional concerns have been raised regarding the use of ketamine. In the USA, physicians retain considerable flexibility in how off-label racemic ketamine is prescribed, and the number of (private) clinics offering this treatment continues to grow. The gap in regulatory supervision became conspicuous during the relaxed restriction phase of the global COVID-19 pandemic during which controlled substances could be prescribed without an in-person examination [21]. Clinics began prescribing various preparations such as ketamine lozenges to take home for oral intake or used telehealth consultations and postal ketamine supply. The potential for overuse, the possibility of patients being exposed to exceedingly high doses with severe acute and long-term medical consequences, and the risk of diversion are therefore of paramount concern. The public's growing interest in ketamine's purported benefits, coupled with the attention it received in the media and popular press, as well as the promotional activities of clinics, has also contributed to much interest in ketamine as a recreational drug. Although the prevalence of illicit ketamine use is relatively low in comparison to other drug abuse, the quantity of ketamine confiscated in the USA has been documented to have increased by 350% from 2017 to 2022 [21]. Additionally, the number of ketamine poisonings reported has been found to have increased by 80% between 2019 and 2021. It is evident that the risks associated with treatment are particularly alarming when patients receive ketamine without the supervision of a mental health professional. The potential for adverse effects on the urinary and liver systems with repeated uses definitely necessitates close patient monitoring. In light of the evolving landscape of ketamine use not only in the USA, there is a pressing need for research on ketamine surveillance and long-term outcomes.

In order to address safety concerns and monitor the effectiveness of ketamine, British expert psychiatrists have proposed to establish mandatory national registries comprising all patients who have undergone or receive ketamine therapy [19]. This would facilitate pharmacovigilance and enable the tracking of patient outcomes across diagnoses, routes, and doses, including the assessment of long-term data. The experts emphasize that any framework which promotes the safe, effective, and equitable use of ketamine is a positive step forward. This notion appears to be particularly valid in view of the positive effects of ketamine on suicidal ideation and TRD [22, 23].

Far beyond, it is possible that improvements in the surveillance and monitoring of ketamine's efficacy in the treatment of mental health disorders may facilitate its use on a global scale. In this regard, expert psychiatrists in New Zealand have indicated that the reluctance to adopt the use of off-label IV ketamine as an antidepressant may be attributed to a dearth of first-hand experience [24]. It is evident that the existing literature on the benefits and safety of ketamine has not been sufficient to prompt a change in their practice. In light of these considerations, experts have proposed a two-pronged approach: initially, intramuscular (IM) ketamine injections would be employed to assess patient response and then a transition to oral intake would be made. However, the absorption of IM ketamine is variable and dependent on the injection technique. In contrast to the intent, varying effects may thus be produced. Oral ketamine may also present certain challenges. For example, it undergoes first-pass metabolism, resulting in only 20% of the administered dose reaching the systemic circulation. Its absorption is also influenced by several factors, including the duration of pre-treatment fasting. A further potential challenge in the proposed guidelines from New Zealand is the suggestion that psychiatrists should observe at least three patient courses receiving IM ketamine in order to become familiar with its psychoactive effects [24]. This has been considered a prerequisite for the prescription of ketamine. All of these endeavors illustrate the necessity for enhanced collaboration between medical specialities and the exchange of knowledge and expertise regarding ketamine in its various applications. For instance, in the context of prehospital and/or perioperative acute and chronic pain management, anesthetists have accumulated extensive experience with ketamine in awake patients over decades and would undoubtedly benefit from sharing their insights with colleagues from other disciplines [25].

In light of the dearth of resources allocated to the management and treatment of depression in low- to middle-income countries, the utilization of ketamine might prove to be a significant contributor to public health. An expert from India has emphasized that an oral racemic ketamine session for antidepressive therapy can be effectively implemented in a public hospital at a cost of approximately half a US dollar in India [26]. The expert also references own observational studies indicating that a fixed dose of 150 mg oral ketamine elicits a moderate antidepressant effect, with higher doses being well-tolerated and more efficacious if used skillfully. When ketamine is administered in a solution over 10–15 min, the risks of adverse effects may be reduced in comparison to IV administration due to slower absorption [26]. The overall effect of authorization of oral, generic ketamine as an antidepressant would be to provide an inexpensive means of improving approaches to depression and depression-related conditions in countries with limited resources if some requirements for use are met. This is particularly the case in settings with restricted healthcare infrastructure and/or patients who have to pay out-of-pocket for therapy. Nevertheless, appropriate randomized controlled trials with a particular focus on the administration routes, comorbidities and unwanted psychoactive effects of ketamine should be conducted at the local level. This may impact policy to gain local regulatory approval for ketamine therapy [26, 27]. Far beyond, such efforts need to include patients struggling with depressive symptoms and suicidality who live in prison on a global scale, as well [28].

To conclude, the following actions should be prioritized to find more global solutions for ketamine treatment in severe depression: (1) the implementation of a surveillance system in all countries; (2) transfer and exchange of knowledge and expertise about pitfalls, benefits, and best possible treatment schedules between medical disciplines for practical ketamine use on an (inter-)national level; and (3) the conductance of high-quality, multicenter clinical trials performed at best with patient-personalized treatment approaches. These endeavors need to be flanked by (4) implementing frameworks for legal ketamine administration to impact healthcare policy changes. Nonetheless, there will be patients who will not respond to ketamine, but for those who respond, ketamine may be a life-changer.

Acknowledgments

We would like to thank Professor G. Sanacora, Yale Depression Research Program, Department of Psychiatry, Yale School of Medicine, Connecticut, and Interventional Psychiatry Service, Yale-New Haven Hospital, New Haven, Connecticut, USA, for expert suggestions on the manuscript. We are also grateful to Professor Sooraj Baijnath, Integrated Molecular Physiology Research Initiative, School of Physiology, University of the Witwatersrand, Johannesburg, South Africa, for his insightful comments.

Conflicts of Interest

SH, CTN and AMEM do not have any conflicts of interest.

References

1. GBD 2021 Diseases and Injuries Collaborators (2024) Global incidence, prevalence, years lived with disability (YLDs), disability-adjusted life-years (DALYs), and healthy life expectancy (HALE) for 371 diseases and injuries in 204 countries and territories and 811 subnational locations, 1990-2021: a systematic analysis for the global burden of disease study 2021. Lancet 403:2133–2161
2. Herrman H, Patel V, Kieling C et al (2022) Time for united action on depression: a lancet-world psychiatric association commission. Lancet 399:957–1022
3. Deng Q, Parker E, Wu C et al (2025) Repurposing ketamine in the therapy of depression and depression-related disorders: recent advances and future potential. Aging Dis 16: 804–840
4. Vilarim M, Rebelo F, Vieira I et al (2024) Prevalence of postpartum depression symptoms in high-income, and low- and middle-income countries in the Covid-19 pandemic: a systematic review with meta-analysis. Braz J Psychiatry 46:e20233453
5. O'Connor RC, Wothman CM, Abanga M et al (2023) Gone too soon: priorities for action to prevent premature mortality associated with mental illness and mental distress. Lancet Psychiatry 10:452–464
6. Tang T, Jiang J, Tang X (2022) Prevalence of depression among older adults living in care homes in China: a systematic review and meta-analysis. Int J Nurs Stud 125:104114
7. Kasturi S, Oguoma VM, Grant JB et al (2023) Prevalence rates of depression and anxiety among young rural and urban Australians: a systematic review and meta-analysis. Int J Environ Res Public Health 20:800
8. Petarli GB, Cattafesta M, Viana MCM et al (2024) Depression in Brazilian farmers: prevalence and associated factors. J Ment Health 33: 127–135
9. Millen AME, Daniles WMU, Baijnath S (2024) Depression, an unmet health need in Africa: understanding the promise of ketamine. Heliyon 10:e28610
10. Seshadri A, Prokop LJ, Singh B (2024) Efficacy of intravenous ketamine and intranasal esketamine with dose escalation for major depression: a systematic review and meta-analysis. J Affect Disord 356:379–384
11. McIntyre RS, Jain R (2024) Glutamatergic modulators for major depression from theory to clinical use. CNS Drugs 38:869–890
12. Gaynes BN, Asher G, Gartlehner G et al (2018) Definition of treatment-resistant depression in the medicare population. Agency for Healthcare Research and Quality, Rockville

13. Johnston JN, Kadriu B, Kraus C et al (2024) Ketamine in neuropsychiatric disorders: an update. Neuropsychopharmacology 49:23–40
14. Johnston JN, Zarate CA, Kvarta MD (2024) Esketamine in depression: putative biomarkers from clinical research. For practical ketamine use. Eur Arch Psychiatry Clin Neurosci. https://doi.org/10.1007/s00406-024-01865-1
15. Glue P, Loo C, Fam J et al (2024) Extended-release ketamine tablets for treatment-resistant depression: a randomized placebo-controlled phase 2 trial. Nat Med 30:2004–2009
16. Price MZ, Price RL (2024) Benefits and risks of esketamie nasal spray continuation in treatment-resistant depression. Biomarkers Neuropsychiatry 11:100104
17. Swainson J, Klassen LJ, Brennan S et al (2022) Non-parenteral ketamine for depression: a practical discussion on addiction potential and recommendations for judicious prescribing. CNS Drugs 36:239–251
18. Le TT, Cordero IP, Jawad MY et al (2022) The abuse liability of ketamine: a scoping review of preclinical and clinical studies. J Psychiatr Res 151:476–496
19. Jelen LA, McShane R, Young AH (2024) Guidelines for ketamine use in clinical psychiatry practice. BJPsych Open 10(e107):1–3
20. McIntyre RS, Bitter I, Buyze J et al (2024) Safety and tolerability of esketamine nasal spray versus quetiapine extended release in patients with treatment resistant depression. Eur Neuropsychopharmacol 85:58–65
21. Wilkinson ST, Palamar JJ, Sanacora G (2024) The rapidly shifting ketamine landscape in the US. JAMA Psychiatry 81:221–222
22. Shen ZY et al (2024) A meta-analysis of the effects of ketamine on suicidal ideation in depression patients. Transl Psychiatry 14:248
23. Kumari S, Chaudhry HA, Sagot A et al (2024) Exploring esketamine's therapeutic outcomes as an FDA-designated breakthrough for treatment-resistant depression and major depressive disorder with suicidal intent: a narrative review. Cureus 16:e53987
24. Beaglehole B, Glue P, Clarke M, Porter R (2023) Multidisciplinary development of guidelines for ketamine treatment for treatment-resistant major depression disorder for use by adult specialist mental health services in New Zealand. BJPsych Open 9(e191):1–7
25. Brenna CTA, Goldstein BJ, Zarate CA, Orser BA (2024) Repurposing general anesthetic drugs to treat depression: a new frontier for anesthesiologists in neuropsychiatric care. Anesthesiology 141:222–237
26. Andrade C (2023) Ketamine for depression – knowns, unknowns, possibilities, barriers, and opportunities. JAMA Psychiatry 80:1189–1190
27. Włodarczyk A, Dywel A, Cubala WJ (2023) Safety and tolerability of the acute ketamine treatment in treatment-resistant depression: focus on comorbities interplay with dissociation and psychomimetic symptoms. Pharmaceuticals 16:173
28. Bayrhammer-Savel M, Ortner M, van Hout MC, Komorowski A (2024) Psychiatric and legal considerations for ketamine treatment within prison settings. Front Psych 15: 1316108

Chapter 12

Key Considerations for Practical Administration of Ketamine in Severe Depression, and Its Use in Community Care in Edmonton, AB, Canada

Carson Chrenek and Jennifer Swainson

Abstract

Although both intranasal esketamine and intravenous ketamine have proven efficacy and tolerability in treatment-resistant depression (TRD), patient access to these treatments is often limited due to cost, availability, and/or monitoring requirements. This article is intended to provide a synthesis of the current literature with regard to the pharmacologic use of ketamine in TRD and describe its application to clinical programming in Edmonton, AB, Canada. Topics discussed include rationale for maintenance ketamine treatments, monitoring of long-term safety concerns, and alternative options to intranasal esketamine and intravenous ketamine as creative solutions to improve patient access. Considerations for cautious and judicious prescribing of ketamine outside of highly monitored clinical settings are reviewed.

Key words IV ketamine, Depression, Treatment-resistant depression, Community program, Practical considerations, Intranasal esketamine, Sublingual ketamine, At-home treatment

1 Introduction

The impact of treatment-resistant depression (TRD) has been increasingly recognized and occurs with a higher frequency than originally thought [1]. A commonly accepted definition includes a failure to respond to two (or more) first-line antidepressants of adequate dose and duration [2]. While intranasal esketamine has been approved by Health Canada and the Food and Drug Administration (FDA) in the United States, due to cost and monitoring requirements in North America, access is limited for many patients, and prescribing is impractical for many clinicians. While off-label, IV racemic ketamine has also demonstrated efficacy for TRD and is included in the Canadian Practice Guidelines for depression [3]. Head-to-head randomized controlled trials (RCTs) between IV racemic ketamine and esketamine are lacking, but one meta-analysis of observational studies showed no difference in efficacy up

Sabine Himmelseher (ed.), *Ketamine: Pain Management, Critical Care, Depression, and Resource-Poor Settings Volume II*, Neuromethods, vol. 225, https://doi.org/10.1007/978-1-0716-4599-4_12,

to 1 month [4]. However, a more recent meta-analysis found an efficacy/tolerability profile in favor of racemic ketamine [5], later supported by a 2023 observational study that demonstrated similar rates of response/remission, but with IV ketamine requiring fewer treatments to achieve this outcome [6]. Comparison with electroconvulsive therapy (ECT) in systematic reviews and meta-analysis has shown ketamine to be similar, or slightly inferior, but with better tolerability than ECT, with the caveat being such studies have been small in sample size [7, 8]. However, private clinics offering ketamine treatment are financially out of reach for many individuals with TRD, and publicly funded programs are rare.

Two hospital sites in Edmonton, Canada, have offered publicly funded IV ketamine for TRD since 2015. Guidance was initially from treatment protocols published in randomized and open-label trials in depression with ketamine. While initially considered "experimental" treatment and only reserved for the individuals who had exhausted most other treatment options, as evidence has accumulated, these programs have evolved to offer treatment to less treatment-resistant individuals. Our clinical experience has been commensurate with reports in the literature that ketamine's initial antidepressant effect will dissipate, though greater efficacy and more prolonged benefit have been noted with repeated treatments compared to a single infusion [9]. As such, there has been demand for ongoing maintenance treatments to sustain benefit. The growing demand for these resources within a public mental healthcare system have prompted the need to explore other models of delivery and treatment settings. Here, the structure of this public ketamine program is reviewed within the context of the most recent literature, and models for ketamine treatment delivery are discussed. It is expected that such models could be readily adopted in clinical practice in other countries around the world. Of course, local regulatory bodies may vary with regard to prescribing restrictions and monitoring requirements as it pertains to the use of therapeutic ketamine.

2 Clinical Programming

2.1 Population, Inclusion, and Exclusion

Our program treats adults over the age of 18 with bipolar or unipolar depression, who have failed to respond to trials of at least two antidepressants and one adjunctive medication. Safety and efficacy data supports ketamine use in patients over 65, so these individuals are eligible for treatment [10, 11]. There is early, but limited evidence in adolescent population [12].

The population treated commonly has multiple psychiatric comorbidities, the presence of which is known to increase the level of treatment resistance. Greater degrees of treatment resistance are less likely to remit, but not less likely to respond to treatment [13].

Exclusion criteria were initially designed to minimize both psychiatric and medical morbidity that could arise iatrogenically from treatment. While initially considered absolute, they are now considered relative, more in keeping with common ECT practices. These include primary psychotic disorder, uncontrolled hypertension, central aneurysmal disease, significant valvular disease, recent cardiovascular event, class III heart failure, or other unstable medical condition [3]. Ketamine crosses the placenta and is contraindicated during pregnancy and breastfeeding due to unknown risks [14].

2.2 Dosing and Frequency

Controlled studies have demonstrated that repeat dosing is more effective than single dosing, with an extended mean time to relapse [9]. Doses of 0.5–1 mg/kg are safe and effective, and the higher 1 mg/kg doses have demonstrated a longer duration of response, but only if the patient is tolerating the treatments well [15]. We recommend starting at 0.5 mg/kg and increasing up to 1.0 mg/kg only if well tolerated and deemed necessary. While a number of clinical studies have offered six treatments as the standard, dose optimization is better supported by a course of 8 twice-weekly treatments, as doses can be increased as tolerated after every few treatments, allowing for a greater number of treatments at the optimal determined dose. Clinical discretion is used if it is believed the patient will benefit from extending the initial course of eight treatments. For example, if a patient only began to respond to ketamine at the 1.0 mg/kg dose, additional treatments may be considered.

2.3 Assessing Response

Clinical assessment should include both subjective and objective assessments of the patient. It is established that the degree of dissociation experienced does not correlate with eventual response to treatment [16]. Notable benefits should be seen in the days following treatment and not just for the duration of the dissociative symptoms or the day of. Such would imply an enjoyment of the treatment experience, but not demonstrate antidepressant efficacy per se.

2.4 Concomitant Medications

Literature has found that while GABAergic medications do not substantially interfere with treatment [17], concurrent benzodiazepine use may delay treatment response or increase the likelihood of relapse afterward [18]. As such, benzodiazepines are reduced (or discontinued) prior to ketamine treatment only when possible, if not likely to significantly jeopardize patient well-being. There are conflicting reports regarding whether naltrexone reduces ketamine's antidepressant response, so this is avoided if possible [19]. While lamotrigine may reduce the dissociative effects of ketamine, systematic reviews have not suggested a negative effect when lamotrigine is prescribed with ketamine, so it may be continued during treatment [20]. Most antidepressants and mood stabilizers can be safely prescribed with ketamine, but caution is advised with monoamine oxidase inhibitors (MAOIs) [3]. Our program does not suggest oral prescribing of ketamine, in part due to unpredictable changes in bioavailability from first-pass metabolism and Cytochrome P450 enzyme variability (either genetic or drug-induced).

2.5 Side Effects

Ketamine is safe and well tolerated when used at the doses suggested. Transient side effects can include psychotomimetic and dissociative experiences, blurred vision, dizziness, anxiety, irritability, headaches, nausea, tachycardia, and elevated blood pressure, among others [21]. These peak within 30–60 min and abate within 1–2 h. They are typically dose-dependent [22]. We provide IV ondansetron for those that struggle with nausea but do not typically provide any other medication during the treatment.

An average increase in blood pressure of 9–19 mmHg is expected, typically returning to baseline within 2–4 h [23]. While earlier consensus statements suggested to pause the ketamine infusion in blood pressure increases above 180/100, and not to proceed with treatment if baseline blood pressure is 160/100 [3], consideration of these transient effects within a broader medical context suggests these universal cautions are unnecessary, and aggressively or unnecessarily treating transient hypertension carries its own risk [24]. For example, transient hypertension in healthy individuals without the relative exclusion criteria mentioned is common in many settings (including exercise, etc.) [25], and asymptomatic hypertension is not typically treated in the emergency room setting. As such, elevated blood pressure from ketamine is not treated unless there are objective symptoms of a hypertensive emergency, or a known risk of subarachnoid hemorrhage such as pre-existing aneurysm or arteriovenous malformation (85% of SAH clinical presentations) [26]. Objective evidence of a hypertensive emergency include crushing chest pain, syncope, severe abdominal pain, decreased (not simply altered) level of consciousness, or shortness of breath. Chronic hypertension should be adequately treated before initiation of ketamine treatment. If blood pressure is ever of concern, staff are required to notify the most responsible physician and ask for advice.

3 Treatment Settings

Intravenous ketamine is delivered in an acute care hospital setting with appropriately trained nursing staff present. An anesthetist is not present to administer ketamine in this program, but a hospital rapid response team is accessible in the case of an emergency. Ketamine is administered in a dimly lit and quiet room, as patient sensory experiences may be amplified during treatment. It is suggested for patients to consider wearing an eye mask and to use headphones to listen to soothing sounds such as mindfulness/relaxation exercises or soft music, depending on their preference. It is important that nursing staff be properly trained in supporting the patient's experience through the treatment. The nurses monitor the patient's experiences and will astutely provide comfort and support when requested, or if a patient appears to be in discomfort or distress. Nurses monitor for approximately 90 min after the infusion is done, discharging the patient only when dissociative symptoms have abated and vitals are stable. Patients are instructed not to drive until the following day after ketamine treatment.

4 Maintenance Treatments

As there is a risk of relapse following ketamine treatment, consideration is invariably given regarding the utility of maintenance treatments for individuals. This is analogous to maintenance ECT treatments [27], as well as treatments with standard antidepressants [28]. A growing body of evidence suggests that maintenance treatments with racemic ketamine are both safe and effective, with one study reported on patients receiving maintenance ketamine treatments (multiple routes of administration) for up to 18 months [29]. Even more robust long-term data is available with intranasal esketamine, with reassuring efficacy and safety data now available for up to 4 years after beginning treatment [30]. Frequency/duration is determined on a case-by-case basis, though the intranasal esketamine model suggests weekly maintenance treatment, decreased to every 2 weeks if there is continued stability. The need for continued treatment should be periodically evaluated.

5 Long-Term Safety

Long-term safety concerns with ketamine originate from data available from chronic ketamine abusers. Of note, in these reports, ketamine was used in doses as much as 20–30 times that of therapeutic doses in depression and frequently used daily [31]. These individuals have demonstrated a risk of ulcerative urinary cystitis, hepatobiliary effects, and cognitive decline, among other negative health effects. Similar effects have not been observed in studies or clinical practice related to therapeutic ketamine in depression. Ketamine has been deemed neutral or pro-cognitive when used to treat depression [32]. Urinary effects are rare, but patients receiving ketamine should be periodically screened for urinary symptoms, and long-term use should be accompanied by periodic urinalysis to screen for microscopic hematuria. Documented cases of hepatobiliary effects in therapeutic ketamine use have been limited to individuals treated for various pain conditions, with doses far in excess of what is used in the treatment of depression [33]. Comparatively, Janssen's vigorous clinical program for esketamine nasal spray did not note significant concerns with urinary, liver, or cognitive side effects, and monitoring for these is not required by the esketamine clinical program [30]. Our practice is to do baseline laboratory investigations prior to treatment, then every 6–12 months thereafter, or sooner if clinically indicated.

While ketamine is a known potential substance of abuse, the prevalence rate is reassuringly low in North America [34]. While clinical studies of ketamine and esketamine have not specifically screened for ketamine misuse, there are no reports of ketamine addiction or misuse in clinical populations using ketamine for depression [35, 36]. Early studies in treatment settings have suggested an extremely low, but not negligible risk of ketamine abuse [37], so particularly with long-term ketamine prescribing, clinicians should screen for signs and risk factors for ketamine misuse or abuse.

6 Other Routes of Administration

Ketamine can be delivered in intravenous (IN), intramuscular (IM), subcutaneous (SC), intranasal (IN), sublingual (SL), and oral (PO) formulation. Specifically, non-parenteral forms of ketamine including IN, SL, and PO ketamine are also reported as safe and efficacious [38]. While the majority of literature is in support of IV ketamine, the Canadian Network for Mood and Anxiety Treatments (CANMAT) has recently acknowledged the utility of "non-IV" forms of ketamine as a third-line adjunctive treatment for depression [39]. Data to guide optimal dosing of these formulations, however, remains limited. Doses reported in the literature

Table 1
Previously reported "non-IV" ketamine doses for depression

Route of administration	Dose ranges reported in the depression literature	Bioavailability (%)
IM	0.5 mg/kg	64–95
SC	0.1–1.0 mg/kg	64–95
IN	50–150 mg	30–50
SL	50–450 mg	20–30
PO	50–500 mg	10–20

Table 2
Suggested dosing for intranasal and sublingual ketamine

Mode of administration	Intravenous	Intranasal	Sublingual
Bioavailability (%)	100	30–50	20–30
Minimum dose suggested (mg/kg)	0.5	1.0	1.5
Maximum dose suggested (mg/kg)	1.0	3.0	5

have been previously reviewed and are summarized in Table 1 [40–42].

As Table 1 notes, doses reported for non-IV forms of ketamine vary significantly. Studies have generally been small and with limitations, and many were not published at the time the Edmonton programs began to utilize SL and IN ketamine. Please note that our own expertise has been with pharmacy-compounded SL/IN racemic ketamine formulations, rather than liquid or tablet oral ketamine, or other commercially prepared products. Consideration was given to the differing bioavailability and pharmacokinetics of each formulation, which potentially affects efficacy and tolerability. Considering these factors, physicians in this program elected to extrapolate dosing from the IV ketamine literature, with consideration of bioavailability to calculate dose adjustments. See Table 2 for a summary of suggested SL and IN dosing. Use of SL or IN ketamine has significantly improved access to ketamine treatment, as pharmacy-compounded preparations typically cost about C$ 150 CAD per month of treatment, substantially less than the cost of IN esketamine or private clinic administered IV ketamine.

SL and IN ketamine were initially integrated into ketamine programming as an option for maintenance treatments to support ongoing treatment where it was deemed necessary, while preserving space in the IV program for patients receiving acute treatment courses. When prescribed this way, both the final IV ketamine dose received and bioavailability of the new agent are considered to calculate an appropriate dose. More recently, SL or IN ketamine have been used for initial treatment courses, in patients who had not received IV ketamine previously. In these cases, tolerability should be established by starting a calculated dose equivalent to the starting IV dose of 0.5 mg/kg, again considering bioavailability as per Table 2. Doses may be increased as tolerated for efficacy toward the maximal dose suggested (approximated as equivalent to 1.0 mg/kg when given IV).

7 Alternative Treatment Settings

Both IV ketamine and IN esketamine require administration and monitoring in a healthcare setting. Significant adverse events have been rare in monitored settings, and the Edmonton programs have not once required assistance from medical specialists (anesthesia, internal or emergency medicine) after an estimated 10,000 infusions administered.

Some patients may be considered appropriate candidates for treatment outside of monitored settings. Use of non-parenteral forms of ketamine (IN and SL) provides options for use in a variety of different settings, which can increase access for patients. Prescribing clinicians should be informed of current literature and medical education on ketamine to learn and adjust prescribing practices as new data become available. It should be highlighted that some jurisdictions may have differing regulations and medico-legal considerations of at-home ketamine use and that both physician and patient comfort will depend on several circumstances. Assessment of potential risks and benefits should be discussed as they apply to each patient, and informed consent should be documentated. Some practitioners may elect to create explicit treatment contracts with patients.

Considerations for at-home use of ketamine administration would include [35]:

- High level of treatment resistance.
- No history or antisocial behavior or drug misuse.
- In the case of maintenance ketamine, a previous positive response to an acute course of ketamine.
- In the case of an acute course of ketamine, limited ability to access more evidence-based IV ketamine or IN esketamine treatments.
- Patient reliability to attend follow-up appointments.
- Suitability for ketamine treatment in a non-medical setting, including stable cardiovascular status and controlled blood pressure.

Practical suggestions for clinicians prescribing of ketamine for home use would include:

- Obtain and document informed consent—review potential risks/benefits.
- Prescribe ketamine in limited quantities and limited refills, and specify interval between refills.

- Consider observing first treatments or dose changes in office to monitor blood pressure in medically at-risk patients and/or dissociation in ketamine-naïve individuals.
- A blood pressure monitor should be available at home for the patient to self-monitor.
- Administer ketamine at night when used at home, and advise patients not to drive until next day after use:
  - Advise a quiet and calm environment.
  - For ketamine-naïve patients, consider the presence of another responsible adult for the first several doses, if not observed in office.
  - Wait until dissociative/sedative effects of ketamine dissipate before using other potentially sedating bedtime medications.
- Screen for bladder toxicity:
  - Check urinalysis at baseline and as indicated thereafter.
  - Ask about urinary symptoms (e.g., frequency, urgency, hematuria).
- Monitor for potential drug liking/signs or symptoms of misuse:
  - Lost prescriptions.
  - Requests for early refills.
  - Requests for dose escalation or increased frequency.

8 Conclusions

This clinical experience with ketamine prescribing has utilized a step-down approach, where patients with a positive IV ketamine response are "stepped down" to IN or SL ketamine when maintenance treatment is required. This has supported patient flow through the program to accommodate acute courses of ketamine, while also offering a form of maintenance treatment. Frequency of ketamine is monitored with close clinical follow-up, with potential consideration of reduced frequency, and ideally, eventual tapering off of ketamine.

As demand for ketamine has increased, a step-up approach has been increasingly utilized, first offering acute courses of SL or IN ketamine to patients due to limited IV resources in the public system. In this approach, patients are "stepped up" to an IV ketamine trial only if there is a lack of response to a SL or IN formulation. While there is limited comparative data available, it has been observed in this program that some patients respond well to SL or IN ketamine and do not require use of IV ketamine resources, while others who fail to respond to SL or IN ketamine may later respond to a course of IV ketamine. With the support of updated treatment guidelines [38], this approach may be more scalable as public systems look to integrate ketamine into their mental health treatment programming.

As the body of evidence for ketamine in multiple delivery models expands, there have been financial and logistical barriers for many individuals preventing them from accessing treatment. Similarly, the use of ketamine is likely to require maintenance treatments for many individuals, so practical solutions such as non-IV forms of ketamine must be available for them. Strong awareness of the potential risks of ketamine is essential, as well as the current limitations with regard to long-term use. These must be considered in the risk-benefit analysis, as with any medical treatment, and considered in the context of the individual patient.

Disclosures

Carson Chrenek has received honoraria from Abbvie.

Jennifer Swainson has received honoraria from Abbvie, Bausch Health, Eisai, Idorsia, Janssen, Lundbeck, Otsuka, and has previously served as a medical advisor for the Newly Institute, a private clinic that utilizes SL ketamine.

References

1. McIntyre RS, Alsuwaidan M, Baune BT et al (2023) Treatment-resistant depression: definition, prevalence, detection, management, and investigational interventions. World Psychiatry 22:394–412
2. Rush A, Fava M, Wisniewski S et al (2004) Sequenced treatment alternatives to relieve depression (STAR*D): rationale and design. Control Clin Trials 25:119–142
3. Swainson J, McGirr A, Blier P et al (2021) The Canadian network for mood and anxiety treatments (CANMAT) task force recommendations for the use of racemic ketamine in adults with major depressive disorder: recommendations. Can J Psychiatr 66:537–539
4. McIntyre RS, Carvalho IP, Lui L et al (2020) The effect of intravenous, intranasal, and oral/esketamine in mood disorders: a meta-analysis. J Affect Disord 276:576–584
5. Bahji A, Vazquez GH, Zarate CA Jr (2021) Comparative efficacy of racemic ketamine and esketamine for depression: a systematic review and meta-analysis. J Affect Disord 278:542–555. https://doi.org/10.1016/j.jad.2020.09.071
6. Singh B, Kung S, Pazdernik V et al (2023) Comparative effectiveness of intravenous ketamine and intranasal esketamine in clinical practice among patients with treatment-refractory depression: an observational study. J Clin Psychiatry 84(2):22m14548. https://doi.org/10.4088/JCP.22m14548
7. Menon V, Varadharajan N, Faheem A, Andrade C (2023) Ketamine vs electroconvulsive therapy for major depressive episode: a systematic review and meta-analysis. JAMA Psychiatry 80: 639–642
8. Rhee TG, Shim SR, Forester BP et al (2022) Efficacy and safety of ketamine vs electroconvulsive therapy among patients with major depressive episode: a systematic review and meta-analysis. JAMA Psychiatry 79:1162–1172
9. Kryst J, Kawalec P, Mitoraj AM et al (2020) Efficacy of single and repeated administration of ketamine in unipolar and bipolar depression: a meta-analysis of randomized clinical trials. Pharmacol Rep 72:543–562
10. Lipsitz O, Di Vincenzo JD, Rodrigues NB et al (2021) Safety, tolerability, and real-world effectiveness of intravenous ketamine in older adults with treatment-resistant depression: a case series. Am J Geriatr Psychiatry 29:899–913

11. Oughli H, Gebara M, Ciarleglio A et al (2023) Intravenous ketamine for late-life treatment-resistant depression: a pilot study of tolerability, safety, clinical benefits, and effect on cognition. Am J Geriatr Psychiatry 31:210–221
12. Dwyer JB, Landeros-Weisenberger A, Johnson JA et al (2021) Efficacy of intravenous ketamine in adolescent treatment-resistant depression: a randomized midazolam-controlled trial. Am J Psychiatry 178:352–362
13. Wilkinson ST, Toprak M, Turner MS et al (2017) A survey of the clinical, off-label use of ketamine as a treatment for psychiatric disorders. Am J Psychiatry 174:695–696
14. KETALAR (2020) Product Information: KETALAR® (ketamine HCl) for intravenous and intramuscular injection. Par Pharmaceutical (FDA.gov), Chestnut Ridge
15. Fava M, Freeman MP, Flynn M et al (2020) Double-blind, placebo-controlled, dose-ranging trial of intravenous ketamine as adjunctive therapy in treatment-resistant depression (TRD). Mol Psychiatry 25:1592–1603
16. Grabski M, Borissova A, Marsh B et al (2020) Ketamine as a mental health treatment: are acute psychoactive effects associated with outcomes? A systematic review. Behav Brain Res 392:1126–1129
17. Alnefessi Y, Chen-Li D, Krane E et al (2022) Real-world effectiveness of ketamine in treatment-resistant depression: a systematic review and meta-analysis. J Psychiatr Res 151: 693–709
18. Andrashko V, Novak T, Brunovsky M et al (2020) The antidepressant effect of ketamine is dampened by concomitant benzodiazepine. Front Psych 11:844
19. Marton T, Barnes DE, Wallace A et al (2019) Concurrent use of buprenorphine, methadone, or naltrexone does not inhibit ketamine's antidepressant activity. Biol Psychiatry 85:e75–e76
20. Wilkowska A, Wiglusz MS, Jakuszkowiak-Wojten K, Cubała WJ (2022) Ketamine and lamotrigine combination in psychopharmacology: systematic review. Cells 11:645. https://doi.org/10.3390/cells11040645
21. Wan L, Levitch CF, Perez AM et al (2015) Ketamine safety and tolerability in clinical trials for treatment-resistant depression. J Clin Psychiatry 76:247–252
22. Rodrigues NB, McIntyre RS, Lipsitz O et al (2020) Safety and tolerability of IV ketamine in adults with major depressive disorder or bipolar disorder: results from the Canadian rapid treatment center of excellence. Expert Opin Drug Safety 19:1031–1040
23. Szarmach J, Cubala W, Wlordarczyk A et al (2019) Short-term ketamine administration in treatment-resistant depression: focus on cardiovascular safety. Psychiatr Danub 31(suppl 3):585–590
24. Yip R, Swainson J, Khullar A et al (2022) Intravenous ketamine for depression: a clinical discussion reconsidering best practices in acute hypertension management. Front Psych 13: 1017504. https://doi.org/10.3389/fpsyt.2022.1017504
25. Daida H, Allison TG, Squires RW et al (1996) Peak exercise blood pressure stratified by age and gender in apparently healthy subjects. Mayo Clin Proc 71:445–452
26. Macdonald RL, Schweizer TA (2017) Spontaneous subarachnoid haemorrhage. Lancet 389: 655–666
27. Navarro V, Gastó C, Torres X et al (2008) Continuation/maintenance treatment with nortriptyline versus combined nortriptyline and ECT in late-life psychotic depression: a two-year randomized study. Am J Geriatr Psychiatry 16:498
28. Kennedy SH, Lam RW, McIntyre RS et al (2016) Canadian network for mood and anxiety treatments (CANMAT) 2016 clinical guidelines for the management of adults with major depressive disorder. Can J Psychiatr 61: 506–603
29. Smith-Apeldorn S, Veraart J, Spijker J et al (2022) Maintenance ketamine treatment for depression: a systematic review of efficacy, safety, and tolerability. Lancet Psychiatry 9: 907–921
30. Zaki N, Chen LN, Lane R et al (2023) Long-term safety and maintenance of response with esketamine nasal spray in participants with treatment-resistant depression: interim results of the SUSTAIN-3 study. Neuropsychopharmacology 48:1225–1233. https://doi.org/10.1038/s41386-023-01577-5
31. Morgan CJ, Curran HV (2012) Independent scientific committee on drugs. Ketamine use: a review. Addiction 107:27–38
32. Gill H, Gill B, Rodrigues NB et al (2021) The effects of ketamine on cognition in treatment-resistant depression: a systematic review and priority avenues for future research. Neurosci Biobehav Rev 120:78–85
33. Cotter S, Wang J, Gada N et al (2021) Repeated or continuous medically supervised ketamine administration associated with hepatobiliary adverse events: a retrospective case series. Drug Saf 44:1365–1374
34. Bokor G, Anderson PD (2014) Ketamine: an update on its abuse. J Pharm Pract 27:582–586

35. Swainson J, Klassen LJ, Brennan S et al (2022) Non-parenteral ketamine for depression: a practical discussion on addiction potential and recommendations for judicious prescribing. CNS Drugs 36:239–251. https://doi.org/10.1007/s40263-022-00897-2
36. Le TT, Cordero IP, Jawad MY et al (2022) The abuse liability of ketamine: a scoping review of preclinical and clinical studies. J Psychiatr Res 151:476–496. https://doi.org/10.1016/j.jpsychires.2022.04.035
37. Chubbs B, Wang J, Archer S et al (2022) A survey of drug liking and cravings in patients using sublingual or intranasal ketamine for treatment-resistant depression: a preliminary evaluation of real-world addictive potential. Front Psych 13:1016439. https://doi.org/10.3389/fpsyt.2022.1016439
38. McIntyre RS, Carvalho IP, Lui LMW et al (2020) The effect of intravenous, intranasal, and oral ketamine in mood disorders: a meta-analysis. J Affect Disord 276:576–584
39. Lam RW, Kennedy SH, Adams C et al (2024) Canadian network for mood and anxiety treatments (CANMAT) 2023 update on clinical guidelines for management of major depressive disorder in adults. Can J Psychiatr 69:641. https://doi.org/10.1177/07067437241245384
40. Chrenek C, Duong B, Khullar A et al (2024) Use of ketamine for treatment resistant depression: updated review of literature and practical applications to a community ketamine program in Edmonton, Alberta, Canada. Front Psych 14:1283733. https://doi.org/10.3389/fpsyt.2023.1283733
41. Zanos P, Moaddel R, Morris P et al (2018) Ketamine and ketamine metabolite pharmacology: insights into therapeutic mechanisms. Pharmacol Rev 70:621–660
42. Abuhelwa AY, Somogyi AA, Loo CK et al (2022) Population pharmacokinetics and pharmacodynamics of the therapeutic and adverse effects of ketamine in patients with treatment-refractory depression. Cin Pharmacol Ther 112:720–729

Chapter 13

Comparing Intravenous Racemic Ketamine and Intranasal Esketamine for Treatment-Resistant Depression: A Clinical Dilemma

Balwinder Singh, Jennifer L. Vande Voort, and Simon Kung

Abstract

The decision-making process for or against intravenous (IV) racemic ketamine versus intranasal (IN) esketamine is a critical choice in therapeutic approaches for treatment-resistant depression (TRD). This chapter draws upon real-world data gathered from a specialized center for depression treatment at the Mayo Clinic in Rochester, Minnesota, USA, to examine important factors which may determine this selection. In the acute phase, IV racemic ketamine and IN esketamine demonstrated similar response and remission rates; however, IV racemic had an accelerated response compared to IN esketamine. Additionally, both treatments were well-tolerated by patients.

Key words Depression, Ketamine, Esketamine, Treatment-resistant depression, Efficacy, Side effects, Anhedonia, Suicidal ideation

1 Introduction

Depression is a global health concern, affecting 300 million individuals [1]. The COVID-19 pandemic and subsequent economic crisis further escalated the incidence of major depressive disorder (MDD) and suicidality [2]. The majority of antidepressants take a considerable amount of time to work, and often patients do not even respond to these medications. This time delay leads to significant suffering among patients and families, placing a considerable burden on caregivers. Depression is associated with high morbidity and mortality. Electroconvulsive therapy (ECT) is one effective therapeutic option but induces significant adverse effects such as neurocognitive impairment. It is not considered a first-line treatment. Other interventions, such as partial wake and/or light therapy [3], have shown rapid antidepressant effects, but questions persist regarding their sustained efficacy.

Sabine Himmelseher (ed.), *Ketamine: Pain Management, Critical Care, Depression, and Resource-Poor Settings Volume II*, Neuromethods, vol. 225, https://doi.org/10.1007/978-1-0716-4599-4_13,

Intravenous (IV) racemic ketamine, an analgesic and anesthetic agent used since the 1970s, has extensively been studied since first randomized controlled trials (RCTs) reported antidepressant effects: in a RCT published in 2000, Berman et al. [4] and, in a cross-over RCT published in 2006, Zarate et al. [5] demonstrated significant antidepressant efficacy of ketamine. Meanwhile, ketamine has emerged as "an examplar" of rapid-acting antidepressants worldwide. Initial studies showed response rates of up to 70% with a single infusion, but "real-world studies" have reported lower response rates. Nevertheless, ketamine may efficiently reduce depressive symptomatology in a significant number of patients where traditional antidepressants have failed.

A problem with IV racemic ketamine is the lack of official US Food and Drug Administration (FDA) approval for clinical use in depressive disorders. This is associated with legal issues and—as any off-label drug use—may become a serious problem in case of litigation and liability questions. It leads to lack of coverage by many insurance providers and often costs for patients and financial disparities. Esketamine, the left-handed enantiomer of worldwide used racemic ketamine, shows a three to four times higher affinity at the N-methyl-D-aspartate (NMDA) receptor compared to right-handed arketamine [6]. This results—at least in part—in a two times greater analgesic, anesthetic, and likely antidepressant efficacy of esketamine as compared to racemic ketamine [7]. In addition to other advantages, esketamine may offer the benefits of shorter drug emergence and (neurocognitive) drug recovery. In 2019, a patented intranasal (IN) formulation of esketamine has been granted FDA approval for adjunctive use in adults with unipolar treatment-resistant depression (TRD) receiving a traditional oral antidepressant. IN esketamine is currently more expensive than IV racemic ketamine but tends to be covered by insurances more frequently than IV racemic ketamine.

Usually IV racemic ketamine is administered at sub-anesthetic dose of 0.5 mg/kg actual body weight as an infusion over 40 min. It may thus exert an antidepressant effect that begins even within 40 min and can last up to a week or longer. However, for a sustained antidepressant response, and/or remission, most patients require ongoing or repeated infusions [8], and there is a lack of consensus regarding the duration of ketamine therapy for maintaining treatment gains [9]. In addition to approval in 2019 [10], esketamine also received FDA authorization for therapy of MDD with suicidal ideation (SI) or behavior in 2020 and as a monotherapy for adults with TRD in 2025.

Despite numerous studies on the efficacy of IV racemic ketamine and IN esketamine, there is currently no head-to-head RCT comparing the efficacy and side effect profiles of the two ketamine formulations [11]. However, there are a few studies that made comparisons in other study designs: in a double-blind, non-inferiority RCT, Correia-Melo et al. [12] allocated participants with TRD to a single 40-min IV infusion of racemic ketamine at a dose of 0.5 mg/kg or to IV esketamine at dose of 0.25 mg/kg. At equipotent dose and equal bioavailability, IV esketamine was effective as antidepressant as IV racemic ketamine. Both treatments were deemed safe and well tolerated. In a meta-analysis published in 2021, Bahji et al. evaluated initial RCTs that utilized either IV racemic ketamine or IN esketamine for therapy of TRD [13]. The results from the meta-analysis showed that IV racemic ketamine was associated with higher efficacy and fewer patient dropouts compared to IN esketamine. In the following, observational studies in real-world settings were performed to offer more insights for improving patient treatment, guide healthcare providers, and provide hypotheses-generating data for controlled trials. In one study, Nikayin et al. [14] found similar remission and response rates for both ketamine formulations, but secondary outcomes favored IV racemic ketamine. Our group also examined a cohort of patients with TRD who received either IV racemic ketamine or IN esketamine in a first short-term observation. In this chapter, we present interim results from this recently published retrospective study [15] and will discuss challenges with comparisons of the two ketamine preparations.

2 Study Patients and Methods

2.1 Study Participants

We compared IV racemic ketamine with IN esketamine for TRD in a retrospective, "real world environment," short-term evaluation. Patients received up to 6 IV ketamine infusions (each 0.5 mg/kg over 40–100 min) or up to 8 IN esketamine treatments with a dose of 56 or 84 mg at the Mayo Clinic Depression Center, Rochester, MN, USA.

2.2 Methods

The Institutional Review Board (IRB)-approved evaluation included consenting adults above 18 years of age with TRD. Data were collected from electronic health records by trained psychiatrists and fellows. The Mayo Clinic Depression Center offers IV racemic ketamine therapy for 18- to 64-year-old patients with MDD or bipolar disorder with TRD. IN esketamine was administered according to FDA guidelines for patients with treatment-resistant MDD without an upper age limit. The inclusion criteria for both treatments were failure to respond to at least two adequate trials of established antidepressant therapies (including conventional antidepressants, mood stabilizers, atypical antipsychotic drugs with indication for bipolar depression, ECT, or transcranial magnetic stimulation). Exclusion criteria included psychosis, recent substance abuse (excluding nicotine and caffeine), and cognitive and non-mood primary psychiatric disorders [16]. Study details have been published earlier [15, 17–19].

Depression symptoms were measured with the 16-Item Quick Inventory of Depressive Symptomatology-Self Report (QIDS-SR) immediately before and 24 h after treatment [20]. Of note, some patients from the Mayo Clinic Depression Center also participated in an open-label, non-randomized, multicenter study investigating the role of ketamine and biomarker development in patients with TRD [21, 22]. If a patient's data was collected using the Montgomery Asberg Depression Rating Scale (MADRS) [23] as part of that study, we converted it to QIDS-SR as recommended [24]. Treatments were generally administered twice or thrice per week, with QIDS-SR depression scale scores recorded 24 h post-treatment and before the next treatment. Outcome assessments were based on changes from baseline to the lowest score within this period. Response was defined as a reduction $\geq$ 50% and remission as a reduction $\leq$ 5 in QIDS-SR score, respectively.

Statistical analysis included Wilcoxon rank sum and Fisher exact tests for demographic, clinical, and medication differences between IV racemic and IN esketamine-treated patients. Cox proportional hazard models evaluated associations between number of treatments to response/remission and treatment group, adjusting for age, sex, body mass index (BMI), duration of depressive episode, and baseline QIDS-SR score. Among patients with baseline SI data, Cox proportional hazard models evaluated the association between number of treatments to response with any baseline SI adjusting for age, sex, BMI, treatment group, and modified baseline QIDS-SR score and, among those with baseline SI, between number of treatments to 50% reduction in SI with treatment group adjusting for age, sex, BMI, and modified baseline QIDS-SR score. Analyses were conducted using R version 4.0.3 and SAS Studio 3.81, with p-values <0.05 considered significant.

3 Results

A total of 44 patients with treatment-resistant MDD and 3 with treatment-resistant bipolar depression received IV racemic ketamine (14/33 males/females, mean age 46 $\pm$ 12 years, 25 employed/22 unemployed). 15 patients with treatment-resistant MDD received IN esketamine (8/7 males/females, mean age 50 $\pm$ 13 years, 8 employed, 7 unemployed). The dose of IN esketamine was 54 mg in 6 patients, and 84 mg in 9 patients. In the acute phase, 25 patients needed 1–3 treatments of IV racemic ketamine, and 22 patients needed 4–6 treatments. One patient needed 1 treatment with IN esketamine, and 14 patients needed 7–8 treatments. BMI was higher in IN esketamine patients (median 32 vs 28; $p = 0.05$), and depressive episode was longer (8.0 vs

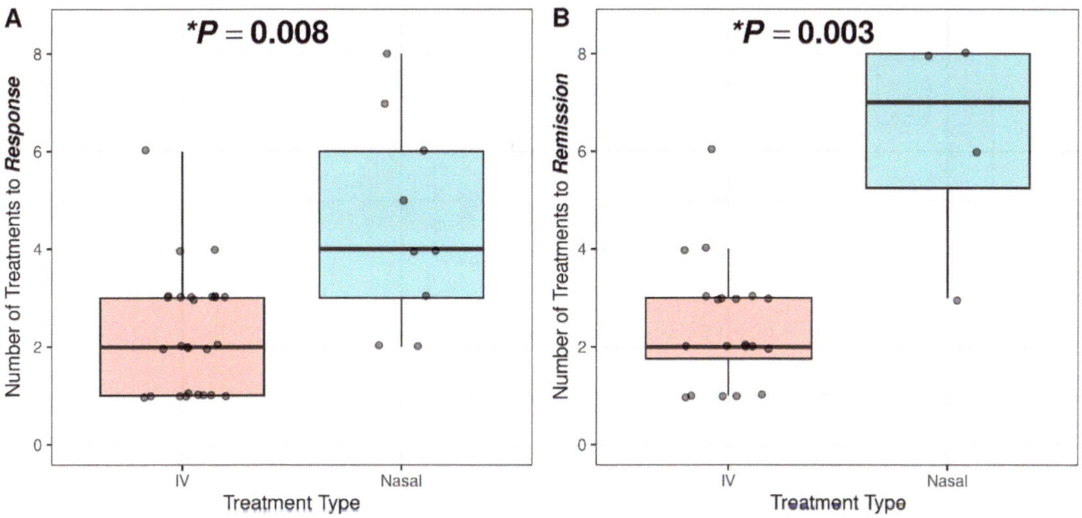

Fig. 1 Boxplots of number of treatments to response (**a**) and remission (**b**) with IV racemic ketamine (red color) versus IN esketamine (blue color). Boxes display the median and interquartile range (IQR), whiskers extend from the hinge to the most extreme value that is no further than 1.5 times the IQR from the hinge, and dots show individual points

2.3 years; $p = 0.01$). Fibromyalgia and chronic pain were more prevalent among the responders compared to nonresponders (19% vs 0.0%; $p = 0.04$), and there was a trend suggesting a higher prevalence of these conditions among remitters compared to non-remitters (21% vs 5%; $p = 0.098$). No significant associations were found between response/remission and other comorbidities, employment status, or the evaluated demographic/clinical characteristics in all patients.

The baseline depression QIDS-SR score was 17.30 ± 3.79 for IV racemic ketamine and 19.07 ± 1.87 for IN esketamine. The end-point QIDS-SR score was 8.57 ± 5.54 for IV racemic ketamine and 9.8 ± 5.14 for IN esketamine. Response rates were 57% and 60%, and remission rates were 43% with IV racemic and 27% with IN esketamine, respectively. Among those who achieved response and remission, respectively, patients receiving IV racemic ketamine compared to IN esketamine required fewer treatments to achieve response (median 2.0 vs 4.0) and remission (median 2.0 vs 7.0) ($p < 0.01$) (Fig. 1). After adjusting for possible confounders (age, sex, BMI, duration of depressive episode, and baseline QIDS-SR score), both, the time—defined as the number of treatments—to response (Hazard ratio, HR = 2.65, $p = 0.06$) and remission were faster with IV racemic ketamine (HR = 6.35, $p = 0.01$). Figure 2 shows Kaplan-Meier and model-adjusted curves of percentage of patients without response (A) and remission (B) after IV racemic ketamine versus IN esketamine. Model-adjusted curves were obtained by averaging individual predictions from the Cox

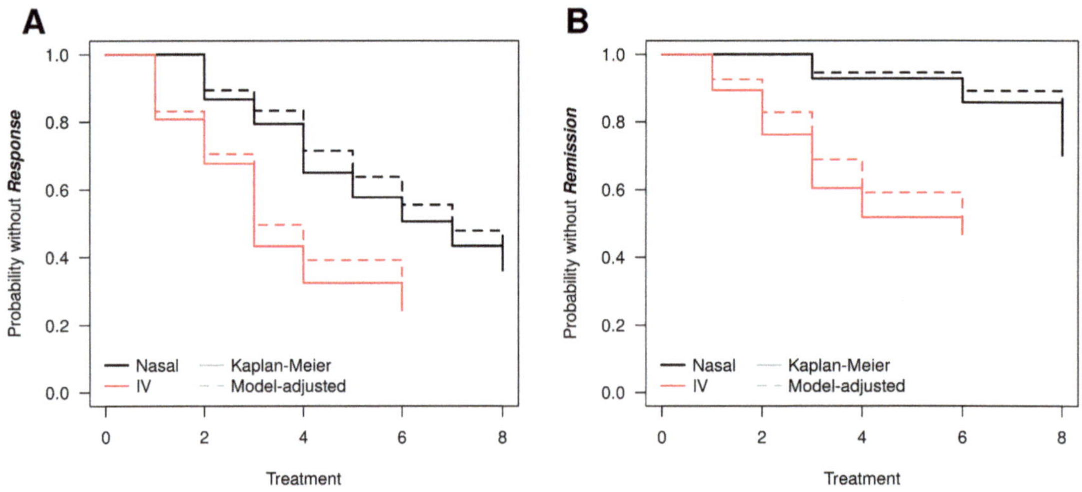

Fig. 2 Kaplan-Meier and model-adjusted curves of percentage of patients without response (**a**) and remission (**b**) after IV racemic ketamine (red lines) versus IN esketamine (black lines). Model-adjusted curves were obtained by averaging individual predictions from the Cox model including main effects for age, sex, BMI, duration of depressive episode, and baseline QIDS-SR score with treatment group as a strata

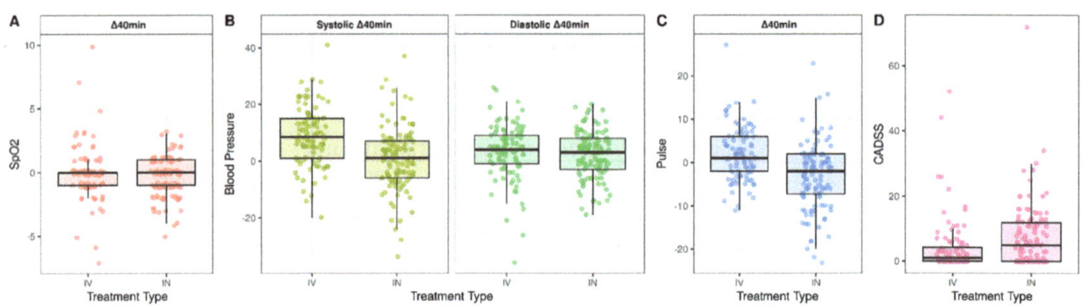

Fig. 3 Boxplots of different vital signs after IV racemic ketamine versus IN esketamine indicating $\Delta =$ change from baseline to 40 min. *SpO*$_2$, partial arterial oxygen saturation measured with pulse oximetry (red color) (**a**), systolic (yellow color) and diastolic blood pressure (green color) (**b**), *pulse* heart rate (blue color) (**c**), and *CADSS* clinician-administered dissociative states scale scores (pink color) (**d**). Boxes display the median and interquartile range (IQR), whiskers extend from the hinge to the most extreme value that is no further than 1.5 times the IQR from the hinge, and dots show individual points

model including main effects for age, sex, BMI, duration of depressive episode, and baseline QIDS-SR score with treatment group as a strata.

The evaluation of adverse effects shows that IV racemic ketamine was associated with a greater increase in systolic blood pressure and heart rate (pulse) at 40 min after start of therapy compared to IN esketamine. No differences were observed in oxygen saturation, diastolic blood pressure, or unwanted psychoactive effects as assessed with the Clinician-Administered Dissociative States Scale (CADSS) (Fig. 3).

In addition, data about SI were available from 37 patients who received IV racemic and 15 receiving IN esketamine [17] (84% of overall patient cohort). Before treatment, 81% of these reported SI, and after therapy, 60% reported improved SI scores. 38% of patients did not report a change. After adjusting for covariates, the HR for response was significantly lower in those with baseline SI (HR = 0.36, 95% CI 0.14–0.92, $p = 0.03$). However, the number of treatments required to achieve a 50% reduction in SI scores did not differ significantly between therapy with IV racemic and IN esketamine ($p = 0.44$).

4 Discussion

The main outcome of this short-term, retrospective evaluation is that IV racemic ketamine induced an earlier response to therapy than IN esketamine, but both treatments were associated with similar response and remission rates at the end of the acute phase observation period. These findings align with those of a recent publication by Nikayin et al. [14]. In a new post hoc analysis of pooled real-world data, d'Andrea et al. [25] compared response and remission rates of 171 patients with IV racemic ketamine treatment at the Canadian Rapid Treatment Center of Excellence in Toronto, Canada, with those of 140 patients treated at multiple TRD centers in Italy. Both sites included patients with insufficient response to two antidepressant trials. The analysis found that there was a significant difference in response rates being 36% for IV racemic versus 25% for IN esketamine ($p = 0.04$). Remission rates were 13% for IV racemic and 12% for IN esketamine. Discontinuation and drop-out rates were similar with both treatments. After IV racemic ketamine, a higher incidence of dissociative symptoms, increased *incidence of* dizziness, and treatment-emergent hypertension were noted.

These three observational, partially retrospective studies [14, 15, 25] demonstrate the dilemma of clinical caregivers in daily practice when decision-making for choice of IV racemic versus IN esketamine in patients with MDD or TRD is necessary. Table 1 summarizes remarkable formal differences between IV racemic ketamine versus IN esketamine [8, 26].

From a medical efficacy perspective, pooled data from the studies show a non-significant difference between IV racemic and IN esketamine in short-term response (odds ratio (OR): 1.34, 95% CI: 0.94–1.92, $p = 0.11$) and remission (OR: 1.29, 95% CI: 0.83–2.00, $p = 0.25$) [11]. However, IV ketamine can provide faster relief, which may be beneficial in patients who need quick symptom improvement [11].

Table 1
Different conditions for IV racemic ketamine versus IN esketamine use for therapy of major depressive disorder and treatment-resistant depression

IV racemic ketamine	IN esketamine
Legal issues	
Off-label use	FDA-approved
No FDA authorization	
Monitoring open to physician discretion	2-h monitoring time required after administration
Coverage/costs	
Insurance coverage often difficult	Mostly covered by insurance with prior authorization
Patients may need to pay out-of-pocket	
Financial concerns	
Only patients with financial means may be able to afford therapy, especially maintenance treatment	More expensive
Dosage unclarities	
Established efficacy at 0.5 mg/kg body weight	Likely better efficacy with doses >28 mg
No additional benefits at 1 mg/kg body weight	Less depressive symptoms with 56 or 84 mg
Social concerns	
No car driving/no operation of heavy machines on day of use	No car driving/no operation of heavy machines on day of use

5 Limitations

There are severe limitations to this work: for example, overall trial evidence is small and of very low certainty. A formal risk-of-bias analysis is not possible, and retrospective evaluation has many confounders. Our study included a limited sample size of 62 real-world participants, who continued their antidepressant treatment. The presented interim data covered only a very small period of time after therapy initiation, and results do not provide trajectory to long-term outcomes. Therapeutic efficacy has been assessed with very few scales which may not depict the entire range of psychic change, and no placebo or active comparator was or could be included. Psychoactive effects were only measured with CADSS scales and not consistently among the studies. While patients received pre-education by mood psychiatrists regarding potential adverse effects of ketamine/esketamine at our site, individual responses to these treatments can vary. The administration of medications for nausea or hypertension was permitted; however, consistent reporting of exact data among the published studies was lacking. Patients were not characterized for their individual disposition to metabolize ketamine, such as with regard to age, genetic differences, ethnicity, medication or food inducing or inhibiting CYP ketamine metabolizing enzymes [27–31]. However, there is currently inconsistent evidence regarding the impact of genetic differences on patient outcomes with ketamine and esketamine. There were no blood concentration measurements of ketamine and/or its metabolites, as this was an observational study and not an RCT. Currently, there is minimal data on ketamine levels and their correlation with antidepressant response in humans. Future studies are necessary to explore the optimal ketamine level for therapeutic response.

Moreover, a significant challenge is to prevent "an unfair"—if ever possible direct—comparison between IV racemic and IN esketamine with regard to dose and effect: at present, the NMDA receptor is still considered as first and essential site of ketamine action [32], and esketamine shows a three to four greater NMDA receptor affinity than arketamine [6], with the result of esketamine's clinical potency being approximately twice that of racemic ketamine [7]. This difference has also been confirmed with respect to antidepressant efficacy when IV half-dose 0.25 mg/kg esketamine was compared to IV 0.5 mg/kg racemic ketamine infused over the same time period resulting in similar improvement [11]. In addition, there is a difference in absolute bioavailability between IV and IN drug administration: IV ketamine has an approximate bioavailability > 98% and IN esketamine applied with the new patented device and skillful use >50% [28]. However, as with every IN drug administration, a part may get uncontrollably swallowed. This will undergo hepatic first-pass metabolism, and ~ 20% of the oral intake will reach the systemic circulation [29]. Moreover, esketamine is metabolized more rapidly than the racemate [29, 33]. Major metabolites of (es)-ketamine, such as nor (es)-ketamine, are detected within ~3 min after IV administration, and different application routes may result in different, multiple ketamine metabolites with pharmacologic effects of their own [34, 35]. Unfortunately, as yet, there are only very few pharmacokinetic/pharmacodynamic (PK/PD) paradigms modeling at least some of these multiple ketamine effects and interactions [36–38]. For a fair comparison, racemic and esketamine would have to be compared at equipotent dose although dose change deficit did not show a significant interaction with treatment in our cohort. Additionally, animal model studies suggest that R-ketamine may exhibit greater potency and longer-lasting antidepressant effects compared to esketamine [39].

6 Conclusion

Our preliminary findings need to be replicated in RCT highlighting the need to assess dose and use schedules for therapy of patients presenting with SI [17]. The impact of repeated ketamine treatment on neurocognitive functions in MDD and TRD also requires further investigation. A RCT reported no difference in pro-cognitive benefit directly comparing IV racemic with versus IV esketamine [40], and a new multicenter trial points out to cognitive improvement, not deterioration, with serial IV racemic ketamine in TRD in short-term evaluation [21].

In conclusion, both IV racemic ketamine and IN esketamine demonstrate efficacy for TRD with reasonable tolerability. IV racemic ketamine was associated with an accelerated response which may have been caused by higher blood concentrations after IV racemic as compared to IN esketamine. This has clinical implications when helping patients make informed decisions with regard to choosing IV racemic or IN esketamine as therapeutic approach to severe depression in combination with the different formal conditions. In view of the gap of knowledge about the above-named interfering PK/PD factors, only pragmatic comparisons between the two ketamine formulations can be performed at present.

Acknowledgments

We would like to extend our sincere gratitude to Vanessa K. Pazdernik, MS, for the help with data analysis and the creation of figures for this book chapter, and to Sabine Himmelseher, MD, for the assistance in reviewing this chapter and providing insightful feedback and thorough critique, which significantly enhanced the quality and clarity of our work.

Conflict of Interest

Dr. Singh has received research grant support from Mayo Clinic, the National Network of Depression Centers Momentum grant, and Breakthrough Discoveries for thriving with bipolar disorder (BD2). He is a KL2 Mentored Career Development Program scholar, supported by CTSA grant KL2TR002379 from the National Center for Advancing Translational Science (NCATS). Dr. Singh has received honoraria (to Mayo Clinic) from Elsevier for editing a Clinical Overview on Treatment-Resistant Depression. Dr. Vande Voort has received grant-in-kind support from Assurex Health (unrelated to this study). Dr. Kung received payment for a recorded CME presentation about transcranial magnetic stimulation by Psychopharmacology Institute. Funding sources had no influence on the work reported.

References

1. Institute of Health Metrics and Evaluation (2024) Global Health Data Exchange (GHDx). https://vizhub.healthdata.org/gbd-results. Accessed 23 Apr 2024
2. Klomek AB (2020) Suicide prevention during the COVID-19 outbreak. Lancet Psychiatry 7(5):390
3. Singh B, Kung S, Poppe KA, Brown SL, Erickson C, Hale AM et al (2024) Partial wake therapy and luminotherapy as an adjunct to usual care in hospitalized patients with depression – a pilot study. Psychiatry Res 337: 115921
4. Berman RM, Cappiello A, Anand A, Oren DA, Heninger GR, Charney DS et al (2020) Antidepressant effects of ketamine in depressed patients. Biol Psychiatry 47(4):351–354
5. Zarate CA Jr, Singh JB, Carlson PJ, Brutsche NE, Ameli R, Luckenbaugh DA et al (2006) A randomized trial of an N-methyl-D-aspartate antagonist in treatment-resistant major depression. Arch Gen Psychiatry 63(8):856–864
6. Oye I, Paulsen O, Maurset A (1992) Effects of ketamine on sensory perception: evidence for a role of N-methyl-D-aspartate receptors. J Pharmacol Exp Ther 260(3):1209–1213
7. Mion G, Himmelseher S (2024) Esketamine: less drowsiness, more analgesia. Anesth Analg 139:78–91. https://doi.org/10.1213/ANE.0000000000006851
8. Seshadri A, Prokop LJ, Singh B (2024) Efficacy of intravenous ketamine and intranasal esketamine with dose escalation for major depression: a systematic review and meta-analysis. J Affect Disord 356:379–384
9. Nash AI, Borentain S, O'Hara M, Zhang Q, Barbreau S, Haughey SM, et al (2021) National network of depression centers annual conference. Virtual meeting
10. Kim J, Farchione T, Potter A, Chen Q, Temple R (2019) Esketamine for treatment-resistant depression – first FDA-approved antidepressant in a new class. N Engl J Med 381(1):1–4
11. Singh B, Kung S, Vande Voort JL (2024) Intravenous (IV) ketamine versus intranasal esketamine for depression– advantage IV ketamine? J Affect Disord 56:564–567
12. Correia-Melo FS, Leal GC, Vieira F, Jesus-Nunes AP, Mello RP, Magnavita G et al (2020) Efficacy and safety of adjunctive therapy using esketamine or racemic ketamine for adult treatment-resistant depression: a randomized, double-blind, non-inferiority study. J Affect Disord 264:527–534
13. Bahji A, Vazquez GH, Zarate CA Jr (2021) Comparative efficacy of racemic ketamine and esketamine for depression: a systematic review and meta-analysis. J Affect Disord 278:542–555
14. Nikayin S, Rhee TG, Cunningham ME, de Fontnouvelle CA, Ostroff RB, Sanacora G (2022) Evaluation of the trajectory of depression severity with ketamine and esketamine treatment in a clinical setting. JAMA Psychiatry 79(7):736–738
15. Singh B, Kung S, Pazdernik V, Schak KM, Geske J, Schulte PJ et al (2023) Comparative effectiveness of intravenous ketamine and intranasal esketamine in clinical practice among patients with treatment-refractory depression: an observational study. J Clin Psychiatry 84(2):22m14548
16. Singh B, Bobo WV, Rasmussen KG, Stoppel CJ, Rico JA Jr, Schak KM et al (2019) The association between body mass index and remission rates in patients with treatment-resistant depression who received intravenous ketamine. J Clin Psychiatry 80(6):19l12852
17. Singh B, Vande Voort JL, Pazdernik VK, Frye MA, Kung S (2024) Treatment-resistant depression patients with baseline suicidal ideation required more treatments to achieve therapeutic response with ketamine/esketamine. J Affect Disord 351:534–540
18. Patarroyo-Rodriguez L, Pazdernik V, Vande Voort JL, Kung S, Singh B (2024) Hypersomnia as a predictor of response to intravenous ketamine/intranasal esketamine in treatment resistant depression. J Affect Disord 349:272–276
19. Joseph B, Nunez NA, Kung S, Vande Voort JL, Pazdernik VK, Schak KM et al (2023) Efficacy of ketamine with and without lamotrigine in treatment-resistant depression: a preliminary report. Pharmaceuticals (Basel) 16(8):1164
20. Rush AJ, Trivedi MH, Ibrahim HM, Carmody TJ, Arnow B, Klein DN et al (2003) The 16-item quick inventory of depressive symptomatology (QIDS), clinician rating (QIDS-C), and self-report (QIDS-SR): a psychometric evaluation in patients with chronic major depression. Biol Psychiatry 54(5):573–583
21. Singh B, Parikh SV, Voort JLV, Pazdernik VK, Achtyes ED, Goes FS et al (2024) Change in neurocognitive functioning in patients with treatment-resistant depression with serial intravenous ketamine infusions: the bio-K multicenter trial. Psychiatry Res 335:115829

22. Parikh SV, Vande Voort JL, Yocum AK, Achtyes E, Goes FS, Nykamp L et al (2024) Clinical outcomes in the biomarkers of ketamine (bio-K) study of open-label IV ketamine for refractory depression. J Affect Disord 348: 143–151
23. Montgomery SA, Asberg M (1979) A new depression scale designed to be sensitive to change. Br J Psychiatry 134:382–389
24. Carmody TJ, Rush AJ, Bernstein IH, Brannan S, Husain MM, Trivedi MH (2006) Making clinicians lives easier: guidance on use of the QIDS self-report in place of the MADRS. J Affect Disord 95(1–3):115–118
25. d'Andrea G, Pettorruso M, Di Lorenzo G, Rhee TG, Chiappini S, Carullo R et al (2024) The rapid antidepressant effectiveness of repeated dose of intravenous ketamine and intranasal esketamine: a post-hoc analysis of pooled real-world data. J Affect Disord 348: 314–322
26. Fava M, Freeman MP, Flynn M, Judge H, Hoeppner BB, Cusin C et al (2020) Double-blind, placebo-controlled, dose-ranging trial of intravenous ketamine as adjunctive therapy in treatment-resistant depression (TRD). Mol Psychiatry 25(7):1592–1160
27. Yamoune S, Müller JP, Langmia IM et al (2024) Uncoupling of cytochrome P450 2B6 and stimulation of reactive oxygen species production in pharmacogenomic alleles affected by interethnic variability. Biochim Biophys Acta Gen Subj 1868:130595
28. Perez-Ruixo C, Rossenu S, Zannikos P et al (2022) Population pharmacokinetics of esketamine nasal spray and its metabolite noresketamine in healthy subjects and patients with treatment-resistant depression. J Clin Pharm Ther 60:501–516
29. Peltoniemi MA, Hagelberg NM, Olkkola KT, Saari T (2016) Ketamine: a review of clinical pharmacokinetics and pharmacodynamics in anesthesia and pain therapy. Clin Pharmacokinet 55:1059–1071
30. Nguyen TML, McGowan JC, Gardier AM (2022) CYP 450 enzymes influence (R,S)-ketamine brain delivery and its antidepressant activity. Neuropharmacology 206:108936
31. Langmia IM, Just KS, Yamoune S et al (2022) Pharmacogenetic and drug interaction aspects on ketamine safety in its use as antidepressant – implications for precision dosing in a global perspective. Br J Clin Pharmacol 88:5149–5165
32. Zhang Y, Ye F, Zhang T et al (2021) Structural basis of ketamine action on human NMDA receptors. Nature 596:301–305
33. Ihmsen H, Geisslinger G, Schüttler J (2001) Stereoselective pharmacokinetics of ketamine: R(−)-ketamine inhibits the elimination of S (+)-ketamine. Clin Pharmacol Ther 70:431–438
34. Glue P, Russell B, Medlicott NJ (2021) Influence of formulation and route of administration on ketamine's safety and tolerability: a systematic review. Eur J Clin Pharmacol 77: 671–676
35. Zanos P, Moaddel R, Morris PJ et al (2018) Ketamine and ketamine metabolite pharmacology: insights into therapeutic metabolites. Pharmacol Rev 70:621–660
36. Henthorn TK, Avram MJ, Dahan A et al (2018) Combined recirculatory-compartmental population pharmacokinetic modeling of arterial and venous plasma S(+)- and R(−) ketamine concentrations. Anesthesiology 129:260–270
37. Abuhelwa AY, Somogyi A, Loo CK et al (2022) Population pharmacokinetics and pharmacodynamics of the therapeutic and adverse effects of ketamine in patients with treatment-refractory depression. Clin Pharmacol Ther 112:720–729
38. Willemin ME, Zannikos P, Mannens G et al (2022) Prediction of drug-drug interactions after esketamine intranasal administration using a physiologically based pharmacokinetic model. Clin Pharmacokinet 61:1115–1128
39. Yang C, Shirayama Y, Zhang JC, Ren Q, Yao W, Ma M et al (2015) R-ketamine: a rapid-onset and sustained antidepressant without psychotomimetic side effects. Transl Psychiatry 5(9):e632
40. Araujo-de-Freitas L, Santos-Lima C, Mendonca-Filho E, Vieira F, Franca R, Magnavita G et al (2021) Neurocognitive aspects of ketamine and esketamine on subjects with treatment-resistant depression: a comparative, randomized and double-blind study. Psychiatry Res 303:114058

Chapter 14

Intranasal Esketamine for Management of Treatment-Resistant Depression: Evidence of Efficacy and Safety and Use in Clinical Practice

Ludovic Samalin, Ludivine Boudieu, and Pierre-Michel Llorca

Abstract

Esketamine, the only intranasal (IN) antidepressant approved by the FDA and EMA for therapy of treatment-resistant depression (TRD) has demonstrated significant depressive symptom reduction in short-term, randomized, placebo-controlled study. It induced superior remission rates compared to extended-release quetiapine after week 8 of therapy in TRD as well. Providing rapid onset of action post-administration, its efficacy can be sustained with repeated use. Long-term study also indicates increased rates of response and remission, with reduced risk of relapse. Real-world observational data corroborate these findings, showing significant symptom improvement at various time points after IN esketamine therapy. In general, IN esketamine has been well tolerated, with mild to moderate adverse events resolving within hours of administration. This chapter narratively reviews available evidence of the efficacy and safety of IN esketamine as well as important considerations for clinical practice when applied as therapy for TRD.

Key words Esketamine, Intranasal formulation, Antidepressant, Major depressive disorder, Treatment-resistant-depression, Trials, Real-world experience

1 Introduction

Despite different options for management of treatment-resistant depression (TRD) including antidepressant switching or combination strategies (such as adding another antidepressant or adjunctive medication like lithium or quetiapine), the prevalence of TRD remains high. In France, it reaches up to 37.6 per 10,000 inhabitants [1]. TRD is associated with an elevated risk of suicide, increased physical and psychic comorbidity, and poor quality of life [2, 3]. Many patients with TRD suffer from heavy disease, socio-economic burden, and healthcare costs [4, 5].

Research on the complex pathophysiology of depression highlighted abnormalities in glutamatergic transmission in neural circuits involved in emotional behavior and mood regulation

Sabine Himmelseher (ed.), *Ketamine: Pain Management, Critical Care, Depression, and Resource-Poor Settings Volume II*, Neuromethods, vol. 225, https://doi.org/10.1007/978-1-0716-4599-4_14,

[6]. These data led to the study of antidepressant effects of agents such as intravenous (IV) ketamine which blocks glutamatergic N-methyl-D-aspartate receptors (NMDAR) as first mechanism and exerts multiple neuroplastic downstream effects. Worldwide used ketamine is a racemic mixture of two stereoisomers, with dextrorotatory esketamine (ESKET) showing a four times greater affinity for NMDAR than levorotatory arketamine. This may in part explain why ESKET is twice as potent as analgesic and anesthetic as racemic ketamine [7]. Although rarely studied so far, a few trials have similarly shown comparable antidepressant efficacy of half-dose IV ESKET as compared to IV racemic ketamine (such as with 0.25 mg/kg ESKET versus 0.5 mg/kg IV racemic ketamine [8].

Recently, an intranasal (IN) ESKET application has been developed in a ready-to-use device that delivers a total of 28 mg pure ESKET in two sprays (one spray per nostril) for one-time use [9, 10]. This new IN ESKET delivery system was approved in 2019 by the US Food and Drug Administration (FDA) and by the European Medicines Agency (EMA) for therapy of patients with TRD or for rapid reduction of depressive symptoms in patients with major depressive disorder (MDD) in combination with conventional antidepressant selective serotonin reuptake inhibitors (SSRIs) or serotonin-norepinephrine reuptake inhibitors (SNRIs). For example, acute suicidal ideation or behavior noticed by clinical judgment is regarded as psychiatric emergency in MDD [11] which requires rapid therapy.

In this chapter, we present a narrative review of available data on the efficacy and safety of IN ESKET as well as key considerations for use in clinical practice for therapy of TRD.

2 Literature Selection

A search of the electronic databases PubMed, Embase, MEDLINE, PsycINFO, and Google Scholar was conducted to identify peer-reviewed articles in French and English published to April 2024. The search terms were selected to capture any article with IN ESKET and MDD including TRD: ("depressive disorder, treatment-resistant" [MeSH Terms] OR "depressive disorder, major" [MeSH Terms] AND (resistance) OR (depressive disorder/drug therapy) [MeshTerms] AND ("ketamine") OR ("ESKET") AND ("administration, intranasal" [MeSH Terms] OR "intranasal" [Text Word] OR "nasal sprays" [MeSHTerms].

Articles were considered if they pertained to patients with TRD who were treated with IN ESKET. Randomized controlled trials (RCTs), open-label or observational studies, and guidelines for managing TRD with specific reference to IN ESKET were included.

3 Efficacy and Safety of Intranasal Esketamine in Trials of Treatment-Resistant Depression

A total of 15 studies can be discussed: seven randomized placebo-controlled trials [12–18], one randomized active-controlled trial [19], one open-label study [20], and six observational studies [21–26]. Two recent national guidelines for the management of TRD which include specific recommendations for ESKET use were also taken into account [27, 28].

Depending on the study, patients were treated with their ongoing antidepressant at the time of randomization or received a new oral antidepressant (mainly SSRI or SNRI). IN ESKET was administrated twice weekly during the induction phase (first 4 weeks) in all studies, except two [18, 25]) (Table 1).

Some studies used a fixed dose [13, 16, 18], while in others, the dose was chosen according to the investigator's clinical judgment. After 4 weeks of therapy, the frequency of administration and/or dose of IN ESKET was adjusted [13, 15, 18–21, 24, 29]. During the optimization/maintenance phase, the frequency was decreased to one use per week and/or every other week. Doses were adjusted based on the investigator's clinical judgment of efficacy and tolerability.

The Montgomery-Åsberg Depression Rating Scale (MADRS) was used in most studies to "more objectively" measure change in depressive symptom severity and to allow for comparison of study results. "Response" was commonly defined as 50% or greater reduction in MADRS total score, whereas remission was defined as total score of $\leq$12 or $\leq$10.

3.1 Short-term Efficacy in Randomized, Controlled Trials

The short-term efficacy of IN ESKET was evaluated at different time points after first therapy in six double-blind, RCT [12–14, 16–18]. The first RCT from Daly et al. (2018) evaluated the effects of IN ESKET [18] on day 8 after first use: the mean MADRS total score decreased from baseline to day 8 in all patients with greater improvement after all doses (28, 56, or 84 mg) of ESKET compared to placebo. An ascending dose-response relationship was observed, with the response being significant at 24 h after first use of all ESKET doses.

The six RCT evaluated IN ESKET's efficacy at day 25/28 after first use [12–14, 16–18]. Two of the six RCTs showed a significantly higher improvement in the MADRS score after IN ESKET compared to placebo [17, 18]. In the four other RCTs with the same endpoint analysis, no significant difference between IN ESKET and placebo was found [12–14, 16].

The efficacy of IN ESKET was also evaluated in comparison with quetiapine augmentation therapy in one randomized active-controlled trial (ESCAPE-TRD) in TRD patients receiving an

Table 1
Characteristics of randomized controlled and open-label studies evaluating the efficacy of IN esketamine

Author	Study design	Study population	Primary outcome	Duration	Main findings
Chen et al. [12]	RCT: IN ESK 56–84 mg vs placebo	250 MDD patients with TRD	Change in MADRS total score from baseline to week 4	4 weeks	No statistical difference in MADRS score between ESK and placebo from baseline to week 4
Reif et al. [19]	RCT (ESCAPE-TRD): IN ESK 28–84 mg vs quetiapine	676 MDD patients with TRD	Remission rate at week 8	32 weeks	Significantly greater remission rate after ESK compared to quetiapine at week 8 (27.1 vs 17.6%, $p = 0.003$, respectively)
Takahashi et al. [13]	RCT: IN ESK 28, 56, and 84 mg vs placebo	202 MDD patients with TRD	Change in MADRS score from baseline to week 4	4 weeks (DB phase) 24 weeks (OL phase)	No statistical difference in MADRS score between ESK and placebo from baseline to week 4
Ochs Ross et al. [14]	RCT (TRANS FORM-3): IN ESK 28–84 mg vs placebo	138 MDD patients with TRD and ≥ 65 years old	Change in MADRS score from baseline to week 4	4 weeks	No statistical difference in MADRS score between ESK and placebo from baseline to week 4
Wajs et al. [20]	OL trial (SUSTAIN-2): IN ESK 28–84 mg	802 MDD patients with TRD	Change in MADRS score from baseline to week 4 (IND phase) and from week 5 to week 52 (OP/MAINT phase)	52 weeks	MADRS score decreased in the IND phase (baseline to week 4) (mean [SD] change: −16.4 [8.76]) and persisted during the OP/MAINT phase (week 5 to week 52) (mean [SD] change: 0.3 [8.12])
Daly et al. [15]	RCT: IN ESK 56–84 mg vs placebo	297 MDD patients with TRD	Time to relapse in patients with stable remission after IND phase (week 4)	Until relapse	Significant delay to relapse with ESK compared to placebo (HR 0.49; 95% CI 0.29–0.84; $p = 0.003$)

Fedgchin et al. [16]	RCT (TRANS FORM-1): IN ESK 56 and 84 mg vs placebo	342 MDD patients with TRD	Change in MADRS score from baseline to week 4	4 weeks	No statistical difference in MADRS score between ESK and placebo from baseline to week 4
Popova et al. [17]	RCT (TRANS FORM-2): IN ESK 56–84 mg vs placebo	227 MDD patients with TRD	Change in MADRS score from baseline to week 4	4 weeks	Significantly greater improvement in MADRS score after ESK compared to placebo from baseline to week 4 (LS mean difference [SE]: −4.0 [1.69], $p = 0.02$)
Daly et al. [18]	RCT: IN ESK 28, 56, and 84 mg vs placebo	67 MDD patients with TRD	Change in MADRS score from baseline to week 1 and week 2	2 weeks (DB phase) d15–d74 (OL phase)	Significantly greater improvement in MADRS score after ESK compared to placebo in both periods (mean difference from placebo [SE]: ESK 28 mg: −4.2 [2.09], $p = 0.02$; ESK 56 mg: −6.3 [2.07], $p = 0.001$; ESK 84 mg: −9.0 [2.13], $p < 0.001$

ASI active suicidal ideation, *d* day, *DB phase* double-blind phase, *ESK* esketamine, *HR* hazard ratio, *IN* intranasal, *IND Phase* induction phase, *LS* least-squares, *MADRS* Montgomery-Asberg Depression Rating Scale, *MDD* major depressive disorder, *OL* open-label, *OP/MAINT phase* optimization/maintenance phase, *RCT* randomized controlled trial, *SE/SD* standard error of the mean/standard deviation *TRD* treatment-resistant depression ongoing SSRI/SNRI [19]. The choice of an active comparator is of great clinical significance because quetiapine is a guideline-supported augmentation antipsychotic, commonly used to treat TRD. The primary end point was remission, defined as a score of 10 or less on the MADRS at week 8. The results demonstrated a significantly greater number of patients in remission after IN ESKET versus quetiapine at week 8 (27.1% vs 17.6%, respectively, $p = 0.003$). After this initial treatment, patients continued the study during a 24-week maintenance phase.

3.2 Long-Term Efficacy in Randomized, Controlled Trials

IN ESKET's long-term efficacy in terms of response, remission, and prevention of relapse was evaluated in several studies: one randomized placebo-controlled [15], one randomized active-controlled (ESCAPE-TRD) [19], and one open-label trial [20].

In the ESCAPE-TRD study, remission, response, and change in the MADRS score from baseline were significantly higher after IN ESKET versus quetiapine for 32 weeks follow-up. The study also demonstrated that a higher proportion of patients stayed in remission at week 32 without relapsing after having achieved remission at week 8 (21.7 vs 14.1%) with an adjusted odds ratio of 1.72 (95% confidence interval (CI) 1.15–2.57).

In the open-label study conducted by Wajs et al. (2020), the improvement in MADRS score persisted among patients who responded to IN ESKET and continued for up to 1 year [20], when the percentage of responders and remitters was 76.5 and 58.2%, respectively.

In Daly et al.'s (2019) randomized placebo-controlled trial, patients who received IN ESKET experienced a significantly longer time to relapse compared to those who received placebo. Relapse was defined as a MADRS score of 22 or higher for 2 consecutive assessments separated by 5–15 days or hospitalization for worsening depression or suicide attempt. Analysis of hazard ratios (HR) revealed that combining IN ESKET with an antidepressant reduced the risk of relapse by 51% in patients who achieved stable remission and by 70% in patients who achieved stable response as compared to placebo combined with antidepressant [15].

3.3 Efficacy in Prospective Observational Studies

The observational studies evaluated IN ESKET's efficacy in *TRD patients in real-world conditions* [21–26] (Table 2).

They offered a broader view on IN ESKET's impact, allowing to better assess therapy-relevant factors such as diverse patient response, various comorbidities, and treatment adherence. Individual patient characteristics, circumstances and conditions may affect outcomes differently than a controlled study setting.

Brendle et al.'s (2022) observational study involved 171 patients receiving IN ESKET. They showed a significant ($p < 0.001$) reduction in mean Patient Health Questionnaire-9 (PHQ-9) and Generalized Anxiety Disorder-7 items (GAD-7)

Table 2
Characteristics of observational studies evaluating the efficacy of IN esketamine

Author	Study design	Study population	Main outcomes	Duration	Main findings
Dvorak et al. [25]	OS, IN ESK 28–84 mg	94 patients with TRD	Change in QIDS score	6 months	Significant reduction in QIDS score between baseline and acute phase, i.e., 8 weeks (22.02 ± 5.4 vs 17.45 ± 6.5, $p < 0.0001$) and between baseline and maintenance phase i.e., 4–6 months (22.73 + 4.8 vs 12.15 ± 6.37, $p < 0.001$)
Estrade et al. [21]	OS, IN ESK 28–84 mg	105 MDD patients with TRD	Define trajectories of ESK response using MADRS score	5 months	After two ESK administrations, the MADRS score predicted the 90-day trajectories of response with an accuracy of 80%
Singh et al. [22]	OS, IV racemic Ketamine 0.5 mg/kg vs IN ESK 56–84 mg	62 MDD patients with TRD	Change in QIDS-SR score and time to achieve response and remission	Up to 6 weeks	No significant difference in response and remission rates between IV racemic ketamine and IN ESK. Significant faster time to remission after IV racemic ketamine compared to IN ESK (HR = 5.0, $p = 0.02$)
Brendle et al. [23]	OS, IN ESK 56–84 mg	171 MDD patients	Change in GAD-7 and PHQ-9 scores	1–71 treatment sessions	Significant reduction in PHQ-9 and GAD-7 scores between baseline and last available treatment (PHQ-9 = mean [SD]: 16.7 [5.82] vs 12 [6.38], $p < 0.001$; GAD-7 = mean [SD]: 12.0 [5.8] vs 8.7 [5.62], $p < 0.001$)
Martinotti et al. [24]	OS, IN ESK 28–84 mg	116 MDD patients with TRD	Change in MADRS score and response and remission rates	3 months	Significant improvement in MADRS score at month 1 (T1) and 3 (T2) (student t-test T1 vs baseline: $t - 15.79$, $p < 0.0001$; T2 vs baseline: $t - 18.07$, $p < 0.0001$)
Samalin et al. [26]	OS, IN ESK 28–84 mg	157 MDD patients with TRD	Characteristics of TRD patients receiving IN ESK and change in MADRS score	12 months	During the 12-month follow-up, 64.6% of participants reached a clinical response. Median time to response with ESK was estimated at 5.7 weeks (95% CI [4.1–8.4])

CI confidence interval, *ESK* esketamine, *GAD* general anxiety disorder; *HR* hazard ratio, *i.e.* that is, *IN* intranasal, *IV* intravenous; *MADRS* Montgomery-Asberg Depression Rating Scale; *MDD* major depressive disorder, *OS* observational study, *PHQ-9* patient health questionnaire, *QIDS-SR* quick inventory of depressive symptomatology-self report, *SD* standard deviation, *TRD* treatment resistant depression, *vs* versus scores from baseline to last available reported treatment (1–71 treatment sessions) [23]. Martinotti et al. (2022) observed a significant improvement in depressive symptoms in 117 patients treated with IN ESKET, with reductions in MADRS scores at 1 month and 3 months' follow-up. 64.2% of patients achieved a clinical response, and 40.6% reached remission after 3 months [24]. Estrade et al.'s (2023) observational study of 105 MDD patients with TRD showed that 52.4% responded and 38.1% remitted 4 weeks after starting IN ESKET. Trajectory analysis identified distinct response patterns, with MADRS scores after two administrations predicting the 90-day trajectory [21]. Singh et al.'s (2023) real-world observational study compared IV racemic ketamine and IN ESKET in 47 and 15 patients, respectively. While acute phase outcomes were similar, IV ketamine led to faster remission compared to IN ESKET (HR = 5.0, $p = 0.02$) [22]. However, as the authors mentioned, from a statistical perspective, the IN ESKET group lacked sufficient power, leading to a potential risk of error in the comparison with the ketamine group. Additionally, the lack of randomization in the groups predisposed the study to a higher risk of bias.

Dvorak et al. (2024) observed a significant improvement in depressive symptoms among 94 TRD patients treated with IN ESKET. There were reductions in Quick Inventory of Depressive Symptomatology (QIDS) scores between baseline and the acute (8 weeks) phase, as well as between baseline and maintenance (4–-6 months). Additionally, 20.8% and 46.5% of patients achieved a clinical response at 8 weeks and at the end of the maintenance phase. It is important to note that response was here defined as improvement in depressive symptoms from baseline to 30% (and not 50%) [25]. Samalin et al. (2024) found that among the 144 patients (91.7%) who completed the MADRS at baseline and met criteria to be included in the response analysis, 93 patients (64.6%) reached a clinical response during the 12 month-follow-up period after a median time of 3.1 weeks (interquartile range (IQR) 1.1;5.7) [26].

In summary, short-term study demonstrated improvement in depressive symptoms with IN ESKET in combination with conventional antidepressants in therapy of TRD. The response was rapid in onset with clinically meaningful benefit 24 h after first use.

IN ESKET often showed sustained efficacy in long-term study, with higher remission and response rates compared to placebo. It significantly delayed relapse in patients who achieved stable response and remission.

Observational studies confirmed the findings of RCTs, showing significant depressive symptom reduction and improved response/remission rates in TRD patients after IN ESKET initiation. Different response patterns were identified in diverse settings.

4 Safety and Adverse Events with Intranasal Esketamine as Reported in Trials on Treatment-Resistant Depression

Overall, IN ESKET was tolerated very well. The most common unwanted (psychoactive) effects were treatment-emergent adverse events (TEAEs) with mild to moderate severity. Usually, they occurred within ~40 min after drug administration and resolved on the day of use (within 1.5–2 h). During the induction phase, common (incidence $\geq$ 10% of patients) TEAEs included dizziness, dissociation, nausea, headache, somnolence, increased blood pressure (BP), dysgeusia, vertigo, hypoesthesia, vomiting, vision changes, and paresthesia. Withdrawal symptoms were not observed 1 and 2 weeks after discontinuation of IN ESKET across four RCTs [13, 14, 16, 20]. No case of drug abuse or craving was reported during follow-up phases.

In the study comparing IN ESKET with quetiapine and placebo, adverse events which lead to drug discontinuation occurred in 0.9–4.5% of patients receiving placebo, 11% of patients with quetiapine, and 0.9–12.2% of patients with IN ESKET [19]. Serious TEAEs, including depression- (e.g., worsening of the depressive episode) and suicide-related events (e.g., developing or increasing suicidal ideations or attempting suicide), were reported, but no deaths occurred during the double-blind phase of any study.

The long-term studies revealed no emergence of new side effects and showed favorable long-term safety and manageable tolerability [15, 19, 20]. The most frequent TEAEs were also mild to moderate in severity and resolved on the day of administration.

Real-world evaluations did not report new safety concerns. The adverse effects observed were consistent with those in RCTs: Most common adverse events were transient increases in BP, dissociation, and sedation. They were generally short-lived, occurred on the day of IN ESKET use, and resolved by themselves within 2 h [23, 24, 29].

In summary, the administration of IN ESKET was not associated with critical adverse effects, and mild to moderate unwanted events resolved within hours of use. Common side effects included dizziness, dissociation, nausea, headache, and increased BP. Withdrawal symptoms were not observed, and no case of drug abuse was reported. Long-term trials confirmed esketamine's favorable safety profile, and real-world evaluation echoed these findings without additional safety concerns.

5 Intranasal Esketamine in Clinical Practice

As written above, IN ESKET was approved in the USA and Europe for therapy of TRD in 2019. So far, this approval is valid for indications such as therapy of moderate to severe depression in adult patients who failed to respond to at least two different antidepressant treatments and in combination with SSRI or SNRI [30].

In adults under 65 years, the initial dose of IN ESKET is 56 mg on day 1, followed by doses of either 56 mg or 84 mg twice a week for the first 4 weeks during the induction phase. Choice of dose depends on individual response and tolerability. In the elderly ≥ 65 years, the starting dose is 28 mg, and subsequent doses are 28, 56, or 84 mg.

If there is a satisfactory therapeutic response and favorable tolerability, the frequency of IN ESKET administration will be decreased to one use per week when maintenance therapy begins at the fifth week of treatment. Further reduction to one use every 2 weeks can be considered from the ninth week onward. It is recommended to maintain the dose the patient receives at the end of the induction in the maintenance phase.

Before the indication for IN ESKET is determined, potential contraindications need to be carefully considered. These include unsatisfactorily controlled BP, aneurysmal vascular disease (including intracranial, thoracic, or abdominal aorta, or peripheral arterial vessels), history of intracerebral hemorrhage, recent (within 6 weeks) cardiovascular event, and hypersensitivity to the active substance, ketamine, or to any of the excipients. Precaution must be taken when ketamine is applied together with monoamine oxidase inhibitors, other sympathomimetic drugs, or in hyperthyroidism. Of course, it is contraindicated in disorders such as thyroid toxicosis or pheochromocytoma.

We apply IN ESKET via the officially approved nasal spray device: Patients do that themselves under supervision of a healthcare professional at a healthcare facility or clinic. Administration of IN ESKET at home raises concerns about potential misuse or abuse, as patients may apply ESKET as they like. Monitoring for adverse effects is challenging outside a healthcare facility though remote monitoring via telemedicine has been reported [31]. Nonetheless, despite safety concerns, home use offers convenience for patients and increased accessibility for those with transportation difficulties or limited access to healthcare. Benefits must therefore be carefully weighed against a need for close medical supervision. Monitoring patient response and effectively managing possible side effects are crucial for treatment success. If at-home administration is considered feasible, healthcare providers must implement strict guidelines and protocols to ensure patient safety. With IN ESKET administration, risk assessment, effect evaluation, and getting help in case of emergency need to be taught and trained.

The consensus statement from the Austrian Society for Neuropsychopharmacology and Biological Psychiatry (ÖGPB) regarding TRD recommends the following ten steps for the administration of IN ESKET in clinical practice [32]:

1. Patients shall assume a comfortable position on a recliner or couch and gently blow their nose using a handkerchief.
2. After BP control by medical staff, the first applicator is provided to patients with normotension.
3. Patients shall hold the applicator with 28 mg ESKET with their right hand. The index and middle finger are placed on the finger rests of the device, and the plunger is slightly supported by the thumb.
4. Patients shall then lean their head back for about 45°. They need to insert the tip of the first applicator straight into the right nostril, ensuring that the nasal pads make contact with the skin between the two nostrils.
5. While using the fingers of their left hand to close the left nostril, patients inhale through the nose while pushing the applicator plunger as far as it can go.
6. After the spray is delivered, patients are asked to pull up the air/spray mixture in their nose slightly so that the drug remains inside the nose.
7. Now patients take the same applicator into their left hand, and deliver the spray into the left nostril in the same way as described above.
8. After both spray cans of the first applicator have been dispensed, the operator checks that no green dot is visible in the display window.
9. Patients shall relax and lean back comfortably for 5 min so that ESKET can be absorbed sufficiently through the nose mucous membranes. Patients shall not blow their nose under any circumstances. For a 56 mg or 84 mg treatment dose, a second/third 28 mg applicator is used after a 5-min break, following the same procedure.
10. During the follow-up period, BP should be monitored approximately 40 min after use and subsequently at clinical discretion. Of course, patients are monitored for well-being and potential adverse events.

After IN ESKET therapy, it is recommended to advise patients that they need to refrain from driving a car or operating machinery on the treatment day until the next day. They shall try to get a restful sleep and arrange transportation to home via public transport, taxi, or other way.

The French Society for Biological Psychiatry and Neuropsychopharmacology (AFPBN) guidelines for the management of patients with treatment-resistant depression recommend the following monitoring measures after each administration of IN ESKET:

- Consciousness (using Glasgow scale).
- BP and heart rate.
- Oxygen saturation and respiratory rate.
- Common adverse events: dissociation, somnolence, and sedation.
- Urinary tract and bladder symptoms.
- Continuous monitoring by healthcare provider for at least 40 min after administration or as long as it takes to amend an adverse event.

The expert panel recommends a duration of IN ESKET treatment of at least 6 months in case of a satisfactory clinical response.

With regard to discontinuation of ESKET, the AFPBN guideline suggests to stop treatment not before an administration period of 2 weeks, regardless of the dose.

Figure 1 summarizes the modalities of practical use of IN ESKET in therapy of TRD.

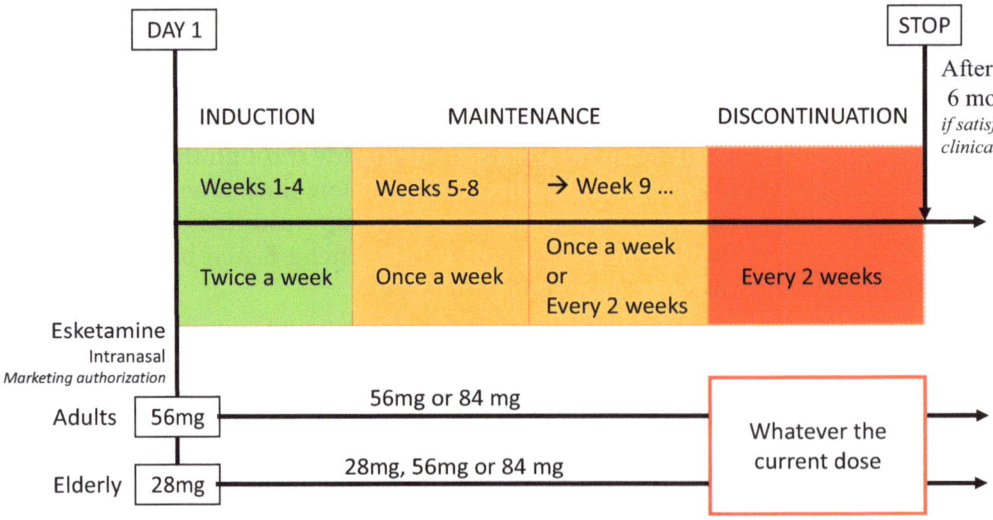

Fig. 1 Modalities of intranasal esketamine therapy for treatment-resistant depression (designed by the authors in Power Point, Microsoft Corp., Redmond, WA, USA)

6 Conclusion

So far, ESKET is the only IN antidepressant approved by the FDA and EMA for use in TRD. The IN application of the S(+)-enantiomer of racemic ketamine has demonstrated significant efficacy in reducing depressive symptoms compared to placebo in various short-term RCT. In addition, ESKET induced superior remission rates at week 8 when compared to extended-release quetiapine in a randomized active-controlled study in TRD patients. Offering a rapid onset of antidepressant action within hours post-administration, its treatment benefits can be maintained over time through repeated use. Long-term studies have shown a notable increase in the percentage of responders and remitters, alongside a decrease in the risk of relapse among patients who achieved stable response or remission. Real-world observational studies have yielded similar outcomes, demonstrating significant improvement in depressive symptoms at various time points after therapy in TRD patients.

The use of IN ESKET as an augmentation strategy complicates the differentiation between intrinsic antidepressant effects of ESKET and benefits of combining ESKET with other oral antidepressants. The results of a double-blind RCT assessing the efficacy of each individual dose of IN ESKET as monotherapy in TRD patients are eagerly awaited [33].

The tolerability of IN ESKET was generally deemed acceptable. The most common TEAEs were mild to moderate, transient, and typically occurring and resolving on the treatment day. Additionally, the lack of incidence of ESKET use disorder in the studies suggests that concerns regarding severe sedative, dissociative, or "euphoric" effects as well as risks of misuse or abuse have to be taken serious but that these risks may also be "manageable." Nevertheless, they necessitate patient education and may require supervision of treatment by healthcare providers at healthcare facilities.

Future updates of guidelines for the treatment of MDD shall endorse IN ESKET as treatment option for TRD. They need to outline key considerations for practical use, detailing dosage, frequency of administration, individual patient-tailored approaches, and monitoring protocols.

Conflict of Interest

Ludovic Samalin has received personal fees and/or nonfinancial support from Janssen-Cilag, Lundbeck, Otsuka, Rovi, and Sanofi-Aventis. Pierre-Michel Llorca has received personal fees and/or nonfinancial support from Eisai, Ethypharm, Janssen, Lundbeck, MSD, Neuraxpharm, Novartis, Otsuka, Roche, Rovi, and Sanofi. Ludivine Boudieu declares no conflicts of interest.

References

1. Bosco-Lévy P, Grelaud A, Blin P et al (2021) Treatment resistant depression incidence and prevalence using the French nationwide claims database. Pharmacoepidemiol Drug Safe 30: 169–177. https://doi.org/10.1002/pds.5082
2. McIntyre RS, Alsuwaidan M, Baune BT et al (2023) Treatment-resistant depression: definition, prevalence, detection, management, and investigational interventions. World Psychiatry 22:394–412. https://doi.org/10.1002/wps.21120
3. Rathod S, Denee T, Eva J et al (2022) Health-related quality of life burden associated with treatment-resistant depression in UK patients: quantitative results from a mixed-methods non-interventional study. J Affect Disord 300: 551–562. https://doi.org/10.1016/j.jad.2021.12.090
4. Heerlein K, De Giorgi S, Degraeve G et al (2022) Real-world evidence from a European cohort study of patients with treatment resistant depression: healthcare resource utilization. J Affect Disord 298:442–450. https://doi.org/10.1016/j.jad.2021.11.004
5. Karkare S, Zhdanava M, Pilon D et al (2022) Characteristics of real-world commercially insured patients with treatment-resistant depression initiated on esketamine nasal spray or conventional therapies in the United States. Clin Ther 44:1432–1448. https://doi.org/10.1016/j.clinthera.2022.09.005
6. Duman RS, Sanacora G, Krystal JH (2019) Altered connectivity in depression: GABA and glutamate neurotransmitter deficits and reversal by novel treatments. Neuron 102:75–90. https://doi.org/10.1016/j.neuron.2019.03.013
7. Mion G, Himmelseher S (2024) Esketamine: less drowsiness, more Analgesia. Anesth Analg 139:78–91. https://doi.org/10.1213/ANE.0000000000006851
8. Correia-Melo FS, Leal GC, Vieira F et al (2020) Efficacy and safety of adjunctive therapy using esketamine or racemic ketamine for adult treatment-resistant depression : a randomized, double-blind, non-inferiority study. J Affect Disord 264:527–534. https://doi.org/10.1016/j.jad.2019.11.086
9. Garay RP, Zarate CAJ, Charpeaud T et al (2017) Investigational drugs in recent clinical trials for treatment-resistant depression. Expert Rev Neurother 17:593–609. https://doi.org/10.1080/14737175.2017.1283217
10. Boudieu L, Mennetrier M, Llorca P-M, Samalin L (2023) The efficacy and safety of intranasal formulations of ketamine and esketamine for the treatment of major depressive disorder: a systematic review. Pharmaceutics 15:2773. https://doi.org/10.3390/pharmaceutics15122773
11. Traynor K (2019) ESKET nasal spray approved for treatment-resistant depression. Am J Health Syst Pharm 76:573. https://doi.org/10.1093/ajhp/zxz065
12. Chen X, Hou X, Bai D et al (2023) Efficacy and safety of flexibly dosed esketamine nasal spray plus a newly initiated oral antidepressant in adult patients with treatment-resistant depression: a randomized, double-blind, multicenter, active-controlled study conducted in China and USA. Neuropsychiatr Dis Treat 19:693–707. https://doi.org/10.2147/NDT.S391096
13. Takahashi N, Yamada A, Shiraishi A et al (2021) Efficacy and safety of fixed doses of intranasal esketamine as an add-on therapy to oral antidepressants in Japanese patients with treatment-resistant depression: a phase 2b randomized clinical study. BMC Psychiatry 21: 526. https://doi.org/10.1186/s12888-021-03538-y
14. Ochs-Ross R, Daly EJ, Zhang Y et al (2020) Efficacy and safety of esketamine nasal spray plus an oral antidepressant in elderly patients with treatment-resistant depression-TRANSFORM-3. Am J Geriatr Psychiatry 28:121–141. https://doi.org/10.1016/j.jagp.2019.10.008
15. Daly EJ, Trivedi MH, Janik A et al (2019) Efficacy of esketamine nasal spray plus oral antidepressant treatment for relapse prevention in patients with treatment-resistant depression: a randomized clinical trial. JAMA Psychiatry 76: 893–903. https://doi.org/10.1001/jamapsychiatry.2019.1189
16. Fedgchin M, Trivedi M, Daly EJ et al (2019) Efficacy and safety of fixed-dose esketamine nasal spray combined with a new oral antidepressant in treatment-resistant depression: results of a randomized, double-blind, active-controlled study (TRANSFORM-1). Int J Neuropsychopharmacol 22:616–630. https://doi.org/10.1093/ijnp/pyz039
17. Popova V, Daly EJ, Trivedi M et al (2019) Efficacy and safety of flexibly dosed ESKET nasal spray combined with a newly initiated oral antidepressant in treatment-resistant depression: a randomized double-blind active-controlled study. Am J Psychiatry 176:428–

438. https://doi.org/10.1176/appi.ajp.2019.19020172
18. Daly EJ, Singh JB, Fedgchin M et al (2018) Efficacy and safety of intranasal esketamine adjunctive to oral antidepressant therapy in treatment-resistant depression: a randomized clinical trial. JAMA Psychiatry 75:139–148. https://doi.org/10.1001/jamapsychiatry.2017.3739
19. Reif A, Bitter I, Buyze J et al (2023) Esketamine nasal spray versus quetiapine for treatment-resistant depression. N Engl J Med 389:1298–1309. https://doi.org/10.1056/NEJMoa2304145
20. Wajs E, Aluisio L, Holder R et al (2020) Esketamine nasal spray plus oral antidepressant in patients with treatment-resistant depression: assessment of long-term safety in a phase 3, open-label study (SUSTAIN-2). J Clin Psychiatry 81:19m12891. https://doi.org/10.4088/JCP.19m12891
21. Estrade I, Petit A-C, Sylvestre V et al (2023) Early effects predict trajectories of response to esketamine in treatment-resistant depression. J Affect Disord 342:166–176. https://doi.org/10.1016/j.jad.2023.09.030
22. Singh B, Kung S, Pazdernik V et al (2023) Comparative effectiveness of intravenous ketamine and intranasal ESKET in clinical practice among patients with treatment-refractory depression: an observational study. J Clin Psychiatry 84:22m14548. https://doi.org/10.4088/JCP.22m14548
23. Brendle M, Robison R, Malone DC (2022) Cost-effectiveness of esketamine nasal spray compared to intravenous ketamine for patients with treatment-resistant depression in the US utilizing clinical trial efficacy and real-world effectiveness estimates. J Affect Disord 319:388–396. https://doi.org/10.1016/j.jad.2022.09.083
24. Martinotti G, Vita A, Fagiolini A et al (2022) Real-world experience of esketamine use to manage treatment-resistant depression: a multicentric study on safety and effectiveness (REAL-ESK study). J Affect Disord 319:646–654. https://doi.org/10.1016/j.jad.2022.09.043
25. Dvorak L, Bloemhof-Bris E, Shelef A et al (2024) Efficacy of esketamine among patients with treatment resistant depression in a "real world" health-care setting in Israel. J Psychiatr Res 174:66–72. https://doi.org/10.1016/j.jpsychires.2024.04.008
26. Samalin L, Mekaoui L, Rotharmel M et al (2024) Use of esketamine nasal spray in patients with treatment-resistant depression in routine practice: a real-world French study. Depress Anxiety 2024:7262794
27. Yrondi A, Javelot H, Nobile B et al (2024) French Society for Biological Psychiatry and Neuropsychopharmacology (AFPBN) guidelines for the management of patients with partially responsive depression and treatment-resistant depression: update 2024. Encéphale S0013-7006(24):00019–00018. https://doi.org/10.1016/j.encep.2023.11.029
28. Kasper S, Cubała WJ, Fagiolini A et al (2021) Practical recommendations for the management of treatment-resistant depression with esketamine nasal spray therapy: basic science, evidence-based knowledge and expert guidance. World J Biol Psychiatry 22:468–482. https://doi.org/10.1080/15622975.2020.1836399
29. Samalin L, Rothärmel M, Mekaoui L et al (2022) Esketamine nasal spray in patients with treatment-resistant depression: the real-world experience in the french cohort early-access programme. Int J Psychiatry Clin Pract 26:352–362. https://doi.org/10.1080/13651501.2022.2030757
30. EMA (2023) Summary of product characteristics of spravato[R] (last update 08 dec 2023) https://www.ema.europa.eu/en/documents/product-information/spravato-epar-product-information_en.pdf. Accessed 29 Apr 2024
31. Mathai DS, Hull TD, Vando L, Malgaroli M (2024) At-home, telehealth-supported ketamine treatment for depression: findings from longitudinal, machine learning and symptom network analysis of real-world data. J Affect Disord 361:198–208. https://doi.org/10.1016/j.jad.2024.05.131
32. Austrian Society for Neuropsychopharmacology and Biological Psychiatry (2022) Treatment-resistant depression: diagnosis and treatment, state of the art. https://oegpb.at/wp-content/uploads/2022/06/10_J_Neuro_OGPB-Konsensus_Therapy-resistant-depression.pdf. Accessed 29 Apr 2024
33. Janssen Research & Development, LLC (2024) A randomized, double-blind, multicenter, placebo-controlled study to evaluate the efficacy, safety, and tolerability of ESKET nasal spray, administered as monotherapy, in adult participants with treatment-resistant depression. clinicaltrials.gov https://classic.clinicaltrials.gov/ct2/show/NCT04599855. Accessed 29 Apr 2024

Chapter 15

Arketamine: Unapproved but Promising Antidepressant with Multiple Brain Benefits

Xiayun Wan and Kenji Hashimoto

Abstract

Racemic ketamine, at subanesthetic doses, has shown rapid and prolonged antidepressant effects in treatment-resistant patients with major depressive disorder, bipolar disorder, and other severe psychiatric conditions. This has led to an increase in off-label use in the USA. Its *(S)*-enantiomer, esketamine, has been approved as a new patented intranasal treatment for treatment-resistant depression in the USA, Canada, and several European countries. However, there are concerns about esketamine's potential adverse effects, including long-term efficacy, addiction risks, and the possibility of increased suicide risk in clinical settings when using nasal spray. Conversely, in rodent models of depression, arketamine, the (*R*)-enantiomer of ketamine, has demonstrated more effective and longer-lasting antidepressant action with fewer side effects than esketamine. However, research on the efficacy and safety of arketamine in patients with depression remains limited. This chapter provides an overview of arketamine's history and its potential as a treatment for various psychiatric and neurological disorders, including depression, cognitive impairment, Parkinson's disease, and stroke.

Key words Arketamine, Bone disease, Cognition, Depression, Enantiomer, Esketamine, Parkinson's disease, Stroke, (*S*)-Norketamine

1 Introduction

Ketamine is a racemic mixture of equal amounts of *R*(−) and *S*(+) stereoisomers, commonly referred to as (*R*,*S*)-ketamine (Fig. 1). It is used as an analgesic, sedative, amnestic, and anesthetic due to its distinct pharmacological characteristics and favorable safety profile, which typically maintain or increase blood pressure, heart rate, and spontaneous ventilation without affecting laryngopharyngeal reflexes [1, 2]. In recent years, subanesthetic doses of ketamine have garnered significant attention in psychiatry for their rapid-onset and sustained antidepressant effects, particularly in the treatment of treatment-resistant depression (TRD) in patients with major depressive disorder (MDD), bipolar disorder (BD), post-traumatic stress disorder (PTSD), and other serious

Sabine Himmelseher (ed.), *Ketamine: Pain Management, Critical Care, Depression, and Resource-Poor Settings Volume II*, Neuromethods, vol. 225, https://doi.org/10.1007/978-1-0716-4599-4_15,

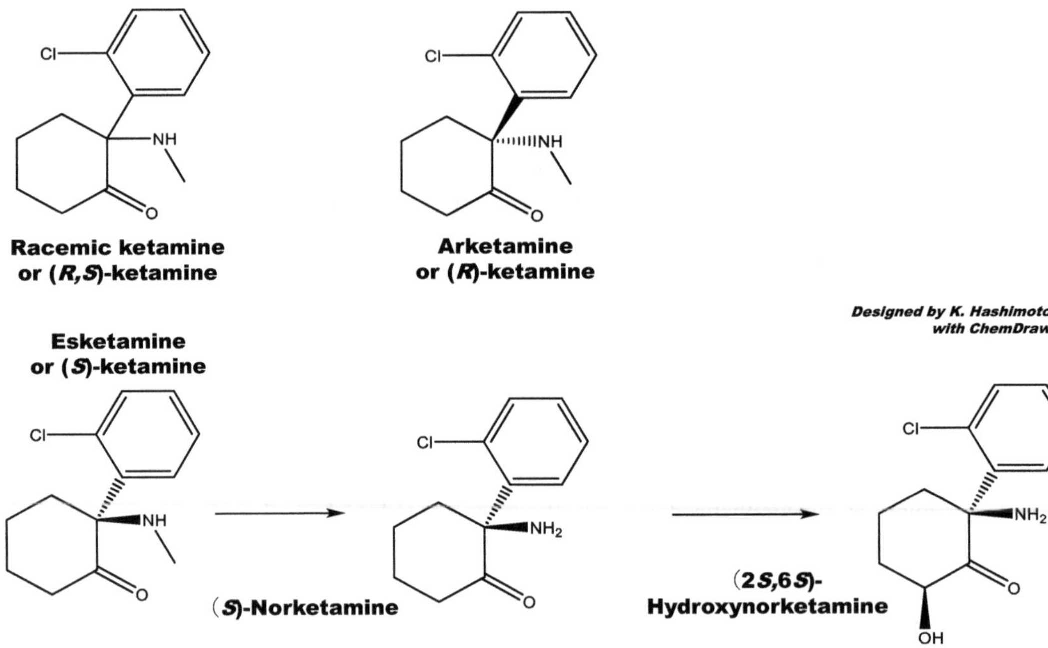

Fig. 1 Chemical structure of racemic, arketamine, esketamine, and esketamine-derived metabolites (*S*)-norketamine and (2*S*,6*S*)-hydroxynorketamine (HNK) (created by K. Hashimoto in free ChemDraw software (revvitysignals))

neuropsychiatric conditions [3–11]. Ketamine can lead to notable improvements in depressive symptoms and suicidal thoughts within hours of a single intravenous (IV) injection. However, its use requires careful patient guidance and monitoring due possible unwanted psychoactive effects, such as dissociation, hallucinations, and psychomimetic adverse reactions [2].

Ketamine consists of two enantiomers in equal amounts: (*S*)-ketamine (esketamine) and (*R*)-ketamine (arketamine) (Fig. 1). These enantiomers have different affinities for the *N*-methyl-D-aspartate receptor (NMDAR) [2, 3, 12]. In 2019, the US Food Drug Administration (FDA) and European authorities approved a patented formulation of esketamine as an intranasal (IN) spray for TRD. However, the effects of IV racemic ketamine administered at 0.5 mg/kg cannot directly be compared with those of IN esketamine used at 28 mg, 56 mg, or 84 mg because the bioavailability of these routes of administration is completely different. IV ketamine has a bioavailability of nearly 100%, whereas the IN patented device has been reported to show a bioavailability of more than 50% [3, 13]. Nevertheless, as with every IN application of a liquid, a portion is uncontrollably swallowed. This portion undergoes hepatic first-pass metabolism and enters the systemic circulation with a bioavailability of less than 20%. Additionally, ketamine metabolites with pharmacological effects of their own may differ after IV versus IN or oral ketamine administration. Unfortunately, the literature has reported and interpreted effects from direct comparisons between IV racemic ketamine and IN esketamine.

Worldwide interest in arketamine has recently increased, as it may be a novel antidepressant with fewer unwanted psychoactive adverse effects compared to racemic ketamine and esketamine [2]. In 1997, Vollenweider and colleagues [14] observed that administrating psychotomimetic IV doses of esketamine significantly elevated glucose metabolism in the frontal cortex of healthy volunteers. These changes in metabolism were linked to experience of ego disintegration and hallucinations. Conversely, similar IV doses of arketamine appeared to reduce glucose metabolism in various brain regions and did not induce psychotic symptoms; instead, it produced a state of relaxation [14]. A randomized, cross-over study evaluated the effects of IV injections of 0.5 mg/kg racemic ketamine, 0.25 mg/kg esketamine, and 1 mg/kg arketamine on neurocognitive functions in healthy volunteers [15]. Both ketamine enantiomers caused less tiredness and cognitive impairment compared to racemic ketamine. Esketamine was also associated with less decline in concentration capacity and memory than racemic ketamine [15].

Keeping these challenges in mind, our chapter provides a concise historical review of arketamine as a treatment approach for psychiatric and neurological disorders. In this work, the term "ketamine" refers to the racemic mixture, while esketamine and arketamine refer to its two stereoisomers.

2 Effects of Ketamine and Esketamine in Patients with Major Depressive Disorder or Treatment-Resistant Depression

Berman and co-workers were among the first to report that a single subanesthetic dose of IV ketamine at 0.5 mg/kg elicited a rapid antidepressant response within hours in medication-free patients with MDD [16]. This effect lasted up to 3 days post-treatment. Numerous subsequent studies have confirmed that both single and repeated IV infusions of ketamine at 0.5 mg/kg are effective in reducing depressive symptoms and suicidal ideation in patients with TRD or BD [17–22]. Additionally, ketamine has shown therapeutic value in TRD, BD, PTSD, and obsessive-compulsive disorder [23–25].

Recognizing the important involvement of NMDARs in ketamine's antidepressant effects, Janssen pharmaceuticals developed esketamine as a new antidepressant. In 2016, Singh and co-authors [26] reported that a single low dose of 0.2 mg/kg and a subanesthetic dose of 0.4 mg/kg of IV esketamine, infused over 40 minutes, resulted in rapid and sustained antidepressant effects in TRD patients. Unwanted medical side effects such as headache, nausea, and adverse psychoactive effects like dissociation were more frequent at the higher dose. However, all side effects were transient and resolved within 4 hours [26]. Further studies, including one by Zhang and colleagues, presented a case series showing that repeated IV esketamine injections at 0.4 mg/kg alleviated depressive symptoms in Chinese TRD patients who were also taking oral antidepressants [27].

As noted earlier, IN esketamine was not approved by the US FDA and European authorities until 2019. To ensure safety, it is available only through a restricted distribution system under a risk and mitigation strategy, which includes a mandatory post-treatment observation period of 2 hours within a hospital setting and reporting of adverse events. Interestingly, a recent analysis of the US FDA Adverse Event Reporting System database identified new adverse reactions, such as flashbacks, associated with the IN spray of esketamine [28]. The clinical significance of these reactions is still unclear, and it is not known if they are linked to specific effects of the IN route or specific ketamine metabolites following IN use. In contrast, multicenter trials have reported on the safety, tolerability, and efficacy of IN esketamine, in TRD patient samples [29]. Long-term clinical outcome studies are needed to fully understand the effects and safety profile of IN esketamine [30]. However, it is important to remember that the use of racemic ketamine via any route is off-label, and in the USA, physicians have considerable freedom in their prescriptions [2, 31]. There is no surveillance system monitoring private clinics, nor is there a legal obligation to report adverse effects. Consequently, it is challenging to assess the incidence and severity of adverse impacts associated with the off-label use of racemic ketamine for depression.

When administered at subanesthetic, analgesic doses in awake patients with acute and chronic pain or during emergence from anesthesia as plasma levels decline, IV esketamine at half dose of IV racemic ketamine shows a slightly lower or comparable incidence of psychoactive events. However, these events are generally more pleasant in quality with esketamine than with racemic ketamine [32]. Additionally, esketamine is metabolized more rapidly than the racemate, leading to noticeably quicker recovery times with esketamine [33].

3 Comparison of Ketamine Enantiomers in Animal Models of Depression

Given the primary target of ketamine's pharmacological antidepressant effects, the NMDAR, various clinical trials have explored NMDAR antagonists as antidepressant agents. However, none of these compounds has exhibited ketamine-like antidepressant efficacy in individuals with MDD [34, 35]. Most experts now believe that unwanted psychoactive effects of ketamine und esketamine, including dissociative symptoms, do not correlate with their antidepressant effects in TRD patients [36]. Considering the significant role of NMDAR antagonism in ketamine-induced dissociation, these clinical findings suggest that NMDARs do not play a major role in ketamine's antidepressant effects [36].

Our research has shown that arketamine exhibited more potent and more enduring antidepressant-like effects than esketamine in multiple rodent depression models, including chronic social defeat stress (CSDS) [37–44]. Importantly, preclinical studies have indicated that arketamine produces fewer behavioral abnormalities, including hyperlocomotion, abuse potential, and prepulse inhibition deficits, when compared to esketamine [38, 42, 45–48]. Additionally, a positron emission tomography study showed that a single subanesthetic IV dose (0.5 mg/kg) of esketamine led to a significant release of dopamine in the monkey striatum, whereas an equimolar dose (0.5 mg/kg) of arketamine did not produce this dopamine release [49]. Our findings suggest that arketamine may be a novel antidepressant with fewer unwanted psychoactive effects [3, 4, 7].

4 Antidepressant-Like Effects of Esketamine Metabolites

Esketamine is metabolized into its primary metabolite, (*S*)-norketamine, which has a lower affinity for the NMDAR (Fig. 1).

Similar to esketamine [38], (*S*)-norketamine improved depression-like behaviors and alterations in dendritic spine density in brain regions of CSDS-susceptible mice [50]. Interestingly, α-amino-3-hydroxy-5-methyl-4-isoxazole propionic acid receptor (AMPAR) antagonist 2,3-dioxo-6-nitro-1,2,3,4- tetrahydrobenzo [f]quinoxaline-7-sulfonamide (NBQX) did not block the antidepressant-like effects of (*S*)-norketamine in CSDS-susceptible mice, whereas it did block the antidepressant-like effects of esketamine in the same model [50]. Similar to esketamine [38], the tropomyosin receptor kinase B (TrkB) antagonist ANA-12 inhibited the antidepressant-like effects of (*S*)-norketamine in mice susceptible to CSDS [50]. Additionally, Yokoyama et al. reported sustained antidepressant-like effects of (*S*)-norketamine in a chronic corticosterone-induced depression model [51] and a social isolation-reared model [52]. Unlike esketamine, (*S*)-norketamine did not cause behavioral, pathological, or neurophysiological abnormalities, such as hyperlocomotion, prepulse inhibition deficits, rewarding effects, loss of parvalbumin (PV)-immunoreactivity in the medial prefrontal cortex, or increased baseline gamma band oscillation [50]. Given their similar antidepressant-like effects, esketamine might act as a prodrug for (*S*)-norketamine [2, 53]. However, it is important to note that (*S*)-norketamine's antidepressant-like effects are less potent than those of arketamine [2].

(*S*)-norketamine is further metabolized into (2*S*,6*S*)-hydroxynorketamine (HNK) (Fig. 1). Contrary to Zanos et al. [54], Yokoyama et al. [52] reported that (2*S*,6*S*)-HNK exhibited antidepressant-like effect in mice treated with chronic corticosterone. In 2024, Kawatake-Kuno et al. [55] demonstrated that (2*S*,6*S*)-HNK produced sustained antidepressant-like effects in mice exposed to repeated stress, accompanied by neurobiological changes in the anterior paraventricular nucleus of the thalamus (aPVT). Additionally, (2*S*,6*S*)-HNK was found to increase the expression of extrasynaptic gamma-amino-butyric acid $(GABA)_A$ receptors and enhance tonic GABAergic currents. These findings suggest that the aPVT plays a role in the sustained antidepressant-like effects of (2*S*,6*S*)-HNK in mice [55].

5 Effects of Arketamine in Other Rodent Models

Cognitive impairment is prevalent among patients with various psychiatric and neurological disorders. Notably, ketamine has demonstrated positive effects on cognitive impairments in TRD patients [56–58]. Additionally, there is a report indicating that esketamine may have effects on neurocognition in TRD patients [59]. Our research group found that arketamine, but not esketamine, could improve cognitive deficits induced by phencyclidine in mice [60], and it also mitigated cognitive deficits in mice following maternal immune activation [61]. These findings suggest that arketamine may be a promising therapeutic option for cognitive impairment associated with various psychiatric and neurological disorders [62, 63].

Osteoporosis is commonly observed in patients with psychiatric conditions such as MDD. In 2018, Kadriu et al. [64] demonstrated that a single subanesthetic dose of ketamine at 0.5 mg/kg significantly increased the osteoprotegerin (OPG)/receptor activator of NF-κB ligand (RANKL) ratio and plasma osteopontin (OPN) levels in TRD patients while decreasing RANKL levels. Our group reported that arketamine, but not esketamine, significantly reduced elevated plasma RANKL levels in CSDS-susceptible mice and improved the decreased ratio of OPG/RANKL in the plasma of these mice [65, 66]. Additionally, decreased bone mineral density in ovariectomized (OVX) mice was ameliorated following repeated, intermittent IP administration of arketamine (10 mg/kg/day), but not esketamine (10 mg/kg/day) [67, 68]. Interestingly, the gut-bone-brain axis may play a role in the beneficial effects of arketamine on bone mineral density in OVX mice [68] and CSDS-susceptible mice [69]. These findings collectively suggest that arketamine could be a potential therapeutic agent for addressing bone metabolism abnormalities in patients with MDD or osteoporosis [7].

Additionally, our research team reported that arketamine, but not esketamine, could improve outcomes in rodent model mice with Parkinson's disease [70], cerebral ischemic stroke [71], and ulcerative colitis [72]. Furthermore, arketamine was able to reduce demyelination in the spinal cord and brain of an experimental autoimmune encephalomyelitis model [73], cuprizone-treated mice [74], and CSDS model [75]. Given the potent anti-inflammatory effects of arketamine, it could be a promising therapeutic option beyond psychiatric disorders [7, 63]. However, it is worth noting that esketamine also demonstrates anti-inflammatory benefits, effectively counteracting inflammatory imbalances in high-quality patient trials across various perioperative settings [76, 77]. Further study is needed to compare the anti-inflammatory effects of the two enantiomers in clinical settings.

6 Antidepressant Effects of Arketamine in Humans

Research on the antidepressant effects of arketamine in MDD or TRD patients remains relatively limited. An open-label pilot study with a small sample size of seven patients found that a single IV dose of 0.5 mg/kg arketamine resulted in rapid and lasting antidepressant effects in female TRD patients [78]. Furthermore, the same research team found that IV injections of arketamine at a dose of 0.5 and 1.0 mg/kg led to rapid-acting antidepressant effects in six depressed patients [79]. In both studies, arketamine was associated with minimal side effects, including dissociation [78, 79].

In January 2023, Perception Neuroscience announced that the phase 2a trial of arketamine (PCN-101) in TRD patients did not meet its primary endpoint of statistical significance [80]. However, further analysis highlighted differences between the US and European cohorts. Specifically, the US subgroup showed clinically significant improvements in depression scores lasting up to 2 weeks after a single 60 mg dose of arketamine. There were no significant differences in sedation or dissociation between the two groups [2]. Case analyses from the trial highlighted notable reductions in symptoms and improvements in vocational and social functioning, including fewer hospitalizations and sick leaves following treatment [81]. These findings suggest that arketamine does not cause dissociative side effects in humans when used at effective doses for depression treatment [2, 36].

Currently, there are no reports directly comparing the antidepressant effects of arketamine and esketamine in patients with depression. While esketamine at antidepressant doses can cause severe side effects like dissociation and hallucination, it appears that higher IV doses of arketamine, such as 0.8 mg/kg, can be administered due to its lower side-effect profile compared to IV esketamine at 0.4 mg/kg [26]. Therefore, it is important to compare the antidepressant effects of these two ketamine enantiomers in TRD or MDD patients, and other psychiatric disorders.

7 Conclusion and Future Directions

As highlighted in this chapter, there is a growing interest in the application of arketamine in major psychiatric and neurologic disorders [82–90]. Preclinical evidence suggests that arketamine could offer more potent and longer-lasting benefits than esketamine with fewer adverse effects. However, current research on the efficacy and safety of arketamine in treating TRD or MDD is limited. Larger, high-quality studies are needed to determine its antidepressant effects in patients with MDD, BD, and other psychiatric and neurological disease. It is crucial to directly compare the efficacy and safety of these two ketamine enantiomers in TRD patients. Equal important are comparisons of the efficacy and safety of esketamine and its metabolites, such as (*S*)-norketamine and (2*S*,6*S*)-HNK, as antidepressants in patients with TRD or MDD.

Acknowledgments

The authors would like to thank our collaborators who are listed as the co-authors of our papers in the reference list. Dr. Wan is supported by JST SPRING (JPMJSP2109) (Tokyo, Japan).

Disclosure Statement

Dr. Hashimoto is the inventor of filed patent applications on "The use of *R*-ketamine in the treatment of psychiatric diseases," "(*S*)-norketamine and salt thereof as pharmaceutical," "*R*-ketamine and derivative thereof as prophylactic or therapeutic agent for neurodegeneration disease or recognition function disorder," "Preventive or therapeutic agent and pharmaceutical composition for inflammatory diseases or bone diseases," "*R*-ketamine and its derivatives as a preventive or therapeutic agent for a neurodevelopmental disorder," and "TGF-β1 in the treatment of depression" by the Chiba University. Dr. Hashimoto also declares that he has received research support and consultant from Otsuka. Dr. Wan has no conflict of interest.

References

1. Domino EF (2010) Taming the ketamine tiger. 1965. Anesthesiology 113:678–684. https://doi.org/10.1097/ALN.0b013e3181ed09a2
2. Hashimoto K, Zhao M, Zhu T et al (2024) Ketamine and its two enantiomers in anesthesiology and psychiatry: a historical review and future directions. J Anes Transl Med 3:65–75. https://doi.org/10.1016/j.jatmed.2024.07.001
3. Hashimoto K (2019) Rapid-acting antidepressant ketamine, its metabolites and other candidates: a historical overview and future perspective. Psychiatry Clin Neurosci 73:613–627. https://doi.org/10.1111/pcn.12902
4. Yang C, Yang J, Luo A et al (2019) Molecular and cellular mechanisms underlying the antidepressant effects of ketamine enantiomers and its metabolites. Transl Psychiatry 9:280. https://doi.org/10.1038/s41398-019-0624-1
5. Hashimoto K (2020) Molecular mechanisms of the rapid-acting and long-lasting antidepressant actions of (*R*)-ketamine. Biochem Pharmacol 177:113935. https://doi.org/10.1016/j.bcp.2020.113935
6. Wei Y, Chang L, Hashimoto K (2020) A historical review of antidepressant effects of ketamine and its enantiomers. Pharmacol Biochem Behav 190:172870. https://doi.org/10.1016/j.pbb.2020.172870
7. Wei Y, Chang L, Hashimoto K (2022) Molecular mechanisms underlying the antidepressant actions of arketamine: beyond the NMDA receptor. Mol Psychiatry 27:559–573. https://doi.org/10.1038/s41380-021-01121-1
8. Hashimoto K (2022) Ketamine: anesthetic, psychotomimetic, antidepressant, or anthelmintic? Mol Psychiatry 27:3116–3118. https://doi.org/10.1038/s41380-022-01587-7
9. Ma L, Hashimoto K (2022) The role of hippocampal KCNQ2 channel in antidepressant actions of ketamine. Neuron 110:2201–2203. https://doi.org/10.1016/j.neuron.2022.05.027
10. Zhang JC, Yao W, Hashimoto K (2022) Arketamine, a new rapid-acting antidepressant: a historical review and future directions. Neuropharmacology 218:109219. https://doi.org/10.1016/j.neuropharm.2022.109219
11. Zhang K, Yao Y, Hashimoto K (2023) Ketamine and its metabolites: potential as novel treatments for depression. Neuropharmacology 222:109305. https://doi.org/10.1016/j.neuropharm.2022.109305
12. Oye I, Paulsen O, Maurset A (1992) Effects of ketamine on sensory perception: evidence for a role of N-methyl-D-aspartate receptors. J Pharmacol Exp Ther 260:1209–1213
13. Perez-Ruixo C, Rossenu S, Zannikos P et al (2022) Population pharmacokinetics of esketamine nasal spray and its metabolite noresketamine in healthy subjects and patients with treatment-resistant depression. J Clin Pharm Ther 60:501–516. https://doi.org/10.1007/s40262-020-00953-4
14. Vollenweider FX, Leenders KL, Oye I et al (1997) Differential psychopathology and patterns of cerebral glucose utilisation produced by (*S*)- and (*R*)-ketamine in healthy volunteers using positron emission tomography (PET). Eur Neuropsychopharmacol 7:25–38. https://doi.org/10.1016/s0924-977x(96)00042-9
15. Pfenninger EG, Durieux ME, Himmelseher S (2002) Cognitive impairment after small dose ketamine isomers in comparison to equianalgesic racemic ketamine in healthy volunteers. Anesthesiology 96:357–366. https://doi.org/10.1097/00000542-200202000-00022
16. Berman RM, Cappiello A, Anand A et al (2000) Antidepressant effects of ketamine in depressed patients. Biol Psychiatry 47:351–354. https://doi.org/10.1016/s0006-3223(99)00230-9
17. Zarate CA Jr, Singh JB, Carlson PJ et al (2006) A randomized trial of an *N*-methyl-D-aspartate antagonist in treatment-resistant major depression. Arch Gen Psychiatry 63:856–864. https://doi.org/10.1001/archpsyc.63.8.856
18. Diazgranados N, Ibrahim L, Brutsche NE et al (2010) A randomized add-on trial of an *N*-methyl-D-aspartate antagonist in treatment-resistant bipolar depression. Arch Gen Psychiatry 67:793–802. https://doi.org/10.1001/archgenpsychiatry.2010.90
19. Su TP, Chen MH, Li CT et al (2017) Dose-related effects of adjunctive ketamine in Taiwanese patients with treatment-resistant depression. Neuropsychopharmacology 42:2482–2492. https://doi.org/10.1038/npp.2017.94
20. Wilkinson ST, Ballard ED, Bloch MH et al (2018) The effect of a single dose of intravenous ketamine on suicidal ideation: a systematic review and individual participant data meta-analysis. Am J Psychiatry 175:150–158.

https://doi.org/10.1176/appi.ajp.2017.17040472

21. Zheng W, Zhou YL, Liu WJ et al (2018) Rapid and longer-term antidepressant effects of repeated-dose intravenous ketamine for patients with unipolar and bipolar depression. Psychiatry Res 106:61–68. https://doi.org/10.1016/j.jpsychires.2018.09.013
22. Fava M, Freeman MP, Flynn M et al (2020) Double-blind, placebo-controlled, dose-ranging trial of intravenous ketamine as adjunctive therapy in treatment-resistant depression (TRD). Mol Psychiatry 25:1592–1603. https://doi.org/10.1038/s41380-018-0256-5
23. Nikolin S, Rodgers A, Schwaab A et al (2023) Ketamine for the treatment of major depression: a systematic review and meta-analysis. EClinicalMedicine 62:102127. https://doi.org/10.1016/j.eclinm.2023.102127
24. Almeida TM, Lacerda da Silva UR, Pires JP et al (2024) Effectiveness of ketamine for the treatment of post-traumatic stress disorder – a systematic review and meta-analysis. Clin Neuropsychiatry 21:22–31. https://doi.org/10.36131/cnfioritieditore20240102
25. Bandeira ID, Lins-Silva DH, Cavenaghi VB et al (2022) Ketamine in the treatment of obsessive-compulsive disorder: a systematic review. Harv Rev Psychiatry 30:135–145. https://doi.org/10.1097/HRP.0000000000000330
26. Singh JB, Fedgchin M, Daly E et al (2016) Intravenous esketamine in adult treatment-resistant depression: a double-blind, double-randomization, placebo-controlled study. Biol Psychiatry 80:424–431. https://doi.org/10.1016/j.biopsych.2015.10.018
27. Zhang K, Yang Y, Yuan X et al (2022) Efficacy and safety of repeated esketamine intravenous infusion in the treatment of treatment-resistant depression: a case series. Asian J Psychiatry 68:102976. https://doi.org/10.1016/j.ajp.2021.102976
28. Jiang Y, Du Z, Shen Y, et al (2023) The correlation of esketamine with specific adverse events: a deep dive into the FAERS database. Eur arch psychiatry Clin Neurosci 2023 Dec 16. Doi:10.1007/s00406-023-01732-5. Online ahead of print
29. Martinotti G, Vita A, Fagiolini A et al (2022) Real-world experience of esketamine use to manage treatment-resistant depression: a multicentric study on safety and effectiveness (REAL-ESK study). J Affect Disord 319:646–654. https://doi.org/10.1016/j.jad.2022.09.043
30. Kaur U, Pathak BK, Singh A et al (2021) Esketamine: a glimmer of hope in treatment-resistant depression. Eur Arch Psychiatry Clin Neurosci 271:417–429. https://doi.org/10.1007/s00406-019-01084-z
31. Wilkinson ST, Palamar JJ, Sanacora G (2024) The rapidly shifting ketamine landscape in the US. JAMA Psychiatry 81:221–222. https://doi.org/10.1001/jamapsychiatry.2023.4945
32. Mion G, Himmelseher S (2024) Esketamine: less drowsiness, more analgesia. Anesth Analg 139:78–91. https://doi.org/10.1213/ANE.0000000000006851
33. Ihmsen H, Geisslinger G, Schüttler J (2001) Stereoselective pharmacokinetics of ketamine: R(−)-ketamine inhibits the elimination of S(+)-ketamine. Clin Pharmacol Ther 70:431–438. https://doi.org/10.1067/mcp.2001.119722
34. Newport DJ, Carpenter LL, McDonald WM et al (2015) Ketamine and other NMDA antagonists: early clinical trials and possible mechanisms in depression. Am J Psychiatry 172:950–966. https://doi.org/10.1176/appi.ajp.2015.15040465
35. Kishimoto T, Chawla JM, Hagi K et al (2016) Single-dose infusion ketamine and non-ketamine N-methyl-D-aspartate receptor antagonists for unipolar and bipolar depression: a meta-analysis of efficacy, safety and time trajectories. Psychol Med 46:1459–1472. https://doi.org/10.1017/S0033291716000064
36. Hashimoto K (2024) Are "mystical experiences" essential for antidepressant actions of ketamine and the classic psychedelics? Eur Arch Psychiatry Clin Neurosci (2024 Feb 27). https://doi.org/10.1007/s00406-024-01770-7
37. Zhang JC, Li SX, Hashimoto K (2014) R (−)-ketamine shows greater potency and longer lasting antidepressant effects than S (+)-ketamine. Pharmacol Biochem Behav 116:137–141. https://doi.org/10.1016/j.pbb.2013.11.033
38. Yang C, Shirayama Y, Zhang JC et al (2015) R-ketamine: a rapid-onset and sustained antidepressant without psychotomimetic side effects. Transl Psychiatry 5:e632. https://doi.org/10.1038/tp.2015.136
39. Fukumoto K, Toki H, Iijima M et al (2017) Antidepressant potential of (*R*)-ketamine in rodent models: comparison with (*S*)-ketamine. J Pharmacol Exp Ther 361:9–16. https://doi.org/10.1124/jpet.116.239228
40. Yang C, Qu Y, Abe M et al (2017) (*R*)-ketamine shows greater potency and longer lasting

antidepressant effects than its metabolite (2*R*,6*R*)-hydroxynorketamine. Biol Psychiatry 82:e43–e44. https://doi.org/10.1016/j.biopsych.2016.12.020

41. Yang C, Ren Q, Qu Y et al (2018) Mechanistic target of rapamycin-independent antidepressant effects of (*R*)-ketamine in a social defeat stress model. Biol Psychiatry 83:18–28. https://doi.org/10.1016/j.biopsych.2017.05.016
42. Chang L, Zhang K, Pu Y et al (2019) Comparison of antidepressant and side effects in mice after intranasal administration of (*R,S*)-ketamine, (*R*)-ketamine, and (*S*)-ketamine. Pharmacol Biochem Behav 181:53–59. https://doi.org/10.1016/j.pbb.2019.04.008
43. Zhang K, Yang C, Chang L et al (2020) Essential role of microglial transforming growth factor-β1 in antidepressant actions of (*R*)-ketamine and the novel antidepressant TGF-β1. Transl Psychiatry 10:32. https://doi.org/10.1038/s41398-020-0733-x
44. Yao W, Cao Q, Luo S et al (2022) Microglial ERK-NRBP1-CREB-BDNF signaling in sustained antidepressant actions of (*R*)-ketamine. Mol Psychiatry 27:1618–1629. https://doi.org/10.1038/s41380-021-01377-7
45. Yang C, Han M, Zhang JC et al (2016) Loss of parvalbumin-immunoreactivity in mouse brain regions after repeated intermittent administration of esketamine, but not *R*-ketamine. Psychiatry Res 239:281–283. https://doi.org/10.1016/j.psychres.2016.03.034
46. Tian Z, Dong C, Fujita A et al (2018) Expression of heat shock protein HSP-70 in the retrosplenial cortex of rat brain after administration of (*R,S*)-ketamine and (*S*)-ketamine, but not (*R*)-ketamine. Pharmacol Biochem Behav 172:17–21. https://doi.org/10.1016/j.pbb.2018.07.003
47. Tan Y, Hashimoto K (2020) Risk of psychosis after repeated intermittent administration of (*S*)-ketamine, but not (*R*)-ketamine, in mice. J Affect Disord 269:198–200. https://doi.org/10.1016/j.jad.2020.03.040
48. Bonaventura J, Lam S, Carlton M et al (2021) Pharmacological and behavioral divergence of ketamine enantiomers: implications for abuse liability. Mol Psychiatry 26:6704–6722. https://doi.org/10.1038/s41380-021-01093-2
49. Hashimoto K, Kakiuchi T, Ohba H et al (2017) Reduction of dopamine $D_{2/3}$ receptor binding in the striatum after a single administration of esketamine, but not R-ketamine: a PET study in conscious monkeys. Eur Arch Psychiatry Clin Neurosci 267:173–176. https://doi.org/10.1007/s00406-016-0692-7
50. Yang C, Kobayashi S, Nakao K et al (2018) AMPA receptor activation-independent antidepressant actions of ketamine metabolite (*S*)-norketamine. Biol Psychiatry 84:591–600. https://doi.org/10.1016/j.biopsych.2018.05.007
51. Yokoyama R, Higuchi M, Tanabe W et al (2024) Long-lasting anti-despair and anti-anhedonia effects of (*S*)-norketamine in social isolation-reared mice. J Pharmacol Sci 154:72–76. https://doi.org/10.1016/j.jphs.2023.12.005
52. Yokoyama R, Higuchi M, Tanabe W et al (2020) (*S*)-norketamine and (2*S*,6*S*)-hydroxynorketamine exert potent antidepressant-like effects in a chronic corticosterone-induced mouse model of depression. Pharmacol Biochem Behav 191:172876. https://doi.org/10.1016/j.pbb.2020.172876
53. Hashimoto K, Yang C (2019) Is (*S*)-norketamine an alternative antidepressant for esketamine? Eur Arch Psychiatry Clin Neurosci 269:867–868. https://doi.org/10.1007/s00406-018-0922-2
54. Zanos P, Moaddel R, Morris PJ et al (2016) NMDAR inhibition-independent antidepressant actions of ketamine metabolites. Nature 533:481–486. https://doi.org/10.1038/nature17998
55. Kawatake-Kuno A, Li H, Inaba H et al (2024) Sustained antidepressant effects of ketamine metabolite involve GABAergic inhibition-mediated molecular dynamics in aPVT glutamatergic neurons. Neuron 112:1265–1285.e10. https://doi.org/10.1016/j.neuron.2024.01.023
56. Zhou Y, Zheng W, Liu W et al (2018) Neurocognitive effects of six ketamine infusions and the association with antidepressant response in patients with unipolar and bipolar depression. J Psychopharmacol 32:1118–1126. https://doi.org/10.1177/0269881118798614
57. Shiroma PR, Thuras P, Wels J et al (2020) Neurocognitive performance of repeated versus single intravenous subanesthetic ketamine in treatment resistant depression. J Affect Disord 277:470–477. https://doi.org/10.1016/j.jad.2020.08.058
58. Gill H, Gill B, Rodrigues NB et al (2020) The effects of ketamine on cognition in treatment-resistant depression: a systematic review and priority avenues for future research. Neurosci Biobehav Rev 120:78–85. https://doi.org/10.1016/j.neubiorev.2020.11.020
59. Pepe M, Bartolucci G, Marcelli I et al (2023) Reduction in cognitive symptoms following intranasal esketamine administration in patients with chronic treatment-resistant

depression: a 12-week case series. J Psychiatr Pract 29:325–332. https://doi.org/10.1097/PRA.0000000000000723

60. Tan Y, Fujita Y, Qu Y et al (2020) Phencyclidine-induced cognitive deficits in mice are ameliorated by subsequent repeated intermittent administration of (*R*)-ketamine, but not (*S*)-ketamine: role of BDNF-TrkB signaling. Pharm Biochem Behav 188:172839. https://doi.org/10.1016/j.pbb.2019.172839
61. Tan Y, Fujita Y, Pu Y et al (2022) Repeated intermittent administration of (*R*)-ketamine during juvenile and adolescent stages prevents schizophrenia-relevant phenotypes in adult offspring after maternal immune activation: a role of TrkB signaling. Eur Arch Psychiatry Clin Neurosci 272:693–701. https://doi.org/10.1007/s00406-021-01365-6
62. Hashimoto K (2023) Arketamine for cognitive impairment in psychiatric disorders. Eur Arch Psychiatry Clin Neurosci 273:1513–1525. https://doi.org/10.1007/s00406-023-01570-5
63. Wang X, Yang J, Hashimoto K (2022) (R)-ketamine as prophylactic and therapeutic drug for neurological disorders: beyond depression. Neurosci Biobehav Rev 139:104762. https://doi.org/10.1016/j.neubiorev.2022.104762
64. Kadriu B, Gold PW, Luckenbaugh DA et al (2018) Acute ketamine administration corrects abnormal inflammatory bone markers in major depressive disorder. Mol Psychiatry 23:1626–1631. https://doi.org/10.1038/mp.2017.109
65. Zhang K, Ma M, Dong C et al (2018) Role of inflammatory bone markers in the antidepressant actions of (*R*)-ketamine in a chronic social defeat stress model. Int J Neuropsychopharmacol 21:1025–1030. https://doi.org/10.1093/ijnp/pyy065
66. Xiong Z, Fujita Y, Zhang K et al (2019) Beneficial effects of (*R*)-ketamine, but not its metabolite (2*R*,6*R*)-hydroxynorketamine, in the depression-like phenotype, inflammatory bone markers and bone mineral density in a chronic social defeat stress model. Behav Brain Res 368:111904. https://doi.org/10.1016/j.bbr.2019.111904
67. Fujita Y, Hashimoto K (2020) Decreased bone mineral density in ovariectomized mice is ameliorated after subsequent repeated intermittent administration of (*R*)-ketamine, but not (*S*)-ketamine. Neuropsychopharmacol Rep 40:401–406. https://doi.org/10.1002/npr2.12132
68. Wan X, Eguchi A, Fujita Y et al (2022) Effects of (*R*)-ketamine on reduced bone mineral density in ovariectomized mice: a role of gut microbiota. Neuropharmacology 213:109139. https://doi.org/10.1016/j.neuropharm.2022.109139
69. Wan X, Eguchi A, Chang L et al (2023) Beneficial effects of arketamine on the reduced bone mineral density in susceptible mice after chronic social defeat stress: role of the gut-microbiota-bone-brain axis. Neuropharmacology 228:109466. https://doi.org/10.1016/j.neuropharm.2023.109466
70. Fujita A, Fujita Y, Pu Y et al (2020) MPTP-induced dopaminergic neurotoxicity in mouse brain is attenuated after subsequent intranasal administration of (*R*)-ketamine: a role of TrkB signaling. Psychopharmacol 237:83–92. https://doi.org/10.1007/s00213-019-05346-5
71. Xiong Z, Chang L, Qu Y et al (2020) Neuronal brain injury after cerebral ischemic stroke is ameliorated after subsequent administration of (*R*)-ketamine, but not (*S*)-ketamine. Pharmcol Biochem Behav 191:172904. https://doi.org/10.1016/j.pbb.2020.172904
72. Fujita Y, Hashimoto Y, Hashimoto H et al (2021) Dextran sulfate sodium-induced inflammation and colitis in mice are ameliorated by (*R*)-ketamine, but not (*S*)-ketamine: a role of TrkB signaling. Eur J Pharmcol 897:173954. https://doi.org/10.1016/j.ejphar.2021.173954
73. Wang X, Chang L, Tan Y et al (2021) (R)-ketamine ameliorates the progression of experimental autoimmune encephalomyelitis in mice. Brain Res Bull 177:316–323. https://doi.org/10.1016/j.brainresbull.2021.10.013
74. Wang X, Chang L, Wan X et al (2022) (R)-ketamine ameliorates demyelination and facilitates remyelination in cuprizone-treated mice: a role of gut-microbiota-brain axis. Neurobiol Dis 165:105635. https://doi.org/10.1016/j.nbd.2022.105635
75. Huang C, Wu Z, Wang D et al (2023) (2024) myelin-associated oligodendrocytic basic protein-dependent myelin repair confers the long-lasting antidepressant effect of ketamine. Mol Psychiatry. https://doi.org/10.1038/s41380-023-02288-5
76. Hublet S, Galland M, Navez J et al (2022) Opioid-free versus opioid-based anesthesia in pancreatic surgery. BMC Anesthesiol 22:9. https://doi.org/10.1186/s12871-021-01551-y
77. Qiu D, Wang XM, Yang JJ et al (2022) Effect of intraoperative esketamine infusion on postoperative sleep disturbance after gynecological laparoscopy: a randomized clinical trial. JAMA

Netw Open 5:e2244514. https://doi.org/10.1001/jamanetworkopen.2022.44514

78. Leal GC, Bandeira ID, Correia-Melo FS et al (2021) Intravenous arketamine for treatment-resistant depression: open-label pilot study. Eur Arch Psychiatry Clin Neurosci 271:577–582. https://doi.org/10.1007/s00406-020-01110-5
79. Bandeira ID, Leal GC, Correia-Melo FS et al (2023) Arketamine for bipolar depression: open-label, dose-escalation, pilot study. J Psychiatr Res 164:229–234. https://doi.org/10.1016/j.jpsychires.2023.06.028
80. Press release from perception neuroscience: PCN-101 P2a - ASCP 2023 — perception neuroscience (perceptionns.com) Access 27 July 2024
81. Włodarczyk A, Słupski J, Szarmach J et al (2024) Single arketamine in treatment resistant depression: presentation of 3 cases with regard to sick-leave duration. Asian J Psychiatry 96: 104016. https://doi.org/10.1016/j.ajp.2024.104016
82. He T, Wang D, Wu Z, Huang C, Xu X, Xu X, Liu C, Hashimoto K, Yang C (2022) A bibliometric analysis of research on (*R*)-ketamine from 2002 to 2021. Neuropharmacology 218:109207. https://doi.org/10.1016/j.neuropharm.2022.109207
83. Scotton E, Casa PL, de Abreu FP et al (2023) Differentially regulated targets in the fast-acting antidepressant effect of (*R*)-ketamine: a systems biology approach. Pharmacol Biochem Behav 223:173523. https://doi.org/10.1016/j.pbb.2023.173523
84. Zhang S, Pu Y, Liu J et al (2024) Exploring the multifaceted potential of (*R*)-ketamine beyond antidepressant applications. Front Pharmacol 15:1337749. https://doi.org/10.3389/fphar.2024.1337749
85. Yokoyama R, Ago Y, Igarashi H et al (2024) (R)-ketamine restores anterior insular cortex activity and cognitive deficits in social isolation-reared mice. Mol Psychiatry 29: 1406–1416. https://doi.org/10.1038/s41380-024-02419-6
86. Pothorszki D, Koncz S, Török D et al (2024) Unique effects of (*R*)-ketamine compared to (*S*)-ketamine on EEG theta power in rats. Pharmaceuticals (Basel) 17:194. https://doi.org/10.3390/ph17020194
87. Shafique H, Demers JC, Biesiada J et al (2024) (R)-(-)-ketamine: the promise of a novel treatment for psychiatric and neurological disorders. Int J Mol Sci 25(12):6804. https://doi.org/10.3390/ijms25126804
88. Chang L, Wei Y, Qu Y et al (2024) Role of oxidative phosphorylation in the antidepressant effects of arketamine via the vagus nerve-dependent spleen-brain axis. Neurobiol Dis 199:106573. https://doi.org/10.1016/j.nbd.2024.106573
89. Zhao LY, Zhang GF, Lou XJ et al (2024) Ketamine and its enantiomers for depression: a bibliometric analysis from 2000 to 2023. Eur Arch Psychiatry Clin Neurosci. https://doi.org/10.1007/s00406-024-01809-9
90. Hashimoto K (2023) Neuroinflammation through the vagus nerve-dependent gut-microbiota-brain axis in treatment-resistant depression. Prog Brain Res 278:61–77. https://doi.org/10.1016/bs.pbr.2023.01.003

Chapter 16

Ketamine or Electroconvulsive Therapy as Treatment Approach to Severe Depression: The Necessity of More Personalized Management

Pouya Movahed Rad and Akihiro Takamiya

Abstract

Depression is a debilitating psychiatric disorder. Electroconvulsive therapy (ECT) and ketamine are effective treatments for severe depression, but target populations may differ. Systematic reviews on the few available trials show higher remission rates with ECT than with ketamine. Nonetheless, ketamine's rapid onset of action is comparable to ECT but with less cognitive adverse effects. This makes ketamine a valuable option, particularly in outpatients or when ECT is unavailable. ECT and ketamine differ in side effects: ECT induces transient cognitive impairments, while ketamine commonly causes dissociative symptoms. Clinical factors can guide choices for ECT versus ketamine. Older patients and those with psychotic symptoms tend to respond better to ECT, while younger patients with longer episode duration may benefit similarly from ketamine. Although biological variables and brain imaging results hold promise as biomarkers to predict treatment outcomes, implementation in everyday practice remains limited due to inconsistent findings and technical challenges. This review underscores the importance of personalized approaches to managing severe depression, advocating the integration of clinical predictors with patient-specific characteristics to improve treatment outcomes.

Key words Racemic ketamine, Electroconvulsive therapy (ECT), Major depressive disorder, Antidepressant effects, Cognitive side effects, Personalized treatment, Clinical predictors

1 Introduction

Depression is a common and debilitating psychiatric disorder characterized by persistent sadness, loss of interest, and a range of physical and cognitive symptoms that heavily impair daily life. In research settings, treatment-resistant depression (TRD) is defined as a failure to respond adequately to at least two different antidepressants, but this definition varies and provides limited insight into the condition itself. TRD poses significant challenges in psychiatric care due to its complex and multifaceted nature and may in fact fathom other conditions coexisting with depressive symptoms. In

Sabine Himmelseher (ed.), *Ketamine: Pain Management, Critical Care, Depression, and Resource-Poor Settings Volume II*, Neuromethods, vol. 225, https://doi.org/10.1007/978-1-0716-4599-4_16,

clinical settings, some patients require rapid relief. This group, referred to as patients with "severe depression," often presents with melancholic, psychotic, catatonic symptoms, or suicidality, necessitating urgent and comprehensive treatment. Moreover, severe depression does not necessarily overlap with TRD.

The treatment landscape for severe depression and TRD includes various pharmacological and non-pharmacological approaches. Two prominent treatments are ketamine and electroconvulsive therapy (ECT). While ECT remains the gold standard [1], both treatments have shown efficacy in severe depression and TRD, though each has distinct benefits and drawbacks. The increasing recognition of personalized treatment in psychiatry underscores the importance of identifying "the right treatment for the right patient," tailored to the individual clinical profile. In this chapter, we review the current research on ECT and ketamine, followed by evidence comparing the two treatment options.

2 Predictors for Response to Electroconvulsive Therapy

2.1 Clinical and Demographic Predictors

ECT has been extensively studied to identify clinical and demographic predictors of treatment outcomes, with several consistent factors emerging from the literature. Age is one of the most robust predictors, with older patients generally achieving higher remission rates. Kellner and Espinoza [2] concluded that older adults with major depressive disorder (MDD) are more likely to benefit from ECT, potentially due to age-related conditions, such as psychotic depression and psychomotor disturbances, which mediated the relationship between age and clinical outcomes [3, 4].

Psychotic features have also been consistently associated with superior ECT outcomes. A large clinical trial reported higher remission rates in patients with psychotic depression compared to those with non-psychotic depression [5], a finding further confirmed by a meta-analysis [6]. The severity of the depressive episode is another predictor, with more severe depression correlating with a better response to ECT, especially in cases requiring rapid symptom resolution. Using a large dataset (i.e., approximately 1900 participants) from the Dutch ECT Consortium, Loef and colleagues identified factors such as older age, shorter episode duration, psychotic symptoms, better cognitive functioning, fewer prior psychotherapy failures, and lower antidepressant resistance as positive predictive features of ECT response [7]. However, as their model explained only a small portion of the variance in outcomes, Loef and co-authors underscored the need for additional biological markers to enhance predictive accuracy, aligning with ongoing efforts to improve personalized ECT approaches. Nonetheless, it is clinically important to note that ECT is also effective for moderate depression [8].

Some clinical predictors, such as melancholic features, have shown inconsistent results. For instance, while melancholic symptoms are often associated with ECT response, van Diermen and colleagues noted that the heterogeneity of definitions and the variable presentation of melancholia limit its predictive reliability [6].

Similarly, Fink et al. criticized the Diagnostic and Statistical Manual of Mental Disorders (DSM)-defined melancholic features, arguing that they do not capture the classical criteria of melancholia, particularly profound psychomotor disturbances [9]. Consistent with this, Hickie et al. (1996) found psychomotor disturbances to be predictive primarily when paired with psychotic features [10]. Veltman et al. further suggested that psychotic features, rather than psychomotor symptoms alone, more reliably predict ECT response, especially in older patients with melancholic depression [11].

2.2 Physiological Variables as Biomarkers

The search for reliable biomarkers that can predict ECT response has expanded into multiple biological domains, including neurotrophic factors, inflammatory markers, and epigenetic modifications. However, the current evidence remains mixed and often lacks the necessary robustness for clinical application.

Neurotrophic factors like brain-derived neurotrophic factor (BDNF) and vascular endothelial growth factor (VEGF) have been studied extensively. ECT has been shown to increase levels of these factors, which are associated with neurogenesis and neuroplasticity, but baseline levels have not reliably predicted clinical outcomes [12].

Inflammatory markers such as interleukin (IL)-6 and tumor-necrosis factor (TNF)-α show an acute increase following ECT, which later subsides, potentially contributing to the therapeutic effect. However, the relationship between baseline inflammatory markers and treatment response remains unclear [12].

Epigenetic markers are a promising area of research, with some studies indicating that methylation levels in the promoter regions of specific genes such as of BDNF and p11 could predict ECT response. Lower baseline methylation in the BDNF promoter has been associated with better outcomes, suggesting that epigenetic modifications could serve as potential biomarkers [13]. Despite these findings, more research is needed to validate such variables as markers across diverse patient populations.

2.3 Models Using Clinical Data and their Performance

Predictive modeling in ECT research has advanced with the integration of clinical data and computational techniques. For example, van der Does et al. (2022) developed a Bayesian network model that achieved an area under the curve (AUC) value of 0.686 in predicting remission, highlighting older age, psychotic symptoms, and absence of personality disorders as strong predictors of remission. However, the model's performance in predicting nonresponse was limited, suggesting the need for further refinement and the inclusion of additional variables, such as biomarkers [14].

Nakajima et al. utilized machine learning to build a prediction model for clinical remission following ECT. Their model achieved 71% accuracy with high sensitivity (86%) and a specificity of 46% [15]. Moreover, they identified a shorter duration of the current episodes as one of the most important predictive features, which is consistent with findings from a previous meta-analysis [16] and data from a large international consortium [17].

Semple et al. developed a machine learning model using a large Scottish dataset ($n = 2074$) that included only clinical variables [18]. The predictive accuracy of their model was 61%. In addition, their statistical analysis identified several clinical variables associated with ECT response, including older age, psychotic features, psychomotor retardation, previous positive response to ECT, lack of medication resistance, severe distress, and the need for urgent intervention.

2.4 Results from Brain Imaging

Functional magnetic resonance imaging (fMRI) has provided valuable insights into the neural mechanisms underlying ECT's effects, identifying brain regions where functional changes are associated with ECT response. For example, the subgenual anterior cingulate cortex (sgACC) has emerged as a key region in predicting ECT outcomes; higher baseline connectivity between the sgACC and other brain regions, such as the superior frontal gyrus, was associated with better clinical outcomes [19]. In addition, multimodal assessments combining structural MRI and fMRI reported an AUC above 0.80 for predicting remission after ECT [20]. Recent research on effective resting-state connectivity by Doesschate et al. further illuminates these neural mechanisms, highlighting specific changes in connectivity within the default mode, salience, and central executive networks that correlate with ECT response. Notably, changes in connectivity from the posterior default mode network to the insula were found to be associated with treatment effectiveness, suggesting a role for these network dynamics in mediating ECT's clinical impact [21].

The influence of individual patient factors on neuroplasticity is another area of active research. Opel et al. demonstrated that elevated body mass index (BMI) may modulate ECT-induced changes in subcortical gray matter volume, particularly within the thalamus and putamen. These findings indicate that body weight could affect the extent of neuroplastic responses to ECT, providing a personalized context for treatment response [22].

However, the complexity and cost of brain imaging techniques limit their widespread use in clinical settings. Adding to the complexity, Wade et al. emphasized that symptom heterogeneity influences neuroimaging findings, showing that separate symptom dimensions, such as somatic disturbances and core mood/anhedonia, are associated with distinct neural networks. This specificity suggests that ECT's impact on brain structure and function may vary according to symptom profiles, highlighting the potential for more targeted, symptom-informed neuroimaging approaches to optimize treatment predictions [23].

3 Study Limitations in Electroconvulsive Therapy Findings

Despite all of these promising findings, the research on ECT predictors is fraught with several limitations:

- **Variability in Study Design:** Differences in study design, including the definitions of clinical features like melancholia, the timing of biomarker assessments, and the specific imaging modalities used, contribute to inconsistent findings across studies. This variability complicates the interpretation of results and hampers the development of standardized predictive models.
- **Lack of Longitudinal, Long-Term Observation:** Many studies focus on immediate or short-term outcomes, with fewer addressing long-term remission and relapse rates. Longitudinal studies are needed to assess the durability of ECT's effects and the long-term predictive value of identified biomarkers.
- **Technical and Practical Barriers:** The implementation of advanced biomarkers and imaging techniques in routine clinical practice faces technical challenges, such as the need for specialized equipment and expertise, as well as the associated cost and time. For example, while MRI has provided valuable insights into ECT's neural mechanisms, its widespread use in predicting individual patient outcomes is currently impractical [19].
- **Ethical and Stigma-Related Issues:** The stigma associated with ECT limits both its application and the willingness of patients to participate in related research. It is crucial to educate the public, patients, their families, and even psychiatrist that ECT is generally safe under general anesthesia and muscle relaxant and that it is not associated with excessive mortality or an increased risk of developing dementia.

4 Conclusions with Respect to Electroconvulsive Therapy

Several clinical factors, such as older age and the presence of psychotic symptoms, are now often successfully used in daily clinical practice to predict ECT outcomes. The identification of reliable predictors holds significant promise for advancing personalized medicine. The search for predictive biomarkers should particularly focus on identifying those that can pinpoint patients who may benefit from ECT, even if they belong to patient categories with typically low response rates. For instance, it would be valuable to know in advance which patients with comorbid emotional instability personality disorder could benefit, in contrast to those with psychotic depression, where response rates may hardly increase.

Future research should focus on large-scale, multicenter studies that integrate clinical, genetic, and neuroimaging data to develop comprehensive predictive models. Multi-omics approaches, integrating genomic, proteomic, and metabolomic data, may offer a deeper understanding of the biological mechanisms underlying ECT response and help identify novel biomarkers. However, the complexity, cost, and technical requirements of these emerging tools currently limit their widespread implementation.

5 Predictors for Response to Ketamine

5.1 Clinical and Demographic Predictors

Ketamine has emerged as a rapid-acting option compared to traditional antidepressant agents for treatment-resistant depression (TRD) and severe depression. Identifying clinical and demographic predictors of ketamine response is crucial for optimizing its clinical use. The following potential clinical predictors have emerged:

- **A Positive Family History of Alcohol Use Disorder (FHP)**: A positive FHP was associated with a favorable response to ketamine, with patients often experiencing greater symptom improvement. This connection may reflect shared neurobiological mechanisms between alcohol use disorder and ketamine's mode of action [24].
- **BMI**: Higher Pretreatment BMI was associated with better acute responses, though its relationship with sustained response remains unclear. This suggests that metabolic factors may play a role in ketamine's antidepressant effects [25]. However, despite higher BMI in the ELEKT-D study compared to the KetECT trial, the ELEKT-D study had numerically lower remission rates [26, 27].
- **History of Suicide Attempts**: Patients without a history of suicide attempts are more likely to experience sustained antidepressant effects possibly due to different underlying psychopathology or resilience factors that influence treatment outcomes [24].
- **Subtype of Depression**: Evidence suggests that patients with anxious depression may respond better to ketamine than those with non-anxious depression [28]. Findings from the KetECT and ELEKT-D trials indicate that anxious features among

patients correlated with exceptionally poor responses to ECT, highlighting that ketamine is a potentially better alternative for this subtype of depression [29].

5.2 Physiological Variables as Biomarkers

Research into biological variables predicting ketamine response spans physiological, biochemical, and genetic factors, though none meets the reliability required for routine clinical use as yet. Those biological variables include:

- **Inflammatory Markers**: Elevated baseline levels such as of C-reactive protein (CRP) and pro-inflammatory cytokines correlated with better ketamine response, suggesting an anti-inflammatory effect, particularly in patients with high levels of inflammation [28]. However, a meta-analysis did not find one single inflammatory marker with consistent predictive value, though some studies reported that lower levels of CRP were linked to greater symptom reduction [30].
- **Neurotrophic Factors**: Brain-derived neurotrophic factor (BDNF) has been among the most studied neurotrophic markers related to ketamine response. Plasma levels of BDNF increase moderately in ketamine responders, implying a role in ketamine effects. Baseline BDNF levels do not reliably predict outcomes [30].
- **Metabolites of the Tryptophan-Kynurenine Pathway**: A lower kynurenic acid to quinolinic acid ratio was associated with greater improvement in depression scores, underscoring the involvement of metabolic pathways [30].
- **Amino Acids and Derivatives**: Certain amino acids, such as D-serine, showed potential as marker for ketamine response, though baseline associations are inconsistent [31, 32]
- **Levels of Ketamine and Metabolites**: A nonsignificant correlation between blood levels of ketamine or its metabolites like norketamine and response suggests limited clinical utility for predicting response [30].
- **Adiponectin Levels**: Lower baseline adiponectin, an anti-inflammatory marker, was associated with rapid response, supporting the notion that inflammatory and metabolic states may influence ketamine's efficacy [24].
- **Neurochemical Markers**: Lower pretreatment glutamine to glutamate ratios predicted better symptomatic improvement, highlighting the importance of glutamatergic neurotransmission in ketamine's mechanism of action [25].
- **Genetic Factors**: The BDNF Val66Met polymorphism may influence ketamine response, with Val/Val genotype responding more favorably than Met carriers. This aligns with BDNF's role in synaptic plasticity and neurogenesis [24].

5.3 Models Using Clinical Data and their Performance

While there is growing interest in developing predictive models for ketamine response, the field is still in its early stages. However, some promising approaches have emerged:

- **Machine Learning Models**: Miller et al. used logistic regression models to predict nonresponse to ketamine based on baseline depression symptoms in military veterans. Their models identified specific symptoms, such as suicidal ideation and sleep difficulties, as predictors of nonresponse, achieving a negative predictive value (NPV) of over 91% for a subset of patients. This approach highlights the potential for using symptom-level data to guide clinical decision-making [33].
- **Longitudinal Models**: Kane et al. (2022) developed a Bayesian hierarchical model to predict individual responses to ketamine intravenous therapy (KIT) for depression. This model accounts for patient-level variability and uses longitudinal symptom data, demonstrating potential for predicting long-term response trajectories. It offers a more nuanced understanding of patient responses to ketamine [34].

5.4 Findings in Brain Imaging

Neuroimaging has provided valuable insights into the neural mechanisms underlying ketamine's antidepressant effects, with several brain regions and connectivity patterns emerging as potential predictive markers:

- **Anterior Default Mode Network (DMN) and Posterior Insular Connectivity**: Increased connectivity between the right medial prefrontal cortex/anterior cingulate cortex (ACC) and the posterior insula predicted significant reductions in depressive symptoms following serial ketamine infusions. This suggests that pretreatment functional connectivity in these regions could serve as a biomarker for ketamine response [35].
- **Pregenual Anterior Cingulate Cortex (pgACC)**: Higher baseline pgACC activity during emotional processing was associated with greater reductions in depressive symptoms, particularly when combined with increases in glutamate levels post-treatment. This multimodal approach using fMRI and magnetic resonance spectroscopy (MRS) holds promise for predicting outcomes [36].
- **Superior Longitudinal Fasciculus (SLF)**: Lower kurtosis in the right SLF, reflecting lower white matter integrity, was associated with greater reductions in depressive symptoms. It suggested a role for structural changes in white matter in ketamine response [35].

6 Study Limitations in Findings from Ketamine Trials

Despite promising findings, research on predictors of ketamine response faces several limitations:

- **Heterogeneity of Patient Populations**: Many studies include diverse patient populations with varying degrees of treatment resistance, comorbid conditions, and medication history. This heterogeneity complicates the interpretation of results and limits the generalizability of findings.
- **Small Sample Sizes**: Many studies on ketamine response predictors have relatively small sample sizes, which limits the statistical power to detect significant effects and increases the risk of type I and type II errors. Larger, multicenter studies are needed to validate existing findings and explore new predictive markers.
- **Variability in Ketamine Administration**: Differences in dosing, administration routes (IV vs. intranasal), and treatment protocols (single vs. serial infusions) introduce outcome variability. Standardized protocols are necessary to improve the reliability of predictive models.
- **Short-term Follow-up**: Most research focuses on short-term outcomes, typically assessing response within hours to days after treatment. Extended follow-up is needed to assess the durability of ketamine's efficacy and the stability of identified predictors over time.
- **Challenges in Translating Biomarkers to Clinical Practice**: While several biomarkers and imaging techniques (e.g., fMRI and MRS) have shown promise in research settings, they face technical, logistical, and cost-related challenges, limiting their routine use.
- **Ethical and Safety Concerns**: Along with unclear long-term safety, risks of ketamine misuse and dependency raise ethical concerns about widespread use, particularly in vulnerable populations. This highlights the need for further research on ketamine's safety profile.

7 Conclusions with Respect to Ketamine Therapy

In conclusion, clinical factors such as family history of alcohol use disorder, BMI, and baseline symptoms like suicidal ideation and sleep difficulties have been suggested as potential predictors of ketamine response, but evidence remains limited. Further research should focus on validating emerging biomarkers and neuroimaging variables like anterior DMN connectivity and pgACC activity, despite their current constraints of complexity, cost, and technical requirements. The future of personalized ketamine treatment lies in developing comprehensive predictive models that integrate clinical, genetic, and neuroimaging data. Advances in multi-omics and artificial intelligence will help identify novel predictors and facilitate personalized treatment approaches.

8 The Synthesis: Clinical Predictors for Selecting Electroconvulsive Therapy or Ketamine in Depression Treatment

The comparative efficacy and safety of ketamine and ECT have been the subject of clinical trials and meta-analyses, although those meta-analyses include the same five to seven RCTs all the time. Two systematic reviews and meta-analyses [37, 38] consistently showed that ECT was more effective than ketamine for major depressive episodes, until one large RCT [27] was published. Reviewing 5 RCTs with 278 patients, Menon et al. concluded that ECT had superior efficacy in depression ratings, response rates, and remission rates at 1-week post-treatment [37]. Reviewing 6 clinical trials (5 RCTs and 1 clinical trial) with 340 patients, Rhee et al. found a higher response rate for ECT, though ketamine's rapid onset of action may benefit those needing immediate relief [38]. Then, Shafiee et al., [39] adding Anand et al. [27] to the six clinical trials, reported ECT was not superior to ketamine for TRD. In contrast, Petrucci et al. reported that ketamine did not meet the non-inferiority margin, especially in inpatient settings. They reinforced the preference for ECT in clinical practice based on the analysis of six RCTs [40].

While meta-analysis provide evidence with respect to greater sample sizes and defined evaluation criteria, the most significant contributions to this comparative analysis come from the KetECT (Ekstrand et al., 2021; n = 186) [26] and ELEKT-D (Anand et al., 2023; n = 403) [27] trials. Interestingly, their conclusions differ: KetECT found ECT superior to ketamine, while ELEKT-D concluded that ketamine was non-inferior to ECT. This discrepancy is primarily due to the significant difference in the remission rates of the ECT arm (63% in KetECT vs. 22% in ELEKT-D), while the ketamine arm showed comparable rates (46% vs. 38%) based on Montgomery Åsberg Depression Rating Scale (MADRS) scores. Therefore, clinicians should be cautious of factors that may influence the divergent clinical outcomes of ECT. A comprehensive assessment of these trials was reported in Ekstrand et al. which discussed the reasons for the discrepant results [29]. Briefly, clinical characteristics of the participants (e.g., age, psychotic features, episode duration, and psychiatric comorbidities) and ECT parameters (e.g., the number of ECT sessions, pulse width, and electrode placement) can significantly impact ECT outcomes.

Variations in such parameters substantially limit the interpretation of these studies' results. In addition, better preparation of patients and more experienced staff might have resulted in higher remission rates for ketamine. As noted, KetECT delivered better results for ketamine than ELEKT-D, likely due to differences in patient selection. This highlights the need for further investigations to optimize ketamine treatment schedules and fully explore its potential in antidepressant use.

Selecting between these two treatments requires careful consideration of multiple clinical factors. The following discussion is primarily based on two key head-to-head studies, KetECT and ELEKT-D.

8.1 Severity of Depression and Psychotic Features

As discussed in Sect. 2, ECT is highly effective for psychotic depression and is recommended as a first-line treatment in many international guidelines at present. This goes along with combining antidepressants and antipsychotics, especially in patients with severe suicidality or life-threatening somatic conditions [5, 41, 42]. Ketamine may be considered as an alternative when ECT is relatively contraindicated or unavailable. The KetECT trial reported that while ECT achieved a higher remission rate of 79% in psychotic depression, ketamine had a remission rate of 50% in a post hoc analysis [26, 29]. These findings suggest that ketamine may be a viable alternative, especially when ECT is less accessible or suitable.

8.2 Duration of Depression

The duration of the depressive episode plays a critical role in determining the optimal treatment. ECT may be less effective in patients with longer-duration episode [15, 16]. The higher remission rates of ECT in the KetECT trial may be explained by depression with a shorter median duration of 3.5 months as compared to the ELEKT-D trial with a median of 24 months episode duration [26, 27]. This suggests that ECT may be preferable in patients with shorter episode durations, while ketamine could be more effective in patients experiencing prolonged or treatment-resistant depression [40, 43].

8.3 Inpatient Versus Outpatient Setting

The setting of treatment whether inpatient or outpatient—significantly affects the choice of treatment modality. ECT has been shown to be more effective in hospitalized patients, particularly those with severe depressive symptoms. This is supported by both the KetECT trial and meta-analyses, which found that ECT was superior in reducing depressive symptoms among inpatients [26, 40, 43]. Conversely, ketamine has shown non-inferiority to ECT in outpatient settings, making it an alternative for outpatients or those with less severe symptoms [27, 29, 40]. However, the distinction between inpatient and outpatient settings likely reflects underlying factors such as depression subtype, severity, and differing healthcare traditions that shape hospitalization criteria and treatment decisions.

8.4 Age

Age is another key factor influencing treatment outcomes. Older patients are known to respond better to ECT [4, 6, 18, 41], as observed in a post hoc analysis of the KetECT trial, where those over 50 had significantly higher remission rates with ECT (77%) compared to ketamine (37%) [26]. Conversely, younger patients (under 50) may respond equally well- or slightly better-to ketamine, making it a preferred option for this age group [26, 40, 43]. Rhee et al. also suggest that ECT remains superior for older patients, while ketamine may be better suited for younger populations [38]. However, the relationship between older age and higher clinical response to ECT may be mediated by the presence of psychotic and psychomotor symptoms, which are more common in older patients [3, 4]. Therefore, age alone should not be used as the only determinant for treatment selection.

8.5 Cognitive Concerns

Transient cognitive adverse effects are well-documented risks associated with ECT. The ELEKT-D trial found that ECT led to a greater temporary decline in memory function, although cognitive recovery was observed at follow-up [27]. In contrast, ketamine is associated with less cognitive impairment. This may make it preferable in patients concerned about cognitive function [26, 37, 40]. Rhee et al. support these findings, suggesting that cognitive concerns should influence treatment selection, particularly in patients vulnerable to memory issues [38]. However, these transient adverse effects should not override the recommendation of ECT when clearly indicated.

8.6 Comorbid Anxiety

For patients with comorbid anxiety disorders, ketamine may be a better option. The ELEKT-D trial highlighted the rapid antidepressant effects of ketamine, which can be particularly beneficial for patients with acute suicidality or significant anxiety [27]. While not all studies show a definitive benefit of ketamine in patients with anxiety, its rapid onset makes it a compelling choice in urgent conditions [26]. However, clinicians should avoid using ketamine indiscriminately for short-term symptom relief, particularly in patients with marked distress when underlying personality disorders or chronic affective dysregulation may be present.

9 Side Effects and Safety Profiles

Despite concerns about transient cognitive impairments, ECT is used when potential benefits are deemed to outweigh risks. Headaches and muscle pain are common side effects of ECT, though they are self-limiting [27, 38]. Nevertheless, ECT use varies significantly by country, influenced by stigma and media portrayal [44–46].

Ketamine's side effects include dissociative symptoms—such as detachment from reality, hallucinations, and perceptual disturbances and psychoactive events [26]. These may distress patients, particularly those prone to psychiatric side effects. It can transiently increase blood pressure and heart rate, warranting monitoring in patients with cardiovascular concerns. Ketamine's popularity and potential for misuse raise dependency concerns and may contribute to depression overdiagnosis. Systematic reviews and meta-analyses [37, 38, 40, 43] found that ketamine is associated with fewer adverse effects like headaches and musculoskeletal pain compared to ECT. Post-treatment cognition scores often favor ketamine [26, 27, 37, 43], and the ELEKT-D trial suggests it may be better tolerated in outpatient settings.

10 Conclusions

In the treatment of severe depression and TRD, both ECT and ketamine are valuable options. However, the growing emphasis on personalized approaches underscores the need for reliable predictors to guide treatment decisions. While biomarkers like BDNF, inflammatory cytokines, and advanced neuroimaging techniques show potential, their predictive value remains uncertain due to inconsistent findings and small sample size investigations [12]. Until validated in large-scale studies, these parameters cannot be used as reliable predictors for treatment response.

In contrast, clinical variables, such as age, psychotic features, episode duration, and psychiatric comorbidities, offer more reliable and consistent guidance, particularly for ECT. These markers have been replicated in several clinical trials and retrospective studies with larger sample sizes. These factors are typically suitable and readily applicable in clinical practice.

Factors such as episode duration, comorbid anxiety, severity of depression, psychotic features, and treatment setting often reflect the overall severity of depression. For instance, severe depression often leads to hospitalization, while comorbid anxiety tends to prolong depressive episodes, complicating diagnosis and treatment. Recognizing these interdependent factors allows for more nuanced and personalized treatment decisions. Patients with prolonged depressive symptoms, often with comorbid personality disorders, may respond less favorably to conventional treatments, including ECT and pharmacotherapy. They may require additional behavioral or augmented psychotherapy. Comorbid anxiety, possibly driven by underlying pathological personality, with prolonged depression complicates treatment [47, 48].

Ketamine's potential benefit may be underestimated due to limited research on dose-response effects and therapeutic context. While the standard dose of 0.5 mg/kg is widely used, some studies have explored higher doses, such as 1.0 mg/kg, to assess possible additional benefits [49, 50]. Although these studies did not find clear evidence that 1.0 mg/kg provided greater antidepressant effects than 0.5 mg/kg, further studies are warranted to explore this question. Due to ketamine's dissociative properties, structured psychological support by experienced healthcare professionals is routinely provided before, during, and after treatment in centers with expertise in its administration. Further research will assess whether treatment at these expert centers offers significantly better outcomes compared to other options, at least for certain patient subgroups.

In conclusion, while ECT remains the gold standard for severe depression, ketamine offers a valuable alternative, especially for younger or outpatient populations [29,40,43,38]. Ketamine should be considered when ECT is contraindicated, inappropriate due to high somatic risk, or for ECT non-responders. Treatment goals should go beyond rapid symptom relief emphasizing sustained remission, improved functions in daily life, and overall well-being. Clinicians should consider the broader clinical context—including comorbidities, episode duration, and patient preferences—when choosing between ketamine and ECT. Future research into clinical and biological predictors may help to further personalize treatment strategies and improve patient outcomes.

Conflict of Interest

Pouya Movahed Rad has received one-time honoraria from TEVA (for expert opinion on tardive dyskinesia) and Lundbeck (for an educational lecture to company staff); Akihiro Takamiya has no potential conflicts of interest.

References

1. Zilles-Wegner D, Kellner CH, Sartorius A (2024) Electroconvulsive therapy and its new competitors: ECT remains the gold standard. J ECT 40:e31. https://doi.org/10.1097/YCT.0000000000001034
2. Espinoza RT, Kellner CH (2022) Electroconvulsive therapy. N Engl J Med 386(7): 667–672. https://doi.org/10.1056/NEJMra2034954
3. Heijnen W, Kamperman AM, Tjokrodipo LD, Hoogendijk WJG, van den Broek WW, Birkenhager TK (2019) Influence of age on ECT efficacy in depression and the mediating role of psychomotor retardation and psychotic features. J Psychiatr Res 109:41–47. https://doi.org/10.1016/j.jpsychires.2018.11.014
4. van Diermen L, Poljac E, Van der Mast R, Plasmans K, Van den Ameele S, Heijnen W et al (2020) Toward targeted ECT: the interdependence of predictors of treatment response in depression further explained. J Clin Psychiatry 82(1). https://doi.org/10.4088/JCP.20m13287
5. Petrides G, Fink M, Husain MM, Knapp RG, Rush AJ, Mueller M et al (2001) ECT remission rates in psychotic versus nonpsychotic depressed patients: a report from CORE. J ECT 17(4):244–253. https://doi.org/10.1097/00124509-200112000-00003
6. van Diermen L, van den Ameele S, Kamperman AM, Sabbe BCG, Vermeulen T, Schrijvers D et al (2018) Prediction of electroconvulsive

therapy response and remission in major depression: meta-analysis. Br J Psychiatry 212(2):71–80. https://doi.org/10.1192/bjp.2017.28

7. Loef D, Hoogendoorn AW, Somers M, Mocking RJT, Scheepens DS, Scheepstra KWF et al (2024) A prediction model for electroconvulsive therapy effectiveness in patients with major depressive disorder from the Dutch ECT consortium (DEC). Mol Psychiatry. https://doi.org/10.1038/s41380-024-02803-2
8. Ostergaard SD, Speed MS, Kellner CH, Mueller M, McClintock SM, Husain MM et al (2020) Electroconvulsive therapy (ECT) for moderate-severity major depression among the elderly: data from the pride study. J Affect Disord 274:1134–1141. https://doi.org/10.1016/j.jad.2020.05.039
9. Fink M, Rush AJ, Knapp R, Rasmussen K, Mueller M, Rummans TA et al (2007) DSM melancholic features are unreliable predictors of ECT response: a CORE publication. J ECT 23(3):139–146. https://doi.org/10.1097/yct.0b013e3180337344
10. Hickie I, Mason C, Parker G, Brodaty H (1996) Prediction of ECT response: validation of a refined sign-based (CORE) system for defining melancholia. Br J Psychiatry 169(1):68–74. https://doi.org/10.1192/bjp.169.1.68
11. Veltman EM, de Boer A, Dols A, van Exel E, Stek ML, Sienaert P et al (2019) Melancholia as predictor of electroconvulsive therapy outcome in later life. J ECT 35(4):231–237. https://doi.org/10.1097/YCT.0000000000000579
12. Maffioletti E, Carvalho Silva R, Bortolomasi M, Baune BT, Gennarelli M, Minelli A (2021) Molecular biomarkers of electroconvulsive therapy effects and clinical response: understanding the present to shape the future. Brain Sci 11(9). https://doi.org/10.3390/brainsci11091120
13. Feng T, Youssef NA (2020) Can epigenetic biomarkers lead us to precision medicine in predicting treatment response and remission for patients being considered for ECT? Psychiatry Res 284:112659. https://doi.org/10.1016/j.psychres.2019.112659
14. van der Does Y, Turner RJ, Bartels MJH, Hagoort K, Metselaar A, Scheepers F et al (2023) Outcome prediction of electroconvulsive therapy for depression. Psychiatry Res 326:115328. https://doi.org/10.1016/j.psychres.2023.115328
15. Nakajima K, Takamiya A, Uchida T, Kudo S, Nishida H, Minami F et al (2022) Individual prediction of remission based on clinical features following electroconvulsive therapy: a machine learning approach. J Clin Psychiatry 83(5). https://doi.org/10.4088/JCP.21m14293
16. Haq AU, Sitzmann AF, Goldman ML, Maixner DF, Mickey BJ (2015) Response of depression to electroconvulsive therapy: a meta-analysis of clinical predictors. J Clin Psychiatry 76(10):1374–1384. https://doi.org/10.4088/JCP.14r09528
17. Blanken M, Oudega ML, Hoogendoorn AW, Sonnenberg CS, Rhebergen D, Klumpers UMH et al (2023) Sex-specifics of ECT outcome. J Affect Disord 326:243–248. https://doi.org/10.1016/j.jad.2022.12.144
18. Semple DM, Suveges S, Steele JD (2024) Electroconvulsive therapy response and remission in moderate to severe depressive illness: a decade of national Scottish data. Br J Psychiatry:1–9. https://doi.org/10.1192/bjp.2024.126
19. Porta-Casteras D, Cano M, Camprodon JA, Loo C, Palao D, Soriano-Mas C et al (2021) A multimetric systematic review of fMRI findings in patients with MDD receiving ECT. Prog Neuro-Psychopharmacol Biol Psychiatry 108:110178. https://doi.org/10.1016/j.pnpbp.2020.110178
20. Bruin WB, Oltedal L, Bartsch H, Abbott C, Argyelan M, Barbour T et al (2024) Development and validation of a multimodal neuroimaging biomarker for electroconvulsive therapy outcome in depression: a multicenter machine learning analysis. Psychol Med 54(3):495–506. https://doi.org/10.1017/S0033291723002040
21. Ten Doesschate F, Bruin W, Zeidman P, Abbott CC, Argyelan M, Dols A et al (2023) Effective resting-state connectivity in severe unipolar depression before and after electroconvulsive therapy. Brain Stimul 16(4):1128–1134. https://doi.org/10.1016/j.brs.2023.07.054
22. Opel N, Narr KL, Abbott C, Argyelan M, Espinoza R, Emsell L et al (2021) Elevated body weight modulates subcortical volume change and associated clinical response following electroconvulsive therapy. J Psychiatry Neurosci 46(4):E418–E426. https://doi.org/10.1503/jpn.200176
23. Wade BSC, Hellemann G, Espinoza RT, Woods RP, Joshi SH, Redlich R et al (2021) Accounting for symptom heterogeneity can improve neuroimaging models of antidepressant response after electroconvulsive therapy. Hum Brain Mapp 42(16):5322–5333. https://doi.org/10.1002/hbm.25620

24. Rong C, Park C, Rosenblat JD, Subramaniapillai M, Zuckerman H, Fus D et al (2018) Predictors of response to ketamine in treatment resistant major depressive disorder and bipolar disorder. Int J Environ Res Public Health 15(4). https://doi.org/10.3390/ijerph15040771
25. Cinderella MA, Nichols NA, Munjal S, Yan J, Kimball JN, Gligorovic P (2022) Antiepileptics in electroconvulsive therapy: a mechanism-based review of recent literature. J ECT 38(2):133–137. https://doi.org/10.1097/YCT.0000000000000805
26. Ekstrand J, Fattah C, Persson M, Cheng T, Nordanskog P, Akeson J et al (2022) Racemic ketamine as an alternative to electroconvulsive therapy for unipolar depression: a randomized, open-label, non-inferiority trial (KetECT). Int J Neuropsychopharmacol 25(5):339–349. https://doi.org/10.1093/ijnp/pyab088
27. Anand A, Mathew SJ, Sanacora G, Murrough JW, Goes FS, Altinay M et al (2023) Ketamine versus ECT for nonpsychotic treatment-resistant major depression. N Engl J Med 388(25):2315–2325. https://doi.org/10.1056/NEJMoa2302399
28. Yavi M, Lee H, Henter ID, Park LT, Zarate CA Jr (2022) Ketamine treatment for depression: a review. Discov Ment Health 2(1):9. https://doi.org/10.1007/s44192-022-00012-3
29. Ekstrand J, Takamiya A, Nordenskjold A, Kirov G, Sienaert P, Kellner CH et al (2024) Ketamine or ECT? What have we learned from the KetECT and ELEKT-D trials? Int J Neuropsychopharmacol 27(1). https://doi.org/10.1093/ijnp/pyad065
30. Medeiros GC, Gould TD, Prueitt WL, Nanavati J, Grunebaum MF, Farber NB et al (2022) Blood-based biomarkers of antidepressant response to ketamine and esketamine: a systematic review and meta-analysis. Mol Psychiatry 27(9):3658–3669. https://doi.org/10.1038/s41380-022-01652-1
31. Moaddel R, Shardell M, Khadeer M, Lovett J, Kadriu B, Ravichandran S et al (2018) Plasma metabolomic profiling of a ketamine and placebo crossover trial of major depressive disorder and healthy control subjects. Psychopharmacology 235(10):3017–3030. https://doi.org/10.1007/s00213-018-4992-7
32. Moaddel R, Luckenbaugh DA, Xie Y, Villasenor A, Brutsche NE, Machado-Vieira R et al (2015) D-serine plasma concentration is a potential biomarker of (R,S)-ketamine antidepressant response in subjects with treatment-resistant depression. Psychopharmacology 232(2):399–409. https://doi.org/10.1007/s00213-014-3669-0
33. Miller EA, Afshar HT, Mishra J, McIntyre RS, Ramanathan D (2024) Predicting non-response to ketamine for depression: an exploratory symptom-level analysis of real-world data among military veterans. Psychiatry Res 335:115858. https://doi.org/10.1016/j.psychres.2024.115858
34. Kane GA, WM J, Oldenkamp AA, McInnes LA (2023) Using a longitudinal model to predict outcomes to ketamine intravenous therapy for depression. [Poster presentation.]. Conference Poster, American Society of Clinical Psychopharmacology (ASCP) Annual Meeting 2023, Tampa, FL, USA
35. Wade BSC, Loureiro J, Sahib A, Kubicki A, Joshi SH, Hellemann G et al (2022) Anterior default mode network and posterior insular connectivity is predictive of depressive symptom reduction following serial ketamine infusion – CORRIGENDUM. Psychol Med 52(12):2399. https://doi.org/10.1017/S0033291722001659
36. Weigand A, Gartner M, Scheidegger M, Wyss PO, Henning A, Seifritz E et al (2022) Predicting antidepressant effects of ketamine: the role of the Pregenual anterior cingulate cortex as a multimodal neuroimaging biomarker. Int J Neuropsychopharmacol 25(12):1003–1013. https://doi.org/10.1093/ijnp/pyac049
37. Menon V, Varadharajan N, Faheem A, Andrade C (2023) Ketamine vs electroconvulsive therapy for major depressive episode: a systematic review and meta-analysis. JAMA Psychiatry 80(6):639–642. https://doi.org/10.1001/jamapsychiatry.2023.0562
38. Rhee TG, Shim SR, Forester BP, Nierenberg AA, McIntyre RS, Papakostas GI et al (2022) Efficacy and safety of ketamine vs electroconvulsive therapy among patients with major depressive episode: a systematic review and meta-analysis. JAMA Psychiatry 79(12):1162–1172. https://doi.org/10.1001/jamapsychiatry.2022.3352
39. Shafiee A, Soltani Abhari F, Jafarabady K, Bakhtiyari M (2023) Ketamine versus electroconvulsive therapy for treatment-resistant depression: an updated meta-analysis of randomized clinical trials. Asian J Psychiatr 88:103720. https://doi.org/10.1016/j.ajp.2023.103720
40. Petrucci ABC, Fernandes JVA, Reis IA, da Silva GHS, Recla BMF, de Mendonca JC et al (2024) Ketamine versus electroconvulsive therapy for major depressive episode: an updated systematic review and non-inferiority meta-

analysis. Psychiatry Res 339:115994. https://doi.org/10.1016/j.psychres.2024.115994

41. van Diermen L, Vanmarcke S, Walther S, Moens H, Veltman E, Fransen E et al (2019) Can psychomotor disturbance predict ect outcome in depression? J Psychiatr Res 117:122–128. https://doi.org/10.1016/j.jpsychires.2019.07.009
42. Leadholm AK, Rothschild AJ, Nolen WA, Bech P, Munk-Jorgensen P, Ostergaard SD (2013) The treatment of psychotic depression: is there consensus among guidelines and psychiatrists? J Affect Disord 145(2):214–220. https://doi.org/10.1016/j.jad.2012.07.036
43. Liu Y, Yang J, Liu Y (2024) Ketamine and electroconvulsive therapy for severe depression: a network meta-analysis of efficacy and safety. J Psychiatr Res 175:218–226. https://doi.org/10.1016/j.jpsychires.2024.05.022
44. Maughan D, Molodynski A (2016) An international perspective on the acceptability and sustainability of electroconvulsive therapy. BJPsych Int 13(1):10–12. https://doi.org/10.1192/s2056474000000891
45. Rohde P, Noorani R, Feuer E, Lisanby SH, Regenold WT (2024) Electroconvulsive therapy across nations: a 2022 survey of practice. J ECT 40(2):96–104. https://doi.org/10.1097/YCT.0000000000000980
46. Abaatyo J, Kaggwa MM (2024) The electroconvulsive therapy story of Africa, a systematic review. Discov Ment Health 4(1):31. https://doi.org/10.1007/s44192-024-00085-2
47. Mulder RT (2002) Personality pathology and treatment outcome in major depression: a review. Am J Psychiatry 159(3):359–371. https://doi.org/10.1176/appi.ajp.159.3.359
48. Newton-Howes G, Tyrer P, Johnson T (2006) Personality disorder and the outcome of depression: meta-analysis of published studies. Br J Psychiatry 188:13–20. https://doi.org/10.1192/bjp.188.1.13
49. Fava M, Freeman MP, Flynn M, Judge H, Hoeppner BB, Cusin C et al (2020) Double-blind, placebo-controlled, dose-ranging trial of intravenous ketamine as adjunctive therapy in treatment-resistant depression (TRD). Mol Psychiatry 25(7):1592–1603. https://doi.org/10.1038/s41380-018-0256-5
50. Dai D, Miller C, Valdivia V, Boyle B, Bolton P, Li S et al (2022) Neurocognitive effects of repeated ketamine infusion treatments in patients with treatment resistant depression: a retrospective chart review. BMC Psychiatry 22(1):140. https://doi.org/10.1186/s12888-022-03789-3

Chapter 17

Therapeutic Ketamine in Youth with Major Depression Versus Illicit Use: Initial Clinical Findings with Short-Term Benefits and Long-Term Unknowns

Malgorzata Zaremba and Patrycja Kleczkowska

Abstract

Severe depression in adolescents significantly contributes to global disease burden due to disability and lost productivity. Major depressive disorder (MDD) and treatment-resistant depression (TRD) are often underdiagnosed and undertreated in this age group. Standard treatments, such as selective serotonin reuptake inhibitors (SSRIs) and psychotherapy, show inadequate response in 30–40% of adolescents, increasing the risk of suicidal behavior. Ketamine, an N-methyl-D-aspartate (NMDA) receptor antagonist, has emerged as a potential treatment option for adolescent MDD and TRD due to its rapid onset of action and ability to reduce suicidal ideation. However, concerns about its neurotoxic effects, potential for abuse, and long-term safety in the developing adolescent brain remain unresolved. Early clinical trials suggest ketamine's effectiveness in youth is comparable to that in adults, but data are limited by small sample sizes and short observation periods. More extensive randomized clinical trials (RCTs) are needed to better understand optimal dosing, administration schedules, and the long-term impacts of ketamine use in adolescents.

Key words Ketamine, Adolescence, Major depressive disorder, Treatment-resistant depression, Efficacy, Adverse events, Abuse, Clinical trials

1 The Burden of Depression in Adolescence

Especially in children and adolescents, mental disorders represent a significant global challenge. Adolescence is a critical developmental period that begins with the onset of puberty and marks the transition from childhood to adulthood. Three stages are described: early (10–13 years (yrs.) of age), middle (14–17 yrs. of age), and late adolescence/young adulthood (18–21 yrs. of age and beyond) [1]. In 2021, the World Health Organization (WHO) underscored that nearly 14% of 10- to 19-year-old individuals carry a burden of severe mental health issues [2]. Depression manifests in approximately 1.1% of 10- to 14-year-old and 2.8% of 15- to 19-year-old persons [2]. According to the National Institutes of Mental Health,

Sabine Himmelseher (ed.), *Ketamine: Pain Management, Critical Care, Depression, and Resource-Poor Settings Volume II*, Neuromethods, vol. 225, https://doi.org/10.1007/978-1-0716-4599-4_17,

approximately 3.2 million 12- to 17-year-old adolescents experienced at least one major depressive episode in the United States in 2017, constituting 13.3% of people within this age group. Notably, there was an elevated prevalence among US females with 20% of the adolescent population being affected [3].

There has also been a significant increase in mood disorders among 11- to 19-year-old people. This has been exacerbated by the COVID-19 pandemic but is also attributed to cyberbullying, education-related pressure, and recent new risk factors, such as social media use [4, 5]. Adolescents from pathological environments characterized by aggression, lack of acceptance, insufficient support, or emotional neglect are particularly vulnerable. While fluctuations in mood, occasional sadness, and outbursts of anger are inherent facets of normal development, an increased incidence of such events coupled with self-harm and suicidal ideation signal potential maladjustment. Depression is usually linked to poor social and educational performance, physical illness, and risky health behavior (e.g., with the result of early pregnancy and substance abuse) [6] and is considered to be the most significant contributor to global disease burden in adolescents because of lost productivity and disability.

Given the potentially lifelong devastating outcomes of adolescent depression, reliable evidence-based guidance for treatment and a better understanding of long-term neurobiological sequelae are an urgent need [7–10]. Timely intervention is imperative, because delayed therapeutic measures may fail to address the emergency of the mental health disorder.

The contemporary approach to adolescent psychiatric disorders encompasses evidence-based pharmacologic and nonpharmacologic interventions and is often extrapolated from adult patients. Therapy of depression in adolescents is always a challenge, as young people seeking treatment are more likely to have comorbidities often accompanied by anxiety and attention-deficit/hyperactivity disorder (ADHD) [10, 11]. An ongoing extensive debate surrounds pharmacotherapeutic interventions for major depressive disorder (MDD) and treatment-resistant depression (TRD) in adolescents (TRD is defined as an unsatisfactory response to two or more adequate trials of antidepressants [12]). Dwyer et al. [13] described TRD in adolescents as a form of MDD in which significant depressive symptoms persist despite the patient receiving an adequate trial of first-line antidepressant, such as SSRI (e.g., fluoxetine, escitalopram, or sertraline) at a dose equivalent to 40 mg of fluoxetine and/or 8–16 sessions of cognitive-behavioral or interpersonal therapy [13]. MDD is prevalent among young individuals, with an estimated point incidence of 2.8% in 6- to 12-year-old, school-age children, and 5.6% in 13- to 18-year-old adolescents [14]. Compared to adults, MDD in children and adolescents is underdiagnosed and undertreated. Young people suffering from depression are often not likely to seek help. According to recent reports, only 35% of adolescents accessed mental health resources, and only 33% received adequate treatment [15]. This underdiagnosis and undertreatment may relate to the fact that younger people present with less differentiated depressive symptoms, such as irritability, aggressive behaviors, and school refusal [14]. In addition, adolescents who do not respond to initial interventions for MDD have an increased rate of suicide, poor academic and social performances, and more interpersonal conflicts in family and peer relationships [16].

Treatment options for adolescent MDD and TRD are very limited and not comprehensively evaluated concerning efficacy and safety. In contrast to adults, the majority of conventional first-line antidepressants do not appear to be suitable as routine therapy [14, 17, 18]. For example, the current National Institute for Health and Care Excellence (NICE) recommends that fluoxetine be administered at least 8 weeks at or above the minimal recommended dose [10, 19]. Adequate titration to maximum tolerance shall be considered when treatment response is insufficient [13]. The starting dose is 10 mg per day, with an increase to 20 mg after 1 week if deemed necessary. Lower doses shall be considered in children with lower body weight. Although evidence for use of doses above 20 mg per day is little, higher doses shall be considered for older, heavier children or in severe cases where a rapid clinical response is prioritized [19]. NICE states that fluoxetine is regarded as the best available choice when pharmacologic treatment of moderate to severe depression is indicated in patients younger than 18 years of age who do not have access to psychotherapy or did not respond to nonpharmacologic intervention [19]. Adequate response to first-line therapy is defined as at least 50% symptom reduction as evaluated by well-validated rating scales [17, 18]. However, response rates to standard antidepressant monotherapy vary from 60% to 70% in pediatric and adolescent patients at best [20]. 30–40% of adolescents with depression show an inadequate response to first-line adult treatment (such as with SSRI or psychotherapy) [13, 17]. Youth patients often do not achieve full remission and/or experience residual symptoms such as anxiety, impaired cognition, fatigue, sleep disturbance, or anhedonia (sadness, loss of gratification, and loss of feelings and affection toward loved ones) [21]. Two-thirds of those treated with conventional antidepressants even do not fully recover [22].

During the initial phase of antidepressant therapy (usually 1–2 weeks), heightened vigilance is indispensable because of an elevated risk of suicidal ideation and behavior [10]. This is not valid for the old drugs, such as monoamine oxidase A inhibitors (IMAO-A) and lithium. Although lithium shows promise in reducing self-harm and suicidality, its use in adolescents is limited by its narrow therapeutic window and potential adverse effects, particularly on kidney and thyroid function [13]. According to a comprehensive systematic review and network meta-analysis comparing conventional antidepressants in the acute treatment of MDD in children and adolescents (mean age 9–18 yrs.), venlafaxine was associated with an increased risk of suicidality (suicidal behavior or ideation) [14]. Suicide is the leading cause of death among adolescents worldwide, and nonfatal behavior (suicidal thoughts and intentional self-harm) is a significant cause of morbidity and the strongest suicide predictor in this group [23]. The US Food and Drug Administration (FDA) has issued a black box warning advising healthcare providers to be very cautious when prescribing antidepressants to patients under 18 yrs. of age [10, 24]. Adolescents taking antidepressants need to be closely monitored, particularly at the beginning of treatment.

All of these circumstances and conditions drastically demonstrate that there is an urgent need for novel, safe, and fast-acting therapy with better response and remission rates in adolescent MDD and TRD. This is especially valid given the few available pharmacologic options and the fact that classical antidepressants require repeated use over days or weeks before benefits manifest [25]. In addition, often several therapeutic attempts are even necessary to find a "working" treatment at all [26]. New strategies must therefore include the provision of rapid-acting pharmacologic agents, potentially combining different mechanisms of action, the prevention of delay in effect, and the improvement of treatment response and/or suicidal behavior [27]. Ketamine may offer all these beneficial actions for the therapy of adolescents with MDD and especially TRD [1]. However, ketamine-related safety and efficacy issues have not been clarified and are completely understudied in these settings [28].

2 Ketamine: A Promising Fast-Acting Candidate Antidepressant for Adolescent Depression

Ketamine is classified as a Schedule III controlled substance because of its potential for abuse but could recently be established as a therapeutic option for therapy-resistant depression at subanesthetic doses (~ 0.5 mg/kg racemic ketamine) in adults. Ketamine is an old/new drug with a decades-long history in anesthesia and is known for its dose-related analgesic, sedative, amnestic, and anesthetic properties. It is used in prehospital emergency therapy via IV or IM administration with a large safety profile in short-term use because of a lack of cardiorespiratory depression [29]. At an anesthetic dose, ketamine rapidly induces a trance-like state (often called dissociation) with memory loss [30]. When intranasal esketamine was used as an antidepressant for up to 12 months in adults, some kind of tolerance for psychoactive effects gradually developed, and no permanent impairment of neurocognitive functions was observed (which is contrary to ketamine abuse at higher doses) [31].

The discovery of racemic ketamine's rapid antidepressant effect with subanesthetic dose IV use (typically 0.5 mg/kg body weight infused over 40 min) introduced an effective option for treating adults with MDD and those with TRD [32–34]. As written, traditional antidepressants are associated with a latency period of about 3–4 weeks until they take a full spectrum of action, and about 30% of patients with depression are completely resistant to their effects [35]. In contrast, ketamine may induce remission in MDD patients within 2 h after IV use which may last for a week [34] or longer with short-course therapy [36]. Ketamine additionally offers much broader therapeutic effects in a variety of mood and mental health disorders than conventional antidepressants and may alleviate MDD refractory to SSRIs [34].

Besides NMDA receptor antagonism, it acts on multiple other targets, resulting in its unique clinical effects (Fig. 1) (*please also see* pharmacology chapters in this book).

However, ketamine's full range of impact remains poorly understood. For example, it has been postulated that ketamine's primary molecular targets and its oscillatory dynamics may facilitate gamma activity among neurons expressing a peptide designated as vasoactive intestinal peptide (VIP) [37]. This exerts health-promoting effects (such as reducing inflammation) which last much longer than ketamine's action on NMDA receptors and beyond antidepressant effects. However, a potential relationship between VIP and better mental health is speculative and necessitates further validation [37].

Ketamine reduces suicidal thoughts [38] and is effective in bipolar disease [39]. Upon initiation of therapy, suicidal thoughts, imagination, or attempts are not induced, which contrasts a significant increase in suicidal ideation in 18- to 25-year-old youth with conventional antidepressants [40]. These discoveries led to a paradigm shift in the study and treatment of depression in adults [41]. Meanwhile, ketamine has emerged as a last-chance treatment option and alternative to electroconvulsive therapy (ECT), which has been the most effective and rapid nonpharmacological treatment for TRD and MDD for nearly 80 years [42–44].

Response to ketamine therapy and potential adverse effects are currently under investigation in adolescents: clinical benefits, safety, and tolerability in 18- to 25-year-old patients were comparable to those in a matched adult sample with MDD [45], suggesting no age-related differences. The NCT02579928 study by Lineham et al. examined possible predictors of response to ketamine among adolescents with TRD [46] and reported that patients with fewer previous antidepressant and adjunctive treatment

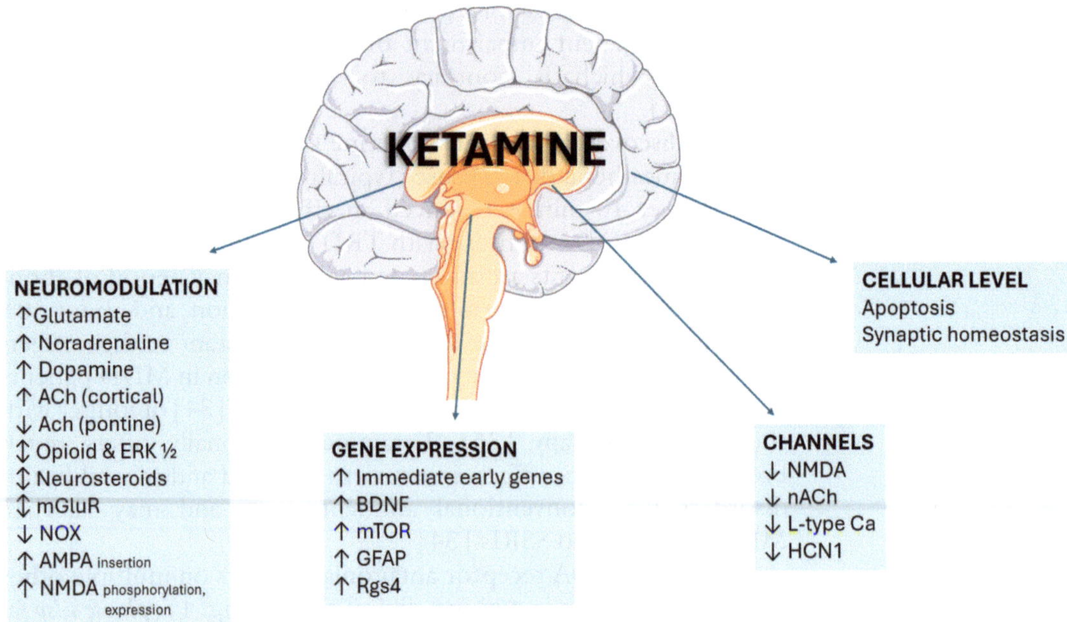

Fig. 1 Comprehensive visualization of ketamine's pharmacological targets and actions in mammalian brain (designed by M. Zaremba and P. Kleczkowska with Servier Medical Art (https:/smart.servier.com) free icon licensed under CC BY 4.0 http:/creativecommons.org/licenses/by/4.0/). *Ach* acetylcholine, *AMPA* a-amino-3-hydroxy-5-methylisoxazole-4-propionic acid, *BDNF* brain-derived neurotrophic factor, *ERK1/2* extracellular signal-regulated kinases, *GFAP* glial fibrillary acidic protein, *HCN1* hyperpolarization-activated cyclic nucleotide channels, *L-type Ca2* L-type calcium channels, *mGluR* metabotropic glutamate receptors, *mTOR* mammalian target of rapamycin, *NMDA* N-methyl-D aspartate, *nACh* nicotinic acetylcholine receptors, *NOX* NADPH oxidase, *Rgs4* regulator of G protein signaling 4

attempts had a greater likelihood of improvement in depressive symptoms with ketamine. A shorter duration of the current depressive episode was also associated with better outcomes. Symptom improvement with ketamine was observed in patients treated with SSRIs and sometimes not with serotonin-norepinephrine reuptake inhibitors (SNRIs). Less severe depressive symptoms according to the Children's Depression Rating Scale (CDRS) but not the Montgomery-Åsberg Depression Rating Scale (MADRS) and a comorbid diagnosis of ADHD were also linked to an increased ketamine response. However, this must be regarded as preliminary and exploratory because of the small trial sample size and multiple secondary analyses. Interestingly, a higher body mass index (BMI) has been associated with better improvement in MDD, and categorization of body weight as obese correlates with greater response to single dose IV ketamine. Although the underlying mechanism is not known, ketamine's anti-inflammatory effects may have played a major role. Obesity and MDD are associated with high levels of inflammation [47].

In 2019, the FDA and European Medicines Agency (EMA) approved a patented nasal spray formulation of the pure S-enantiomer of ketamine, called esketamine. Worldwide racemic ketamine is a mixture of two enantiomers, *R(−)*–ketamine and *S (+)*–ketamine (arketamine and esketamine) which combines the two enantiomers at equal amounts. Esketamine has a fourfold increased affinity for NMDA receptors, which results in a two times higher clinical potency versus racemic ketamine. Intranasal esketamine is officially approved for adult patients with depressive symptoms in MDD or TRD with co-occurring acute suicidal thoughts or behaviors in combination with an oral antidepressant [48]. Intranasal esketamine may offer a potentially more pragmatic route of administration as compared to IV racemic ketamine [49–51]. A recent systematic review has suggested that IV ketamine is more effective in treating MDD than intranasal esketamine [52]. However, this conclusion is not valid because 0.5 mg/kg IV racemic ketamine cannot directly be compared to often variably dosed intranasal esketamine (used at 28, 56, or 84 mg) because of huge differences in potency between racemic and esketamine and huge differences in bioavailability between an IV infusion compared to intranasal administration with the patented device. Esketamine is twice as potent as an analgesic and anesthetic and likely as an antidepressant as compared to racemic ketamine. It offers the advantage of shorter neurocognitive recovery and better maintains logical functions. Psychoactive events are of better quality with a slightly smaller incidence at a 50% dose reduction compared to the racemate [53]. Intravenous ketamine administration shows a bioavailability >98%, whereas intranasal esketamine applied with the new device has a bioavailability >50%. Nevertheless, as with every intranasal application, a part gets uncontrollably swallowed. Because of first-pass metabolism, the bioavailability of oral ketamine is approximately 20%. Some researchers are even of the opinion that ketamine metabolites are different after use via different administration routes, and many of the different ketamine metabolites have pharmacologic properties of their own. A randomized, double-blind study involving 63 adults with TRD assessed the effectiveness and safety of adjunctive therapy with either IV esketamine or IV racemic ketamine. It was shown that esketamine at half-dose racemic ketamine was comparable to racemic ketamine in terms of efficacy 24 h post-infusion. Both treatments demonstrated effectiveness, safety, and good tolerability [54].

Likewise, IV racemic ketamine and intranasal esketamine is not free of adverse effects such as sedation and dissociation [55]. Side effects are similar in incidence but of different quality if racemic and esketamine are compared at equipotent doses. Nonetheless, esketamine also shows a potential for abuse and misuse. Black box warning cautions that patients are at risk for sedation and difficulty with attention, judgment and thinking, abuse and misuse, and suicidal thoughts and behaviors. But esketamine is only available through a restricted distribution system, drug programs under medical supervision, and at a certified doctor's office or clinic. This is part of a risk evaluation and mitigation strategy (REMS) mandated by the FDA [56]. REMS is a safety program monitoring drugs with significant safety concerns, ensuring benefits outweigh the risks. Healthcare facilities certified under REMS monitor patients for at least 2 hours after (es-) ketamine administration including, for example, blood pressure measurements for at least 40 minutes. General observation continues until the patient is stable and side effects have resolved (usually within 2 h) [31]. However, patients should not drive or operate heavy machinery for the rest of the day when receiving ketamine.

Numerous trials have confirmed ketamine's advantages as a fast-acting antidepressant and antisuicidal agent in adults [38, 57–59], but reliable evidence for ketamine use and long-term observations in pediatric and adolescent patients are lacking [28]. There is always a delay between the approval of a therapy for adults and subsequent authorization for use in adolescents, but there are also serious concerns regarding possibly neurotoxic effects and increased odds of abuse of ketamine as a recreational drug [3]. Repeated, nonmedical ketamine intake is known to cause long-term neuroadaptations that can contribute to addiction [60]. The age of initial drug exposure has additionally been shown to be one of the most important risk factors contributing to future drug abuse [61]. More research into ketamine exposure and associated risk/benefit ratios in children and adolescents with a focus on long-term sequelae is therefore indispensable. Published data on ketamine use in youth psychiatric patients are presented in the following sections and Table 1.

3 Risks of Recreational Ketamine Use by Adolescents

Ketamine is acknowledged as an antidepressant, analgesic, and anesthetic with good safety profile in adults. Anticipated therapeutic benefits outweigh potential risks when administered at the appropriate dosage, approved indication, and administration routes for medical purposes. However, risks become evident when ketamine is used recreationally, especially by adolescents: ketamine is classified as a Schedule III controlled substance due to its potential for abuse [62, 63]. It is relatively cheaper than other recreational drugs and does not appear to produce immediate, serious side effects or severe withdrawal symptoms [64]. At lower doses, ketamine produces stimulant effects, mild dissociation, hallucinations, and a distortion of time and space. These effects typically resolve 2 h after ketamine intake [29]. Ketamine's psychoactive properties have always limited its clinical use and may occur even in the

Table 1
The evidence: Ketamine trials in children and adolescents with major depressive disorder and/or treatment-resistant depression

Study design	Patient age [years]	N	Indication	Intervention/dose regimen	Outcomes efficacy/safety	Ref.
Case study	16	1 M	MDD Suicidal ideation, ADHD, and Crohn's disease	7 x KET IV (0.5 mg/kg) infusions, 8 weeks	Significant improvement in depressive symptoms (MADRS and CDRS-R) and suicidal ideation (SSI) dissociative symptoms, nausea	[131]
Open-label	12–18	13 M = 8 F = 5	TRD	6 x KET IV (0.5 mg/kg) infusions, 2 weeks	Significant reduction of depression (CDRS score) in 38% of subjects; 3 went into remission dysphoria, transient blood pressure, heart rate, and oxygen saturation changes, nausea	[132]
Case study	14	1 F	MDD	1 x KET IV (0.42 mg/kg/h), 24 h per day, 5 days	Significant decrease in depressive symptoms (MADRS) Nystagmus, visual changes	[133]
Case study	15	1	TRD psychotic depression	6 x KET IV (0.5 mg/kg) infusions, 3 weeks	Significant decrease in depressive symptoms (MADRS and CDRS-R) and suicidal ideation (SSI) Derealization, nausea	[134]
RCT	13–17	17 M = 4 F = 13	TRD	1 x KET IV (0.5 mg/kg) infusion vs midazolam (0.045 mg/kg) over 40 min	Significant improvement in depressive symptoms (MADRS and CDRS-R) 24 h after infusion compared with midazolam dissociative symptoms reduction, systolic blood pressure increase	[135]

(continued)

Table 1
(continued)

Study design	Patient age [years]	N	Indication	Intervention/dose regimen	Outcomes efficacy/safety	Ref.
Open-label	12–18	13 M = 8 F = 5	TRD	6 x KET IV (0.5 mg/kg) infusion over 2 weeks, fMRI was conducted before and after KET series	Depression reduction correlated with increased nucleus accumbens entropy (fMRI) Dysphoria, transient blood pressure, heart rate, oxygen saturation changes, nausea	[136]
Case study	17	1 F	TRD and chronic suicidal ideation	2 x nasal spray ESKET (each 28 mg) per week with 2-day intervals for 4 weeks, then once a week for 3 weeks	No clinically significant changes dizziness, nausea, fatigue	[137]
Case study	14–23	4 M = 2 F = 2	TRD and other psychiatric disorders (anxiety, panic attacks, eating disorders)	KET–assisted psychotherapy; KET SL (100–200 mg), then KET IM (50–100 mg)	Symptomatic and functional improvements (BDI and HAMD) dissociative effects: Numbness, walking difficulty, and sleep disturbances, nausea	[133]
RCT	13–18	54	MDD anxious and non-anxious group	3 x ESKET IV (0.25 mg/kg) vs midazolam (0.045 mg/kg), 5 days	Significant antidepressant effects in both groups; higher remission rate in the non-anxious group 4 weeks posttreatment (MADRS), less suicidal ideation in the non-anxious group (C-SSRS) moderate dissociative effects (greater in an anxious group)	[138]

RCT	13–18	54	MDD and suicidal ideation	3 x ESKET IV (0.25 mg/kg) vs midazolam (0.02 mg/kg)	Significant antidepressant effects (MADRS) and lower SSI-5 score for suicide ideation from baseline to days 6 and 12. Improvement in processing speed (MCCB) infusions did not harm cognition	[129]
RCT	13–18	54	MDD and suicidal ideation	3 x ESKET IV (0.25 mg/kg) vs midazolam IV (0.02 mg/kg) 5 days	Significant decrease in depressive symptoms (MARDS), reduction in suicidal ideation (C-SSRS) dissociative symptoms, dizziness, headache, dry mouth, nausea	[139]
Open label	15	20	MDD	1x KET IV (0,5 mg/kg) infusion over 40 min	Significant decrease in depressive symptoms (MADRS, CDI), plus EDA recording No significant adverse effects	[140]

ADHD attention-deficit/hyperactivity disorder, *BDI* Beck's Depression Inventory, *CDI* Children's Depression Inventory, *CDRSR* Children's Depression Rating Scale-Revised, *C-SSRS* Columbia-Suicide Severity Rating Score, *EDA* electrodermal activity, *ESKET* esketamine, *F* female, *fMRI* functional magnetic resonance imaging, *HAMD* Hamilton Depression Rating Scale, *IM* intramuscular, *IV* intravenous, *KET* ketamine, *M* male, *MADRS* Montgomery-Åsberg Depression Rating Scale, *MCCB* MATRICS Consensus Cognitive Battery, *MDD* major depressive disorder, *N* number of participants, *RCT* randomized controlled trial, *SSI* scale for suicide ideation, *SL* sublingual, *TRD* treatment-resistant depression, vs. versus subanesthetic dose range (i.e., 0.1–0.4 mg/kg). However, when, for example, combined with a GABA agonist in awake patients, unwanted psychoactive effects are prevented or enormously reduced [29]. In contrast, ketamine's psychedelic effects made it a popular recreational drug among adolescents in the 1970s with an upward trend in the early twenty-first century. According to the United Nations Office on Drugs and Crime World Drug Reports, ketamine has been detected in illicit drug markets in over 50 countries [65]. In the 2017/2018 Crime Survey for England and Wales, ketamine abuse was most pronounced among 16- to 24-year-old young people with usage increasing from 1.2% to 3.1% [66]. Although the number of individuals seeking treatment for ketamine-related abuse issues remains relatively low overall, several EU Member States have reported increases. Current rates of ketamine abuse by young adults (15–34 yrs. of age) in the EU have been reported to range from 0.4% in Denmark (2021) to 0.8% in Romania (2019) [67]. However, it is alarming to note that Belgium, Spain, France, and Italy have witnessed an increase in the number of patients seeking treatment for ketamine use disorders in 2021. This is evidenced by the fact that the total number of documented cases has increased from 93 in 2015 to 414 in 2021. The Netherlands reports that ketamine abuse has especially increased among young people in nightclubs [67]. The popularity of ketamine in dance clubs and party settings enhances risks of misuse and abuse [68, 69], and for example, ketamine intake at electronic dance music events in New York City increased from 5.9% to 15.3% [3]. Nevertheless, despite the increase within the dance club culture, overall rates of nonmedical ketamine administration by young people remain relatively low [3]. Routes of ketamine intake include IV, IM, through snorting, or smoking together with other materials. A typical "recreational" drug dose is 60 to 250 mg [70]. Since about 2018, an increasing number of samples of mixtures containing ketamine; 3,4-methylenedioxymethamphetamine (MDMA), commonly known as ecstasy; and cocaine have been reported to the EU Early Warning System. Some of these substances are marketed as "pink cocaine" or "tucibi," and those consuming them may be unaware that ketamine is included [67]. Given the widespread clinical use, theft and diversion of ketamine from legitimate purposes may become a bigger problem [67].

According to the REMS, using ketamine without proper medical supervision can be quite challenging: there has been a recent growth in interest to use compounded ketamine products, including oral formulations which were intended for psychiatric treatment and are facilitated by telemedicine platforms. In April 2023, the FDA received an adverse event report regarding a patient who experienced severe, unusual respiratory depression after ingesting a compounded oral ketamine product without medical monitoring

[71]: the patient's blood concentrations were found to be approximately twice the ketamine level normally used in general anesthesia [71]. In addition, the doses of sublingual and oral ketamine-containing products sold by pharmaceutical companies and telemedicine platforms vary, which makes it difficult to predict product risks. Mostly, product information about appropriate use is also lacking. Administering ketamine-containing products without a healthcare professional monitoring sedation, "drowsiness," psychoactive effects, and vital signs (such as blood pressure and heart rate) will expose patients to a higher risk of serious adverse events.

Ketamine's rewarding effects often led to increased drug consumption and dose escalation. In some individuals, ketamine abuse causes repeated, compulsive behavior indicative of addiction. Siegel [72] conducted a seminal study documenting physical and psychological ketamine addiction in four populations, and Jansen [73], along with Bokor and Anderson [74], discuss ketamine's potential to induce dependence in numerous illustrative case reports [60]. Rapid tolerance develops quite commonly, and ketamine cessation may result in all kinds of withdrawal symptoms, including anxiety, dysphoria, depression, tremors, diaphoresis, palpitations, increased appetite, and even respiratory depression in severe cases.

Some of the most common subjective ketamine effects in recreational abusers are presented by Hansen et al. [75] who listed the following: experiences related to altered "body consistency" (e.g., being made of wood, rubber, or plastic), feelings of unreal shape or size of body parts, a sensation of floating or being weightless, timelessness, visions and hallucinations, sudden insight into the self, and experiences of oneness with the universe. Dissociative states enjoyed by some users typically include a sense of detachment from the physical body and the external world, often referred to as the "K-hole" [75]. It has also been shown that regular use of ketamine causes cognitive impairment, particularly related to difficulty with working memory [76].

Most ketamine overdose symptoms are self-resolving, and a specific antidote is not available (as it is for opioids), thus only supportive, symptom-based interventions are possible [29]. Unlike with opioids, deaths solely attributed to ketamine are rare and typically involve accidental poisonings, drownings, traffic accidents, or suicide. For example, a review of ketamine-related deaths in the United Kingdom from 1993 to 2006 identified only four cases where ketamine was the sole drug detected [77]. However, ketamine increases the risk of fatal accidents such as motor vehicle collisions, so driving a car after ketamine intake is strongly discouraged [78]. This is likely because of its dissociative and sedative effects, which reduce awareness and realization of the real environment and may decrease hand-eye coordination and balance [30]. Long-term ketamine abuse has been associated with various lower urinary tract pathologies, such as detrusor overactivity,

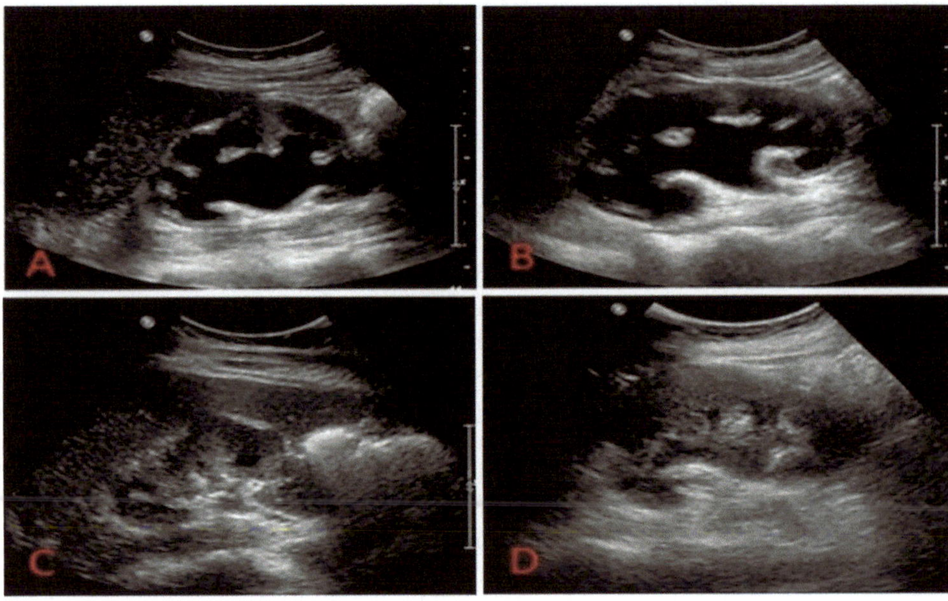

Fig. 2 Renal ultrasound scans showing significant bilateral hydronephrosis in the (**a**) right and (**b**) left kidney, with a slightly more pronounced effect on the left after 10 years of ketamine abuse. There are signs of renal atrophy/scarring and likely underlying medical renal disease. Nearly complete resolution of bilateral hydronephrosis is noted in the (**c**) right and (**d**) left kidney 7 days after bilateral nephrostomy (from Manongi N et al. [81] under a CC-BY 4.0 commercial international license; http:/creativecommons.org/licenses/by/4.0/)

difficulty passing urine, hydronephrosis, and kidney failure [30, 79, 80]. Ketamine-induced uropathy is becoming more prevalent as ketamine abuse becomes widespread. The computed tomography (CT) scan image below documents one of the first few published cases of severe ketamine-induced uropathy with cholangiopathy necessitating bilateral nephrostomies in an Asian 29-year-old female with a 10-year history of ketamine abuse (Figs. 2 and 3).

The CT scan image shows a thickened bladder wall and ureter with hydronephrosis. The patient underwent a percutaneous bilateral nephrostomy in the acute phase of management to improve urine output and prevent chronic renal disease [81]. The mechanism of both ketamine-induced uropathy and cholangiopathy remains unknown; however, ureteral inflammation, stricture, hydronephrosis, and impaired renal function were reported. Studies have shown that ketamine and its metabolites directly impact urothelial cells, causing microvascular changes that induce autoimmune responses [82].

In addition, this case, along with others, also highlights that chronic recreational ketamine use can lead to structural damage in the hepatobiliary system, including biliary dysfunction, epigastric pain, and hepatic injury [83–86]. The liver biopsy (Fig. 4) shows an

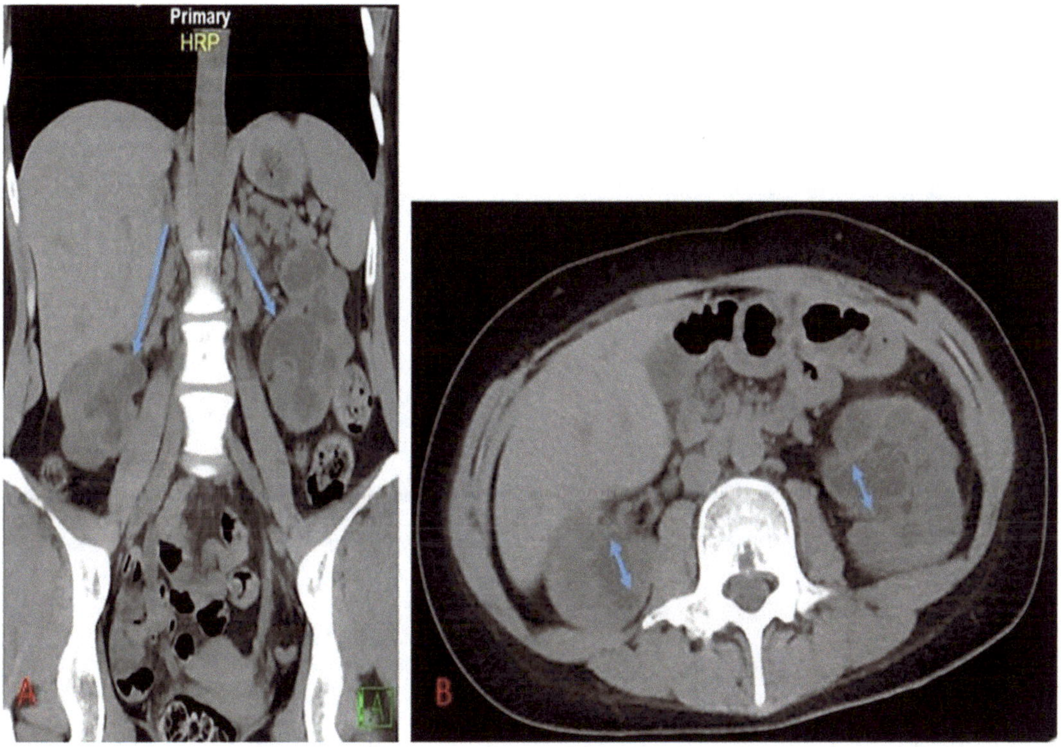

Fig. 3 Non-contrast CT scan of the abdomen and pelvis showing the ureteric wall thickening and hydroureteronephrosis (indicated with blue arrows), with no evidence of nephrolithiasis in (**a**) coronal view and (**b**) axial section after 10 years of ketamine abuse (from Manongi N et al. [81] under a CC-BY 4.0 commercial international license; http:/creativecommons.org/licenses/by/4.0/)

example of bile duct proliferation with ketamine-induced neutrophilic infiltration, along with rare focal hepatocytic ballooning within the hepatic lobule [81]. These complications may be reversible with ketamine cessation but may even cause hepatobiliary failure [85–87].

Given a possibly long-term treatment with ketamine in mental illness, it seems necessary to consider the risks mentioned earlier and especially its addictive potential when elucidating ketamine's safety profile for medical use in children and adolescents. Unfortunately, the precise mechanisms through which ketamine produces psychoactive and addictive effects are not fully understood [88].

4 The Possible Impact of Ketamine on Adolescent Brain

Exploring ketamine's brain impact is a complex and evolving field of research with a notable lack of data on long-term safety issues. Understanding the neurobiological consequences of ketamine

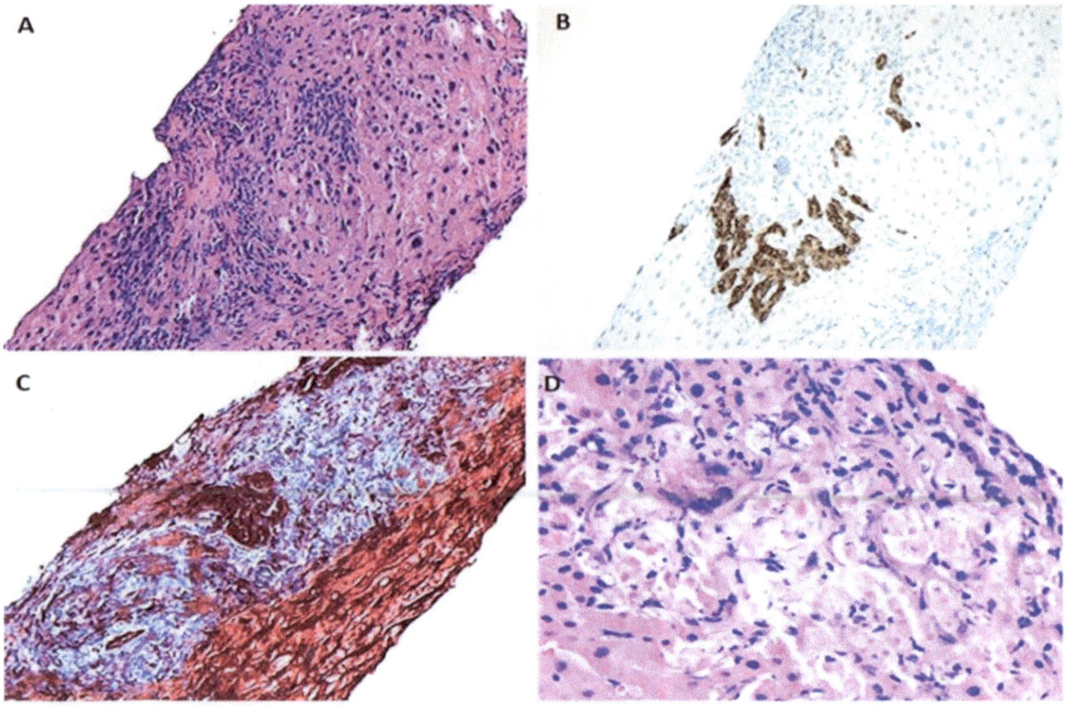

Fig. 4 (**a**) Needle liver biopsy revealing an expanded portal tract with mixed lymphocytic and neutrophilic inflammation, ductular reaction, and mild fibrosis; cholestasis is observed in hepatocytes and bile canaliculi within the periportal region after 10 years of ketamine abuse (H&E stain, x200); (**b**) proliferating bile ductules (CK7 immunostain, x200), (**c**) mild portal fibrosis (Mallory trichrome stain, x200), and (**d**) liver damage showing hepatocyte ballooning and neutrophilic infiltration (H&E stain, x400) (from Manongi N et al. [81] under a CC-BY 4.0 commercial international license; http:/creativecommons.org/licenses/by/4.0/). *H&E* hematoxylin and eosin, *CK7* cytokeratin, *x* magnification factor

applied before adulthood is "fragmented" and primarily originates from abuse case reports, neurocognitive observations after abstinence, or chronic ketamine administration in animal addiction models. Adolescence is critical for brain and behavioral development; therefore a deepening of knowledge on ketamine effects in the young brain and how these differ from the adult is essential [89].

Ketamine induces physiological and behavioral change associated with altered brain structures which is in part related to the modulation of the excitatory amino acid glutamate via NMDA and α-amino-3-hydroxy-5-methyl-4-isoxazolepropionic acid (AMPA) receptor channels [33]. Ketamine's initial rapid antidepressant action necessitates complex neuromodulatory interactions involving NMDA and possibly opioid receptors [33, 90]. Naltrexone, a mu-opioid receptor antagonist, can reduce ketamine's antidepressant effects without affecting dissociation [91].

The adolescent response to ketamine varies from that observed in adults, particularly with efficacy and safety issues. Adolescence is characterized by numerous developmental changes, including modification of brain regions associated with drug use and abuse [60]. The primary target of ketamine, NMDA receptor subunits, increases in expression during brain development in childhood, peaks during adolescence (including within the hippocampus), and remains high throughout life or progressively decline into adulthood [92]. Because NMDA receptors mature relatively late in life, ketamine will likely elicit different responses in adolescents as compared to adults [92, 93]. The higher density of NMDA receptors during adolescence could theoretically lead to enhanced effects of NMDA antagonists.

The significant NMDA receptor developmental shift may also explain why certain ketamine effects, such as locomotor stimulation, are more pronounced in adolescents than in adult rats [60]. Pharmacokinetic factors are likely not responsible, because drug metabolism tends to be higher in adolescents than adults, and ketamine elimination is faster in adolescence [94]: ketamine clearance is ~79 l/h/70 kg in an adult [95], whereas it can reach values of up to 90 l/h/70 kg in children [96]. Adolescents may require higher doses of ketamine to achieve antidepressant effects than adults, and the same ketamine/esketamine dosage would have weaker effects on cognitive function in adolescents compared with adults [97]. Younger individuals may therefore have less propensity to substance abuse compared to older ones [60]. Animal data also show less bioavailability of ketamine in younger adults compared to older rats [98]. The need for a higher dose of ketamine in adolescents may also be related to different developmental stages in cerebral circulation and metabolism, neural myelination, and cardiovascular differences, such as increased cardiac output [99, 100].

Pathophysiologically, NMDA receptors are involved in long-term potentiation [101], a physiological process related to learning, memory, and synaptic plasticity (i.e., the ability to adapt after experience) [102]. Ketamine's pharmacological effects may therefore have long-lasting brain impact, especially in developing adolescent brains [30]. Its interaction with a multitude of other neurotransmitter systems, such as inhibition of muscarinic acetylcholine receptors [103] or induction of dopamine release [104], may further contribute to these phenomena (Fig. 1).

The most important finding in an animal model with adult and adolescent rats was that repeated single-dose ketamine (25 mg/kg i.p.) did not produce deficits in novel object recognition or the Barnes maze in either age group [60]. Adult rats who received a moderate dose of ketamine (30 mg/kg i.p.) twice daily for 7 days showed a decrease in object exploration; however, treatment once daily for 5 days did not result in relevant memory deficits [105]. Significant deficits in attention and/or learning are likely to occur at higher ketamine doses. A correlation between dose and emergence of cognitive deficits is in line with clinical reports of poorer verbal fluency with higher ketamine dose [106] and correlations between heavy lifetime ketamine abuse and verbal learning and memory decline [107]. Ketamine's disruptive effects on memory appear to depend on the duration of intake and the time elapsed since drug cessation. Data from a study on adolescent cynomolgus monkeys show that brain function remained unchanged after 1 month of ketamine, while it was significantly impaired after 6 months of use [108]. Whether cognitive deficits are reversible or not after ketamine discontinuation remains unknown, there is a trend that longer abstinence may result in better cognitive performance [106].

Studies in mice examining spatial (such as the Morris water maze and the T-maze) and other tasks like novel object recognition showed that adolescent mice exposed to ketamine had deficits in working memory as adults, which was not observed when adult animals were exposed to ketamine [60]. Unmedical human use of ketamine at high doses was also found to impair cognition and memory, with results such as deficient working memory [109, 110]. Ketamine may thus interfere with developmental processes involved in spatial working memory.

There is also a lack of adequate reflection on ketamine's cognitive effects. Studies in long-term illicit ketamine users show altered structure in brain regions related to cognition. For instance, less frontal gray matter, multiple cortical atrophy, widely reduced axial diffusivity of the right prefrontal cortex (anterior corona radiata and forceps minor), less connectivity between the thalamus and cortices, and decreased brain volume over the right insula have been reported in neuroimaging study [64, 111–115] (Fig. 5).

Many experimental data support the thesis of neurotoxic and neurodegenerative ketamine effects on developing adolescent brains. Onaolapo et al. reported neurodegeneration in the cerebral cortex, hippocampus, cerebellum, and pons in adolescent rats after 8 weeks of ketamine administration up to higher doses [116]. It is also well-known that high doses (especially above clinical routine) or extended use periods may readily induce pathologic neural change [117]. Harmful synaptic and extra-synaptic NMDA receptor blockade may inhibit neuronal and behavioral plasticity and new brain cell development [101, 102, 118]. Although various animal models reported toxic effects with suppression of neural excitability and/or induction of hippocampal apoptosis [119–122], it must not be forgotten that there are tremendous species differences, for example, between rodents and human brains. It is therefore not possible to directly translate these results to clinical settings.

Differences in harmful effects on the brain between sexes have also been found. Histological study differentiating between the

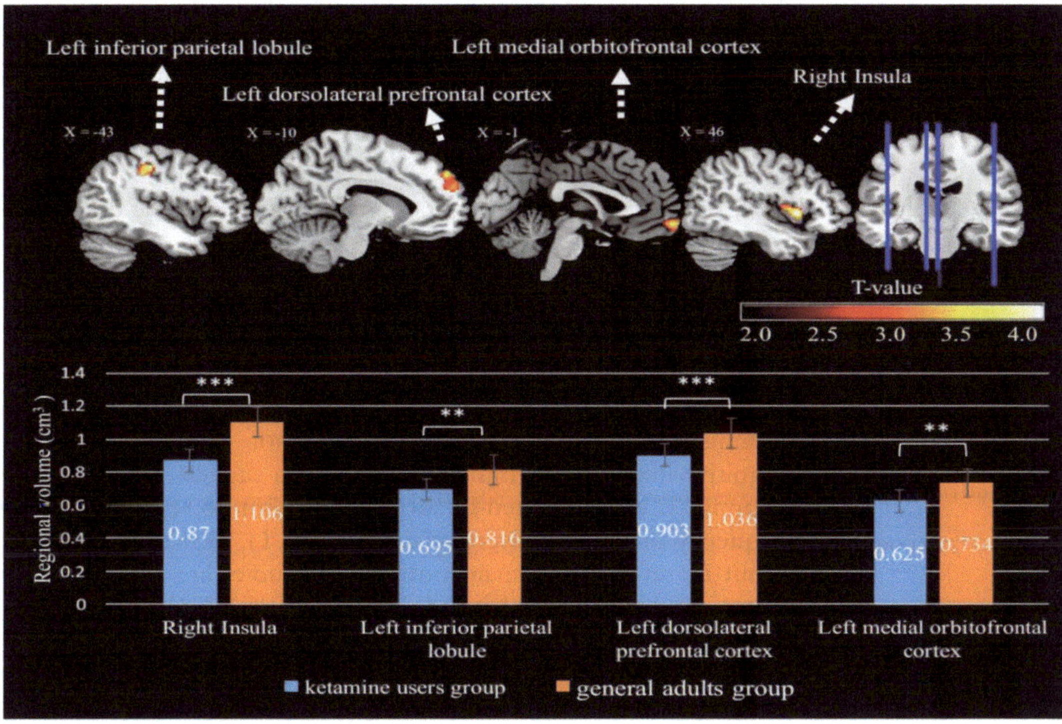

Fig. 5 The T-score map above reveals significantly reduced gray matter volume in ketamine users compared to the general population, with clusters determined by Monte Carlo simulation and significance corrected to $p < 0.05$. The bar graph below shows reduced regional volumes in the right insula, left inferior parietal lobule, left dorsolateral prefrontal cortex, and left medial orbitofrontal cortex in ketamine users (blue bars). The analyses were adjusted for age, gender, and years of education. ***LSD-corrected $p < 0.001$, **LSD-corrected $p < 0.01$, following ANOVA with post hoc tests (from Hung CC et al. [64] under a CC-BY 4.0 commercial international license; http:/creativecommons.org/licenses/by/4.0/). *LSD* least significant difference, *ANOVA* analysis of variance

effects of ketamine in male and female rats [123] revealed that males experienced cell death throughout the hippocampus. In contrast, females localized cell death to the dentate gyrus after a lower dose of ketamine. Because gonadal hormones can influence drug sensitivity, this difference may be the result of interactions between circulating gonadal hormones and ketamine [123]. 17-β-estradiol has been reported to provide neuroprotection in numerous brain injury models in both sexes and alleviates ketamine-induced neurodegeneration [124–126].

The developmental shift of NMDA receptors also alters mesolimbic dopamine levels which leads to variable hypersensitivity to drugs and drug-related stimuli. Behavioral drug sensitization results in the orientation of attention toward drugs and may induce the pathological motivation that is characteristic of addiction [60]. This implies that younger adolescents may be sensitized to drugs, whereas older teens with more developed brains may repel this sensitization and respond less. The different odds for addiction in the heterogeneous group of adolescents may thus in part be explained. Risk assessments in children and adolescents may therefore not be based on ketamine data extrapolated from adults or ketamine abusers.

Cognitive deficits in frequent and chronic ketamine users were typically noted after more than 20 months of use [76]. Average daily doses ranged from 2.77 to 3.5 g, with a usage frequency of 6 to 6.5 days per week [76, 127, 128]. In contrast, patients treated for depression received much lower doses (0.2 to 0.75 mg/kg) with infusions typically given two to three times a week over 2 weeks. This suggests that such subanesthetic doses of ketamine, administered at low frequencies, may not lead to significant cognitive impairment. Limited clinical data suggest that low-dose ketamine is well tolerated in adolescents, with no reports of high-frequency or serious adverse events (Table 1). However, concerns about its safety will remain until larger studies are conducted. A recent randomized trial in adolescents with MDD showed that three esketamine infusions at a 0.25 mg/kg subanesthetic dose did not impair cognitive function over 12 days [129]. Adolescent patients treated with esketamine showed improved processing speed, but no significant changes in working memory, verbal, or visual learning were observed (see Fig. 6). Baseline cognitive performance was not linked to its antidepressant or antisuicidal effects.

In summary, present findings of ketamine effects on the adolescent brain have primarily been obtained in animal studies with a focus on rodents and primates. Because there are huge species and experimental paradigm differences as compared to clinical reality, long-term clinical observations are warranted to assess ketamine effects on the adolescent human brain, especially with repeated use. Physiological mechanisms underlying the differences between adolescent and adult responses to ketamine need to be elucidated, as well as mechanisms responsible for lasting behavioral effects. It is essential to define dose and use schedules that are beneficial without inducing unwanted long-term consequences on the brain.

5 The Evidence: Insights on Efficacy and Safety of Ketamine in Trials of Adolescents with MDD and TRD

From the time that ketamine has shown fast-acting and potentially long-lasting antidepressant efficacy in adults [130], healthcare practitioners have begun with its off-label use in depressive adolescents. Previous data reported ketamine's effective and safe utility with use in children and adolescents when skillfully administered in multimodal analgesia, sedation, and anesthesia. Nevertheless, the current

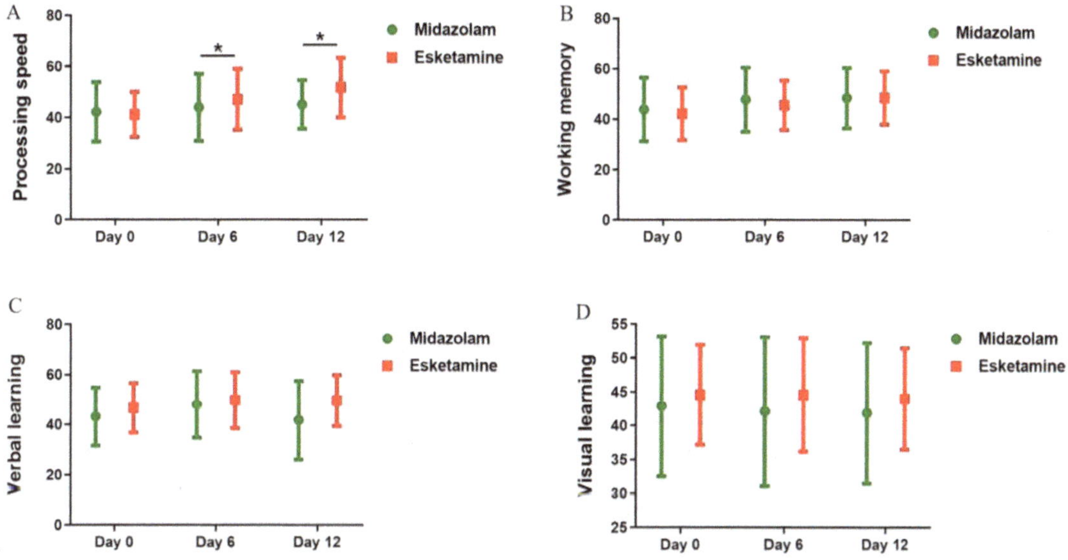

Fig. 6 Cognitive performance changes on days 0, 6, and 12, showing significant differences in processing speed (**a**) between esketamine-treated adolescents (red) versus midazolam-treated adolescents (green) on days 6 and 12. No significant changes in working memory (**b**), and verbal (**c**) or visual (**d**) learning were observed. Asterisks (*) indicate significant group differences (*$p < 0.05$) (from Lan et al. [129] under a CC-BY 4.0 commercial international license; http:/creativecommons.org/licenses/by/4.0/)

information about ketamine's profile for therapy in adolescent depression is still limited, possibly because of concerns about the negative impact on the adolescent, developing brain. To date, relatively few RCTs are available that study the effects of ketamine exposure in depressive adolescents (see Table 1). However, several NIH-funded clinical registered trials are ongoing (NCT02401139, NCT04613453, NCT04592809, NCT05045378, NCT05217706, NCT04834427), so more data are expected soon, also from the Chinese Clinical Trials Registry (http://www.chictr.org.cn).

In the literature, among publications from 2017 to 2024, 12 clinical trials involving 233 participants aged 12–23 years (range through early/middle/late adolescence to young adult) were found (Table 1).

However, common study limitations were, for example, a relatively small sample size, as indicated by the number of the largest cohort consisting of 54 patients [138, 139]. The studies varied in settings and methodologies. Of the ten, only four studies were randomized clinical trials (RCT) [131, 135, 138], five were case reports [131, 133, 134, 137, 141], and three were open-label trials [132, 136]. Most studies administered IV ketamine at a dose of

0.5 mg/kg, except for three RCTs that used 0.25 mg/kg and one study that utilized 0.42 μg/kg/h in both single and repeated dosing paradigms (see Table 1). Limited data report the use of non-IV ketamine formulations: one case report investigated 100–200 mg sublingual ketamine followed by 50–100 mg IM ketamine [133], while another used the approved nasal spray device with two doses of 28 mg esketamine [137]. The duration of ketamine treatment ranged from a single administration to over 5 days and a maximum of 8 weeks.

The trial findings revealed that ketamine caused a significant reduction in MDD symptoms, decrease in acute suicidality, and a reduction in mood lability (assessed with the following scales: *Montgomery-Åsberg Depression Rating Scale* (MADRS), *Children's Depression Rating Scale-Revised* (CDRS), *Beck's Depression Inventory* (BDI), *Children's Depression Inventory* (CDI), *Children's Depression Rating Scale-Revised* (CDRS), and the *Hamilton depression rating scale* (HAMD)). One study demonstrated a substantial decrease in affective and anhedonic symptoms in adolescents with depression following IV ketamine therapy [140]. This anti-anhedonic effect is thought to originate from downstream dopaminergic regulation through the glutamatergic system [142], and there is a correlation between suicidality and the presence of anhedonia.

The majority of study data indicate that even a single ketamine infusion may significantly reduce depressive symptoms in adolescents within 24 hours post-infusion, with therapeutic benefits lasting up to 14 days [135]. Ketamine was found to reduce suicidal ideation and hopelessness after an initial IV infusion [131]. This result from a randomized controlled trial (RCT) demonstrates that incorporating a three-dose regimen of IV esketamine into standard inpatient care and treatment effectively reduced depression and suicidal thoughts in adolescents with MDD [139]. Notably, at the 4-week follow-up, antisuicidal and antidepressant effects were still evident in 69.2% and 61.5% of patients [139]. In another study, some patients experienced substantial improvement in depressive symptoms on the first day of use, with symptom relief lasting up to 5 months after the first ketamine infusions [141]. Of note, the results demonstrate that participants who received repeated and higher ketamine doses had better responses. However, in one study, the response rate was 38% which is less than noticed in adults. Ketamine may therefore also show reduced efficacy in some adolescents [132]. One case report failed to demonstrate the effectiveness of intranasal esketamine in a 17-year-old female with TRD [137], which may have been caused by inappropriate therapy duration, likely too small drug dose, or inappropriate route of use. In one study, three infusions of esketamine demonstrated a more pronounced immediate posttreatment antisuicidal effect in adolescents with non-anxious MDD compared to those with anxious

MDD [138]. However, this benefit was temporary and did not persist over time. Children and adolescents often suffer from anxiety and depression, and approximately 50% develop at least one lifetime anxiety disorder [138]. Additionally, patients with anxious depression are more likely to experience recurrent depressive episodes.

Published data evaluating the safety of multiple doses of ketamine in children and adolescents indicate that ketamine is well tolerated and safe for treatment in these age groups, even when used in a regimen similar to those for antidepressant treatment in adults. Noteworthy, observations were limited to a maximum of 8 weeks. Ketamine was associated with brief, self-limiting adverse events that did not lead to patient drop-out. Overall, ketamine infusions were well tolerated yet with transient dissociative and hemodynamic adverse effects [136]. Adolescent patients sometimes experience cardiovascular side effects with transient blood pressure, HR, and oxygen saturation change (Table 1). The underlying mechanisms for ketamine's sympathomimetic stimulation on the central nervous system (CNS) are still incompletely understood [143, 144]. Ketamine is also thought to inhibit the production of centrally synthesized nitric oxide by decreasing NMDA receptor activity [145]. Other proposed mechanisms of sympathomimetic effects include systemic catecholamines release, norepinephrine release from sympathetic ganglion neurons, inhibition of vagal nerve and centrally localized muscarinic receptors, and inhibition of norepinephrine reuptake in peripheral nerve endings and cardiac muscle [146–148]. There is only one study on acute ketamine effects on sympathetic regulation specifically shown by electrodermal activity (EDA). The trial showed improved depressive symptomatology without a significant increase in HR and change in EDA parameters [140]. A multimodal analysis of cardiocirculatory variability such as with blood pressure change together with EDA variables may therefore improve knowledge about ketamine and the sympathetic regulatory network.

The most common psychoactive adverse effects of ketamine treatment were psychedelic and psychotomimetic symptoms such as dissociation, derealization, and dysphoria (Table 1). In the study by Lan et al., post hoc analysis revealed that patients with anxiety experienced more severe dissociative symptoms 30 min after the initial esketamine infusion compared to non-anxious patients. This difference was not observed in participants treated with midazolam [138].

These adverse reactions are generally irrespective of drug administration routes, but IV ketamine appears to be associated with more psychotic events than other administration routes (e.g., oral, subcutaneous) [149]. This can readily be explained by higher drug blood concentration levels achieved with IV ketamine. Unblinded or unmasked patients may also influence the incidence and quality of psychotic effects according to their expectations [150]. It has long been known that awake patients who receive ketamine are more prone to suggestions and easier to hypnotize. In many competent centers, patients are therefore prepared for psychoactive effects before ketamine infusion and suggestions such as "now you will have beautiful dreams" help a lot to reduce the unwanted quality of psychic events. A calm, supportive atmosphere also increases patient well-being and decreases unwanted mental effects. Some studies did not report behavioral or other side effects related to impaired cognitive functions [132]. Adverse reactions from the gastrointestinal tract, such as typical nausea, were the next most often reported adverse event (Table 1). In one female patient, ketamine caused neurologic-ocular symptoms like nystagmus and altered vision [141]. In adult patients, some short-term RCT reported more adverse effects in women than in men [151], but this correlation requires further investigation.

6 Ketamine in Non-depressive Pediatric and Adolescent Mental Disorder

Increasing evidence shows that ketamine has emerged as a promising therapeutic option for a range of psychiatric conditions beyond depressive disorders in pediatric and adolescent patients.

Papolos and colleagues [152] studied treatment-resistant individuals with a diagnosis of the fear-of-harm (FOH) phenotype bipolar disorder, including 12 children and adolescents of both sexes, 6–19 years of age, and 45 patients 6–37 years old. The first study characterized patients with comorbid conditions such as verbal and self-aggression, separation anxiety, and suicidal ideation: all participants received intranasal ketamine within a dose range from 30 to 120 mg every 3–7 days. The treatment was well-tolerated and caused minimal side effects. While aggression decreased and manic/hypomanic and depressive symptoms almost completely resolved, dissociative effects were brief and self-limiting with a duration of less than 60 min. Symptoms of the FOH phenotype, including fear of harm, sleep-onset and offset disturbances, parasomnias, and thermoregulatory disturbances, improved as well [152].

In a subsequent longitudinal study, Papolos et al. [153] administered nasal ketamine (50–200 mg/ml) in a repeated use schedule every 3–4 days for periods ranging from 3 months to 6.5 years. The authors did not analyze results based on patient age but published averaged data for the entire cohort: (i) 8 patients were rated as very much better by evaluators, (ii) 26 patients as much better, and (iii) only 5 patients as mildly better or unchanged. The publication reported that adolescent patients who previously struggled with school attendance, poor social skills, lack of friends, and multiple medications showed fewer school attendance issues, better social interaction, and less need for a difficult medication regimen after ketamine. Overall, a significant reduction in all symptom categories was induced by ketamine.

There were no age-related differences in the severity of common ketamine side effects. The potential efficacy of ketamine treatment was investigated in a 7-year-old boy with posttraumatic stress disorder (PTSD), significant aggression, and behavioral dysregulation in a treatment approach by Donoghue et al. [154]. After two doses of 10 mg IV ketamine, the patient presented with a remarkable decrease in behavioral outbursts and aggression. He made significant progress in psychotherapy because of more openness and direct discussion of childhood trauma.

7 Conclusion

First trials with ketamine in youth psychiatric patients have shown treatment success rates comparable to those reported in adults in short-term observation. Ketamine has significantly reduced suicidal ideation, thus decreasing suicide risk in depressive adolescents. In combination with its rapid onset of action, it may therefore serve as a good option for acute antidepressant therapy of adolescents with MDD who are at high risk of suicide.

The promising first findings suggest that ketamine's benefits may outweigh its risks, but current evidence is very limited by small sample size studies and insufficient data on confounding variables. The long-term impact of especially repeated ketamine use during adolescence is not understood, and available data mostly originate from primate and rodent experiments. Properly designed RCTs with long-term study periods are therefore indispensable in child and adolescent populations of comparable developmental stages. Investigation on the best possible dosing and administration schedules that maximize ketamine benefits while minimizing unwanted events are urgently needed.

Without such a database, child and adolescent psychiatrists will have to face a significant dilemma: whether to use ketamine as therapy despite a lack of knowledge about long-term, possibly harmful, sequelae or whether to accept the greater risk of inaction with possibly deleterious and even suicidal consequences in the young patients. It is important not to forget that a delay in therapeutic intervention may not be possible in a mental health problem with emergency.

Conflict of Interests

The authors declare no conflicts of interest.

References

1. Chrisitie D, Viner R (2005) Adolescent development. BMJ 330:301–304
2. World Health Organization (2021) Mental health of adolescents. https://www.who.int/news-room/fact-sheets/detail/adolescent-mental-health. Accessed 23 Mar 2024
3. Acevedo J, Siegel JA (2022) Neurobiological, behavioral, and cognitive effects of ketamine in adolescents: a review of human and pre-clinical research. Behav Brain Res 435:114049
4. Keyes KM, Platt JM (2024) Annual research review: sex, gender, and internalizing conditions among adolescents in the 21st century—trends, causes, consequences. J Child Psychol Psychiatry 65:384–407
5. Racine N, McArthur BA, Cooke JE, Eirich R, Zhu J, Madigan S (2021) Global prevalence of depressive and anxiety symptoms in children and adolescents during COVID-19: a metaanalysis. JAMA Pediatr 175:1142–1150
6. Dunn V, Goodyer IM (2006) Longitudinal investigation into childhood and adolescence-onset depression: psychiatric outcome in early adulthood. Br J Psychiatry 188:216–222
7. Emslie GJ, Mayes TL (2001) Mood disorder in children and adolescents: psychopharmacological treatment. Biol Psychiatry 49:1082–1090
8. Zito JM, Safer DJ, DosReis S et al (2002) Rising prevalence of antidepressants among US youth. Pediatrics 109:721–727
9. Coyle JT, Pine DS, Charney DS et al (2003) Depression and bipolar support alliance consensus statement on the unmet needs in diagnosis and treatment of mood disorders in children and adolescents. J Am Acad Child Adolesc Psychiatry 42:1494–1503
10. Crawford CM, Fallucco E, Fava M, Ingelfinger J, Scott-Vernaglia S (2024) Depression—screening and treating depression in adolescents. N Engl J Med 390:e56
11. Cummings CM, Caporino NE, Kendall PC (2014) Comorbidity of anxiety and depression in children and adolescents: 20 years after. Psychol Bull 140:816–845
12. Gaynes BN, Lux L, Gartlehner G et al (2020) Defining treatment-resistant depression. Depress Anxiety 37:134–145
13. Dwyer JB, Stringaris A, Brent DA et al (2020) Annual research review: defining and treating pediatric treatment-resistant depression. J Child Psychol Psychiatry 61:312–332
14. Cipriani A, Zhou X, Del Giovane C et al (2016) Comparative efficacy and tolerability of antidepressants for major depressive disorder in children and adolescents: a network meta-analysis. Lancet 388:881–890
15. Walter HJ, Abright AR, Bukstein OG, Diamond J, Keable H, Ripperger-Suhler J et al (2023) Clinical practice guideline for the assessment and treatment of children and adolescents with major and persistent depressive disorders. J Am Acad Child Adolesc Psychiatry 62:479–502
16. Maalouf FT, Atwi M, Brent DA (2011) Treatment-resistant depression in adolescents: review and updates on clinical management. Depress Anxiety 28:946–954
17. Brent D, Emslie G, Clarke G et al (2008) Switching to another SSRI or to venlafaxine with or without cognitive behavioral therapy for adolescents with SSRI-resistant depression. JAMA J Am Med Assoc 299:901–913
18. March J, Silva S, Petrycki S et al (2004) Fluoxetine, cognitive-behavioral therapy, and their combination for adolescent with depression: treatment for adolescent with depression study (TADS) randomized controlled trial. JAMA 292:807–820
19. NICE Recommendations. Depression in children and young people: identification and management. Guidance. https://www.nice.org.uk/guidance/ng134/chapter/Recommendations. Accessed 10 Feb 2024
20. Kennard B, Silva S, Vitiello B et al (2006) Remission and residual symptoms after short-term treatment in the treatment of adolescents with depression study (TADS). J Am Acad Child Adolescent Psychiatry 45:1404–1411
21. McIntyre RS, Alsuwaidan M, Baune BT et al (2023) Treatment-resistant depression: definition, prevalence, detection, management, and investigational interventions. World Psychiatry 22:394–412
22. Jawad MY, Vincenzo JD, Ceban F et al (2022) The efficacy and safety of adjunctive intranasal esketamine treatment in major depressive disorder: a systemic review and meta-analysis. Expert Opin Drug Saf 21:841–852
23. Ledesma-Corvi S, Jornet-Plaza J, Gálvez-Melero L et al (2024) Novel rapid treatment options for adolescent depression. Pharmacol Res 201:107085
24. Richmond TK, Rosen DS (2005) The treatment of adolescent depression in the era of the

black box warning. Curr Opin Pediatrics 17: 466–472
25. Abi-Dargham A, Moeller SJ, Ali F et al (2023) Candidate biomarkers in psychiatric disorders: state of the field. World Psychiatry 22: 236–262
26. Eder J, Dom G, Gorwood P et al (2023) Improving mental health care in depression: a call for action. Eur Psychiatry 66:e65
27. Marwaha S, Palmer E, Suppes T et al (2023) Novel and emerging treatments for major depression. Lancet 401:141–153
28. Meshkat S, Rosenblat JD, Ho RC et al (2022) Ketamine use in pediatric depression: a systemic review. Psychiatry Res 317:114911
29. Li L, Vlisides PE (2016) Ketamine: 50 years of modulating the mind. Front Hum Neurosci 10:612
30. Morgan CJA, Curran HV (2012) Ketamine use: A review. Addiction 107:27–38
31. Hope J, Copolov D, Tiller J et al (2023) What clinicians need to know about intranasal esketamine for treatment-resistant depression? Australas Psychiatry 31:841–845
32. Berman RM, Cappiello A, Anand A et al (2000) Antidepressant effects of ketamine in depressed patients. Biol Psychiatry 47:351–354
33. McIntyre RS, Rosenblat JD, Nemeroff CB et al (2021) Synthesizing the evidence for ketamine and esketamine in treatment-resistant depression: an international expert opinion on the available evidence and implementation. Am J Psychiatry 178:383–399
34. Zarate CA Jr, Singh JB, Carldon PJ et al (2006) A randomized trial of an N-methyl-D-aspartate antagonist in treatment-resistant major depression. Arch Gen Psychiatry 63: 856–864
35. Zhang K, Hashimoto K (2019) An update on ketamine and its two enantiomers as rapid-acting antidepressant. Expert Rev Neurother 19:83–92
36. Newport DJ, Carpenter LL, McDonald WM et al (2015) APA council of research task force on novel biomarkers and treatments. Ketamine and other NMDA antagonists: early clinical trials and possible mechanisms in depression. A randomized trial of an N-methyl-D-aspartate antagonist in treatment-resistant major depression. Am J Psychiatry 172:950–966
37. Adam E, Kowalski M, Akeju O et al (2024) Ketamine can produce oscillatory dynamics by engaging mechanisms dependent on the kinetics of NMDA receptors. Proc Natl Acad Sci USA 121:e2402732121
38. Wilkinson ST, Ballard ED, Bloch MH et al (2018) The effect of a single dose of intravenous ketamine on suicidal ideation: a systemic review and individual participation data meta-analysis. Am J Psychiatry 175:150–158
39. Diazgranados N, Ibrahim L, Brutsche NE et al (2010) A randomized add-on trial of an N-methyl-D-aspartate antagonist in treatment-resistant bipolar depression. Arch Gen Psychiatry 67:793–802
40. Masi G (2022) Controversies in the pharmacotherapy of adolescent depression. Curr Pharm Des 28:1975–1984
41. Krystal JH, Abdallah CG, Sanacora G et al (2019) Ketamine: a paradigm shift for depression research and treatment. Neuron 101: 774–778
42. Espinoza RT, Kellner C (2022) Electroconvulsive therapy. N Engl J Med 386:667–672
43. Lisanby SH (2007) Electroconvulsive therapy for depression. N Engl J Med 357:1939–1945
44. Ross EL, Zivin K, Maixner DF (2018) Cost-effectiveness of electroconvulsive therapy vs pharmacotherapy/psychotherapy for treatment-resistant depression in the United States. JAMA Psychiatry 75:713–722
45. Chisamore N, Danayan NB, Rodrigues JD et al (2023) Real-world effectiveness of repeated intravenous ketamine infusions for treatment-resistant depression in transitional age youth. J Psychopharmacol 37:755–846
46. Lineham A, Avila-Quintero VJ, Bloch MH et al (2024) Exploring predictors of ketamine response in adolescent treatment-resistant depression. J Child Adolesc Psychopharmacol 34:73–79
47. Freeman MP, Hock RS, Papakostas GI et al (2020) Body mass index as a moderator of treatment response to ketamine for major depressive disorder. J Clin Psychopharmacol 40:287–292
48. Mahase E (2019) Esketamine is approved in Europe for treating resistant major depressive disorder. BMJ 367:17069
49. Correia-Melo FS, Leal GC, Carvalho MS (2018) Comparative study of esketamine and racemic ketamine in treatment-resistant depression. Medicine 97:e12414
50. Maina G, Adami M, Ascione G et al (2023) Nationwide consensus on the clinical management of treatment-resistant depression in Italy: a Delphi panel. Ann General Psychiatry 22:48
51. Schatzberg AF (2019) A word to wise about intranasal esketamine. Am J Psychiatry 176: 422–424

52. Singh B, Kung S, Pazdernik V et al (2023) Comparative effectiveness of intravenous ketamine and intranasal esketamine in clinical practice among patients with treatment-refractory depression: an observational study. J Clin Psychiatry 84:22m14548
53. Mion G, Himmelseher S (2024) Esketamine: less drowsiness, more analgesia. Anesth Analg 139:78–91. https://doi.org/10.1213/ANE.0000000000006851
54. Correia-Melo FS, Leal GC, Vieira F et al (2020) Efficacy and safety of adjunctive therapy using esketamine or racemic ketamine for adult treatment-resistant depression: a randomized, double-blind, non-inferiority study. J Affect Disord 264:527–534
55. Wlerick L, Devreese M, Peremans K et al (2020) Pharmacokinetics, absolute bioavailability and tolerability of ketamine after intranasal administration to dexmedetomidine sedated dogs. PLoS One 15:e0227762
56. FDA (2019) FDA approves new nasal spray medication for treatment-resistant depression; available only at a certified doctor's office or clinic. https://www.fda.gov/news-events/press-announcements/fda-approves-new-nasal-spray-medication-treatment-resistant-depression-available-only-certified. Accessed 3 Mar 2024
57. Grunebaum MF, Galfalvy HC, Choo T-H et al (2018) Ketamine for rapid reduction of suicidal thoughts in major depression: a midazolam-controlled randomized clinical trial. Am J Psychiatry 175:327–335
58. Pettorruso M, d'Andrea G, Di Carlo F et al (2023) Comparing fast-acting interventions for treatment-resistant depression: an explorative study of accelerated HF-rTMS versus intranasal esketamine. Brain Stimul 16: 1041–1043
59. Zanos P, Moaddel R, Morris PJ et al (2018) Ketamine and ketamine metabolite pharmacology: insights into therapeutic mechanisms. Pharmacol Rev 70:621–660
60. Bates MLS, Trujillo KA (2019) Long-lasting effects of repeated ketamine administration in adult and adolescent rats. Behav Brain Res 369:111928
61. Anthony JC, Petronis KR (1995) Early-onset drug use and risk of later drug problems. Drug Alcohol Depend 40:9–15
62. Freedman R, Brown AS, Cannon TD et al (2018) Can a framework be established for the use of ketamine? Am J Psychiatry 175: 587–589
63. Singh I, Morgan C, Curran V et al (2017) Ketamine treatment for depression: opportunities for clinical innovation and ethical foresight. Lancet Psychiatry 4:419–426
64. Hung CC, Liu YH, Huang CC et al (2020) Effects of early ketamine exposure on cerebral gray matter volume and functional connectivity. Sci Rep 10:15488
65. Yuen SKK, Ng ACF (2020) Management of complications of ketamine abuse. In: Hashimoto K, Ide S, Ikeda K (eds) Ketamine from abused drug to rapid-acting antidepressant. Springer, Singapore. https://doi.org/10.1007/978-981-15-2902-3. Accessed 21 May 2024
66. Home Office (U.K.) NS (2018) Drug misuse: findings from the 2017/18 crime survey for England and Wales. https://assets.publishing.service.gov.uk/government/uploads/system/uploads/attachment_data/file/729249/drug-misuse-2018-hosb1418.pdf. Accessed 3 May 2024
67. European Monitoring Centre for Drugs and Drug Addiction (2023), European Drug Report 2023: Trends and Developments. https://www.emcdda.europa.eu/publications/european-drug-report/2023_en. Accessed 3 May 2024
68. Rocha A, Hart N, Trujillo KA (2017) Differences between adolescents and adults in the acute effects of PCP and ketamine and in sensitization following intermittent administration. Pharmacol Biochem Behav 157:24–34
69. Li F, Liu J, Yip PSF et al (2020) Mortalities of methamphetamine, opioid, and ketamine abusers in Shanghai and Wuhan. China Forensic Sci Int 306:110093
70. European Monitoring Centre for Drugs and Drug Addiction EMCDDA (2002) Report on the risk assessment of ketamine in the framework of the joint action on new synthetic drugs. http://bookshop.europa.eu/uri?target=EUB:NOTICE:TD3901053:EN:HTML. Accessed 14 Mar 2024
71. Drugs.com (2023) FDA warns patients and health care providers about potential risks associated with compounded ketamine products. https://www.drugs.com/fda/fda-warns-patients-health-care-providers-potential-risks-associated-compounded-ketamine-products-14620.html?hash2=7839a07de57eca461c3adda6dd5732b8&utm_source=ddc&utm_medium=email&utm_campaign=FDA+Safety+Alert++FDA+Warns+Patients+and+Health+Care+Providers+About+Potential+Risks+Associated+With+Compounded+Ketamine+Products. Accessed 5 May 2024

72. Siegel RK (1978) Phencyclidine and ketamine intoxication: a study of four populations of recreational users. NIDA Res Monograph 21:119–147
73. Jansen KL (2000) A review of the nonmedical use of ketamine: use, users and consequences. J Psychoactive Drugs 32:419–433
74. Bokor G, Anderson PD (2014) Ketamine: an update on its abuse. J Pharm Pract 27:582–586
75. Hansen G, Jensen SB, Chandresh L et al (1988) The psychotropic effect of ketamine. J Psychoactive Drugs 20:419–425
76. Morgan CJA, Muetzelfeldt L, Curran HV (2010) Consequences of chronic ketamine self-administration upon neurocognitive function and psychological wellbeing: a 1-year longitudinal study. Addiction 105: 121–133
77. Schifano F, Corkery J, Oyefeso A (2008) Trapped in the "K-hole" overview of deaths associated with ketamine misuse in the UK (1993–2006). J Clin Psychopharmacol 28: 114–116
78. Cheng JY, Chan DT, Mok VK (2005) An epidemiological study of alcohol/drugs related fatal traffic crash cases of deceased drivers in Hong Kong between 1996 and 2000. Forensic Sci Int 153:196–201
79. Chu PS, Ma WK, Wong SC et al (2008) The destruction of the lower urinary tracts by ketamine abuse: a new syndrome? BJU Int 102:1616–1622
80. Tsai T-H, Cha T-L, Lin C-M et al (2009) Ketamine-associated bladder dysfunction. Int J Urol 16:826–829
81. Manongi N, Yoon SH, Luo JJ, Trivedi P (2022) Severe ketamine-induced uropathy associated with cholangiopathy requiring percutaneous nephrostomy. Cureus 14(10): e30443
82. Venyo A, Benatar B (2012) Severe lower urinary tract symptoms associated with bilateral hydronephrosis and renal impairment: a case report of ketamine abuse uropathy with a review of the literature. WebmedCentral UROLOGY 3:2969
83. Lo RS, Krishnamoorthy R, Freeman JG et al (2011) Cholestasis and biliary dilatation associated with chronic ketamine abuse: a case series. Singapore Med J 52:e52–e55
84. Poon TL, Wong KF, Chan MY et al (2010) Upper gastrointestinal problems in inhalational ketamine abusers. J Dig Dis 11:106–110
85. Wong GL, Tam YH, Ng CF et al (2014) Liver injury is common among chronic abusers of ketamine. Clin Gastroenterol Hepatol 12: 1759.e1–1762.e1
86. Yu WL, Cho CC, Lung PF et al (2014) Ketamine-related cholangiopathy: a retrospective study on clinical and imaging findings. Abdom Imaging 39:1241–1246
87. Zhou J, Shaw SG, Gilleece Y (2013) Dilated common bile duct and deranged liver function tests associated with ketamine use in two HIV-positive MSM. Int J STD AIDS 24:667–669
88. Sleigh J, Harvey M, Voss L et al (2014) Ketamine—more mechanisms of action than just NMDA blockade. Trends Anaesth Crit Care 4:76–81
89. Spear LP (2013) Adolescent neurodevelopment. J Adolesc Health 52:S7–S13
90. Klein ME, Chandra J, Sheriff S et al (2020) Opioid system is necessary but not sufficient for antidepressive actions ketamine in rodents. Proc Natl Acad Sci USA 117:2656–2662
91. Williams NR, Heifets BD, Bentzley BS et al (2019) Attenuation of antidepressant and antisuicidal effects of ketamine by opioid receptor antagonism. Mol Psychiatry 24: 1779–1786
92. Henson MA, Roberts AC, Salimi K et al (2008) Developmental regulation of the NMDA receptor subunits, NR3A and NR1, in human prefrontal cortex. Cereb Cortex 18: 2560–2573
93. Colwell CS, Cepeda C, Crawford C et al (1998) Postnatal development of glutamate receptor-mediated responses in the neostriatum. Dev Neurosci 20:154–163
94. Cook DR, Stiller R, Dayton P (1982) Pharmacokinetics of ketamine (K) in infants and small children. Anesthesiology 57:A428
95. Kamp J, Olofsen E, Henthorn TK et al (2020) Ketamine pharmacokinetics: a systemic review of the literature, meta-analysis, and population analysis. Anesthesiology 133:1192–1213
96. Herd D, Anderson BJ (2007) Ketamine disposition in children presenting for procedural sedation and analgesia in a children's emergency department. Pediatr Anesth 17:622–629
97. Tuscher JJ, Luine V, Frankfurt M, Frick KM (2016) Estradiol-mediated spine changes in the dorsal hippocampus and medial prefrontal cortex of ovariectomized female mice depend on ERK and mTOR activation in the dorsal hippocampus. J Neurosci 36(5):1483–1489
98. Veilleux-Lemieux D, Castel A, Carrier D et al (2013) Pharmacokinetics of ketamine and xylazine in young and old Sprague-Dawley rats. J Am Assoc Lab Animal Sci 52:567–570

99. Edginton AN, Schmitt W, Voith B (2006) Mechanistic approach for the scaling of clearance in children. Clin Pharmacokinet 45:683–704
100. Lockhart CH, Nelson WL (1974) The relationship of ketamine requirement to age in pediatric patients. Anesthesiology 40:507–508
101. Bliss TV, Lomo T (1973) Long-lasting potentiation of synaptic transmission in the dentate area of the anaesthetized rabbit following stimulation of the perforant path. J Physiol 232:331–356
102. McCabe BJ, Horn G (1988) Learning and memory: regional changes in N-methyl-D-aspartate receptors in the chick brain after imprinting. Proc Natl Acad Sci USA 85: 2849–2853
103. Durieux ME (1995) Inhibition by ketamine of muscarinic acetylcholine receptor function. Anesth Analg 81:57–62
104. Rabiner EA (2007) Imaging of striatal dopamine release elicited with NMDA antagonists: is there anything there to be seen? J Psychopharmacol 21:253–258
105. Schumacher A, Sivanandan B, Tolledo EC et al Different dosing regimens of repeated ketamine administration have opposite effects on novelty processing in rats. Prog Neuropsychopharmacol Biol Psychiatry 69:1–10
106. Wang LJ, Chen CK, Lin SK et al (2018) Cognitive profile of ketamine-dependent patients compared with methamphetamine-dependent patients and healthy controls. Psychopharmacology 235:2113–2121
107. Chan KW, Lee TM, Siu AM et al (2013) Effects of chronic ketamine use on frontal and medial temporal cognition. Addict Behav 38:2128–2132
108. Sun L, Li Q, Li Q et al (2014) Chronic ketamine exposure induces permanent impairment of brain functions in adolescent cynomolgus monkeys. Addict Biol 19:185–194
109. Krystal JH, Karper LP, Seibyl JP et al (1994) NMDA antagonist, ketamine, in humans. Psychotomimetic, perceptual, cognitive, and neuroendocrine responses. Arch Gen Psychiatry 51:199–214
110. Tang WK, Lau CG, Ungvari GS (2019) Recovery of cognitive functioning following abstinence from ketamine. Addict Behav 99: 106081
111. Liao Y, Tang J, Ma M et al (2010) Frontal white matter abnormalities following chronic ketamine use: a diffusion tensor imaging study. Brain 133:2115–2122
112. Liao Y, Tang J, Corlett PR et al (2011) Reduced dorsal prefrontal gray matter after chronic ketamine use. Biol Psychiatry 69:42–48
113. Narendran R, Frankle WG, Keefe R et al (2005) Altered prefrontal dopaminergic function in chronic recreational ketamine users. Am J Psychiatry 162:2352–2359
114. Roberts ER, Curran HV, Friston KJ et al (2014) Abnormalities in white matter microstructure associated with chronic ketamine use. Neuropsychopharmacology 39:329–338
115. Wang C, Zheng D, Xu J et al (2013) Brain damages in ketamine addicts as revealed by magnetic resonance imaging. Front Neuroanat 7:23
116. Onaolapo AY, Ayeni OJ, Ogundeji MO et al (2019) Subchronic ketamine alters behaviour, metabolic indices and brain morphology in adolescent rats: involvement of oxidative stress, glutamate toxicity and caspase-3-mediated apoptosis. J Chem Neuroanat 96: 22–33
117. Ding R, Li Y, Du A et al (2016) Changes in hippocampal AMPA receptors and cognitive impairments in chronic ketamine addiction models: another understanding of ketamine CNS toxicity. Sci Rep 6:38771
118. Chaoxuan D, Anand KJS (2013) Developmental neurotoxicity of ketamine in pediatrics clinical use. Toxicol Lett 220:53–60
119. Brown BP, Kang SC, Gawelek K et al (2015) In vivo and in vitro ketamine exposure exhibits a dose-dependent induction of activity-dependent neuroprotective protein in rat neurons. Neuroscience 290:31–40
120. Cao SE, Tian J, Chen S et al (2015) Role of miR-34c in ketamine-induced neurotoxicity in neonatal mice hippocampus. Cell Biol Int 39:164–168
121. Huang L, Liu Y, Zhang P et al (2013) In vitro dose-dependent inhibition of the intracellular spontaneous calcium oscillations in developing hippocampal neurons by ketamine. PLoS One 8:e59804
122. Jiang XL, Du BX, Chen J et al (2014) MicroRNA-34a negatively regulates anesthesia-induced hippocampal apoptosis and memory impairment through FGFR1. Int J Clin Exp Pathol 7:6760–6767
123. Saland SK, Schiepfer KJ, Kabbaj M (2016) Hedonic sensitivity to low-dose ketamine is modulated by gonadal hormones in a sex-dependent manner. Sci Rep 6:21322
124. Li J, Wu H, Xue G et al (2013) 17β-oestradiol protects primary-cultured rat cortical neurons from ketamine-induced apoptosis by

activating PI3K/Akt/Bcl-2 signalling. Basic Clin Pharmacol Toxicol 113:411–418

125. Li J, Wang B, Wu H et al (2014) 17β-estradiol attenuates ketamine-induced neuroapoptosis and persistent cognitive deficits in the developing brain. Brain Res 1593:30–39
126. Li W, Li H, Wei H et al (2019) 17β-estradiol treatment attenuates neurogenesis damage and improves behavior performance after ketamine exposure in neonatal rats. Front Cell Neurosci 13:251
127. Ke X, Ding Y, Xu K, He H, Wang D, Deng X et al (2018) The profle of cognitive impairments in chronic ketamine users. Psychiatry Res 266:124–131
128. Zhang C, Xu Y, Zhang B, Hao W, Tang WK (2020) Cognitive impairment in chronic ketamine abusers. Psychiatry Res 291:113206
129. Lan X, Wang C, Zhang F, Liu H, Li W, Ye Y et al (2023) Short-term cognitive effects of repeated-dose esketamine in adolescents with major depressive disorder and suicidal ideation: a randomized controlled trial. Child Adolesc Psychiatry Ment Health 17:108
130. Molero P, Ramos-Quiroga JA, Martin-Santos R et al (2018) Antidepressant efficacy and tolerability of ketamine and esketamine: a critical review. CNS Drugs 32:411–420
131. Dwyer JB, Beyer C, Wilkinson ST et al (2017) Ketamine as a treatment for adolescent depression: a case report. J Am Acad Child Adolesc Psychiatry 56:352–354
132. Cullen KR, Amatya P, Roback MG et al (2018) Intravenous ketamine for adolescents with treatment-resistant depression: an open-label study. J Child Adolesc Psychopharmacol 28:437–444
133. Wolfson PE, Andries J, Ahlers D et al (2023) Ketamine-assisted psychotherapy in adolescent with multiple psychiatric diagnoses. Front Psych 14:1141988
134. Zarrinnegar P, Kothari J, Cheng K (2019) Successful use of ketamine for the treatment of psychotic depression in a teenager. J Child Adolesc Psychopharmacol 29:472–473
135. Dwyer JB, Landeros-Weisenberger A, Johnson JA et al (2021) Efficacy of intravenous ketamine in adolescent treatment-resistant depression: a randomized midazolam-controlled trial. Am J Psychiatry 178:352–362
136. Roy AV, Thai M, Klimes-Dougan B et al (2021) Brain entropy and neurotropic molecular markers accompanying clinical improvement after ketamine: preliminary evidence in adolescents with treatment-resistant depression. J Psychopharmacol 35:168–177
137. Skala K, Doganay K, Eder H et al (2023) Intranasal esketamine as therapeutic option: a case report of an adolescent with treatment resistant depression. Front Psych 14: 1118737
138. Lan X, Wang C, Zhang F et al (2023) Efficacy of repeated intravenous esketamine in adolescents with anxious versus non-anxious depression. Gen Psychiatry 36:e101007
139. Zhou Y, Lan C, Wang C et al (2024) Effect of repeated intravenous esketamine on adolescents with major depressive disorder and suicidal ideation: a randomized active-placebo-controlled trial. J Am Acad Child Adolesc Psychiatry 63:P507–P518
140. Kovacova V, Maceiova A, Tonhaizerova I et al (2024) Effect of acute ketamine treatment on sympathetic regulation indexed by electrodermal activity in adolescent major depression. Pharmaceuticals 17:358
141. Weber G, Yao J, Binns S et al (2018) Case report of subanesthetic intravenous ketamine infusion for the treatment of neuropathic pain and depression with suicidal features in a pediatric patient. Case Rep Anesthesiol 2018: 9375910
142. Lally N, Nugent AC, Luckenbaugh DA et al (2015) Neural correlates of change in major depressive disorder anhedonia following open-label ketamine. J Psychopharmacol 29: 596–607
143. Kienbaum P, Heuter T, Pavlakovic G et al (2001) (+)–ketamine increases muscle sympathetic activity and maintains the neural response to hypotensive challenges in humans. Anesthesiology 94:252–258
144. Couch GA, White MP, De Gray LE (2020) Central nervous system stimulants: basic pharmacology and relevance to anaesthesia and critical care. Anaesth Intensive Care Med 21:503–511
145. Okamoto H, Hoka S, Kawasaki T et al (1994) L-arginine attenuates ketamine-induced increase in renal sympathetic nerve activity. Anesthesiology 81:137–146
146. Liebe T, Li S, Lord A et al (2017) Factors influencing the cardiovascular response to subanesthetic ketamine: a randomized, placebo-controlled trail. Int J Neuropsychopharmacol 20:909–918
147. Miller RD, Cohen NH, Eriksson LI et al (2015) Miller's anesthesia, eight edition. Can J Anesth 62:558–559
148. Salt PJ, Barnes PK, Beswick FJ (1979) Inhibition of neuronal and extraneuronal uptake of noradrenaline by ketamine in the isolated perfused rat heart. Br J Anaesth 51:835–838

149. Short B, Fong J, Galvez V et al (2018) Side-effects associated with ketamine use in depression: a systemic review. Lancet Psychiatry 5: 65–78

150. Horowitz MA, Moncrieff J (2020) Are we repeating mistakes of the past? A review of the evidence for esketamine. Br J Psychiatry 219:614–617

151. Jones RR, Freeman MP, Kornstein SG et al (2022) Efficacy and safety of esketamine nasal spray by sex in patients with treatment-resistant depression: finding from short-term randomized, controlled trials. Arch Womens Ment Health 25:313–326

152. Papolos DF, Teicher MH, Faedda GL et al (2013) Clinical experience using intranasal ketamine in the treatment of pediatric bipolar disorder/fear of harm phenotype. J Affect Disord 147:431–436

153. Papolos DF, Frei M, Rossignol D et al (2018) Clinical experience using intranasal ketamine in the longitudinal treatment of juvenile bipolar disorder with fear of harm phenotype. J Affect Disord 225:545–551

154. Donoghue AC, Roback MG, Cullen KR (2015) Remission from behavioral dysregulation in a child with PTSD after receiving procedural ketamine. Pediatrics 136:e694–e696

Chapter 18

Future Directions of Esketamine Nasal Spray Continuation in Treatment-Resistant Depression

Maxwell Z. Price and Richard L. Price

Abstract

Treatment-resistant depression (TRD) is a challenging condition with low response and remission rates. Intranasal esketamine (ESK) is the first regulatory-approved medication specifically designed for difficult-to-treat depression that builds upon the discovery of rapid antidepressant effects of intravenous (IV) racemic ketamine. For more than 6 years now, ESK has demonstrated efficacy, safety, and practicality in acute and long-term management of TRD and major depressive disorder with suicidal ideation or behavior. This chapter builds upon the insights gained in patient trials and clinical experience with ESK and will discuss emerging indications, future research, and paths forward in clinical practice. Potential new indications include approaches to improve neurocognitive performance, even in the elderly, and to provide better care in posttraumatic stress and substance use disorders, in depressed youth as well as in bipolar depression. In everyday care, IV racemic ketamine can be used in acute situations, before a transition to ESK is feasible. For outpatient practice, new combinations such as with dextromethorphan/bupropion and sometimes magnesium have emerged. Ketamine's side effects can induce beneficial subjective experience especially with patient preparation and guidance through psychoactive events and appropriate behavioral or augmented psychotherapy.

Key words Treatment-resistant depression, Esketamine, Intranasal

1 Introduction

After 6.5 years of practical experience with intranasal (IN) esketamine (ESK) for major depressive disorder in adults, it appears that the time has come for some consideration of important findings in clinical study and practical uses including effects on comorbidities and associated conditions. Although we prescribe ESK for its approved indications in treatment-resistant depression (TRD) and for major depression with suicidal ideation (MDSI), we and others have found utility for ESK in addressing cognitive disorders, such as dementia and traumatic brain injury (TBI), post-traumatic stress disorder (PTSD), substance use disorders, bipolar depression, anxiety, and adolescent depression. We have also considered how to

Sabine Himmelseher (ed.), *Ketamine: Pain Management, Critical Care, Depression, and Resource-Poor Settings Volume II*, Neuromethods, vol. 225, https://doi.org/10.1007/978-1-0716-4599-4_18,

optimize ESK efficacy with co-prescribed oral antidepressants along with psychotherapy. Finally, we will propose a paradigm shift in the treatment of severe depression by integrating short-term intravenous (IV) ketamine followed by longer-term ESK maintenance.

2 Emerging Research Areas for Future Exploration

Several, in part, unexpected, new potential indications which need to be explored have emerged: based on evidence of enhanced cognitive function in non-cognitively impaired seniors with TRD taking ESK [1], research on the effects of ESK in (TRD) patients with baseline cognitive deficits, dementia, or following TBI could yield valuable insights. In our clinical practice, ESK has shown benefits for neurocognitive performance in the Mini Mental Status Exam (MMSE), independent of mood improvement measured by the Patient Health Questionnaire (PHQ-9). Among five patients with dementia and three with TBI, MMSE scores unexpectedly continued to improve beyond 4 weeks, even as PHQ-9 scores plateaued.

Research also suggests potential benefits of ESK for TRD in patients with PTSD symptoms [2, 3], though these individuals often experience heightened dissociative responses. Incorporating psychotherapy, either in person or through virtual tools, before or after ESK sessions may help mitigate dissociative effects and support the reprocessing of traumatic memories [2]. Additional research, including a planned trial at Yale, is needed to examine psychotherapy's role in enhancing ESK outcomes [4].

Small studies have also indicated ESK's effectiveness and tolerability in patients with a history of substance use disorders, including alcohol, benzodiazepines, cocaine, and cannabis [5]. These findings align with our experience, including cases involving opioid use disorder. While ESK would not be appropriate for individuals abusing ketamine, most patients do not find ESK's side effects reinforcing due to the low incidence of euphoria [6]. Therefore, effectively treating TRD with ESK may support substance abstinence.

For treatment-resistant bipolar depression, off-label ESK use has shown promising results in improving depression and anxiety, with minimal risk of manic switch [7]. However, insurance reimbursement may be challenging since bipolar depression is excluded from a TRD diagnosis. In practice, distinguishing between TRD with mixed features and a mixed episode of bipolar disorder can be difficult. In such cases, combining olanzapine–fluoxetine, the only Food and Drug Administration (FDA)-approved oral antidepressant for TRD that also treats bipolar I depression, can provide a dual benefit. This combination satisfies insurance requirements in the USA [8]. Emerging research suggests potential future approval of ESK for adolescents. A phase 2 randomized, double-blind study demonstrated significant and clinically meaningful reductions in depressive symptoms at 24 h after the first dose of ESK compared to midazolam in adolescents at imminent suicide risk [9].

In our experience, preparing patients before ESK sessions can significantly improve tolerance and outcomes. The subjective effects of ESK are highly dependent on the patient's mental state at the time of treatment, with negative moods intensifying unpleasant experiences and positive moods enhancing beneficial effects. Encouraging patients to adopt a neutral, open, and hopeful mindset—supported by tools such as music, aromatherapy, mindfulness, or relaxation techniques—can create a more tolerable and effective treatment experience, with lasting benefits between sessions.

3 Directions for Moving Forward in Every-Day Patient Care

For emergency departments (ED) and hospitals, we recommend short-term, off-label IV ketamine use for rapid antidepressant treatment. High cost of branded medications such as ESK often impact the daily compensation rate in these settings [10, 11]. ED physicians who are mostly already familiar with IV ketamine for analgesia, sedation, or even anesthesia can collaborate with psychiatrists to facilitate quicker discharges of patients from overcrowded EDs to appropriate outpatient psychiatric facilities. Here, patients can continue well-tolerated long-term ESK treatment. For patients requiring psychiatric inpatient admission, IV ketamine may reduce prolonged hospital stays, is associated with lower costs, and may free up critical bed space. It can significantly accelerate response times compared to traditional oral antidepressants. ESK is more appropriate for outpatient settings, where necessary resources such as Advanced Cardiac Life Support (ACLS), crash carts, and IV access are less common. Insurance coverage often alleviates treatment costs, improving access to ESK as a long-term therapeutic option.

Controlled short-term TRD trials have shown that the choice of oral antidepressant combined with ESK does not significantly influence its efficacy [12]. A recently FDA-approved oral glutamatergic antidepressant, dextromethorphan/bupropion (DXM/Bup), became available as a single tablet in 2022 in the USA [13] at a dose of 45 mg/105 mg twice daily. Although currently approved for major depressive disorder (MDD) rather than TRD, DXM/Bup demonstrates several advantages. It has a lower N-methyl-D-aspartate (NMDA) receptor half-maximal inhibitory concentration (IC_{50}) than ketamine, though it dissociates from the receptor more quickly. This may account for its greater tolerability and lower risk of nausea, sedation, or dissociation [14]. These attributes make DXM/Bup suitable for daily at-home use without being classified as a controlled substance. DXM/Bup also offers multimodal antidepressant mechanisms beyond NMDA receptor antagonism, including serotonin, norepinephrine, and dopamine reuptake inhibition. It acts as sigma receptor agonist, being superior to ketamine in this regard [14]. Sigma receptor agonism has been associated with antidepressant effects via various intracellular pathways [15]. Similar to ketamine, DXM impacts neuroplasticity-related gene expression through the transcriptional regulator methyl-CpG-binding protein 2 (MeCP2) [16].

Although α-amino-3-hydroxy-5-methyl-4-isoxazolepropionic acid (AMPA) or mu-opioid receptor (MOR) antagonists may theoretically interfere with ketamine's antidepressant effects [15, 17], this has not been conclusively demonstrated with DXM/Bup, likely due to its multimodal actions. For example, while some argue that MOR antagonists, like naltrexone, inhibit ketamine's effects by blocking the mammalian target of rapamycin (mTOR) pathway, studies have found that the mTOR inhibitor rapamycin can actually enhance ketamine's antidepressant effects [17]. This may be due to rapamycin's anti-inflammatory properties and its possible compensatory enhancement of the extracellular signal-regulated kinase (ERK) pathway [18]. The shared glutamatergic mechanisms between ESK and DXM/Bup suggest a potential role for DXM/Bup in sustaining ESK's effects between treatment sessions, offering a promising avenue for future exploration. In outpatient practice, we often combine ESK with daily oral DXM/Bup [13] and sometimes add magnesium to enhance NMDA receptor blockade [19]. This protocol warrants further investigation against alternative oral antidepressant regimens. Dextromethorphan (DXM) has also been shown to extend ketamine's analgesic effects and provide additional benefits for cognition and anxiety [20]. Taken together, although DXM/Bup may have less sustained NMDA receptor antagonism than ketamine, its multimodal mechanism may offer a synergistic antidepressant effect.

Since regulatory guidelines mandate combining ESK with an oral antidepressant at the moment [6], identifying the most effective drug combinations could improve outcomes while reducing treatment time and costs. This is likely also valid for future therapies if Spravato® will get approval as "stand-alone" antidepressant as already officially requested by the drug manufacturer in June 2024.

It is also essential to consider the unique patient experiences with ESK in TRD, beyond what rating scales and statistical analyses capture. Unlike conventional oral antidepressants, ESK has been observed to rapidly promote adaptive thinking in patients feeling "stuck" in depression, likely due to its neuroplastic, neurogenic, and synaptic plasticity-enhancing effects, as well as anti-inflammatory properties [17]. When paired with individualized psychotherapy—whether ongoing or immediately following treatment—patients often report "a feeling of being more connected to themselves and others," regaining a sense of "meaning and purpose" [21]. Such individual insights can extend beyond the treatment sessions, fostering long-lasting positive changes [22]. Although the dissociative side effects of ESK are typically milder than those of IV ketamine and diminish over time for both treatments [23], ESK has demonstrated sustained antidepressant effects over years of use.

Rather than viewing ESK as a competitor to other TRD treatments, we advocate for its integration into comprehensive care plans. ESK effectively augments existing antidepressant regimens, offering rapid symptom relief and sustained benefits without additional daily side effects [24]. For patients unable to afford paying for IV ketamine treatment out of pocket, ESK offers a more accessible alternative with widespread insurance reimbursement from private and government insurers in the USA [25]. At private clinics, IV ketamine can cost over $1000 USD per session. Additionally, ESK's suitability for routine outpatient use minimizes administrative burdens and overhead costs, making it an efficient option in psychiatric care of patients tremendously suffering from depression. In addition, failure of repetitive transcranial magnetic stimulation (rTMS) or electroconvulsive therapy (ECT) to provide relief in TRD is not necessarily predictive of future ESK response. ESK can also be successfully combined with neurostimulation [26].

4 Conclusions

Controlled clinical trials have demonstrated the efficacy and safety of adjunctive ESK in treating TRD across acute, short-term, and long-term periods, with evidence supporting its use in maintenance treatment for over 6 years. In emergency and inpatient settings, IV ketamine remains a practical option due to its lower cost and greater accessibility, with the potential to transition patients to ESK for ongoing outpatient management. A favorable therapeutic response may be associated with a variety of conditions, personal factors, patient histories, and trajectories, such as [1] achieving clinical improvement within 4 weeks, extending intervals between treatments, and requiring fewer concomitant medications (excluding benzodiazepines); [2] bipolarity, a family history of alcohol use disorder, a personal history of substance use disorder, female sex, younger age, and early-onset depression (before age 55 in seniors); and [3] successful reinitiation of prior effective acute ESK regimens, integrating psychotherapy for comorbid PTSD, as well as patient preparation and guidance through psychoactive events during treatment. ESK's therapeutic effects appear to be independent of its dissociative side effects, prior responses to neurostimulation, or outcomes from individual antidepressant trials and atypical antipsychotic augmentation. Notably, ESK has shown superiority over extended-release quetiapine as an augmentation strategy in a multicenter trial.

Future research priorities include exploring its application in patients with comorbid cognitive impairments, anxiety, obsessive–compulsive disorder, and substance use disorders, as well as its use in adolescents. Additional studies are needed to identify biomarkers predictive of response, refine psychotherapy integration, and optimize co-prescribed antidepressant strategies—such as dextromethorphan/bupropion—to enhance and sustain treatment outcomes.

Conflict of Interest

The authors declare the following financial interests/personal relationships which may be considered as potential competing interests: Maxwell Z. Price certifies that there is no conflict of interest with any financial organization regarding the material discussed in the manuscript. Richard L. Price has received honoraria from AbbVie, Alkermes, Axsome, Biogen, Idorsia, Intra-Cellular Therapies, Janssen, Jazz, Lundbeck, Neuronetics, Otsuka, Sage, Supernus, and Vanda and discloses no other conflicts of interest in this work.

References

1. Zaki N, Chen L, Lane R, et al (2023) Long-term safety of esketamine nasal spray in adults with treatment-resistant depression: final results of the SUSTAIN-3 study. Presented at: Psych Congress; September 6–10, 2023; Nashville, TN, USA
2. Rotharmel M, Benosman C, El-Hage W et al (2022) Efficacy and safety of intranasal esketamine in patients with treatment-resistant depression and comorbid chronic post-traumatic stress disorder: open-label single-arm pilot study. Front Psych 13:865466. https://doi.org/10.3389/fpsyt.2022.865466
3. Artin H, Bentley S, Mehaffey E et al (2022) Effects of intranasal (s)-ketamine on veterans with co-morbid treatment-resistant depression and PTSD: a retrospective case series. eClinicalMedicine 48:101439. https://doi.org/10.1016/j.eclinm.2022.101439
4. Kitay BM, Murphy E, Macaluso M et al (2023) Cognitive behavioral therapy following esketamine for major depression and suicidal ideation for relapse prevention: the CBT-endure randomized clinical trial study protocol. Psychiatry Res 330:115585. https://doi.org/10.1016/j.psychres.2023.115585
5. Chiappini S, d'Andrea G, De Filippis S et al (2023) Esketamine in treatment-resistant depression patients comorbid with substance-use disorder: a viewpoint on its safety and effectiveness in a subsample of patients from the real-esk study. Eur Neuropsychopharmacol 74:15–21. https://doi.org/10.1016/j.euroneuro.2023.04.011
6. Medication guide Spravato® – janssenlabels.com. https://www.janssenlabels.com/package-insert/product-patient-information/SPRAVATO-medication-guide.pdf. Accessed 3 Oct 2023
7. Martinotti G, Dell'Osso B, Di Lorenzo G et al (2023) Treating bipolar depression with esketamine: safety and effectiveness data from a naturalistic multicentric study on esketamine in bipolar versus unipolar treatment-resistant depression. Bipolar Disord 25:233–244. https://doi.org/10.1111/bdi.13296
8. Highlights of prescribing information – Eli Lilly and company. Accessed 3 Oct 2023. https://pi.lilly.com/us/symbyax-pi.pdf

9. Janssen Research & Development, LLC (2024) Study to evaluate the efficacy and safety of 3 fixed doses of intranasal esketamine in addition to comprehensive standard of care for the rapid reduction of the symptoms of major depressive disorder, including suicidal ideation, in pediatric participants assessed to be at imminent risk for suicide. Clinicaltrials.gov. Accessed 6 Feb 2024. https://clinicaltrials.gov/study/NCT03185819?cond=NCT03185819&rank=1
10. Singh B, Kung S, Pazdernik V et al (2023) Comparative effectiveness of intravenous ketamine and intranasal esketamine in clinical practice among patients with treatment-refractory depression. J Clin Psychiatry 84(2). https://doi.org/10.4088/JCP.22M14548
11. Nikayin S, Rhee TG, Cunningham ME et al (2022) Evaluation of the trajectory of depression severity with ketamine and esketamine treatment in a clinical setting. JAMA Psychiatr 79:736. https://doi.org/10.1001/jamapsychiatry.2022.1074
12. Adigun M, Sliwa JK, Turkoz I, et al (2019) Impact of concomitant oral antidepressants on the efficacy and safety of esketamine nasal spray in patients with treatment-resistant depression. Poster presented at: US Psych Congress; October 3–6, 2019; San Diego, California, USA
13. Auvelity Label – food and drug administration (FDA). Accessed August 20, 2023. https://www.accessdata.fda.gov/drugsatfda_docs/label/2022/215430s000lblCorrect3.pdf
14. Stahl SM (2019) Dextromethorphan/bupropion: a novel oral NMDA (N-methyl-Daspartate) receptor antagonist with multimodal activity. CNS Spectr 24:461–466. https://doi.org/10.1017/S1092852919001470
15. Fishback JA, Robson MJ, Xu Y et al (2010) Sigma receptors: potential targets for a new class of antidepressant drug. Pharm Ther 127: 271–282. https://doi.org/10.1016/j.pharmthera.2010.04.003
16. Morelli KH, Jin W, Shathe S et al (2022) MECP2-related pathways are dysregulated in a cortical organoid model of myotonic dystrophy. Sci Transl Med 14:651. https://doi.org/10.1126/scitranslmed.abn2375
17. Krystal JH, Kavalali ET, Monteggia LM (2023) Ketamine and rapid antidepressant action: new treatments and novel synaptic signaling mechanisms. Neuropsychopharmacology 49: 41–50. https://doi.org/10.1038/s41386-023-01629-w
18. Dai J, Bercury KK, Macklin WB (2014) Interaction of mTOR and ERK1/2 signaling to regulate oligodendrocyte differentiation. Glia 62:2096–2109. https://doi.org/10.1002/glia.22729
19. Price MZ, Price RL (2023) Intranasal Esketamine: a primer. Curr. Psychiatry 22(4). https://doi.org/10.12788/CP.0352
20. Martin E, Sorel M, Morel V et al (2019) Dextromethorphan and memantine after ketamine analgesia: a randomized control trial. Drug Des Dev Ther 13:2677–2688. https://doi.org/10.2147/DDDT.S207350
21. Mathai DS, Mora V, Garcia-Romeu A (2022) Toward synergies of ketamine and psychotherapy. Front Psychol 13:868103. https://doi.org/10.3389/fpsyg.2022.868103
22. Price RB, Spotts C, Panny B et al (2022) A novel, brief, fully automated intervention to extend the antidepressant effect of a single ketamine infusion: a randomized clinical trial. Am J Psychiatry 179:959–968. https://doi.org/10.1176/appi.ajp.20220216
23. Chubbs B, Wang J, Archer S et al (2022) A survey of drug liking and cravings in patients using sublingual or intranasal ketamine for treatment resistant depression: a preliminary evaluation of real world addictive potential. Front Psych 13:1016439. https://doi.org/10.3389/fpsyt.2022.1016439
24. Anand A, Mathew SJ, Sanacora G et al (2023) Ketamine versus ECT for nonpsychotic treatment-resistant major depression. N Engl J Med 388:2315–2325. https://doi.org/10.1056/nejmoa2302399
25. Megli D (2024) The ketamine economy: new mental health clinics are a "wild west" with few rules. NPR. January 30, 2024. Accessed 6 Feb 2024. https://www.npr.org/sections/health-shots/2024/01/30/1227630630/ketamine-infusion-clinic-mental-health -depression-anxiety-fda-off-label
26. Al-Sabbagh A, Solly A, Bending N et al (2022) Esketamine following transcranial magnetic stimulation for treatment resistant depression. Brain Stimul 15:1327–1328. https://doi.org/10.1016/j.brs.2022.07.025

Chapter 19

Ketamine for Treatment of Addiction in Alcohol, Opioid, and Cocaine Use Disorder

Katarina Savić Vujović, Ana Jotić, Branislava Medić, and Ivana Ćirković

Abstract

Substance use disorders (SUDs), including alcohol, opioid, and cocaine dependencies, present significant challenges in public health. Current treatment modalities often face limitations in efficacy and accessibility, necessitating exploration of novel therapeutic approaches. Ketamine, originally recognized for its anesthetic properties, has garnered attention for its potential in treating SUDs due to its unique pharmacological actions. This chapter synthesizes current literature on the use of ketamine in alcohol, opioid, and cocaine use disorders, highlighting its mechanisms of action, clinical effectiveness, and safety profile. Through its N-methyl-D-aspartate receptor antagonism and rapid-acting antidepressant effects, ketamine offers promise in reducing cravings, withdrawal symptoms, and relapse rates in clinical study of SUDs. Ketamine's ability to modulate glutamatergic transmission and promote neural plasticity may contribute to the therapeutic effects in addiction. Challenges such as optimal dosing, long-term safety, and potential for abuse require further investigation. This chapter underscores the evolving landscape of ketamine as a potential adjunctive therapy in SUDs, emphasizing the need for rigorous clinical trials to establish its place in addiction treatment protocols.

Key words Ketamine, Alcohol use disorder (AUD), Opioid use disorder (OUD), Cocaine use disorder (CUD)

1 Introduction

The medical community's acceptance of addiction has fluctuated throughout history. Viewing addiction as a disease aligns with modern disease models and scientific progress in understanding how neurobiology and environment interact [1]. Definitions of addiction must recognize the complex interaction of brain circuits, genetics, environmental influences, and personal experiences. This perspective aligns with diagnostic classifications of substance use disorders that emphasize broader criteria beyond just tolerance and withdrawal symptoms. Evolving social and political perspectives on addiction continually influence and reflect changes in medical treatments and terminology within the field [2].

Sabine Himmelseher (ed.), *Ketamine: Pain Management, Critical Care, Depression, and Resource-Poor Settings Volume II*, Neuromethods, vol. 225, https://doi.org/10.1007/978-1-0716-4599-4_19,

Substance addiction, also known as drug addiction, is a neuropsychiatric condition characterized by a persistent urge to continue using a substance despite its detrimental effects. This definition highlights different aspects of addictive behaviors [3]. Substance use disorders (SUD) are disorders involving the use of legal and illegal psychoactive substances [4]. SUD typically begin during adolescence or young adulthood and pose significant health risks for this demographic group globally [5]. Vulnerable individuals often initiate harmful substance use during early adolescence, which can progress into chronic use disorders which likely relapse and cause comorbidities evolving over time. Drug addiction is a serious global issue marked by both physical and psychological dependence on drugs. It involves an intensified urge to seek and use drugs, loss of control over drug intake, and the onset of withdrawal symptoms like dysphoria, anxiety, and irritability upon discontinuation of drug use. The shift from occasional drug use to addiction is driven by neurobiological changes that disrupt normal brain function within circuits responsible for regulating motivated behavior [6]. Alcohol use disorder (AUD), opioid use disorder (OUD), and cocaine use disorder (CUD) are commonly encountered SUD.

Research into drug dependence neurobiology attempts to identify specific neuropharmacological targets to facilitate the creation of addiction medications and novel therapy. This research is crucial to enhance understanding of molecular, cellular, and systemic alterations caused by addictive substances so that it can be clarified how they affect human behaviors linked to drug dependence [7]. Decades of neurobiological research has confirmed the involvement of dopamine (DA), glutamate, and other neurotransmitter systems in both the immediate and prolonged effects of addictive drugs. For example, many substances influence the mesocorticolimbic DA system by contributing to their initial reinforcing effects [8]. Chronic use of these substances disrupts other systems as well, such as the corticostriatal and extended amygdala networks. These disruptions ultimately affect emotional regulation and impair higher cognitive functions like impulse control and self-regulation of drug use, leading to a persistent inclination to relapse [9]. Although extensive research has been done on how addiction drugs interact with specific neuronal sites to activate the mesocorticolimbic DA system and induce long-term changes in the brain, efforts to target dysfunction of this system with pharmacotherapy have mostly been unsuccessful. This is still valid despite advances in understanding neuronal mechanisms involved in drug actions.

Recent research has shown that subanesthetic dose racemic ketamine produces fast and long-lasting antidepressant effects. In 2019, the US Food and Drug Administration (FDA) approved intranasal esketamine, the more potent S-enantiomer of ketamine, for treating patients with major depressive disorder and later treatment-resistant depression (TRD) [10]. This approval led to increased interest in exploring ketamine for treating various mental health conditions beyond depression, including substance use disorders (SUDs). Ketamine has intensively been investigated during recent years for its potential to treat various forms of addiction, including alcohol abuse, opioid abuse, and cocaine use disorder. The powerful anesthetic has shown promising results in reducing cravings and withdrawal symptoms in individuals struggling with addiction. In this article, we will explore ketamine's potential as a treatment option for addiction and its effectiveness in addressing alcohol, opioid, and cocaine use disorders.

2 Ketamine: Therapeutic Potential for Use Disorders

Ketamine is a psychoactive anesthetic that is commonly used in medical practice for inducing analgesia, sedation, and anesthesia. By blocking N-methyl-D-aspartate (NMDA) receptors, subanesthetic dose ketamine causes pain relief and dissociative effects [11]. In recent years, there has been growing interest in the potential of ketamine for treatment of severe neuropsychiatric disorders, including depression, post-traumatic stress disorder (PTSD), and addiction. Ketamine's antidepressant effects are primarily mediated by inhibition of NMDA receptors, which leads to subsequent activation of α-amino-3-hydroxy-5-methyl-4-isoxazolepropionic acid (AMPA) receptors. Activation of AMPA receptors triggers downstream effects, including upregulation of brain-derived neurotrophic factor (BDNF) and activation of tropomyosin receptor kinase B (TrkB) signaling [12]. Ketamine also activates the mammalian target of rapamycin (mTOR) pathway, resulting in inhibition of glycogen synthase kinase 3 (GSK-3) and suppression of eukaryotic elongation factor 2 (eEF2) kinase phosphorylation [13]. The active metabolite of ketamine, hydroxynorketamine, does not directly interact with NMDA receptors but can indirectly stimulate AMPA receptors and potentially contribute to the rapid onset of ketamine's antidepressant effects [14, 15]. Recent studies suggest that ketamine's antidepressant effects may also involve inhibition of the lateral habenula, known as the "anti-reward center" [16]. Ketamine additionally blocks muscarinic acetylcholine receptors, voltage-gated calcium channels, and may enhance descending monoaminergic pain pathways [17, 18]. It exhibits partial agonist activity at μ and κ opioid receptors and may enhance gamma-aminobutyric acid (GABA) synaptic inhibition while modulating DA release and presynaptic glutamate release.

The potential mechanisms by which ketamine may treat addiction include the following [19]:

1. Increase in neuroplasticity and neurogenesis: Ketamine may enhance neuroplasticity and promote brain neurogenesis,

which may contribute to reverse some of the changes caused by addiction.

2. Alteration of decisive structures and connectivity: Ketamine's ability to block certain neural pathways and/or alter neural circuitry and connectivity may disrupt addictive behaviors and cravings.
3. Alleviation of symptoms of depression: Because depression often co-occurs with addiction, ketamine's rapid antidepressant effects may indirectly benefit individuals with SUD by mood improvement and reduction of the desire to use substances as a coping mechanism.
4. Interference with memory reconsolidation: Ketamine may disrupt the reconsolidation of memories associated with drug use, potentially weakening the associations between drug-related cues and cravings.
5. Induction of mystical experiences: Some research suggests that ketamine may induce profound mystical and/or transformative experiences, which may lead to changes in perspective and behavior that are beneficial for addiction recovery.
6. Enhancement of efficacy of other therapeutic modalities: Ketamine-assisted therapy sessions may enhance the effectiveness of traditional therapies by facilitating introspection, emotional processing, and insight into underlying psychological issues contributing to addiction.

All these mechanisms emphasize ketamine's potential as a novel treatment approach to addiction, offering multiple avenues through which it may help individuals achieve recovery (Fig. 1).

3 Ketamine and Alcohol Use Disorder

AUD manifests as a psychiatric condition marked by persistent and excessive alcohol consumption despite adverse effects on health, social relationships, and work or school performance. Alcohol abuse is a global, major health problem and the leading risk factor for premature mortality, poor health, and disability in 20–39-year-old individuals, with a 13% death rate in this age group (World Health Organization 2023). Alcohol works by enhancing the activity of the inhibitory GABA receptor while reducing excitatory NMDA receptor functions [20]. It promotes the release of opioid peptides and DA, which leads to craving for alcohol and other substances [21]. These combined effects result in sedation, euphoria, drowsiness, and the development of physiological dependence on alcohol over time. Prolonged alcohol use causes a decrease in GABA receptors and an increase in NMDA receptors, necessitating continued alcohol consumption to maintain a chemical balance and

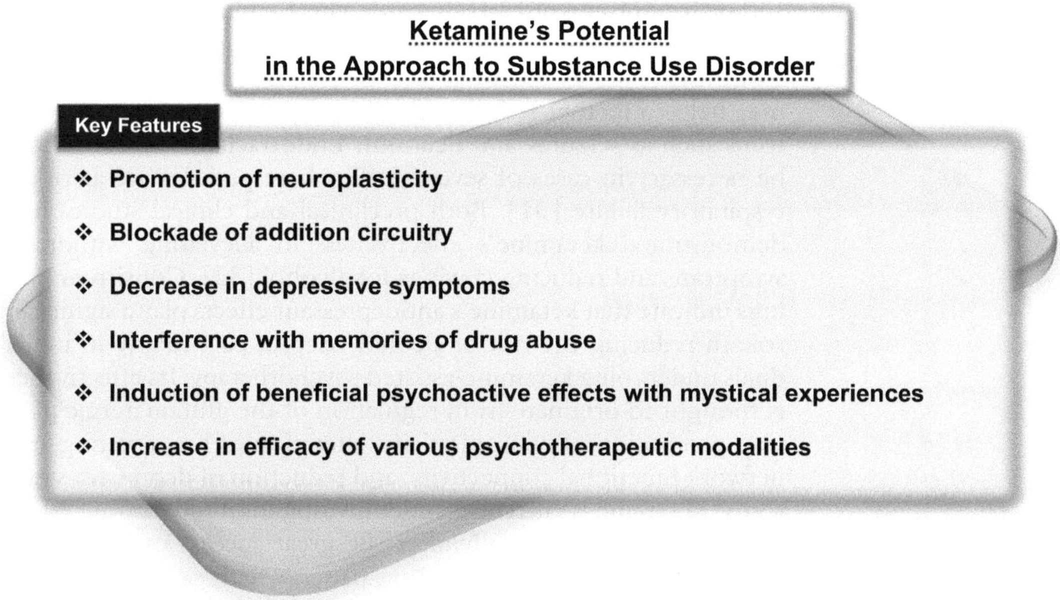

Fig. 1 Ketamine's potential in the approach to substance use disorder (created by the authors and S. Himmelseher in Word, Microsoft Corp., Redmond, WA, USA)

prevent withdrawal experiences [22, 23]. In addition to its complex effects on GABA and glutamate receptors, transient DA increases in the nucleus accumbens (NAc) are induced [24]. In summary, AUD involves a complex pathology that affects multiple neurotransmitter systems, synaptic plasticity, hippocampal neurogenesis, and neural connectivity, all of which contribute to its behavioral manifestations [25]. Depressed mood and stress are prominent factors influencing relapse in AUD. AUD often coexists with depression and stress disorders, indicating possible shared mechanisms for therapeutic approaches [26]. Current treatment options for AUD are limited, and effectiveness and response rates are generally modest [27]. FDA-approved medications include acamprosate, disulfiram, or naltrexone, but there are also off-label used drugs such as baclofen, gabapentin, varenicline, topiramate, prazosin, and ondansetron [28]. The FDA recommended agents naltrexone, acamprosate, and disulfiram are administered in oral or injectable form [29]. For example, systematic review and meta-analysis including 118 clinical trials and 20,976 participants found that both, 50 mg/d oral naltrexone and acamprosate, caused significant reduction in alcohol consumption in comparison to placebo [25].

Alcohol withdrawal syndrome (AWS) is a prevalent condition that occurs when heavy drinking is abruptly stopped or reduced, whether intentionally or unintentionally. It affects approximately 5–10% of individuals who face significant medical and psychiatric challenges, including withdrawal seizures, perceptual disturbances, and delirium tremens [30]. In cases where patients do not respond to benzodiazepines, physicians may need to consider alternative treatments such as phenobarbital or propofol. These medications may help to further sedate patients and prevent dangerous symptoms such as seizures and delirium. Endotracheal intubation may be necessary in cases of severe AWS where patients are at risk of respiratory failure [31]. Both preclinical and clinical studies have demonstrated ketamine's effectiveness in alleviating withdrawal symptoms and reducing cravings for alcohol [32]. Consistent findings indicate that ketamine's antidepressant effects play a significant role in reducing the risk of alcohol relapse, particularly in individuals undergoing ketamine-assisted psychotherapy. Its effectiveness is thought to originate from regulation of the glutamatergic pathway, promotion of neuroplasticity, alteration of brain resting-state network functional connectivity, and reduction of depressive symptoms. Ketamine is currently under investigation for treating AUDs and other SUDs especially after the great success of intravenous (IV) racemic ketamine in treating patients with severe depression and the FDA-approved preparation of intranasal esketamine for TRD [33]. Along with preliminary evidence supporting ketamine's utility in treating SUDs, its effectiveness in managing depression and stress disorders like PTSD points to a real potential in treating AUDs as well [34]. Goldfine et al. identified a total of 204 studies related to alcohol and ketamine, out of which 10 specifically explored its use in AUD or AWS in humans [35]: 7 studies focused on ketamine's value in AUD, and 3 examined AWS. In AUD, ketamine demonstrated beneficial effects including reduced cravings, decreased alcohol consumption, and improved rates of sustained abstinence compared to standard treatment approaches. In all AUD studies, ketamine was applied at subanesthetic doses. In some studies, ketamine was administered either as single IV infusion once a week for 4 weeks or as three infusions per week with varying frequencies and routes. The maximum dose of ketamine used in these studies was 2.5 mg/kg administered as intramuscular (IM) injection. In AWS trials, ketamine was primarily used as adjunct to standard benzodiazepine therapy in severe and refractory cases, including those presenting with delirium tremens. Ketamine was administered as IV infusion at subanesthetic doses (0.15–0.75 mg/kg/h). This adjunctive use showed promising outcomes such as quicker resolution of delirium tremens and AWS symptoms, reduced duration of intensive care unit (ICU) stay, and decreased likelihood of requiring intubation. In AUD and AWS, ketamine induced adverse effects such as oversedation, headache, hypertension, and euphoria, which were manageable. These findings highlight ketamine's potential as a therapeutic option for managing AUD and AWS, particularly in cases where traditional treatments have been insufficient or ineffective [36]. Kelson et al. conducted a systematic review which examined 11 studies with

854 adult patients across the USA, the United Kingdom, and Russia. Study sample sizes varied from 5 to 211 individuals [37]. Among these trials, seven focused on patients with AUD, one on heavy drinkers, and three on alcohol withdrawal. The combination of ketamine and psychotherapy showed the most promising results in terms of patients achieving abstinence and less alcohol consumption. Findings regarding relapse, craving, and withdrawal symptoms varied. Ketamine therefore holds promise as a therapeutic option in conditions without response to FDA-approved treatment. In a three-arm open label trial, Terasaki et al. [38] compared IV ketamine (0.5 mg/kg infused over 40 min), IM naltrexone (380 mg once), and a control group receiving IV placebo (normal saline solution) in 44 high-utilization patients with AUD. The controls received "linkage treatment" alone, which involved outpatient addiction clinic referrals and research stipends, but no pharmacological intervention. Interestingly, there was a trend toward fewer 30-day hospital readmissions after ketamine and naltrexone treatment, and patient immediate acceptability rates were very high in both treatment arms without adverse effects. Tran et al. [39] conducted a rapid review about ketamine in AUD including two previous systematic reviews with long-term observation of a 5-year period [37, 40]. The evidence shows that combining IV ketamine infusion with psychotherapy treatment may effectively promote abstinence and reduce alcohol and cocaine consumption. Still, there were mixed results regarding the effects of ketamine on withdrawal symptoms and cravings. In a recent double-blind placebo-controlled trial, Gent and co-workers randomized 28 individuals with severe AUD to 1 oral esketamine treatment or placebo and 2 weeks of daily mindfulness-based therapy [41]. Esketamine was given in a new oral thin film preparation with good absorption through mucous membranes at a total dose of 115.1 mg. The authors found that esketamine enhanced patient psychological engagement. Combining ketamine with psychotherapy led to a decrease in alcohol cravings and was associated with more beneficial mystical and unwanted dissociative states compared to placebo.

The systematic reviews by Kelson et al. [37] demonstrated that 6 out of 11 studies documented adverse effects linked to ketamine use [42, 43], but 2 studies explicitly reported no observable unwanted effects [44, 45]. Mostly, adverse effects were oversedation, hypertension, sedation, and headache [31, 42, 46]. In a randomized controlled trial (RCT), Grabski et al. categorized 53 adverse events, including hypertension, tachycardia, euphoria, and low mood in 20 participants, as definitively (7 cases), probably (3 cases), or possibly (43 cases) related to ketamine infusion [47].

4 Ketamine and Opioid Use Disorder

Opioid use disorder (OUD) is a complex illness characterized by compulsive opioid drug use. It poses a significant public health challenge and has become an "opioid crisis" in the USA with increasing rates of addiction contributing to recent decreases in life expectancy. OUD carries a substantial disease burden, evident in high rates of hospitalization and deaths from overdose. It also contributes significantly to involvement in criminal acts and punishment by the justice system, as indicated by association with elevated incarceration rates [48, 49]. Overall, there is a significant increase in OUD and opioid overdose deaths which may be attributed to several factors, such as easy access to prescription opioid analgesics, higher purity levels of heroin, and the emergence of potent illicit fentanyl compounds [50].

OUD is associated with acute effects and long-term brain neuroadaptations related to intoxication, tolerance, and dependence, which foster compulsive use. Activation of μ, ∂, and κ opioid receptors modulates the activity of neural stress and reward pathways. Both animal and human studies have identified the orbitofrontal cortex (OFC) as a major player in the initiation and persistence of drug addiction. This brain region is associated with impairments in inhibiting impulsive behaviors and alterations in decision-making processes related to rewards [51]. Several epigenetic modifications regulate gene expression that influences behaviors pertinent to opioid addiction [52]. In contrast to alcohol, opioids inhibit inhibitory interneurons expressing the μ opioid receptor, which leads to disinhibition of ventral tegmental area (VTA) dopamine (DA) neurons. This results in transient increases in DA release in the NAc, as observed in mice injected with heroin using genetically encoded DA sensors. However, the dependence of opioid drug reinforcement on mesolimbic DA has also been questioned because heroin self-administration persists even after DA depletion or blockade of NAc DA receptors. Additionally, individuals addicted to heroin who receive methadone substitution experience a marked euphoric "high" from IV administered heroin but do not show an increase in striatal DA levels [53–55]. OUD is a chronic condition typically treated with both medication and psychosocial intervention. Individuals with OUD are commonly prescribed opioid receptor agonists like methadone, partial agonists such as buprenorphine, or opioid antagonists like extended-release naltrexone [50, 56]. These drugs are effective in supporting recovery. Despite strong evidence of their efficacy, they are still not widely utilized. The severity of the opioid epidemic underscores the importance of exploring every potential new avenue for OUD treatment. In their systematic review, Walsh et al. [40] included findings from a single study assessing the impact of ketamine on treatment with respect to acute withdrawal and long-term opioid consumption. However, results were inconclusive: when the randomized, placebo-controlled, double-blind trial investigated the impact of IV ketamine infused at a dose of 0.5 mg/kg, ketamine reduced the need for additional medication to manage acute opiate withdrawal after 48 h compared to controls. Specifically, lower doses of carbamazepine and clonazepam were required. Nonetheless, no statistically significant difference in opiate use between ketamine and control groups was observed after 4 months. In another RCT with 50 patients, ketamine demonstrated superior control over withdrawal symptoms which persisted beyond the ketamine infusion period [57]. The study showed that subanesthetic IV ketamine was effective in managing acute precipitated opiate withdrawal as adjunctive treatment. However, it did not exhibit long-term effects with respect to opiate dependence. In a prospective, long-term observation of 59 detoxified heroin-dependent patients, repeated sessions of ketamine-assisted psychotherapy resulted in a higher rate of abstinence at 1-year follow-up: 13 out of 26 study patients (50%) with multiple treatment sessions remained abstinent as compared to 6 out of 27 patients (22.2%) receiving only one ketamine session [58].

The treatment of buprenorphine-precipitated opioid withdrawal in a patient using fentanyl demonstrated a synergistic effect when IV ketamine at a dose of 0.6 mg/kg was infused over 1 h [59]. Ketamine also showed promising results in treating precipitated opioid withdrawal by enhancing μ-opioid receptor-mediated signaling [60]. Furthermore, anesthesia-assisted opioid detoxification is an option for patients who received conventional detoxification without success. When 31 patients with long OUD were acutely withdrawn with naltrexone during propofol/clonidine anesthesia and mechanical ventilation, effects of adding 1.5 mg/kg esketamine were studied: esketamine attenuated central nervous system hypersensitivity to naltrexone withdrawal as indicated by beneficial impact on changed electroencephalographic power spectra and nerve-evoked somatosensory potentials [61] (Fig. 2).

5 Ketamine and Cocaine Use Disorder

Cocaine use is an escalating global public health problem, leading to substantial health risks such as cardiovascular disease and deaths. In the USA, increasing deaths from cocaine overdose even outnumbered overuse deaths from heroin and opiate disorder in certain populations [62], and the COVID-19 pandemic exacerbated cocaine misuse. Effective treatments for cocaine use disorder (CUD) primarily involve psychosocial interventions sometimes combined with pharmacotherapy. Two main psychotherapeutic approaches for treatment of CUD are "contingency management"

Ketamine Effects in Treatment of Substance Use Disorder in Clinical Study

Key Features

- **Alcohol use disorder**
  - **Reduction of cravings for alcohol**
  - **Decrease in alcohol consumption**
  - **Improvement in abstinence rates**
  - **Quicker resolution of delirium tremens and withdrawal symptoms**
- **Opioid use disorder**
  - **Reduction of cravings for opioids**
  - **Decrease in opioid consumption**
  - **Better management of precipitated opioid withdrawal with adjunctive use**
  - **Alleviation of central nervous system hyperactivity in rapid opioid detoxification**
- **Cocaine use disorder**
  - **Enhancement of motivation to quit cocaine**
  - **Promotion of abstinence**
  - **Reduction of cravings for cocaine**
  - **Decrease in risks of relapse**
  - **Direct targeting of multiple genes and pathways involved**

Fig. 2 Ketamine effects in treatment of substance use disorder in clinical study (created by the authors and S. Himmelseher in Word, Microsoft Corp., Redmond, WA, USA)

and cognitive behavioral therapy (CBT). Contingency management uses rewards to reinforce abstinence and/or treatment compliance and is sometimes effective in reducing cocaine misuse. CBT focuses on changing behaviors and thought patterns associated with CUD which shall equip individuals with skills and capabilities to prevent relapse. Cocaine stimulates the brain's reward pathway known as the mesolimbic DA system [63]. Its effects are achieved by binding to and inhibiting the function of monoamine transporters responsible for DA, serotonin (5-HT), and norepinephrine (NE) [64]. These transporters normally regulate neuronal communication by facilitating the uptake of neurotransmitters from the synapse back into neighboring neurons. Cocaine leads to DA accumulation with the result of intensified and prolonged signaling between pre- and post-synaptic neurons. This mechanism is mainly thought to produce cocaine's stimulant effects on psychomotor activity and contributes to the pleasurable sensations induced by cocaine, such as euphoria. In addition to DA's central role, other neurotransmitter systems involving NE and 5-HT are also considered as viable cocaine targets which modulate and influence cocaine sensitization, craving, and relapse [65]. The leading pharmacological approaches to treat CUD currently focus on DA agonists like long-acting amphetamine and modafinil, as well as medications targeting glutamatergic and GABA-ergic systems such as topiramate. These approaches show promise in managing cravings and withdrawal symptoms associated with cocaine use, but so far, there is no FDA-approved medication for CUD [66, 67]. A meta-analysis reviewed 66 different drugs or drug combinations across 7 systematic reviews and 48 RCT with respect to pharmacotherapy for CUD [65]. Antidepressants were the most extensively studied class with 38 RCT, yet they did not demonstrate effectiveness in reducing cocaine use or improving treatment retention. Bupropion, psychostimulants, and topiramate were found to have potential to increase abstinence rates in individuals with CUD. In eight RCT, antipsychotics were reported to enhance treatment retention. These findings suggest promising avenues for pharmacological interventions in managing cocaine addiction, but further research is urgently warranted to strengthen evidence based-approaches and improve therapeutic strategies.

Ketamine shows one of the greatest potentials among pharmacotherapy for CUD treatment. In a cross-over, double-blind trial, eight volunteers received a 52-min IV infusion of ketamine at a dose of 0.41 or 0.71 mg/kg which was compared to 2 mg lorazepam in three treatment sessions separated by 48 h. Compared with the active control lorazepam, subanesthetic ketamine demonstrated promising effects in enhancing motivation to quit cocaine and reducing cue-induced cravings when evaluated 24 h after infusions [68]. In another study, 55 cocaine-dependent individuals were randomly assigned to receive either a 40-min IV infusion of ketamine at 0.5 mg/kg or midazolam during a 5-day inpatient stay [69]. A single infusion of ketamine improved several key treatment outcomes in those adults who also participated in mindfulness-based behavioral therapy. Ketamine effects included promotion of abstinence, reduction of cravings, and a decreased risk for relapse. In another study by the same author group with 20 patients, the use of IV ketamine compared to controls resulted in a significant 67% decrease in cocaine self-administration relative to baseline over more than 24 h after ketamine infusion [70].

In a milestone article applying an integrated strategy to identify repurposed FDA-approved drugs for CUD therapy, artificial intelligence predictions and expert insight ranked ketamine as leading candidate for clinical validation [71] (Fig. 3).

This was based on an analysis of electronic health records from more than 90 million people worldwide and CUD patient cohorts who were prescribed ketamine for anesthesia or depression in comparison with matched controls who received non-ketamine anesthesia or antidepressants/midazolam. CUD ketamine-anesthetized patients showed a significantly higher rate of CUD remission compared to matched controls who received other anesthetics. Likewise, CUD patients with antidepressant ketamine prescription

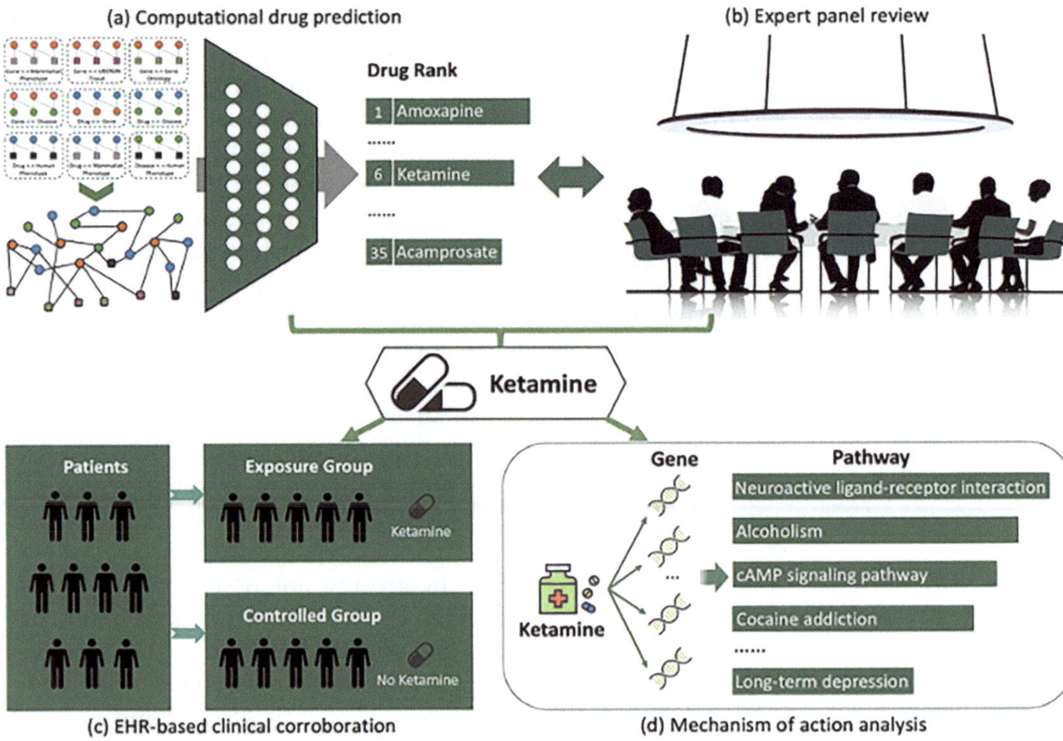

Fig. 3 Pipeline of drug repurposing strategy for ketamine in CUD: artificial intelligence and drug experts ranked ketamine as top candidate for clinical corroboration based on electronic health recordings (**a**), (**b**), and (**c**). Analysis of gene and brain functions showed that ketamine directly targets multiple genes and pathways important in CUD (d) (from Gao et al. [71] with permission from John Wiley and Sons via Copyright Clearance Center, Inc.). *cAMP* cyclic adenosine monophosphate, *CUD* cocaine use disorder, *EHR* electronic health recordings

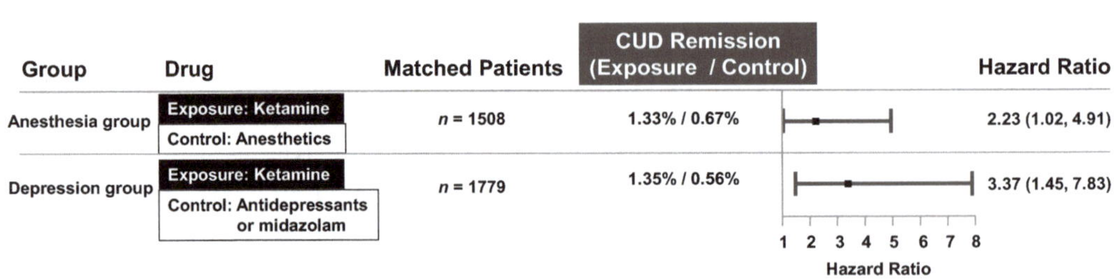

Fig. 4 Clinical efficacy of ketamine in cocaine use disorder: Remission from CUD in two nonoverlapping patient cohorts after ketamine anesthesia or antidepressant ketamine therapy as compared to anesthesia or antidepressant treatment without ketamine (modified from Gao et al. [71] with permission from John Wiley and Sons via Copyright Clearance Center, Inc.). *CUD* cocaine use disorder

exhibited a substantially higher remission rate compared to those matches receiving antidepressants or midazolam without ketamine (Fig. 4). In addition, genetic and pathway enrichment analyses were conducted to elucidate ketamine's potential mechanisms of action. This evaluation revealed multiple genes and pathways associated with cocaine abuse/or dependence as direct ketamine targets.

6 Limitations

Almost all, as yet available, studies regarding ketamine for SUD have small sample sizes, were monocentric, and did not include distinct assessments of specific patient (genetic) dispositions, gender, or other setting-specific conditions and circumstances. Many trials underscore a variability in drug efficacy and response to ketamine in therapeutic approaches to SUD but show remarkable heterogeneity themselves. Depending on the setting and skillful use, ketamine is effective for treatment of SUD when administered as IV infusion at subanesthetic dose alone or as adjuvant to other medications. In rapid opioid detoxication, anesthetic dose ketamine may be useful. Future studies need to assess the effects of repeated or longer-term ketamine administration at different application routes, the value of combination therapies such as psychotherapeutic approaches and ketamine, and ketamine's efficacy compared to active comparators such as benzodiazepines. Overall, there is a great necessity for further research to establish more substantial pharmacologic treatment options for SUD.

7 Conclusions

The emerging research on ketamine's potential in treating alcohol, opioid, and cocaine use disorders presents a promising avenue for addressing these challenging conditions. Ketamine's unique mechanisms of action, particularly its ability to modulate glutamate pathways and enhance neuroplasticity, offers a novel approach in a field where effective treatments are urgently needed. While preliminary findings suggest therapeutic efficacy including rapid reduction of cravings and withdrawal symptoms, future well-controlled clinical trials need to demonstrate long-term ketamine effects, its safety, and potential for integration into comprehensive treatment strategies. Nonetheless, current evidence is reassuring that ketamine will become a valuable addition to SUD treatment.

Acknowledgments

This work was supported by Ministry of Education, Science and Technological Development of Serbia (project N° 451-03-66/2024-03/200110).

Conflicts of Interest

The authors declare no conflict of interest.

References

1. Olsen Y (2022) What Is addiction? History, terminology, and core concepts. Med Clin North Am 106(1):1–12
2. Testino G, Balbinot P, Sampietro L, Bottaro LC (2019) Patologiedelledipendenze: un nuovo management? [Addiction disorders: a new management]. Recenti Prog Med 110(5):230–235
3. Zou Z, Wang H, d'OleireUquillas F, Wang X, Ding J, Chen H (2017) Definition of substance and non-substance addiction. Adv Exp Med Biol 1010:21–41. https://doi.org/10.1007/978-981-10-5562-1_2
4. Thomasius R, Paschke K, Arnaud N (2022) Substance-use disorders in children and adolescents. Dtsch Arztebl Int 119(25):440–450. https://doi.org/10.3238/arztebl.m2022.0122
5. Hoch E, Bonnet U, Thomasius R, Ganzer F, Havemann-Reinecke U, Preuss UW (2015) Risks associated with the non-medicinal use of cannabis. Dtsch Arztebl Int 112:271–278
6. Bachtell RK, Jones JD, Heinzerling KG, Beardsley PM, Comer SD (2017) Glial and neuroinflammatory targets for treating substance use disorders. Drug Alcohol Depend 180:156–170. https://doi.org/10.1016/j.drugalcdep.2017.08.003
7. Winhusen T, Theobald J, Kaelber DC, Lewis D (2020) Increased morbidity and mortality in hypertensive patients with substance use disorders: electronic health record findings. J Stud Alcohol Drugs 81(4):471–478. https://doi.org/10.15288/jsad.2020.81.471
8. Koob GF, Volkow ND (2010) Neurocircuitry of addiction. Neuropsychopharmacology 35(1):217–238. https://doi.org/10.1038/npp.2009.110
9. Kramer J, Dick DM, King A, Ray LA, Sher KJ, Vena A, Vendruscolo LF, Acion L (2020) Mechanisms of alcohol addiction: bridging human and animal studies. Alcohol Alcohol 55(6):603–607. https://doi.org/10.1093/alcalc/agaa068
10. Nikayin S, Murphy E, Krystal JH, Wilkinson ST (2022) Long-term safety of ketamine and esketamine in treatment of depression. Expert Opin Drug Saf 21(6):777–787. https://doi.org/10.1080/14740338.2022.2066651
11. Savić Vujović K, Jotić A, Medić B, Srebro D, Vujović A, Žujović J, Opanković A, Vučković S (2023) Ketamine, an old-new drug: uses and abuses. Pharmaceuticals (Basel) 17(1):16. https://doi.org/10.3390/ph17010016
12. Schwenk ES, Pradhan B, Nalamasu R, Stolle L, Wainer IW, Cirullo M, Olson A, Pergolizzi JV, Torjman MC, Viscusi ER (2021) Ketamine in the past, present, and future: mechanisms, metabolites, and toxicity. Curr Pain Headache Rep 25:57. https://doi.org/10.1007/s11916-021-00977-w
13. Molero P, Ramos-Quiroga JA, Martin-Santos-R, Calvo-Sánchez E, Gutiérrez-Rojas L, Meana JJ (2018) Antidepressant efficacy and tolerability of ketamine and esketamine: a critical review. CNS Drugs 32:411–420. https://doi.org/10.1007/s40263-018-0519-3
14. Zanos P, Thompson SM, Duman RS, Zarate CA, Gould TD (2018) Convergent mechanisms underlying rapid antidepressant action. CNS Drugs 32:197–227
15. Vujović KS, Vučković S, Stojanović R, Divac N, Medić B, Vujović A, Srebro D, Prostran M (2021) Interactions between ketamine and magnesium for the treatment of pain: current state of the art. CNS Neurol Disord Drug Targets 20(5):392–400. https://doi.org/10.2174/1871527320666210121144216
16. Yang Y, Cui Y, Sang K, Dong Y, Ni Z, Ma S, Hu H (2018) Ketamine blocks bursting in the lateral habenula to rapidly relieve depression. Nature 554:317–322

17. Subramanian S, Haroutounian S, Palanca BJA, Lenze EJ (2022) Ketamine as a therapeutic agent for depression and pain: Mechanisms and evidence. J Neurol Sci 434:120152. https://doi.org/10.1016/j.jns.2022.120152
18. Savić Vujović K, Vučković S, Vasović D, Medić B, Stojanović R, Divac N, Srebro D, Prostran M (2019) Involvement of serotonergic and opioidergic systems in the antinociceptive effect of ketamine-magnesium sulphate combination in formalin test in rats. Pharmacol Rep 71(6):1014–1019. https://doi.org/10.1016/j.pharep.2019.05.020
19. Ezquerra-Romano I, Lawn W, Krupitsky E, Morgan CJA (2018) Ketamine for the treatment of addiction: evidence and potential mechanisms. Neuropharmacology 142:72–82. https://doi.org/10.1016/j.neuropharm.2018.01.017
20. Schumann G, Saam C, Heinz A, Mann K, Treutlein J (2005) Identifikation von Risikogenen für Alkoholabhängigkeit. Das NMDA-Rezeptor-System [The NMDA receptor system: genetic risk factor for alcoholism]. Nervenarzt 76:1355–1362. https://doi.org/10.1007/s00115-005-1917-6
21. Pastor R, Aragon CMG (2005) The role of opioid receptor subtypes in the development of behavioral sensitization to ethanol. Neuropsychopharmacology 31:1489–1499. https://doi.org/10.1038/sj.npp.1300928
22. Elvig SK, McGinn MA, Smith C, Arends MA, Koob GF, Vendruscolo LF (2021) Tolerance to alcohol: a critical yet understudied factor in alcohol addiction. Pharmacol Biochem Behav 204:173155. https://doi.org/10.1016/j.pbb.2021.173155
23. DeBellis R, Smith BS, Choi S, Malloy M (2005) Management of delirium tremens. J Intensive Care Med 20:164–173. https://doi.org/10.1177/0885066605275353
24. Abrahao KP, Salinas AG, Lovinger DM (2017) Alcohol and the brain: neuronal molecular targets, synapses, and circuits. Neuron 96:1223–1238
25. McPheeters M, O'Connor EA, Riley S, Kennedy SM, Voisin C, Kuznacic K, Coffey CP, Edlund MD, Bobashev G, Jonas DE (2023) Pharmacotherapy for alcohol use disorder: a systematic review and meta-analysis. JAMA 330(17):1653–1665. https://doi.org/10.1001/jama.2023.19761
26. Worrell SD, Gould TJ (2021) Therapeutic potential of ketamine for alcohol use disorder. NeurosciBiobehav Rev 126:573–589. https://doi.org/10.1016/j.neubiorev.2021.05.006
27. Hilal F, Jeanblanc J, Naassila M (2023) Intérêt et mécanismesd'action de la kétamine dans le traitement de l'addiction à l'alcool—Revue des études cliniques et précliniques [Interest and mechanisms of action of ketamine in alcohol addiction - a review of clinical and preclinical studies]. Biol Aujourdhui 217(3–4):161–182. https://doi.org/10.1051/jbio/2023028
28. Burnette EM, Nieto SJ, Grodin EN, Meredith LR, Hurley B, Miotto K, Gillis AJ, Ray LA (2022) Novel agents for the pharmacological treatment of alcohol use disorder. Drugs 82(3): 251–274. https://doi.org/10.1007/s40265-021-01670-3
29. Zindel LR, Kranzler HR (2014) Pharmacotherapy of alcohol use disorders: seventy-five years of progress. J Stud Alcohol Drugs Suppl 75:79–88. https://doi.org/10.15288/jsads.2014.75.79
30. Choi HY, Seo JS, Lee SK (2020) Assessment and treatment of alcohol withdrawal syndrome. Korean J Gastroenterol 76(2):71–77. https://doi.org/10.4166/kjg.2020.76.2.71
31. Shah P, McDowell M, Ebisu R, Hanif T, Toerne T (2018) Adjunctive use of ketamine for benzodiazepine-resistant severe alcohol withdrawal: a retrospective evaluation. J Med Toxicol 14:229–236. https://doi.org/10.1007/S13181-018-0662-8
32. Farrokh S, Roels C, Owusu KA, Nelson SE, Cook AM (2021) Alcohol withdrawal syndrome in neurocritical care unit: assessment and treatment challenges. Neurocrit Care 34(2):593–607. https://doi.org/10.1007/s12028-020-01061-8
33. McIntyre RS, Rosenblat JD, Nemeroff CB, Sanacora G, Murrough JW, Berk M, Brietzke E, Dodd S, Gorwood P, Ho R, Iosifescu DV, Lopez Jaramillo C, Kasper S, Kratiuk K, Lee JG, Lee Y, Lui LMW, Mansur RB, Papakostas GI, Subramaniapillai M, Thase M, Vieta E, Young AH, Zarate CA Jr, Stahl S (2021) Synthesizing the evidence for ketamine and esketamine in treatment-resistant depression: an international expert opinion on the available evidence and implementation. Am J Psychiatry 178(5):383–399. https://doi.org/10.1176/appi.ajp.2020.20081251
34. Le Daré B, Pelletier R, Morel I, Gicquel T (2022) Histoire de la ketamine; une molecule ancienne qui a toujours la cote [History of ketamine: An ancient molecule that is still popular today]. Ann Pharm Fr 80(1):1–8. https://doi.org/10.1016/j.pharma.2021.04.005
35. Goldfine CE, Tom JJ, Im DD, Yudkoff B, Anand A, Taylor JJ, Chai PR, Suzuki J (2023) The therapeutic use and efficacy of ketamine in alcohol use disorder and alcohol withdrawal

syndrome: a scoping review. Front Psych 14: 1141836. https://doi.org/10.3389/fpsyt.2023.1141836

36. Ungur AL, Neumann T, Borchers F, Spies C (2020) Perioperative management of alcohol withdrawal syndrome. Visc Med 36(3): 160–166. https://doi.org/10.1159/000507595
37. Kelson M, Burnett JM, Matthews A, Juneja T (2023) Ketamine treatment for alcohol use disorder: a systematic review. Cureus 15(5): e38498. https://doi.org/10.7759/cureus.38498
38. Terasaki D, Loh R, Cornell A, Taub J, Thurstone C (2022) Single-dose intravenous ketamine or intramuscular naltrexone for high-utilization inpatients with alcohol use disorder: pilot trial feasibility and readmission rates. Addict Sci Clin Pract 17(1):64
39. Tran K, MacDougall D (2024) Ketamine for adults with substance use disorders: Rapid Review [Internet]. Canadian Agency for Drugs and Technologies in Health, Ottawa. Report No.: RC1523. PMID: 38588371 Accessed 27 June 2024
40. Walsh Z, Mollaahmetoglu OM, Rootman J et al (2021) Ketamine for the treatment of mental health and substance use disorders: comprehensive systematic review. BJPsych Open 8(1):e19
41. Gent EM, Bryan JW, Cleary MA et al (2024) Esketamine combined with a mindfulness-based intervention for individuals with alcohol problems. J Psychopharmacol 38:541–550
42. Wong A, Benedict NJ, Armahizer MJ, Kane-Gill SL (2015) Evaluation of adjunctive ketamine to benzodiazepines for management of alcohol withdrawal syndrome. Ann Pharmacother 49:14–19. https://doi.org/10.1177/1060028014555859
43. Dakwar E, Levin F, Hart CL, Basaraba C, Choi J, Pavlicova M, Nunes EV (2020) A single ketamine infusion combined with motivational enhancement therapy for alcohol use disorder: a randomized midazolam controlled pilot trial. Am J Psychiatry 177:125–133. https://doi.org/10.1176/appi.ajp.2019.19070684
44. Krupitsky EM, Grineko AY, Berkaliev TN, Paley AI, Tetrov UN, Mushkov KA, Borodikin YS (1992) The combination of psychedelic and aversive approaches in alcoholism treatment: the affective contra-attribution method. Alcohol Treat Q 9:99–105. https://doi.org/10.1300/J020V09N01_09
45. Krupitsky EM, Grinenko AY (1997) Ketamine psychedelic therapy (KPT): a review of the results of ten years of research. J Psychoactive Drugs 29:165–183. https://doi.org/10.1080/02791072.1997.10400185
46. Pizon AF, Lynch MJ, Benedict NJ et al (2018) Adjunct ketamine use in the management of severe ethanol withdrawal. Crit Care Med 46: e768–e771. https://doi.org/10.1097/CCM.0000000000003204
47. Grabski M, McAndrew A, Lawn W, Marsh B, Raymen L, Stevens T, Hardy L, Warren F, Bloomfield M, Borissova A, Maschauer E, Broomby R, Price R, Coathup R, Gilhooly D, Palmer E, Gordon-Williams R, Hill R, Harris J, Mollaahmetoglu OM, Curran HV, Brandner B, Lingford-Hughes A, Morgan CJA (2022) Adjunctive ketamine with relapse prevention-based psychological therapy in the treatment of alcohol use disorder. Am J Psychiatry 179(2):152–162. https://doi.org/10.1176/appi.ajp.2021.21030277
48. Rompala G, Nagamatsu ST, Martínez-Magaña JJ, Nuñez-Ríos DL, Wang J, Girgenti MJ, Krystal JH, Gelernter J (2023) Traumatic Stress Brain Research Group; Hurd YL, Montalvo-Ortiz JL. Profiling neuronal methylome and hydroxymethylome of opioid use disorder in the human orbitofrontal cortex. Nat Commun 14(1):4544. https://doi.org/10.1038/s41467-023-40285-y
49. Edinoff AN, Wu NW, Nix CA, Bonin B, Mouhaffel R, Vining S, Gibson W, Cornett EM, Murnane KS, Kaye AM, Kaye AD (2022) Historical pathways for opioid addiction, withdrawal with traditional and alternative treatment options with ketamine, cannabinoids, and noribogaine: a narrative review. Health Psychol Res 10(4):38672. https://doi.org/10.52965/001c.38672
50. Hoffman KA, Ponce Terashima J, McCarty D (2019) Opioid use disorder and treatment: challenges and opportunities. BMC Health Serv Res 19(1):884. https://doi.org/10.1186/s12913-019-4751-4
51. Olausson P, Jentsch JD, Krueger DD, Tronson NC, Nairn AC, Taylor JR (2007) Orbitofrontal cortex and cognitive-motivational impairments in psychostimulant addiction: evidence from experiments in the non-human primate. Ann N Y Acad Sci 1121:610–638. https://doi.org/10.1196/annals.1401.016
52. Browne CJ, Godino A, Salery M, Nestler EJ (2020) Epigenetic mechanisms of opioid addiction. Biol Psychiatry 87:22–33. https://doi.org/10.1016/j.biopsych.2019.06.027
53. Corre J, van Zessen R, Loureiro M, Patriarchi T, Tian L, Pascoli V, Lüscher C (2018) Dopamine neurons projecting to medial shell of the nucleus accumbens drive

heroin reinforcement. Elife 7:e39945. https://doi.org/10.7554/eLife.39945
54. Ting-A-Kee R, van der Kooy D (2012) The neurobiology of opiate motivation. Cold Spring Harb Perspect Med 2(10):a012096. https://doi.org/10.1101/cshperspect.a012096
55. Nutt DJ, Lingford-Hughes A, Erritzoe D, Stokes PR (2015) The dopamine theory of addiction: 40 years of highs and lows. Nat Rev Neurosci 16:305–312
56. Bruneau J, Ahamad K, Goyer MÈ, Poulin G, Selby P, Fischer B, Wild TC, Wood E (2018) CIHR Canadian research initiative in substance misuse. management of opioid use disorders: a national clinical practice guideline. CMAJ 190(9):E247–E257. https://doi.org/10.1503/cmaj.170958
57. Jovaisa T, Laurinenas G, Vosylius S, Sipylaite J, Badaras R, Ivaskevicius J (2006) Effects of ketamine on precipitated opiate withdrawal. Medicina (Kaunas) 42(8):625–634
58. Krupitsky EM, Burakov A, Dunaevsky IV et al (2007) Single versus repeated sessions of ketamine-assisted psychotherapy for people with heroin dependence. J Psychoactive Drugs 39:13–19
59. Hailozian C, Luftig J, Liang A, Outhay M, Ullal M, Anderson ES, Kalmin M, Shoptaw S, Greenwald MK, Herring AA (2022) Synergistic effect of ketamine and buprenorphine observed in the treatment of buprenorphine precipitated opioid withdrawal in a patient with fentanyl use. J Addict Med 16(4):483–487. https://doi.org/10.1097/ADM.0000000000000929
60. Christian NJ, Butner JL, Evarts MS, Weimer MB (2023) Precipitated Opioid Withdrawal Treated With Ketamine in a Hospitalized Patient: A Case Report. J Addict Med 17(4):488–490. https://doi.org/10.1097/ADM.0000000000001151
61. Freye E, Latasch L, Levy JV (2006) S(+)-ketamine attenuates increase in electroencephalograph activity and amplitude height of sensory-evoked potentials during rapid opioid detoxification. Anesth Analg 102(5):1439–1444. https://doi.org/10.1213/01.ane.0000202382.82847.64
62. Kampman KM (2019) The treatment of cocaine use disorder. Sci Adv 5(10):eaax1532. https://doi.org/10.1126/sciadv.aax1532
63. Baik J-H (2013) Dopamine signaling in reward-related behaviors. Front Neural Circuits 7:152
64. Rothman RB, Baumann MH, Dersch CM, Romero DV, Rice KC, Carroll FI et al (2001) Amphetamine-type central nervous system stimulants release norepinephrine more potently than they release dopamine and serotonin. Synapse 39:32–41
65. Shorter D, Domingo CB, Kosten TR (2015) Emerging drugs for the treatment of cocaine use disorder: a review of neurobiological targets and pharmacotherapy. Expert Opin Emerg Drugs 20:15–29
66. Chan B, Kondo K, Freeman M, Ayers C, Montgomery J, Kansagara D (2019) Pharmacotherapy for cocaine use disorder – a systematic review and meta-analysis. J Gen Intern Med 34(12):2858–2873. https://doi.org/10.1007/s11606-019-05074-8
67. Brandt L, Chao T, Comer SD, Levin FR (2021) Pharmacotherapeutic strategies for treating cocaine use disorder-what do we have to offer? Addiction 116(4):694–710. https://doi.org/10.1111/add.15242
68. Dakwar E, Levin F, Foltin RW, Nunes EV, Hart CL (2014) The effects of subanesthetic ketamine infusions on motivation to quit and cue-induced craving in cocaine-dependent research volunteers. Biol Psychiatry 76(1):40–46. https://doi.org/10.1016/j.biopsych.2013.08.009
69. Dakwar E, Nunes EV, Hart CL, Foltin RW, Mathew SJ, Carpenter KM, Choi CJJ, Basaraba CN, Pavlicova M, Levin FR (2019) A single ketamine infusion combined with mindfulness-based behavioral modification to treat cocaine dependence: a randomized clinical trial. Am J Psychiatry 176(11):923–930. https://doi.org/10.1176/appi.ajp.2019.18101123
70. Dakwar E, Hart CL, Levin FR, Nunes EV, Foltin RW (2017) Cocaine self-administration disrupted by the N-methyl-D-aspartate receptor antagonist ketamine: a randomized, crossover trial. Mol Psychiatry 22(1):76–81. https://doi.org/10.1038/mp.2016.39
71. Gao Z, Winhusen TJ, Gorenflo M, Ghitza UE, Davis PB, Kaelber DC, Xu R (2023) Repurposing ketamine to treat cocaine use disorder: integration of artificial intelligence-based prediction, expert evaluation, clinical corroboration and mechanism of action analyses. Addiction 118(7):1307–1319. https://doi.org/10.1111/add.16168

Part IV

Special Medical Situations

Chapter 20

Anesthesia in Low-Income and Lower-Middle-Income Countries and the Important Role of Ketamine for Care Provision

Chimaobi Tim Nnaji

Abstract

Anesthesia care provision remains slender and deficient in most low-income and lower middle-income countries despite the efforts of the World Federation of Societies of Anesthesiologists (WFSA) and the World Health Organization (WHO) and their member states to make safe emergency and essential anesthesia and surgical care a priority and an integral component of Universal Health Coverage and Health Care Systems strengthening. These challenges are extremely demanding in sub-Saharan Africa. Here, the situation is very poor because the growing population and their need for surgery and anesthesia outweighs the insufficient number of anesthesia providers, poor surgical theater infrastructure, and limitations in anesthesia equipment, materials, life-saving drugs, anesthetic agents, and healthcare resources by far. To somehow cope with conditions and external circumstances, anesthesia providers have adapted their practice to the environmental challenges by using ketamine to provide safe anesthesia, pain, and critical care. Although healthcare materials, equipment, manpower, and educational resources are very limited, this has revolutionized the use of ketamine in such regions, harnessing its clinical effects with the goal of improving patients' outcomes, experiences, and healthcare systems effectiveness.

Key words Ketamine, Low- and lower-middle-income countries, Anesthesia care, Healthcare provision, Patient safety, WHO, WFSA, Sub-Saharan Africa

1 Introduction

Anesthesia care provision in the majority of low-income and lower-middle-income countries (LMICs) is quite challenging and demanding. The job description extends beyond perioperative anesthesia care, encompassing cardiopulmonary resuscitation, critical care medicine, transportation of critically ill patients, acute and chronic pain management, sedation services, as well as healthcare system management, advocacy, medical simulation, education, and research [1]. In addition, anesthetists face challenges associated with resource-limited environment like poor healthcare funding;

Sabine Himmelseher (ed.), *Ketamine: Pain Management, Critical Care, Depression, and Resource-Poor Settings Volume II*, Neuromethods, vol. 225, https://doi.org/10.1007/978-1-0716-4599-4_20,

poor state of surgical theater; unavailability of anesthesia and critical care machines; lack of equipment, materials, lifesaving, essential, and anesthesia drugs, and medical gases; poor blood bank and laboratory services; and shortage of healthcare workers [2, 3].

The World Bank defined LMICs as countries with developing healthcare systems (HCS), unstable government, and developing economy [4]. These states are limited in their resources and are often characterized by HCS with ineffectiveness in structure and components, poor healthcare provision, overall shortage, and poorly managed healthcare workforce. Poor governance and funding transcend to anesthesia care provision, which often lacks prioritization by governments. Most HCS in LMICs therefore offer only basic critical care resources like oxygen in life-threatening illnesses and healthcare providers who may not be adequately trained to handle emergencies [5]. Furthermore, anesthesia care provision in some regions of LMICs has been affected by regional armed conflicts leading to many insecurities. This makes the supplies of anesthesia drugs and materials even more difficult, as well as a mobilization of healthcare workers to come to affected regions. In sub-Saharan Africa, HCS are further affected by their growing population and the need for critical, surgical, and anesthesia treatment in numbers that outweigh the insufficient number of physician anesthetists and anesthesia resources by far [2, 6, 7]. Many hospitals in LMICs differ significantly in their access to quality and safe anesthesia care provision, with most of the services being performed by untrained anesthesia providers. Without surprise, this results in avoidable anesthesia mishaps and an anesthesia-related increase in morbidity and mortality [8–10]. Anesthesia providers therefore try to adapt to their environmental hardship by harnessing benefits of clinical effects of ketamine to provide safe anesthesia, pain, and critical care despite all limitations. This involves training of healthcare staff and even providers without direct medical background on how to administer ketamine and manage patients who receive ketamine anesthesia. All these endeavors must be viewed as a revolutionizing use of ketamine with the goal of improving patients' outcomes, perioperative experiences, and HCS effectiveness [2, 7].

In a recent review, the author evaluated the important role of ketamine in enhancing anesthesia care and found a special value with respect to its utility in surgical settings in LMICs [7]. The review highlighted the clinical role of ketamine as safe and mono-anesthetic agent to conduct general anesthesia and sedation for various surgical categories, both in children and adults. The author found that the utilization rate of ketamine ranged from 45% to 100% in surgical procedures in LMICs. Administration of ketamine in anesthesia care provides beneficial effects like analgesia, amnesia, immobility, and loss of consciousness with a wide safety margin because of lack of cardiorespiratory depression. Ketamine is cheap, readily available, and safe. Its administration does not rely heavily on access to costly equipment, monitoring, or intensely trained and skilled physician anesthetists. When it is used alongside with drugs like atropine and diazepam that are also readily available and cheap in this environment, their combination effects help to decrease some of ketamine-related, unwanted actions. These include increased salivation and negative psychoactive effects such as dissociation or hallucinations. It was also reported that ketamine facilitated enhanced recovery after surgery patient management approaches, even in the hands of nonphysician anesthesia providers in LMICs. These benefits were observed in settings without recovery room, machine-equipped healthcare facilities, or physician anesthetists, where patients need to wake up on the ward and in their beds [7].

2 The State of Healthcare Provisions in Low & Lower Middle-Income Countries

Healthcare services should be patient centered, focused on HCS functionality and effectiveness, patients' safety, and good individual experiences. Services should be delivered in a timely, equitable, and efficient manner [11]. The effectiveness of a HCS will also be reflected in healthcare staff working conditions, their level of skill and competence, and how well referral and surveillance systems are integrated and coordinated. Although HCS can be pervasive, complex, and adaptive at individual and organizational levels, many of them are marred with inequitable and ineffective services and poor patient outcomes in LMICs. Causes for this dismal situation consist of unsafe, untimely, and disintegrated services [12–15], which originate from insecurity of lives and properties, chronic underfunding, and lack of all already named necessities. In addition, poverty, malnutrition, use of unclean water, unsafe environment, and no access to basic and advanced healthcare continue to proffer disease transmissions. It has been reported that poor-quality healthcare accounts for up to 15% of overall mortality and imposes costs of $1.4–1.6 trillion each year regarding lost productivity in LMICs [11, 13].

The gap between quality of healthcare and the ability to access healthcare also continues to widen in LMICs, reflecting the inequity between the rich and poor and urban and rural populations. It is estimated that in Nigeria, six out of ten patients lack access to basic healthcare, and only 43% of the general population can access quality healthcare [14]. In Mozambique, patients seeking surgical care travel more than 25 kilometers to gain access to a hospital, and more than 50% of the population live >30 kilometers away from the nearest doctor [15]. In 2018, the National Academies of Sciences, Engineering, and Medicine reported that across the communicable conditions studied, approximately 8.4 million people who had a diagnosis each year were not treated, resulting in 460,000 quality-related deaths (QRD) and 2.5 million years of life lived with disability (YLD) in LMICs annually. Also, 35% of under 5-year-old children did not receive adequate oral rehydration therapy when taken to a medical facility for diarrhea treatment. This resulted in 140,000 QRD annually. Five percent of people diagnosed with tuberculosis were lost to follow-up, resulting in 130,000 QRD and 90,000 YLD annually. Overall, ineffective treatment of communicable diseases led to 800,000 deaths and 2.6 million YLD per year [13]. In 2023, the WHO demonstrated that despite global declining maternal mortality rates (MMR) from 339 to 223 deaths per 100,000 live births with a 2% annual decrease from 2000 to 2020 per year, the LMICs in Sub-Saharan Africa continued to account for 70% of global MMR. Another WHO study showed that the neonatal mortality rate of 27 deaths per 1000 live births is caused by HCS's deficiencies in LMICs [16–18]. The goal of a functional HCS should be to use its resources for achieving long-term and sustainable improvements that are characterized by equity, accessibility, affordability, quality, effectiveness, and efficiency. This requires developing a healthcare policy that can direct and support such endeavors and includes health promotion activities as well as a provision of financial risk protection to patients, particularly among the poor and vulnerable groups like women and children [16, 19]. At present, most of the healthcare expenditures in LMICs is dominated by out-of-pocket (OOP) payments and private voluntary funding. A study on a country in LMICs showed that healthcare expenditure was controlled by OOP pay in 70–90% of patients. It dragged 1,268,800 individuals of the population into poverty and financial hardship. In the same country, the national health insurance scheme which has been established for a decade has barely enrolled <10% of the country population. Among those, 90% were formal sector and government employees, whereas 33% of unemployed and 81% of self-employed people were not adequately covered putting them at high health and financial risks [20–23]. Of course, inadequate health insurance and lack of financial risk protection affects access and use of surgery and anesthesia with the result of poor outcomes from morbidity and injury. Adamu and co-workers noted that preoperative preparation and emergency surgery of patients in LMICs was associated with a high level of delays because of OOP conditions for healthcare service. This resulted in increased perioperative morbidity and mortality [24].

3 Surgical Care Provisions in Low & Lower Middle-Income Countries

To prevent disability or death, some disease conditions require nothing else but surgery as treatment. This scenario has been estimated to occur with an ~11% prevalence, but the global burden of disease because of no surgery has also been associated with a 33% of mortality and 28% of disability-adjusted life years in a multinational survey [25]. Meara and colleagues estimated that 313 million surgeries are performed annually, but LMICs contribute only with 6% to this volume although they make up 35% of the world population. [26]. The Lancet Commission on Global Surgery estimated that 5 billion people in the world do not have access to safe and affordable surgery and anesthesia, and where it is available, it is inequitably distributed. For example, nine of ten people in LMICs are unable to access surgery and anesthesia [27].

Several authors have described why this access is so inequitably distributed [24–30]: in the Democratic Republic of Congo, cost of surgery is not affordable for 73% of the population who rely on OOP for healthcare costs, and 8% of healthcare facilities lack doctors with surgical skills (27). In Afghanistan, safe surgery is inaccessible to most of the population, especially in remote areas. This is caused by costs, poor healthcare coverage, and lack of transportation systems. Premkumar and colleagues reported that 90% of the population in Northern Tanzania do not have access to orthopedic surgery. In Nigeria, preoperative preparation and emergency surgeries are delayed, as patients' caregivers sort for funds to pay for laboratory and blood bank services, surgical consumables, and cost of surgeries. Of course, this results in increased perioperative morbidity and mortality [24–30].

Without surprise, there is still a great disparity regarding outcomes from surgery between high-income countries (HICs) and LMICs, with morbidity and mortality being higher following surgery in LMICs. According to Ariyaratnam and colleagues, the postoperative mortality rate is 9 times higher in LMICs with an incidence of 3.44 per 100 admissions compared to 0.38 per 100 admissions in HICs [31]. A Nigerian surgical outcome study reported that patients have 18.5% odds of developing complications and 6% of death due to surgery, while an African surgical outcome study showed that the risk of mortality following surgery in patients across Africa was twice as high as the global average [32, 33]. LMICs are also marred with poor maternal and neonatal outcomes within caesarean section settings. According to Bishop and co-workers, pregnant women have a 17% odds of developing complications and a 5.43 death risk per 1000 births from caesarean section with a neonatal mortality rate as high as 4.4%. Twenty-five percent of the mothers died because of obstetric hemorrhages, 10% due to anesthesia complications, and 5% after surgery-related infections. This indicates that maternal mortality after caesarean delivery in Africa is 50 times higher than that of HICs being driven by peripartum hemorrhage and anesthesia complications [34]. When surgery is performed in LMICs, several factors have been identified that contribute to poor surgical care provision, with solutions requiring multidimensional approaches as shown in Table 1.

Table 1
The state of surgical care provision in low & lower middle-income countries

Author	Country	Common causes of poor outcomes
Faierman et al. [15]	Mozambique	Surgical patients need to travel >25 kilometers to access surgical care, and > 50% cannot access a doctor due to distance and transportation costs
Adamu et al. [24]	Nigeria	Preoperative preparations and emergency surgeries are delayed because caregivers must search for payment of laboratory, surgical consumables, and surgery costs, which increases perioperative morbidity and mortality
Lin et al. [28]	Democratic Republic of Congo	73% of the population depend on OOP healthcare expenditure, and 8.2% of hospitals do not have a doctor
Trani et al. [29]	Afghanistan	Transportation cost and inequitable distribution of HCS make safe surgical care inaccessible and unaffordable in poor and remote areas
Premkumar et al. [30]	Tanzania	90% of the population in northern Tanzania does not have access to orthopedic surgical care
Endeshaw et al. [35]	Ethiopia	Poor preoperative preparation of surgical patients results in a 3.69% mortality
Mughal et al. [36]	LMICs	Typically, there is a lack of resources, financial and job constraints, shortage of manpower, electricity, team commitment, and high clinical workload burden
Chao et al. [37]	Ethiopia	Shortage of physicians: One shall take care of thousands of patients, ranging from 1:4715 to 1:107,602 people
Linden et al. [38]	Uganda	Lack of human resources and infrastructure to provide adequate, safe surgery. There were 0.2 major operating theater facilities per 100,000 people
Bari et al. [39]	Cambodia	Shortage of financial and human resources. Physician specialist workforce is 0.18 per 10,000 population
Khan & Khan [10]	LMICs	Poor anesthesia services contributed to 11% of direct mortalities and 23% of indirect mortalities, of which 28.5% were considered avoidable
Afsana et al. [40]	Bangladesh	Costs, fear of hospitals, and stigma of "abnormal" birth are major constraints in accepting caesarean section
Bronsard et al. [41]	Tanzania	Gender preference and local health beliefs about cataracts and surgery result in late presentation for cataract surgery
Weiser et al. [42]	LMICs	LMICs make up only 3.5% of the estimated 234 million major surgeries undertaken worldwide
Briesen et al. [43]	Kenya	Patients refuse cataract surgeries due to rumor of blinding after surgery, shame, and fear of missing social support in the community

Note: HCS healthcare system, LMICS, low- and lower-middle-income countries, *OOP* out-of-pocket

Nevertheless, studies found out modifiable factors, such as anesthesia-related morbidity and mortality, availability of laboratory and blood bank services, and provision of critical care personnel and intensive care units [2, 44–46].

In 2014, the WHO called on its member states to ensure that anesthesia and surgery are properly prioritized within the overall context of Universal Health Coverage (UHC) and HCS, as well as to alert that processes of support, follow-up, reporting, and benchmarking are indispensable [46]. In 2015, the Lancet Commission on Global Surgery (LCGS) and the WHO Assembly decided to strengthen essential surgery and anesthesia provision issuing a call to action. They provided an urgently needed platform and framework for key stakeholders [27, 46–48]. The outlined goals for all populations were as follows:

- Access to timely, essential surgery with a minimum of 80% coverage of essential surgical and anesthesia services per country by 2030.
- Increase in specialist surgical workforce density in 100% of countries so that at least 20 surgical, anesthesia, and obstetric physicians per 100,000 population by 2030 are available.
- Increase in surgical volume of 80% by 2020 and 100% by 2030, tracking surgical volume with a minimum of 5000 procedures per 100,000 population by 2030.
- Decrease in perioperative mortality rate of 80% by 2020 and 100% by 2030, tracking perioperative mortality rate; in 2020, evaluation of global data and national target setting for 2030.
- Protection against impoverishing expenditure by 100% from OOP pay for surgical and anesthesia care by 2030.
- Protection against catastrophic expenditure by 100% from OOP pay for surgical and anesthesia care by 2030.

The WFSA through its member national societies and international network supports the LCGS initiatives even in remote regions of the world [48]. The large burden of diseases that require essential surgical and trauma care has thus been brought to the fore within the global health community. The WHO also created the Global Initiative for Essential and Emergency Surgical Care (GIEESC) to address and alleviate the burden of lack of surgery further. This initiative provides an outline for assessing, planning, and organizing surgical and anesthesia care in LMICs and evaluates surgical and anesthesia care capacities [49].

4 Anesthesia Care Provisions in Low & Lower Middle-Income Countries

Anesthesia is valuable in providing a pain- and anxiety-free (surgical) procedure. It offers comfort to the patient even in painful screening and diagnostic tests, tissue sample removal, treatment of certain psychiatric illnesses, and dental work. In most LMICs, anesthesia services also involve resuscitation and critical care, all forms of pain treatment and sedation services, as well as healthcare systems management, advocacy, medical simulation, education, and research [1, 50]. All these tasks must be performed in a timely way and with good quality to prevent and reduce morbidity and/or mortality.

Unfortunately, anesthesia care provision and its impact on safe surgery is still often neglected, rather unpopular, and not prioritized by governments of LMICs. This results in poor funding and eventually poor patients' outcomes [2]. To improve anesthesia care, especially in acute and critical care scenarios, it is necessary to identify the basic problems that underly ineffective services in such settings: regrettably, yet, only a few data originate from LMICS about these problems, and often studies are generated by researchers living and working in HICs. Some of them may have made brief experiences in LMICs as a "missionary anesthesia physician," but some may have no practical knowledge or exposure to conditions of "what it's like to live and work in LMICs" as an anesthetist [2, 7].

In a previous paper, the author described some of the crises and innovations in providing anesthesia care in LMICs that led to improve patient safety and experiences [7]. The basic problems that affect anesthesia care were classified by Bharati and colleagues, including factors related to the population, human and technical resources, educational system, socioeconomic, and other utilities problems. This seems to be especially valid in rural, underprivileged areas of LMICs, where HCS and workforce are usually overburdened by numbers of patient load, infectious diseases, trauma-related injuries, and noncommunicable disorders [8, 51, 52]. Other authors identified problems that affected even the type of anesthesia provided to patients, such as lack of skills, level of training of the provider, shortage of anesthesia physicians, bad payment and welfare packages for healthcare workers, poor theater infrastructure, unavailability of anesthesia equipment, and lack of lifesaving, essential, and anesthetic medication [2, 37, 44, 53]. Problems associated with resource-limited working environments such as unreliable electricity and unavailability of compressed medical gases and laboratory and blood bank services [2, 37] added to the difficult situations. This means that—in this present-day clime for patient safety—people in LMICs continue to suffer. In rural and underprivileged areas of LMICs, the access to safe anesthesia stays very difficult and is often unchanged.

Many LMICs try to meet the high demand for anesthesia care and anesthesia providers by allowing non-anesthesia physicians, nurse anesthetists, anesthesia technicians, and other healthcare personnel with highly variable on-the-job training to provide anesthesia with minimal equipment, drugs, and essential anesthesia resources. These anesthesia providers work alone without any supervision from specialist physician anesthetists in most regions, and they handle cases much beyond their qualifications or training [2, 37, 54]. This makes the safety of anesthesia care in LMICs unaltered questionable, and patients' outcomes have not improved in comparison with HICs. Walker and Wilson reported that anesthesia in a central hospital in Malawi accounts for 1 death per 504 surgical procedures [55]. In a Nigerian teaching hospital, 1 death per 387 anesthetic procedures for caesarean sections occurred, whereas in a Togo teaching hospital, 1 death per 133 anesthesia procedures was found. In a Zimbabwe district hospital, 1 death per 482 anesthesia procedures for surgery had to be counted [55]. A meta-analysis also reported that anesthesia accounted for 2.8% of all maternal deaths during caesarean section [56], and anesthesia experts believe that the quality of anesthesia care correlates highly with surgical and nonsurgical patients' outcomes [2, 7, 8].

The shortage of physician anesthetist in LMICs has resulted in a serious lack of equitable anesthesia care services at all levels of HCS increasing morbidity and mortality. The West African College of Surgeons and National Postgraduate Medical College of Nigeria therefore established an Anesthesia Diploma training program some decades ago [57]. The goal was to educate "middle-level" manpower physician anesthetists, who can work at secondary/district hospitals and rural and remote areas of sub-Saharan Africa. But this initiative has not been fully successful. An audit of the performance of the training showed that most program graduates moved on to acquire anesthesia fellowship. Twenty-three percent of them even migrated to HICs, while 67% remained in the sub-Saharan African zone as physician anesthetists. Nevertheless, only 9% of the graduates work at rural hospitals [57].

There is also a great need for specialist physician anesthetists that proffer effective leadership in anesthesia services, are capable of delivering emergency care, and can educate trainee anesthesia physicians. They should additionally be able to guide trainee physicians from other specialties who rotate through anesthesia and intensive care units, medical students, and allied healthcare workers. Without these physicians, the reported high numbers of avoidable anesthesia-related morbidities and mortalities [51, 52, 55, 58] will not be reducible in LMICs. Workforce density will have to be increased from 0 to approximately 20 per 100,000 people with respect to surgeon, anesthetist, and obstetrician specialists [52]. In 2017, the WFSA Global Anesthesia Workforce Survey stated that at least 5 physician anesthesia providers per 100,000 population would be sufficient to ensure leadership in anesthesia services, delivery of emergency care, and essential services for surgery in LMICs [59]. However, statistics about anesthetists show that 0.3–3 physicians per 100,000 people are still "real-world conditions" [52]. Overall, many referral, district, and rural hospitals face lots of challenges that are not only caused by lack of manpower but also by deficits in all kinds of technical, educational, and financial resources. Even basic equipment, materials, and essential drugs are often missing as summarized in Table 2.

Table 2
The state of anesthesia care provision in low & lower middle-income countries

Author	Country	Common challenges
Hodges et al. [2]	Uganda	Basic equipment and materials are not always available in surgical theater: For example, in 74% unavailability of pulse oximeter, in 22% no oxygen source, and in 21% no appropriately sized tracheal tubes
Chao et al. [37]	Ethiopia	Limited access to continuous electricity, running water, essential and anesthesia medications, and anesthesia monitoring equipment
Hadler et al. [44]	LMICs	Only 43% of hospitals had at least one functional anesthesia machine, 22% have equipment for cricothyroidotomy, and 48% and 42% have a consistent supply of adult and pediatric endotracheal tubes, respectively
Vo et al. [50]	22 LMICs	Only 53.4% of the healthcare facilities studied had functional anesthesia machines, 53% had access to pulse oximetry, and 21–45% lacked basic airway equipment and materials like face masks, laryngoscopes, and endotracheal tubes
Nnaji & Okoli [51]	Nigeria	This study reported that there are only 50 specialist anesthesia physicians in eastern Nigeria that shall serve 21,955,414 people (0.2 per 100,000 population)
Sobhy et al. [56]	LMICs	Anesthesia accounted for 2.8% of all maternal deaths
Tao et al. [58]	Cambodia	Lack of commonly used anesthesia and essential drugs and monitoring equipment, posing major threats to the safety of anesthesia practice
Kingham et al. [60]	Sierra Leone	Only 20% of facilities surveyed in Sierra Leone had access to a functioning anesthesia machine
Petroze et al. [61]	Rwanda	Anesthesia was performed primarily by anesthesia technicians, and 6 of 44 hospitals evaluated had no trained anesthesia provider

(continued)

Table 2 (continued)

Author	Country	Common challenges
Ouro-Bang'na Maman et al. [62]	Togo	Anesthesia-related deaths occurred in fit young patients because of airway problems or hypovolemia. Nearly 50% of deaths occurred in mothers undergoing caesarean section
Lonnée et al. [63]	Zimbabwe	In 42 hospitals studied, anesthesia workforce constituted of 22% medical officers and 77% nurse anesthetists. The only specialist physician anesthetist was part-time in a provincial hospital
Adkoli [64]	Sri Lanka	Shortage of specialist anesthesia physicians, with 37% of Sri Lankan anesthetists migrating to countries such as the United Kingdom and the United States
Dünser et al. [65]	DR Congo and Mongolia	Anesthesia care provision is limited by low educational standards and widespread lack of medical oxygen supplies and equipment
Choo et al. [66]	Ghana	More than 80% of hospitals were reported to have nurse anesthetists to provide anesthesia care, 76% have a functional anesthesia machine, but only 18% have basic anesthesia equipment required for endotracheal intubation and emergency cricothyroidotomy. There were no physician anesthesiologists in any of the facilities studied
Nuhu et al. [67]	Nigeria	No specialist physician anesthetist could be found in 14 public hospitals that undertake major surgeries in northern Nigeria

Note: LMICS low- and lower-middle-income countries

Despite the initiatives of educating specialist physician anesthetists, LMICs must cope with often decreasing numbers of trained personnel. This is partially due to migration of healthcare professionals searching for better wages, working conditions, and securities, sometimes in HICs. In addition, even in countries with a large population of specialist physician anesthetists, professionals are inequitably distributed between urban and rural regions. For example, while there are large numbers of physician anesthetists in Indian cities, they are only a few or no in remote and rural areas [68]. Within the confines of Bhutan, there are only 8 specialist physician anesthetists who shall serve a population of about 700,000 [68]. Afghanistan has likely only ~9 specialist physician anesthetists who shall care for 32 million people, while Uganda has 13 indigenous specialist physician anesthetists "serving" 27 million people [2, 55, 68].

In their quest to improve and enhance the quality of anesthesia provision, patient care, safety, and outcomes in LMICs and globally, the WFSA and WHO jointly developed the International Standards for Safe Practice of Anesthesia (ISSPA). This initiative aims to enhance global safe patient care through advocacy, education, training, safety, and quality improvement measures, innovations, and research efforts [69]. Notwithstanding, to overcome the current deficient care provision in LMICs requires more high-level advocacies including national and political commitments. To meet the present and projected population demands in anesthesia and surgical services does also require investments. These will not be possible without continuous stakeholders' engagement at national and political levels in a coordinated and strategic manner. Healthcare financing mechanisms, government budgetary allocation, and national health insurance will have to be integrated and guided by local and international advocacies [19, 20, 70]. No wonder WHO is calling on its member states to remind that anesthesia and surgery are properly prioritized and included within UHC and HCS strengthening [46, 47, 49].

Most states from LMICs have meanwhile responded to the WHO resolution of prioritizing anesthesia and surgical care provisions in UHC by establishing national surgical, obstetric, anesthesia, and nursing plans (NSOANP) into their national health policy and within the context of their HCS [71–75]. This is visible on established platforms and six frameworks of LCGS vision of year 2030 [26]. The ISSPA targets of WFSA and WHO are still unmet in most regions of LMICs, but, for example, trainee physician anesthetists get educated to be more flexible and adaptable in their work with the available anesthesia resources while sustaining standards of anesthesia care and high-quality treatment. LMICs depend on basic equipment and essential drugs, but if even these are not available, such mishaps are considered in the training of physician anesthetists. Furthermore, trainings have incorporated a culture of rational decision-making, reasoning, and education to perform anesthesia procedures in accordance with local resources and cultures [7, 65].

Public health plays a strategic role in the socioeconomic growth and development of a nation. Nevertheless, little information is available on the effects of anesthesia care capacity within HCS in LMICs to improve a population's health. Vice versa, anesthesia care provision in LMICs is affected by the socioeconomic status of the people, as most depend on OOP pay for healthcare expenditures. Altogether, this situation has persuaded remaining and available anesthesia providers to use simple and effective techniques and methods that are cheap and readily accessible in their daily work [2, 7]. On one hand, they have adapted to their environmental challenges and ensure safe anesthesia and critical care despite all hardships, and on the other hand, they try to train more specialist physician anesthetists and improve quality of care and services through driving research. This has revolutionized the use of ketamine in LMICs for enhancing anesthesia care provision and improve perioperative and critical care outcomes [7].

5 Role of Ketamine in Anesthesia Care Provision in Low & Lower Middle-Income Countries

Ketamine is an important asset in anesthesia, acute and chronic pain therapy, and critical care in LMICs. This is not only true because many anesthesiologic procedures are conducted by nonphysician anesthesia providers but also because ketamine can be administered via several routes and does not depend heavily on oxygen availability, electricity, sophisticated anesthesia and critical care equipment, or expert anesthetists, all of which are scarce in LMICs. The drug is cheap and readily available. Because HCS in LMICs are poorly funded and have only limited resources and all kinds of supplies at their disposal, most hospitals conduct anesthesia or sedation with room air and seldomly deliver oxygen from an oxygen concentrator. Patients' airways are often kept patent by simple patient positioning, chin lift, and jaw thrust. Mucus extractors, rubber bulb suction devices, or rarely foot-pedal manually operated suction machines (because of lack of electricity) are used to suction the airway off secretion. Improvised precordial stethoscopes are used to monitor patients' breath sounds, heart rates, and volumes. Many hospitals do not have anesthesia machines, and where available, there may be no supply of inhalation anesthesia agents (which has been mostly halothane), or patients may not be able to afford them. Isoflurane, sevoflurane, and desflurane are luxury "goods," and many patients cannot pay for them. Unfortunately, halothane has become scarce, because manufacturers are discontinuing its production [2, 3, 7, 8].

Ketamine's specific features as analgesic, antidepressant, sedative, amnestic, anesthetic, and immobility- and loss-of-consciousness-inducing substance make the agent a drug of choice for sedation and general anesthesia in LMICs. It can be administered orally, subcutaneously, intranasally, intravenously (IV), intramuscularly (IM), or intraosseous at different doses calculated for pain treatment, sedation, or anesthesia. IV ketamine administration shows an almost 100% bioavailability and very predictable effects. For general anesthesia, an induction dose of 2 mg/kg IV ketamine in adults or 1 mg/kg in children is administered, followed by an increment of 0.1–0.5 mg/min titrated to patient response for maintenance of anesthesia. This is especially used at places where there is no anesthesia machine, vaporizer, and/or anesthetic vapors. Nevertheless, establishing an IV access for ketamine administration may be difficult in emergencies, children, and obese patients due to arduous peripheral venous cannulation. The IM administration of ketamine for sedation, anesthesia, or pain treatment at a dose of $\leq$5 mg/kg is possible but can have a relatively slow onset of action with still a high bioavailability. IM use is also associated with a longer recovery time. Oral ketamine is a more cost-effective and convenient alternative to other routes but shows a ~ 20%

bioavailability, only. The efficacy of intranasal ketamine has been documented but with variable bioavailability based on particle size and formulation of the drug. Furthermore, the alternative use of subcutaneous ketamine injection at the dose of 0.5 mg/kg for analgesia and sedation or for the purpose of bronchodilation in ICU has been described. When administered through any of the routes outlined, the drug induces analgesia, sedation, amnesia, immobility, and loss of consciousness in a strict dose-dependent manner [76–80]. Furthermore, ketamine exerts sympathomimetic effects with the benefit of hemodynamic stabilization, a preferred option in critically ill and hemodynamically unstable patients. Ketamine has also proved to be beneficial in multimodal analgesia concepts, burn care, chronic pain management, and palliative care. It can be used to sedate patients in the ICU and emergency room and is applied for treatment of depression at subanesthetic dose. It is also useful for managing patients with refractory bronchospasm or refractory status asthmaticus as a bronchodilator of choice at a loading dose of 0.1–0.2 mg/kg followed by an infusion of 0.15–2.5 mg/kg/hr. [76, 77, 81]. Notwithstanding, ketamine may have unwanted effects: For example, the anesthetic dose causes dissociative anesthesia; increases the incidence of postoperative nausea and vomiting; may induce transient apnea, especially when administered rapidly; and increases salivation, which may increase the risk of laryngospasm. Increased salivation can be minimized by co-administration of atropine at best given before ketamine use. The administration of ketamine can provoke imaginative and dissociative states and psychotic symptoms and impairs semantic and episodic memory. Additionally, it can cause emergent phenomena during recovery from anesthesia and from the drug in awake patients. Psychoactive actions are described as floating sensations, vivid pleasant dreams and nightmares, hallucinations, and delirium. Patients receiving ketamine are therefore often premedicated with benzodiazepines like diazepam or midazolam, which prevents or reduces hallucinatory effects and emergence phenomenon. Diazepam is also readily available and cheap in low-resource countries. In combination with atropine and diazepam, ketamine forms a reliable regimen for total IV general anesthesia for different modalities of surgery with room air and minimal equipment [7, 76]. The goal of safe anesthesia care provision is to ensure efficiency, safety, and effectiveness, which should be timely and accessible. The availability of ketamine, diazepam/midazolam, and atropine has made these three drugs a good and reliable combination regimen for the conduct of general anesthesia, sedation of patients, and treatment of pain, with room air or oxygen where its available in LMICs. Minimal equipment is required. This has made ketamine gain lots of credit in anesthesia practice in LMICs and led to utilization rates of 47–100% [7, 82–84] in surgical procedures. Anesthesia providers in LMICs rely heavily on ketamine for the daily demand in their services as indicated in Table 3.

Table 3
The role of ketamine in anesthesia care in low & lower middle-income countries

Author	Country	Background	Role of ketamine
Hodges et al. [2]	Uganda	To quantify and describe the problems faced by anesthetists in developing countries	Anesthesia is largely ketamine-based; in 92%, it is readily available for anesthetists
Nnaji [7]	LMICs	To review the use of ketamine in low-resource countries	Utilization of ketamine ranges from 44.7% to 100% as sole anesthetic agent for general anesthesia due to limited manpower and all resources
Hadler et al. [44]	Ghana	To study anesthesia care capacity in LMICs to identify patterns of deficits and provide useful targets for advocacy and intervention	Ketamine was used in 73% of general anesthesia cases due to critical deficiencies in various anesthesia care capacities
Vo et al. [50]	22 LMICs	To determine the resources available to deliver anesthesia care in LMICs	Majority (71.5%) of hospitals depends on anesthesia with ketamine. Inhalational agents were available in 58.5% of general anesthesia cases
Ogboli-Nwasor et al. [79]	Nigeria	To study the use of oral ketamine during minor orthopedic procedures in the emergency room	Oral ketamine provided analgesia for minor emergency room procedures with reliable onset of action and few side effects. It is accessible and used effectively for pain relief in emergency rooms in LMICs
Oham et al. [80]	Nigeria	To evaluate the efficacy of subcutaneous ketamine on post-operative analgesia in children who had unilateral inguinal herniotomy	Subcutaneous ketamine prolonged the analgesic effect of plain bupivacaine surgical wound site infiltration in children undergoing unilateral inguinal herniotomy
Jochberger et al. [85]	Zambia	To evaluate the status of anesthesia, intensive care medicine, emergency medicine, and pain therapy in Zambia	Ketamine was used for general anesthesia in 50% of cases; endotracheal intubation was performed in 10% of general anesthesia
Burke et al. [86]	Kenya	To evaluate the safety and feasibility of the emergency and essential surgery–ketamine package procedures when no anesthetist is available	Non-anesthetists used ketamine in 1216 surgical procedures in 15 hospitals studied. This was safe and feasible in expanding access to emergency and essential surgeries in rural Kenya where there is no physician anesthetist

Note: LMICS low- and lower-middle-income countries

Taken together, in LMICs, ketamine has been shown to be the most widely used and safest anesthetic drug, as reflected by being "always available" according to 92% of anesthetists surveyed in Uganda [2].

6 Conclusion

Despite the efforts of the WFSA and WHO and their member societies and states to make safe emergency and essential anesthesia and surgical care an integral component of Universal Health Coverage and HCS strengthening, anesthesia care provision in LMICs continues to have to cope with tremendous challenges because of insufficient number of anesthesia providers, poor surgical theater infrastructure, limited anesthesia equipment, materials, life-saving drugs, anesthetic agents, and healthcare resources. Ketamine, a cheap and readily available anesthesia drug, has proved to be beneficial in anesthesia, critical care, multimodal analgesia and acute pain management, burn care, chronic pain therapy, and palliative care. It is also used to sedate patients in the ICU and emergency room and for treatment of depression. Ketamine's specific pharmacological profile offers these opportunities even under austere conditions and circumstances in LMICs with shortage of well-trained healthcare personnel, equipment and materials, and essential and anesthesia drugs.

Conflicts of Interest

The author declares that he has no competing interest.

References

1. Nnaji CT (2013) This medical field called Anaesthesia. Gazette Med 1(2):65–66
2. Hodges SC, Mijumbi C, Okello M, McCormick BA, Walker IA et al (2007) Anaesthesia services in developing countries: defining the problems. Anaesthesia 62(1):4–11. https://doi.org/10.1111/j.1365-2044.2006.04907.x
3. Wirtz VJ, Hogerzeil HV, Gray AL, Bigdeli M, de Joncheere CP et al (2017) Essential medicines for universal health coverage. Lancet 389(10067):403–476. https://doi.org/10.1016/S0140-6736(16)31599-9
4. World Bank (2022) Data. In: World Bank Country and Lending Groups. https://datahelpdesk.worldbank.org/knowledgebase/articles/906519-world-bank-country-and-lending-groups. Accessed 27 March 2024
5. Van Zyl C, Badenhorst M, Hanekom S, Heine M (2021) Unravelling 'low-resource settings': a systematic scoping review with qualitative content analysis. Br Med J 6(6):e005190. https://doi.org/10.1136/bmjgh-2021-005190
6. Geiling J, Burkle FM Jr, Amundson D, Dominguez-Cherit G, Gomersall CD et al (2014) Task force for mass critical care. Resource-poor settings: infrastructure and capacity building: care of the critically ill and injured during pandemics and disasters: CHEST consensus statement. Chest 146(4 Suppl):e156S–e167S. https://doi.org/10.1378/chest.14-0744

7. Nnaji CT (2022) Ketamine and low-resource countries. In: Saiz-Sapena N, Gil MG (eds) Ketamine revisited. IntechOpen, London, GB. https://doi.org/10.5772/intechopen.104651. Accessed 3 July 2024
8. Bharati SJ, Chowdhury T, Gupta N, Schaller B, Cappellani RB et al (2014) Anaesthesia in underdeveloped world: present scenario and future challenges. Niger Med J 55(1):1–8. https://doi.org/10.4103/0300-1652.128146
9. Bainbridge D, Martin J, Arango M, Cheng D (2012) Perioperative and anaesthetic-related mortality in developed and developing countries: a systematic review and meta-analysis. Lancet 380(9847):1075–1081. https://doi.org/10.1016/S0140-6736(12)60990-8
10. Khan M, Khan FA (2007) Anesthetic deaths in a developing country. Middle East J Anesthesiol 19:159–172
11. WHO. Delivering quality health services: a global imperative for universal health coverage 2021. http://www.who.int/publications/i/item/9789241513906. Accessed 22 June 2024
12. Cameron E, Green M (2020) Making sense of change management: a complete guide to the models, tools and techniques of organizational change, 5th edn. Kogan Page Publishers, London, GB
13. National Academies of Sciences, Engineering, and Medicine; Health and Medicine Division; Board on Health Care Services; Board on Global Health; Committee on Improving the Quality of Health Care Globally (2018) Crossing the global quality chasm: improving health care worldwide. The Current State of Global Health Care Quality National Academies Press, Washington, DC, USA. http://wwwncbinlmnihgov/books/NBK535654/. Accessed 3 July 2024
14. Welcome MO (2021) The Nigerian health care system: need for integrating adequate medical intelligence and surveillance systems. J Pharm Bioallied Sci 3(4):470–478. https://doi.org/10.4103/0975-7406.90100
15. Faierman ML, Anderson JE, Assane A, Bendix P, Vaz F et al (2015) Surgical patients travel longer distances than non-surgical patients to receive care at a rural hospital in Mozambique. Int Health 7(1):60–66. https://doi.org/10.1093/inthealth/ihu059
16. Adeyanju O, Tubeuf S, Ensor T (2017) Socioeconomic inequalities in access to maternal and child healthcare in Nigeria: changes over time and decomposition analysis. Health Policy Plan 32(8):1111–1118. https://doi.org/10.1093/heapol/czx049
17. WHO. Trends in maternal mortality: 2000–2020. WHO, Geneva. https://data.unicef.org/topic/maternal-health/maternal-mortality/. Accessed 30 Apr 2024
18. WHO. Newborn mortality (2024). Available at: https://www.who.int/news-room/fact-sheets/detail/newborn-mortality. Accessed 5 May 2024
19. Madu AC, Osborne K (2023) Healthcare financing in Nigeria: a policy review. Int J Soc Determinants Health Health Serv 53(4):434–443. https://doi.org/10.1177/27551938231173611
20. Edeh HC (2022) Exploring dynamics in catastrophic health care expenditure in Nigeria. Health Econ Rev 12:22. https://doi.org/10.1186/s13561-022-00366-y
21. Aregbeshola BS, Khan SM (2018) Out-of-pocket payments, catastrophic health expenditure and poverty among households in Nigeria 2010. Int J Health Policy Manag 7(9):798–806. https://doi.org/10.15171/ijhpm.2018.19
22. Ipinnimo,TM, Durowade KA, Afolayan CA, Ajayi PO, Akande TM (2022) The Nigeria National Health Insurance Authority act and its implications towards achieving universal health coverage. Niger Postgrad Med J 29(4):281–287. doi: https://doi.org/10.4103/npmj.npmj_216_22
23. Alawode GO, Adewole DA (2021) Assessment of the design and implementation challenges of the National Health Insurance Scheme in Nigeria: a qualitative study among sub-national level actors, healthcare and insurance providers. BMC Public Health 21:124. https://doi.org/10.1186/s12889-020-10133-5
24. Adamu A, Maigatari M, Lawal K, Iliyasu M (2010) Waiting time for emergency abdominal surgery in Zaria. Nigeria African Health Sci 10(1):46–53
25. Debas HT, McCord C, Thind A (2006) Surgery. In: Jamison DT, Breman JG, Measham AR, Alleyene G, Claeson M et al (eds) Disease control priorities in developing countries, 2nd edn. Oxford University Press and World Bank, Washington DC, USA, pp 1245–1259
26. Meara JG, Leather AJM, Hagander L, Alkire BC, Alonso N et al (2015) Global surgery 2030: evidence and solutions for achieving health, welfare, and economic development. Lancet 386:569–624 https://doi.org/10.1016/s0140-6736(15)60160-x

27. Shrime MG, Bickler SW, Alkire BC, Mock C (2015) Global burden of surgical disease: an estimation from the provider perspective. Lancet Glob Health 3(Suppl 2):S8–S9. https://doi.org/10.1016/S2214-109X(14)70384-5
28. Lin BM, White M, Glover A, Wamah GP, Trotti DL et al (2017) Barriers to surgical care and health outcomes: a prospective study on the relation between wealth, sex, and postoperative complications in the republic of Congo. World J Surg 1:14–23. https://doi.org/10.1007/s00268-016-3676-x
29. Trani JF, Bakhshi P, Noor AA, Lopez D, Mashkoor A (2010) Poverty, vulnerability, and provision of healthcare in Afghanistan. Soc Sci Med 70(11):1745–1755. https://doi.org/10.1016/j.socscimed.2010.02.007
30. Premkumar A, Ying X, Mack Hardaker W, Massawe HH, Mshahaba DJ et al (2018) Access to Orthopaedic surgical Care in Northern Tanzania: a modelling study. World J Surg 42:3081–3088. https://doi.org/10.1007/s00268-018-4630-x
31. Ariyaratnam R, Palmqvist CL, Hider P, Laing GL, Stupart D et al (2015) Toward a standard approach to measurement and reporting of perioperative mortality rate as a global indicator for surgery. Surgery 158(1):17–26. https://doi.org/10.1016/j.surg.2015.03.024
32. Osinaike B, Ayandipo O, Onyeka T, Alagbe-Briggs O, Nnaji CT et al (2019) Nigerian surgical outcomes study investigators. Nigerian surgical outcomes - report of a 7-day prospective cohort study and external validation of the African surgical outcomes study surgical risk calculator. Int J Surg 68:148–156. https://doi.org/10.1016/j.ijsu.2019.06.003
33. Biccard BM, Madiba TE, Kluyts HL, Munlemvo DM, Madzimbamuto FD et al (2018) Perioperative patient outcomes in the African surgical outcomes study: a 7-day prospective observational cohort study. Lancet 391(10130):1589–1598. https://doi.org/10.1016/S0140-6736(18)30001-1
34. Bishop D, Dyer RA, Maswime S, Rodseth RN, van Dyk D et al (2019) (ASOS investigators). Maternal and neonatal outcomes after caesarean delivery in the African surgical outcomes study: a 7-day prospective observational cohort study. The lancet. Glob Health 7(4): e513–e522. https://doi.org/10.1016/s2214-109x(19)30036-1
35. Endeshaw AS, Kumie FT, Molla MT, Zeru GA, Abera KM et al (2023) Incidence and predictors of perioperative mortality in a low-resource country, Ethiopia: a prospective follow-up study. Br Med J Open 13(5): e069768. https://doi.org/10.1136/bmjopen-2022-069768
36. Mughal NA, Hussain MH, Ahmed KS, Waheed MT, Munir MM et al (2023) Barriers to surgical outcomes research in low-and middle-income countries: a scoping review. J Surg Res 290:188–196. https://doi.org/10.1016/j.jss.2023.04.017
37. Chao TE, Burdic M, Ganjawalla K et al (2012) Survey of surgery and anesthesia infrastructure in Ethiopia. World J Surg 36:2545–2553. https://doi.org/10.1007/s00268-012-1729-3
38. Linden AF, Sekidde FS, Galukande M, Knowlton LM, Chackungal S et al (2012) Challenges of surgery in developing countries: a survey of surgical and anesthesia capacity in Uganda's public hospitals. World J Surg 36:1056–1065. https://doi.org/10.1007/s00268-012-1482-7
39. Bari S, Incorvia J, Ahearn O, Dara L, Sharma S et al (2021) Building safe surgery knowledge and capacity in Cambodia: a mixed-methods evaluation of an innovative training and mentorship intervention. Glob Health Action 14(1):1998996. https://doi.org/10.1080/16549716.2021.1998996
40. Afsana K, Rashid SF (2001) The challenges of meeting rural Bangladeshi women's needs in delivery care. Reprod Health Matters 9(18): 79–89. https://doi.org/10.1016/s0968-8080(01)90094-1
41. Bronsard A, Geneau R, Shirima S, Courtright P, Mwende J (2008) Why are children brought late for cataract surgery? Qualitative Findings from Tanzania Ophthalmic Epidemiol 15(6):383–388. https://doi.org/10.1080/09286580802488624
42. Weiser TG, Regenbogen SE, Thompson KD, Haynes AB, Lipsitz SR et al (2008) An estimation of the global volume of surgery: a modelling strategy based on available data. Lancet 372(9633):139–144. https://doi.org/10.1016/S0140-6736(08)60878-8
43. Briesen S, Geneau R, Roberts H, Opiyo J, Courtright P (2010) Understanding why patients with cataract refuse free surgery: the influence of rumours in Kenya. Trop Med Int Health 15(5):534–539. https://doi.org/10.1111/j.1365-3156.2010.02486.x
44. Hadler RA, Chawla S, Stewart BT, McCunn MC, Kushner AL (2016) Anesthesia care capacity at health facilities in 22 low- and middle-income countries. World J Surg 40(5): 1025–1033. https://doi.org/10.1007/s00268-016-3430-4

45. Geoffroy E, Harries AD, Bissell K, Schell E, Bvumbwe A et al (2014) Bringing care to the community: expanding access to health care in rural Malawi through mobile health clinic. PHA 4(4):252–258. https://doi.org/10.5588/pha.14.0064
46. WHO (2014) Strengthening emergency and essential surgical care and anaesthesia as a component of universal health coverage. Report by the Secretariat. http://apps.who.int/gb/ebwha/pdf_files/WHA68/A68_31-en.pdf. Accessed 14 Mar 2024
47. Price R, Makasa E, Hollands M, World Health Assembly Resolution WHA68.15 (2015) Strengthening emergency and essential surgical care and anesthesia as a component of universal health coverage—addressing the public health gaps arising from lack of safe, affordable and accessible surgical and anesthetic services. World J Surg 39(9):2115–2125. https://doi.org/10.1007/s00268-015-3153-y
48. Barreiro G, Mellin-Olsen J, Gore-Booth J (2018) The role of the WFSA in reaching the goals of the lancet commission on global surgery. Anesth Analg 126(4):1400–1404. https://doi.org/10.1213/ANE.0000000000002543
49. WHO. Global Initiative for Emergency and Essential Surgical Care. https://www.who.int/initiatives/who-global-initiative-for-emergency-and-essential-surgical-care. Accessed 10 June 2024
50. Vo D, Cherian MN, Bianchi S et al (2012) Anesthesia capacity in 22 low and middle income countries. J Anesth Clin Res 3:207. https://doi.org/10.4172/2155-6148.1000207
51. Nnaji CT, Okoli CE (2023) The evaluation of anaesthesia workforce capacity, an index of perioperative care in the South-Eastern Nigeria. J Anesth Clin Res 14:1097. https://doi.org/10.35248/2155-6148.23.14.1096
52. Law T, Lipnick MS, Morriss W, Gelb AW, Mellin-Olsen J et al (2024) The global anesthesia workforce survey: updates and trends in the anesthesia workforce. Anesth Analg 139:15–24. https://doi.org/10.1213/ANE.0000000000006836
53. Nnaji CT, Chikwe K (2017) Anesthesia for abdominal myomectomy – a five-year audit of a Federal Medical Centre in Owerri. Nigeria J Anest Anesthes 1(1):16–19
54. Dubowitz G, Detlefs S, McQueen KA (2010) Global anaesthesia workforce crisis: a preliminary survey revealing shortages contributing to undesirable outcomes and unsafe practices. World J Surg 4:438–444. https://doi.org/10.1007/s00268-009-0229-6
55. Walker IA, Wilson IH (2008) Anaesthesia in developing countries—a risk for patients. Lancet 371:968–969. https://doi.org/10.1016/S0140-6736(08)60432-8
56. Sobhy S, Zamora J, Dharmarajah K, Arroyo-Manzano D, Wilson M et al (2016) Anaesthesia-related maternal mortality in low-income and middle-income countries: a systematic review and meta-analysis. Lancet Glob Health 4(5):e320–e327. https://doi.org/10.1016/S2214-109X(16)30003-1
57. Bode CO, Olatosi J, Amposah G, Desalu I (2013) Has the middle-level anaesthesia manpower training program of the west African College of Surgeons fulfilled its objectives? Anaesth Intensive Care 41:359–362. https://doi.org/10.1177/0310057X1304100313
58. Tao KM, Sokha S, Yuan HB (2020) The challenge of safe anesthesia in developing countries: defining the problems in a medical center in Cambodia. BMC Health Serv Res 20(1):204. https://doi.org/10.1186/s12913-020-5068-z
59. Kempthorne P, Morriss WW, Mellin-Olsen J, Gore-Booth J (2017) The WFSA global anesthesia workforce survey. Anesth Analg 125(3):981–990. https://doi.org/10.1213/ANE.0000000000002258. PMID: 28753173
60. Kingham TP, Kamara TB, Cherian MN, Gosselin RA, Simkins M, Meissner C, Foray-Rahall L, Daoh KS, Kabia SA, Kushner AL (2009) Quantifying surgical capacity in Sierra Leone: a guide for improving surgical care. Arch Surg 144(2):122–127. https://doi.org/10.1001/archsurg.2008.540
61. Petroze RT, Nzayisenga A, Rusanganwa V, Ntakiyiruta G, Calland JF (2012) Comprehensive national analysis of emergency and essential surgical capacity in Rwanda. Br J Surg 99:436–443. https://doi.org/10.1002/bjs.7816
62. Ouro-Bang'na Maman AF, Tomta K, Ahouangbévi S, Chobli M (2005) Deaths associated with anaesthesia in Togo. West Africa Trop Doct 35(4):220–222. https://doi.org/10.1258/004947505774938666
63. Lonnée HA, Madzimbamuto F, Erlandsen ORM, Vassenden A, Chikumba E et al (2018) Anesthesia for cesarean delivery: a cross-sectional survey of provincial, district, and Mission hospitals in Zimbabwe. Anesth Analg 126(6):2056–2064. https://doi.org/10.1213/ANE.0000000000002733. PMID: 29293184
64. Adkoli BV (2006) Migration of health workers: perspectives from Bangladesh, India, Nepal, Pakistan and Sri Lanka. Regional Health Forum 10(1):49–58

65. Dünser M, Baelani I, Ganbold L (2006) The specialty of anesthesia outside Western medicine with special consideration of personal experience in The Democratic Republic of the Congo and Mongolia. Anaesthesist 55(2): 118–1132. German. https://doi.org/10.1007/s00101-006-0979-6
66. Choo S, Perry H, Hesse AAJ et al (2010) Assessment of capacity for surgery, obstetrics and anaesthesia in 17 Ghanaian hospitals using a WHO assessment tool. Trop Med Int Health 15:1109–1115. https://doi.org/10.1111/j.1365-3156.2010.02589.x
67. Nuhu SI, Embu HY, Onoja AA, Dung D (2017) Anaesthesia workforce and infrastructure in a north central state of Nigeria: a survey. Highland Med Res J 17(1):50–54
68. Mavalankar D, Sriram V (2009) Provision of anaesthesia services for emergency obstetric care through task shifting in South Asia. Reprod Health Matters 17:21–31. https://doi.org/10.1016/S0968-8080(09)33433-3
69. Merry AF, Cooper JB, Soyannwo O, Wilson IH, Eichhorn JH (2010) An iterative process of global quality improvement: the international standards for a safe practice of anesthesia 2010. Can J Anesth 57:1021–1026. https://doi.org/10.1007/s12630-010-9380-7
70. Cherian M, Choo S, Wilson I, Noel L, Sheikh M, Dayrit M, Groth S (2010) Building and retaining the neglected anaesthesia health workforce: is it crucial for health systems strengthening through primary health care? Bull World Health Organ 88:637–639. https://doi.org/10.2471/BLT.09.072371
71. Fatima I, Shoman H, Peters AW et al (2020) Pakistan's national surgical, obstetric, and anesthesia plan: an adapted model for a devolved federal-provincial health system. Can J Anesth 67:1212–1216. https://doi.org/10.1007/s12630-020-01708-2
72. Citron I, Jumbam D, Dahm J, Mukhopadhyay S, Nyberger K et al (2019) Towards equitable surgical systems: development and outcomes of a national surgical, obstetric and anaesthesia plan in Tanzania. BMJ Glob Health 4(2):e001282. https://doi.org/10.1136/bmjgh-2018-001282
73. Truché P, Shoman H, Reddy CL et al (2020) Globalization of national surgical, obstetric and anesthesia plans: the critical link between health policy and action in global surgery. Glob Health 16:1. https://doi.org/10.1186/s12992-019-0531-5
74. Seyi-Olajide JO, Anderson JE, Williams OM, Faboya O, Amedu JO et al (2021) National surgical, obstetric, anaesthesia and nursing plan. Nigeria Bull World Health Organ 99(12):883–891. https://doi.org/10.2471/BLT.20.280297
75. Peters AW, Roa L, Rwamasirabo E, Ameh E, Ulisubisya MM et al (2020) National surgical, obstetric, and anesthesia plans supporting the vision of universal health coverage. Glob Health Sci Pract 8(1):1–9. https://doi.org/10.9745/GHSP-D-19-00314
76. Kurdi MS, Theerth KA, Deva RS (2014) Ketamine: current applications in anesthesia, pain, and critical care. Anesth Essays Res 8(3): 283–290. https://doi.org/10.4103/0259-1162.143110
77. Mei Gao M, Rejaei D, Liu H (2016) Ketamine use in current clinical practice. Acta Pharmacol Sin 37:865–872. https://doi.org/10.1038/aps.2016.5
78. Helm M, Hossfeld B, Schlechtriemen T, Braun J, Lampl L, Bernhard M (2011) Use of intraosseous infusion in the German air rescue service: nationwide analysis in the time period 2005 to 2009. Anaesthesist 60:1119–1125. https://doi.org/10.1007/s00101-011-1937-5
79. Ogboli-Nwasor E, Amaefule KE, Audu SS (2014) Use of oral ketamine for analgesia during reduction/manipulation of fracture/dislocation in the emergency room: an initial experience in a low-resource setting. Pain Studies and Treatment 2:17–20. https://doi.org/10.4236/pst.2014.21004
80. Oham A, Ekwere I, Tobi K (2020) Subcutaneous ketamine prolongs the analgesic effect of local infiltration of plain bupivacaine in children undergoing inguinal herniotomy. Afr Health Sci 20(2):806–814. https://doi.org/10.4314/ahs.v20i2.34
81. Goel S, Agrawal A (2013) Ketamine in status asthmaticus: a review. Indian J Crit Care 1793: 154–161. https://doi.org/10.4103/0972-5229.117048
82. Ikechebelu JI, Udigwe GO, Obi RA, Joe-Ikechebelu NN, Okoye IC (2003) The use of simple ketamine anaesthesia for day-case diagnostic laparoscopy. J Obstet Gynaecol 23:650–652. https://doi.org/10.1080/01443610310001604457
83. Makin J, Suarez-Rebling D, Suarez S, Leone A, Burke TF (2021) Operations supported by ketamine anesthesia in resource-limited settings: surgeons' perceptions and recommendations–qualitative study. Int J Surgery Open 29:1–8. https://doi.org/10.1016/j.ijso.2020.12.009
84. Koka R, Chima AM, Sampson JB et al (2016) Anesthesia practice and perioperative outcomes at two tertiary care hospitals in Freetown,

Sierra Leone. Anesth Analg 123:213–227. https://doi.org/10.1213/ANE.0000000000001285

85. Jochberger S, Ismailova F, Lederer W, Mayr VD, Luckner G et al (2008) Anesthesia and its allied disciplines in the developing world: a nationwide survey of the Republic of Zambia. Anesth Analg 106(3):942–948. https://doi.org/10.1213/ane.0b013e318166ecb8

86. Burke TF, Suarez S, Senay A, Masaki C, Rogo K et al (2017) Safety and feasibility of a ketamine package to support emergency and essential surgery in Kenya when no anesthetist is available: an analysis of 1216 consecutive operative procedures. World J Surg 41(12): 2990–2997. https://doi.org/10.1007/s00268-017-4312-0. Erratum in: World J Surg 2018;42(9):3049. https://doi.org/10.1007/s00268-017-4383-y

Chapter 21

Snakebites, Envenomation, and Treatment of Snakebite Pain with Ketamine

Sabine Himmelseher, Georges Mion, and Christopher V. Maani

Abstract

The global incidence of snakebites is estimated to exceed five million annually. Envenomations are responsible for more than 150,000 deaths and over 400,000 permanent disabilities. Although snakebites are a medical emergency, substantial obstacles frequently impede the provision of effective therapy in regions with limited medical facilities. The treatment of acute snakebite pain is often inadequate, which can contribute to the development of post-traumatic stress disorders (PTSD), depression, psychosocial impairments, and chronic, potentially debilitating pain. The antidepressant, analgesic, and resilience- and PTSD-improving effects of ketamine make it a promising addition to snakebite management. Because ketamine shows a wide safety margin and low costs and is available in most low-resource settings, it is a suitable option for therapy in any location. The aim of this work is to improve our understanding of the challenges and therapeutic principles associated with snakebite envenomation. The positive effects of ketamine on acute snakebite pain, as observed in a prospective observational study in Guinea, will be discussed.

Key words Snakebite, Envenomation, Neglected global health crisis, Ketamine, Pain, Post-traumatic stress disorder, Low-resource setting, Austere wilderness medicine

1 Introduction

A few years ago, Kofi Annan very aptly said that "snakebites are (one of) the biggest public health crisis you have likely never heard of"[1]. Due to a lack of reporting, the true scale of the problem is still unknown. However, estimates suggest that there are more than five million bites each year. Envenomations have been linked to over 150,000 deaths and more than 400,000 instances of permanent disability including paralysis and amputations [1–5]. While the likelihood of exposure to snakes is rather low in North America and Europe, the ongoing climate change may alter this situation [6]. Additionally, the trend of keeping snakes as pets and private collections has increased.

Snakebites are a medical emergency. In impoverished, rural communities and marginalized areas of the world, as well as tropical

Sabine Himmelseher (ed.), *Ketamine: Pain Management, Critical Care, Depression, and Resource-Poor Settings Volume II*, Neuromethods, vol. 225, https://doi.org/10.1007/978-1-0716-4599-4_21,

or subtropical regions with limited medical facilities, the obstacles to providing effective treatment are immense. High-risk groups for snakebites include those who engage in outdoor activities and occupations (particularly in agriculture and nature), children, and individuals residing in precarious conditions where environmental factors intersect with snake habitats. Snakebites perpetuate the cycle of poverty, and in their aftermath, victims and families may face substantial financial challenges while bearing a significant medical burden.

After the World Health Organization (WHO) categorized snakebites as a high priority, neglected, potentially life-threatening tropical disease in 2017, international funding increased [3]. In 2021, the Wellcome Trust research fund announced the launch of a £80 million GBP (~$100.560.400 USD) program with the objective of transforming the way snakebite treatments are researched and delivered [7]. In light of the ongoing challenges related to production and international antivenom management [8], the WHO has taken the initiative to enhance quality control measures for antivenom products and target profiles of animal plasma-derived antivenoms [5]. Additionally, the organization has made information campaigns about snakebite prevention and guidance for first aid freely available on official WHO Internet sites. It is well documented that antivenoms are the only effective etiological treatment for snakebites, but accessibility and cost remain significant barriers [9, 10]. There is also a lack of prospective clinical studies and pragmatic trials on snakebite treatment and the course of the disease, and there is an urgent need to reduce morbidity and long-term sequelae for venomous snakebite patients.

Despite the vast, global biodiversity of snake species, the medical principles of care after snakebites are similar across geographical regions [3, 4, 9–13]. For example, in the acute phase after a bite (or spitting), insufficient medical management is frequently observed due to inadequate access to equipped medical facilities, antivenoms, and/or pain medication [3, 4, 11, 12]. Insufficient pain treatment can lead to the development of chronic pain [14], and inadequate analgesia following traumatic injury has been correlated with the onset of post-traumatic stress symptoms in children and adults [15, 16]. After a snakebite, victims frequently experience not only severe pain but also post-traumatic stress disorders (PTSD), depression, and psychosocial impairments [17, 18].

The pharmacological profile of ketamine, coupled with its low cost and widespread availability even in rural and remote regions, makes it suitable for use in low-resource settings. It can be considered a safe, affordable and often accessible therapeutic option [19, 20]. From a medical perspective, its antidepressant, analgesic, and resilience-improving effects, along with its potential for PTSD treatment, make ketamine a promising agent for inclusion in snakebite therapy [21, 22]. This work will briefly delineate the principles of medical care in snakebites and will present two lucky patients as illustrative examples of snakebite disease courses to enhance our understanding of the challenges with snakebite envenomations. It will also describe the findings of a prospective observational study on ketamine in snakebite-poisoned patients with acute pain [23].

2 Principles of Care for Snakebite Envenomation

The majority of medically important snakes are classified within two taxonomic families: Elapidae (elapids, including cobras and mambas) and Viperidae (such as vipers, adders, rattlesnakes) [3, 4, 11, 12, 24] (Fig. 1).

Snakes utilize their venom for predatory and defensive purposes. When biting, the venom is injected via grooved or hollow

Fig. 1 Snake species responsible for high mortality resulting from envenomation: a spectacled cobra (**a**) and a king's cobra (**b**) (family Elapidae) and a common lancehead viper (**c**) and a saw-scaled viper (**d**) (family Viperidae) (photo **a**) from Martin G [24], image courtesy of Kasambe R, Education Conversation Centre, Bombay Natural History Society, Mumbai, India, under a CC BY-SPA 4.0 license; (**b**) from Veto et al. [28] with permission from John Wiley and Sons conveyed via Copyright Clearance Center, Inc.), and (**c**) and (**d**) from Gutiérezz et al. (2017) [4] with permission from Springer Nature via Copyright Clearance Center, Inc.)

teeth, called fangs, which are connected to venomous glands in the upper jaw. The length of the fangs determines whether the venom is introduced subcutaneously or intramuscularly. Snake venoms contain a variety of complex biochemical components, characterized by diversity and interspecies variations. These variations are influenced by a number of factors, the first of which is the snake species. Antivenoms are antibodies that are specific to one (monovalent) or several (polyvalent) species. This is the reason why antivenoms of one species or against snakes at one location may be ineffective in other species or regions [3, 4, 11, 12].

The multifocal toxicity of a venom determines the symptoms of poisoning, which can be classified as either viperine or cobraic syndromes. The volume of venom injected is under the control of the snake and has been associated with a variable degree of clinical effects. Viper venoms mostly induce local tissue damage with necrosis and systemic manifestations such as bleeding, coagulopathies, and hypovolemic shock. Elapide snake venoms predominantly manifest neurotoxic effects, such as neuromuscular paralysis which may result in respiratory failure. After a bite, the adverse effects include edema, cardiotoxicity, rhabdomyolysis, autonomic hyperactivity, local and systemic bleeding, compartment syndrome, superinfection, thrombotic microangiopathy and macrothrombosis, acute kidney injury, shock, and death (for more information, *please see* reviews [3–5, 10–12]). The only effective treatment against snake venoms are antivenoms. Because the prognosis for severely envenomed patients rapidly deteriorates within the first hours, snakebites are always classified as emergencies: Cardiovascular disturbances up to collapse can even develop within seconds and respiratory paralysis within minutes.

First aid focuses on immobilization of a bitten limb and patient transport to the nearest medical facility [3, 4, 11, 12]. Please remove any rings or other tight objects from the affected limb. The use of pressure immobilization is a topic of debate, but research suggests that pressure pads may help to reduce venom spread in viper bites. It is not advisable to use a tourniquet. Any delay in reaching a medical center must be avoided. Reassure patients, who may often be fearful and anxious. Take a medical history. If the situation deteriorates into a life-threatening emergency, resuscitate and stabilize vital signs. At the medical facility, try to identify the causative species. Perform a thorough clinical and laboratory examination (in low-resource settings, use the 20-min whole blood clotting test to detect coagulopathy) [11]. In case of signs of infection, administer appropriate antibiotics [11, 12, 25]. Check for local effects of envenomation, such as severe pain, swelling, blistering, bruising, and lymphangitis, compartment syndrome, and/or venom ophthalmia (Fig. 2).

Check again for systemic effects, such as impaired clotting, shock, muscle dysfunction, paralysis, and kidney injury. In the

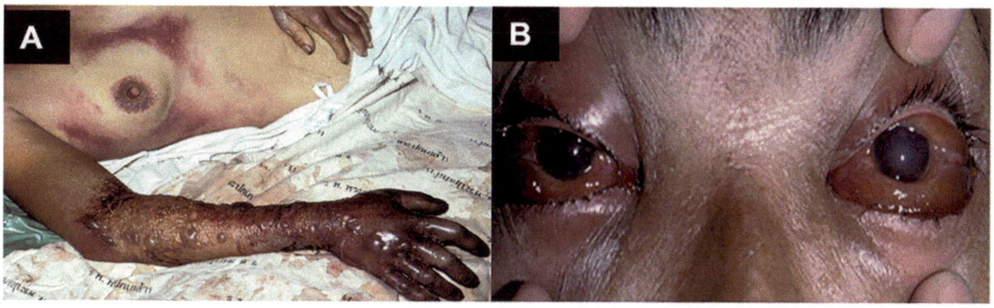

Fig. 2 (**a**) Blistering, swelling, and a gangrenous right hand, which required amputation after a Malayan pit viper bite in Thailand. Extensive ecchymoses (skin discoloration owing to bleeding under the skin) can be seen. (**b**) Bilateral conjunctival edema (chemosis) indicating a generalized increase in capillary permeability after an Eastern Russell's viper bite in Myanmar (from Gutiérezz et al. [4] with permission from Springer Nature via Copyright Clearance Center Inc., images courtesy of DA Williams, WHO Geneva, Switzerland)

event that antivenoms are not available, a prompt decision must be made regarding transfer to a facility with antivenoms.

At present, there is no test that can identify patients with a systemic spread of the venom. When clinically indicated (such as in combined diagnostic, clinical, and systemic envenoming symptoms), antivenoms must be administered as soon as possible (3,4,11,12,26). It is imperative to monitor patients after antivenom treatment for any adverse effects, such as allergic or pyrogenic reactions. Additional medical treatment remains symptom-oriented and supportive. In a "wait-and-see"approach, serial clinical examinations to check for envenomation signs should be repeated for at least 24 h. However, even after several days, delayed manifestations of envenoming have been reported.

Table 1 shows the favorable course of a pregnant patient who correctly identified the snake that had bitten her as a king cobra [27]. Several days after caesarean section delivery, the patient developed a compartment syndrome, necessitating multiple surgical procedures. Table 2 outlines the case of a zoo worker who was also bitten by a king cobra some years ago [28]. The situation was complicated by the development of an anaphylactic shock following the administration of antivenom. Both patients exhibited typical complications associated with snakebites. Antivenoms often are not immediately available, and consultation with poison control and toxicology centers for medical expertise should be sought as early as possible. Additionally, the second course illustrates one of the rare instances where antigenic agents must be reinitiated following anaphylaxis once stability has been restored, and preventive measures have been implemented.

Table 1
Patient report: 26-year-old term pregnant woman with envenomation from spectacled cobra bite complicated by compartment syndrome [27]

Course	Findings, management
Day 1	**Snake bite at dorsal right foot, snake identified as spectacled cobra**
45 min later	**Development of severe pain at bite site**
60 min	**Transport to nearby health facility**
	Administration of 10 vials of polyvalent antivenom
	Referral to All India Institute of Medical Sciences Center at Bhubaneswar
4 h	**Arrival in emergency department, Bhubaneswar**
	Pain in foot: 10/10 NRS—initial treatment with IV paracetamol 1 g
	Pregnancy at 34 weeks 3 days gestational age
	Vitals unremarkable except HR 110 bpm
	Negative "broken neck sign "(snakebite-induced paralysis of neck flexor muscles)
	2 fang marks on dorsal right foot, moderate swelling and tenderness extending to ankle, dorsalis pedis artery pulse present
	Blood tests: Elevated leucocytes ~14,000 x 10^3cumm, neutrophils ~92%, whole blood clotting and coagulation profile within physiological range
	Wound cleaning and dressing (existing tetanus immunization)
	Antibiotic therapy; second dose of 10 vials polyvalent antivenom
	Persistent pain: ultrasound-guided blocks of the superficial peroneal nerve and sural nerve with 3.5 mL 0.375% ropivacaine; result: pain relief, NRS 2/10
Transferral to	Exams: singleton live fetus in longitudinal lie and cephalic presentation
obstetrics	Relaxed uterus, fetal HR 191 bpm; no vaginal bleeding
	Sustained fetal tachycardia: emergency cesarean section
	Uneventful delivery of preterm female baby, 2140 g birth weight
	Managed at neonatal ICU for 48 h because of mild respiratory distress and tachypnea
	Monitoring of mother for changes in foot and limb swelling
Day 3	
	Development of blisters and local tissue necrosis over bite site
	Broad-spectrum antibiotics and anti-inflammatory drugs
Subsequent days	
	Extended foot swelling to below knee and rupture of blisters
	Recording of distal pulses at regular intervals;
	Daily monitoring for compartment syndrome

(continued)

Table 1 (continued)

Course	Findings, management
Day 7	
	Fasciotomy because of fear of compartment syndrome in view of weak peripheral dorsalis pedis artery pulsations
Subsequent days	
	Daily wound dressing changes
Day 14	
Within 1 week	Local pain subsides
Day 21	
Within 2 weeks	**Complete resolution of limb swelling**
21 days after bite	**Uneventful reconstruction of foot defect with free flap from front of thigh**
6 weeks later	
Follow-up	Asymptomatic post-snakebite patient; **mother and baby are doing well**

bpm beats per minute, *cumm* cubic millimeter, *g* grams, h hours, *HR* heart rate, *ICU* intensive care unit, *IV* intravenous, *mL* milliliter, *NRS* numeric rating scale

3 Ketamine in Snakebite Pain Envenomations

In a dedicated snakebite clinic in Guinea, 12 patients, between 14 and 64 years old, were treated by physicians with extensive experience in snakebite therapy in Africa [23]. The protocol implemented for the administration of ketamine was based on standard dose regimens established in high-resource emergency departments, but the doses necessary were lower than those usually needed for out-of-hospital analgesia after severe trauma [29, 30]: on average, 5.75 hours after the snakebite, a dose of 0.05–0.1 mg/kg intravenous ketamine was administered over 2 min in order to treat significant pain, with or without the presence of swelling or blistering of the affected limb. Within 1 min, all patients reported a reduction in pain and felt more comfortable while showing a decrease in hypertension and tachycardia. Nine patients required only a single dose of ketamine. No adverse interactions were observed between ketamine and the applied polyvalent antivenom, which was recently introduced and shown to be effective and well tolerated in Cameroon (Inoserp™ Pan-Africa, Inosan Biopharma, S.A., Spain) [31]. After ketamine and antivenom treatment, paracetamol was the sole analgesic required for the management of

Table 2
Patient report: 22-year-old male zoo shop worker with envenomation from king cobra bite complicated by anaphylactic shock (with a history of asthma treated by inhalers) [28]

Course	Findings, management
Day 1	**Snake bite at lateral aspect of left index finger**
20 min later	**Arrival in accident and emergency room** in Bristol Hospital, United Kingdom
	HR 124 bpm, "dizziness"
	Swollen left index finger with double puncture bite marks, superficial glass lacerations
Within 10 min	**Bilateral eyelid ptosis**
	IV access, 1000 mL saline 0.9%, blood tests
	Contact with Toxbase and Liverpool School of Tropical Medicine:
	Urgent order of specific antivenom from Guy's/St. Thomas's Hospital Poisons Unit, London
	Pressure bandage of entire limb to shoulder for prevention of proximal lymphatic drainage of large molecular weight toxins
20 min later	**Complaints of shortness of breath and inability to swallow saliva**
	Endotracheal intubation per rapid sequence induction
	Sedation and analgesia with propofol and morphine
90 min later	**Transfer to intensive care unit;** arterial BP monitoring, large bore IV access, urinary and nasogastric catheters
	Unremarkable electrocardiography, chest X-ray, and blood tests except for neutrophilia
Intensive care unit	Hypertension (systolic BP 200–220 mm Hg), erratic HR 45–85 bpm
6 h after bite	**20 vials monospecific freeze-dried equine antivenom delivery for *Ophiophagus hannah* envenomation (manufact. by Queen Saovabha Memorial Institute Bangkok, Thailand);** reconstituted in 200 mL water for injection, made up to 500 mL with saline 0.9%, and infused at a rate of 2 mL/h
15 min later	No adverse reactions; **increase in infusion rate to 999 mL/h to administer full dose within 30 min as recommended by manufacturer**
5 min after increase	**Further systolic BP rise up to ~ 250 mm Hg; treated with antihypertensive**
	Development of 2 maculopapular spots on the chest; 200 mg IV hydrocortisone, 10 mg IV chlorphenamine, and 50 mg IV ranitidine
Within 10 min	**Rash progresses to urticaria, finally involving entire chest, flanks, and upper thighs then severe bronchospasm (peak airway pressure 40–50 cm H_2O, maximum tidal volumes 300–350 mL), S_aO_2 drop to 80% with $F_iO_2 = 1$**
Within 1 min	**Systolic BP decrease to 80 mm Hg**
	STOP of antivenom infusion; tentative/probable diagnosis: ANAPHYLAXIS

(continued)

Table 2 (continued)

Course	Findings, management
	10 μg IV adrenaline boluses up to 100 μg and start of 0.1 μg/kg/min adrenaline infusion
	Rapid infusion of 500 mL hydroxyethyl starch and administration of bronchodilators
After 25 min	**Continued anaphylaxis therapy; resolution of bronchospasm and hypotension, rash invisible**
	Further advice from Liverpool; re-start of antivenom infusion at lower rate of 100 mL/h and additional 200 mg hydrocortisone in 24 h
	Result: Rash reappeared, but no more serious adverse reactions occurred
9.5 h post-bite	**Antivenom infusion finished; cessation of adrenaline**; removal of pressure dressings
17 h post-bite	Clinical examinations after reducing sedation: **regain of motor function, head lift from pillow, no ptosis, strong limb power, tidal volumes > 10 mL/kg, good cough reflex**
	Stop of sedation and tracheal extubation
Day 2	
	Discharge from intensive care unit
	Transferral to plastic surgery for care of isolated ischemia of bitten finger

BP blood pressure, *bpm* beats per minute, *cumm* cubic millimeter, F_iO_2 fraction of inspired oxygen, *kg* kilograms, *h* hour (s), *HR* heart rate, *IV* intravenous, *mL* milliliter, S_aO_2 oxygen saturation

residual pain in all patients. Apart from a slightly sedative effect at a dose of 0.1 mg/kg, no other ketamine-related adverse effects were observed. The authors conclude that ketamine can be employed as an efficacious analgesic for the management of pain associated with snakebites. They also propose that it can be administered as a bridge for pain treatment until antivenom therapy is available [23]. In addition to providing adequate analgesia, ketamine may reduce the development of post-bite PTSD, depression, and other neuropsychiatric problems. In non-snakebite settings, ketamine has been shown to improve such disorders in diverse populations when they have become chronic as well [22, 32, 33].

In impoverished areas, snakebite victims may not seek help in medical centers out of fear of high treatment costs. However, in certain regions, even traditional healers are nowadays open for collaboration with health services to administer antivenom to snakebite victims [34]. When it comes to ketamine, the cost is relatively low in many low- to low-middle-income countries. At the clinic in Guinea where the pain study patients were treated, the average costs per effective ketamine dose were $0.03–0.06 USD

[23]. The lead study author from the United States, who utilizes ketamine in everyday emergency medicine, therefore stated that he believes that "withholding a safe and effective (analgesic) therapy despite availability at the local pharmacy at an affordable price, might be considered as dereliction of professional duty as a physician" [23].

4 Take-Home Message

Antivenoms represent the primary treatment against venomous snakebites. Snakebite envenomation constitutes a medical emergency that requires immediate treatment. The initial symptoms of envenomation may manifest as severe, localized pain, which can result in significant, long-term complications. In addition to the physical disabilities that often result from snakebites, victims frequently develop serious psychological problems, with a high incidence of PTSD and depression. Ketamine is a safe, effective, and affordable analgesic for the management of pain associated with snakebites. It is often available even in resource-poor settings. In addition to providing acute pain control, ketamine may prevent post-snakebite psychological disorders and may be useful for the prevention of pain chronification. Further scientific investigation is required to clarify the role of ketamine in physical and psychological snakebite traumas.

Conflicts of interest

The authors declare no conflicts of interest.

Disclaimer

The opinions or assertions contained herein are the private views of the authors and are not to be construed as official or as reflecting the views of the Department of the Army or the Department of Defense of the USA.

References

1. Kofi A (2018) The biggest public health crisis you have never heard of. In: Kofi Annan Foundation, Geneve, Switzerland. Snakebite: The biggest public health crisis you've never heard of (kofiannanfoundation.org). Accessed 01 Oct 2024
2. World Health Organization (2023) Snakebite envenoming. In: News-Room / Factsheets / Details. Available at Snakebite envenoming (who.int). Accessed 1 Oct 2023
3. Warrell DA, Williams DJ (2023) Clinical aspects of snakebite envenomings and its treatment in low-resource settings. Lancet 10385: 1382–1398

4. Gutiérezz JM, Calvete JJ, Habib AG et al (2017) Snakebite envenoming Nat Rev Dis Primers 3:17063
5. World Health Organization (2023) Snakebite envenoming. In: Health topics. Available at Snakebite (who.int). Accessed 10 Oct 2024
6. Martinez PA, da Fonseca Teixeira IB, Siquera-Silva T et al (2024) Climate change related distributional range of shifts of venomous snakes: a predictive modelling study of effects on public health and biodiversity. Lancet Planet Health 8:e163–e171
7. Wellcome Trust (wellcome.org) (2021) Press release / Snakebites—it's time to solve the world's biggest health crisis. Available at Snakebites – it's time to solve the world's biggest hidden health crisis (wellcomeorg) Accessed 5 Sep 2024
8. Kumar R, Rathore AS (2024) Snakebite management: the need for reassessment, international relations, and effective economic measures to reduce the considerable SBE burden. J Epidemiol Glob Health 14:586–812
9. Abouyannis M, Esmail H, Hamaluba M et al (2023) A global core outcome measurement set for snakebite clinical trials. Lancet Glob Health 11:e296–e300
10. Bhaumik S, Beri D, Lassi ZS, Jagnoor J (2020) Interventions for the management of snakebite envenoming: an overview of systematic reviews. PLoS Negl Trop Dis 14:e0008727
11. Ralph R, Faiz MA, Sharma SK et al (2022) Managing snakebite. BMJ 376:e057926
12. Seifert SA, Armitage JO, Sanchez EE (2022) Snake envenomation. N Engl J Med 386:68–78
13. Nielsen VG, Wagner MT (2022) Review of the mechanisms of snake venom induced pain: it's all about location, location, location. Int J Mol Sci 23:2128
14. Ma X, Yan J, Jiang H (2023) Application of ketamine in pain management and the underlying mechanism. Pain Res Manag 2023: 1928969
15. van Meijel EPM, Gigengack MR, Verlinden E et al (2019) The association between acute pain and posttraumatic stress symptoms in children and adolescents 3 months after accidental injury. JCPMS 26:88–96
16. Hua R, Yang Y, Yao A et al (2021) Effect of intensive analgesia on post-traumatic stress disorder in patients with acute trauma. Chin J Emerg Med 12:217–220
17. Bhaumik S, Kallakuri S, Kaur A et al (2020) Mental health conditions after snakebite: a scoping review. BMJ Glob Health 5:e004131
18. Habib ZG, Salihu AS, Hamza M et al (2023) Post-traumatic stress disorder and psychosocial impairment following snakebite in Northeastern Nigeria. Int J Psychiatry Med 56:97–115
19. Nnaji CT (2022) Ketamine and low-resource countries. In: Saiz-Sapena N, Gil MG (eds) Ketamine revisited. IntechOpen, London, GB. Accessed 3 Oct 2024
20. Ooms GI, Usman MA, Reed T et al (2024) The impact of scheduling ketamine as an internationally controlled substance on anaesthesia care in Sub-Saharan Africa: a case study and key informant interviews. BMC Health Serv Res 24:598
21. Magruder T, Isenhart M, Striepe MV et al (2024) Ketamine—an imperfect wonder drug? Biochem Pharmacol 229: 116516
22. Feder A, Costi S, Rutter SB et al (2023) A randomized controlled trial of repeated ketamine administration for chronic posttraumatic stress disorder. Focus (Am Psychiatr Publ) 21: 296–305
23. Brandehoff N, Benjamin JM, Balde C, Chippaux JP (2020) Ketamine for pain control of snake envenomation in Guinea: a case series. Toxicon 187:82–85
24. Martin G (2021) Spectacled cobra: an unwilling item. In: RoundGlass sustain—an encyclopedia of Indian's wildlife. Available via RoundGlass sustain. Available at https://roundglasssustain.com/species/spectacled-cobraSpectacled Cobra: Accessed 27 April 2025
25. Larréché S, Bousquet A, da Silva L et al (2023) Antibiotic susceptibility of cultivable microbiota from the oral cavity of captive *Bothrops atrox* and *Bothrops lanceolatus*: implications for the treatment of snakebite-associated infections in the French departments of America. Infect Dis Now 53:104721
26. Mutricy R, Heckmann X, Douine M et al (2018) High mortality due to snakebites in French Guiana: time has come to re-evaluate medical management protocols. PLoS Negl Trop Dis 12:e0006482
27. Panda R, Singh S, Radhakrishnan RV et al (2023) A case of cobra bite in a term pregnant woman: the obstetric and wound management challenges. Wilderness Environ Med 34:571–575
28. Veto T, Price R, Silsby JF, Carter JA (2007) Case report: Treatment of the first known case of king cobra envenomation in the United Kingdom, complicated by severe anaphylaxis. Anaesthesia 62:75–78

29. Guo J, Zhao F, Bian J et al (2024) Low-dose ketamine versus morphine in the treatment of acute pain in the emergency department: a meta-analysis of 15 randomized, controlled trials. Am J Emerg Med 76:140–149
30. Le Cornec C, Le Pottier M, Broch H et al (2024) Ketamine compared with morphine for out-of-hospital analgesia for patients with traumatic pain: a randomized clinical trial. JAMA Netw Open 7:e2352844
31. Chippaux JP, Ntone R, Benhammou D et al (2023) Real life condition evaluation of Inoserp PAN-AFRICA antivenom effectiveness in Cameroon. PLoS Negl Trop Dis 17:e0011707
32. Liu JJW, Ein N, Gervasio J et al (2024) Ketamine in the effective management of chronic pain, depression, and posttraumatic stress disorder for Veterans: a meta-analysis and systematic review. Front Psych 15:1338581
33. Rohde J, Hickmann E, Buchmann M et al (2024) Combined effects of nasal ketamine and trauma-focused psychotherapy in treatment-resistant post-traumatic stress disorder: a pilot case series. Behav Sci (Basel) 14:717
34. Chuat M, Alcoba G, Eyong J et al (2021) Dealing with snakebite in rural Cameroon: a qualitative investigation among victims and traditional healers. Toxicon X Jul 15:9–10

Chapter 22

Ketamine: Anesthesia and Analgesia in War Zones and Disasters and the Scheduling Issue in Resource-Limited Countries

Jannicke Mellin-Olsen and Sabine Himmelseher

Abstract

Worldwide, 1.8 billion people live in conflict-affected areas. Providing safe anaesthesia and surgery to victims of war and natural disasters in resource-limited settings remains a significant global problem. Ketamine has the advantage of providing analgesia and anesthesia while maintaining upper airway reflexes and spontaneous breathing. With skilled use, this allows some surgical procedures to be performed without advanced airway management. Ketamine is the drug of choice when access to anesthesia machines is not available. It does not affect hemodynamic stability which is important in severe bleeding situations. Because ketamine is sometimes abused as recreational drug, some countries have requested that ketamine be scheduled under international drug control conventions. However, the World Health Organization's Expert Committee on Drug Dependence has repeatedly not recommended such restrictions, and in 2016, the country requesting scheduling withdrew the proposal before it came to a vote at the United Nations Commission on Narcotic Drugs. If such scheduling were to occur, the availability of ketamine would decrease dramatically, with the greatest impact on the health of people living in low- and middle-income countries.

Key words Ketamine, War, Man-made and natural disasters, Anesthesia, Analgesia, Care provision, Safety, Scheduling

1 Introduction

Surgery of war and natural disaster victims often takes place in resource-strained infrastructure, and collapse of supply chains might require improvisations and compromises in patient management [1]. In humanitarian crises, caregivers will often face very different geographical, cultural, social, and many other local factors [2] than they are accustomed to. Electricity, oxygen supply, and monitoring might be unreliable or absent, and patient numbers might by far outweigh the capacity of caregivers. Furthermore, patients will present with different injury patterns and other

Sabine Himmelseher (ed.), *Ketamine: Pain Management, Critical Care, Depression, and Resource-Poor Settings Volume II*, Neuromethods, vol. 225, https://doi.org/10.1007/978-1-0716-4599-4_22,

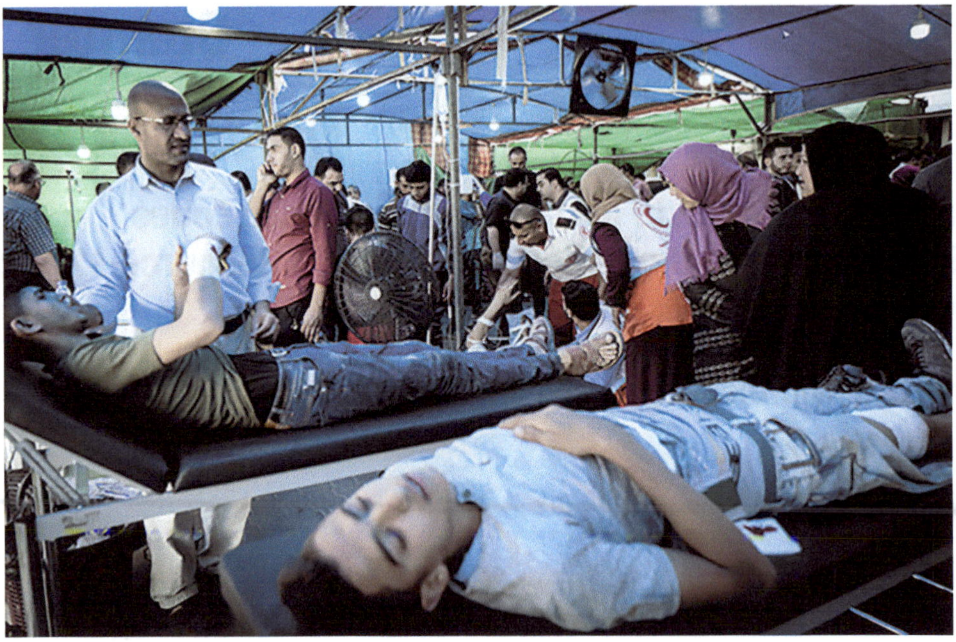

Fig. 1 Photo showing a busy, overcrowded 30-bed makeshift emergency room erected by the International Red Cross (ICRC) in front of the Shifa Hospital, Gaza, in 2018. The capacity to treat patients who need surgical treatment due to gunshot wounds is increased (provided by and with permission from the ICRC, Geneva, Switzerland, without publication restrictions; photographer: Alyona Synenko, May 14, 2018, ICRC Number V-P-IL-E-03156)

disorders than in "normal" civilian prehospital emergencies [3] (Fig. 1).

2 War Zones and Disasters

When ketamine first was approved by the US Food and Drug Administration in 1970, it was for use as a field anesthetic in the Vietnam war [4, 5]. It was soon considered to be a "good, safe anesthetic" mainly because it did not destabilize hemodynamics in trauma patients. However, it was also observed to exert undesirable psychoactive events, such as hallucinations, especially during emergence from ketamine monoanesthesia. If medically feasible and available, a benzodiazepine such as diazepam was therefore combined with ketamine, reducing or preventing adverse psychic effects. After safe and effective use by the British Army in an armed conflict in the Middle East a short time later, ketamine became the "drug of war" [5], due to its ease of application via various routes, its analgesic and hypnotic action, and with skilled use, relatively wide safety margin maintaining upper airway reflexes and spontaneous respiration.

Achieving analgesia at subanesthetic dose, ketamine exerts opioid-sparing effects or may abolish opioid requirements throughout the continuum of war care provision, from Level 1 outpatient emergency care, via Level 2 inpatient surgical field hospital emergency care, to Level 3 inpatient referral hospital and intensive care [1, 5–12]. In a patient safety perspective, this is extremely important, as the range between hypoventilation and apnea is narrow. Less experienced caregivers unable to detect and manage airway and ventilatory hazards might not recognize and treat hypoventilation and apnea until it is too late. The same is valid if the caregiver does not have access to the patient's airways in entrapment situations, if airway equipment is not available, or if there is a lack of caregivers and monitoring, e.g., during recovery from anesthesia. In addition, opioid availability in conflict and disaster areas may be difficult due to scheduling or import restrictions [1].

Another safety aspect of ketamine, especially in resource-strained situations, is the fact that respiratory reflexes and function are maintained even with anesthetic doses. Therefore, ketamine anesthesia can be used for several procedures, including Cesarean section, which normally would require endotracheal intubation. However, in ketamine anesthesia, these procedures can be performed without advanced airway management [1] (Fig. 2).

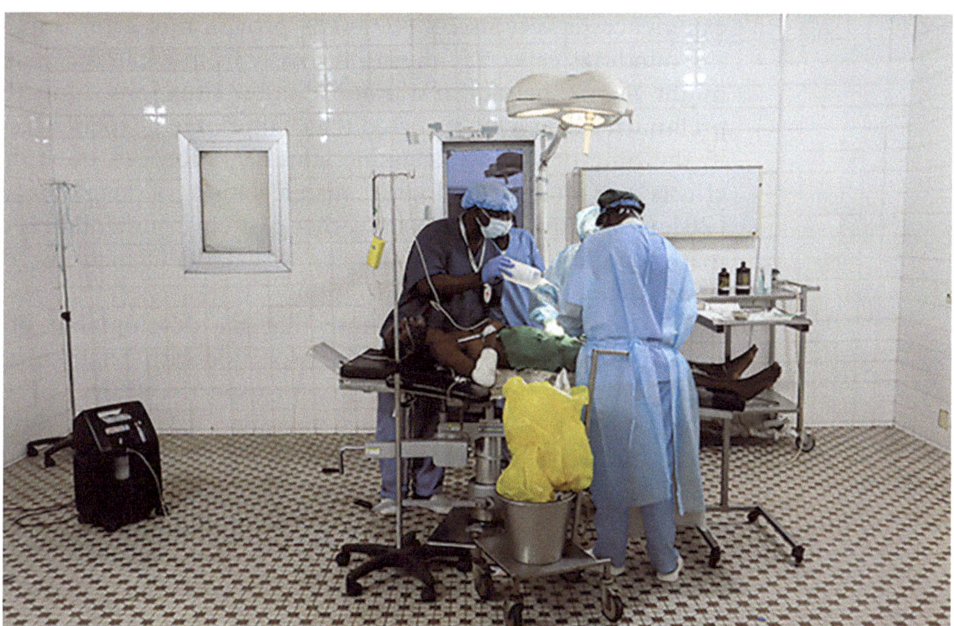

Fig. 2 Photo depicting the conditions under which a medical team has to operate at the university hospital centre, Adré, Abéché, Sudan, Chad. No anesthesia machine can be seen; the patient receives oxygen via nasal cannula from an oxygen concentrator, only (provided by and with permission from the International Red Cross (ICRC), Geneva, Switzerland, without publication restrictions; photographer: Denis Sassou, July 7, 2023, ICRC Number V-P-TD-E-00737)

For instance, after the Haiti earthquake, disaster response teams from the humanitarian organization Doctors without Borders administered general anesthesia with ketamine in approximately 90% of interventions [13].

Furthermore, in severe bleeding situations, including hemorrhagic shock in obstetric emergencies, ketamine does not worsen the hemodynamic situation [14]. Sometimes, trauma patients who undergo damage control surgery may or cannot be extubated during transport to the next level center [1]. Ketamine infusion is indicated for patients that need sedation and analgesia for transport and transfer, including air evacuation, as it requires less equipment and logistics [15, 16]. In the post-acute and intensive care phase, ketamine is very useful for painful procedures such as frequent burn wound dressing changes [17, 18].

As delineated, ketamine dose-dependently can induce undesirable medical and psychoactive effects. These may manifest as nausea, dizziness, hallucinations, nightmares, and even "badtrips" or war experience flashbacks with panic attacks in awake patients and those recovering from especially ketamine monoanesthesia. If medically possible and available, such events should be prevented or reduced by co-administration of a gamma-aminobutyric acid (GABA) agonist. In civilian conditions, patient guidance, positive suggestions, and mental support by care providers during such experiences is very useful [19]. The provision of calm, supportive surroundings especially during recovery from ketamine anesthesia might also be helpful in war and disaster situations. Nonetheless, preliminary data have reported that "imprinted" experiences made in the time preceding ketamine administration can exert delayed effects on phenomenological manifestation of ketamine action [20]. Qualitative data also show that past environmental exposures may manifest as, for example, visual hallucinations during subsequent ketamine administration.

In this context, it is of note that the development of posttraumatic stress disorders was not associated with ketamine administration during hospitalization in a retrospective analysis of combat casualties and wounded service members [21–23]. Conversely, when the development of psychological trauma after extrication from the rubble during the 2016 earthquake in Amatrice, Italy, was examined 3 years later, the administration of 0.3–0.5 mg/kg ketamine compared to 0.1–0.15 mg/kg morphine was associated with less trauma-related psychological disorders [24]. Subanesthetic dose ketamine has been proposed to offer stress-resilience-enhancing effects [25], and in a recent meta-analysis, perioperative esketamine has been demonstrated to prevent and reduce postpartum depression after Cesarean section [26]. Furthermore, another meta-analysis published in 2024 reported that ketamine can be effective in approaching chronic pain, depression, and posttraumatic stress disorder in veterans [27].

3 Resource-Limited Countries and the Scheduling Issue

Around 1.8 billion people live in conflict-affected areas worldwide [2]. Conflicts disrupt the healthcare system and increase the exodus of healthcare workers in resource-limited countries. Safe anaesthesia and surgery continue to be a significant problem under such conditions. In 2015, the Lancet Commission on Global Surgery reported that five out of seven billion people in this world lack access to safe, affordable anaesthesia and life-saving surgical procedures [28] (*please see* more details in Chapter **20** by Nnaji). This causes more than four times as many deaths than HIV, malaria, and tuberculosis combined. A vast majority of these people live in low- and lower-middle-income countries (LMICs). In a 2021–2023 survey, the World Federation of Anaesthesiologists (WFSA) demonstrated that there are still more than 75 member countries with a physician anesthesia provider density <5 per 100,000 inhabitants [29]. Anesthesia services from LMICs also reported that there remains a huge lack of accessibility to well-maintained anesthesia machines and equipment due to financial constraints and infrastructure challenges [30] and that there is still inconsistent supply of electricity, oxygen, and running water especially in rural areas. Therefore, these workforce and material resource-scarce settings depend heavily on racemic ketamine as anesthetic drug of choice for surgeries [31] (Fig. 3).

However, ketamine is and has been abused as illicit recreational drug due to its psychic side effects. Many countries regulate the trade and possession of narcotic drugs by law. Without special license or prescription, their use is considered illegal, or the drugs may not be allowed to be prescribed at all. The United Nations Commission on Narcotic Drugs (UNCND) is the political body which decides whether substances (such as ecstasy) are brought under the control set of the UN drug control conventions. Some years ago, a few countries had requested to place ketamine under these international drug control conventions claiming that this would be decisive to counter its illicit use. In contrast, several medical associations and civil society groups argued that this would severely limit the availability, accessibility, and affordability of ketamine in those countries who need this substance the most. From a medical care provision and human rights perspective, such efforts are therefore of serious concern [32–34]. As yet, the WHO Expert Committee on Drug Dependence (ECDD) has evaluated data on ketamine abuse several times, every time concluding that its misuse does not pose "significant global public health risks" of such an extent that would warrant regulations of ketamine as a globally scheduled drug [35]. Until now, previous scheduling proposals have not been brought to a vote by the UNCND [32, 33, 35, 36]. The significance and associated problems of such scheduling is

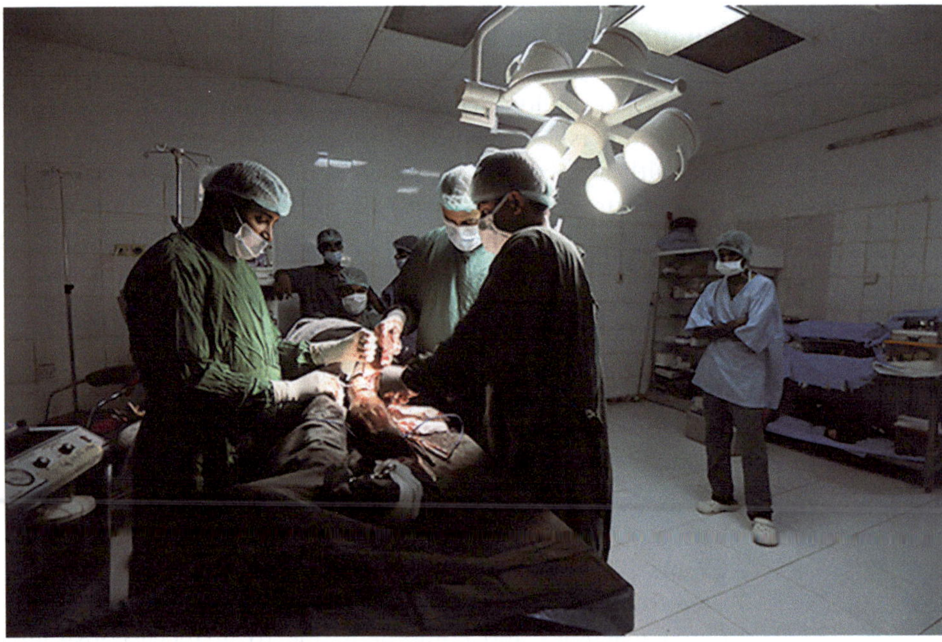

Fig. 3 Photo illustrating "poor light" circumstances in Al Thawra Hospital, Hodeidah, Yemen, under which an International Red Cross (ICRC) surgical team has to work. The ICRC team had been deployed to provide technical and operational support to the local hospital staff (provided by and with permission from the ICRC, Geneva, Switzerland, without publication restrictions; photographer: Abduliabbar Zeyad, March 29, 2017, ICRC Number V-P-YE-E-01644)

illustrated in the experiences made after morphine scheduling in India in 1985: the massively increased bureaucratic hurdles and exuberant rules led to a 97% reduced prescription over 10 years [37]. Meanwhile, improved access to opioids has been re-found in some Indian states, but the United Nations Office on Drugs and Crime published an interview in April 2024 in which a lead physician complained that, e.g., palliative care patients are still massively undertreated with opioids [38]. Clearly, it is expected that such a scenario will also develop with ketamine should it be designated a globally scheduled drug. For public health in LMICs, this would be a catastrophe and put millions of patients at risk for suffering and losing their lives (Fig. 4).

4 Conclusion

The World Health Organization (WHO) has estimated that 1.6 billion people live in areas with conflicts worldwide [2]. The provision of safe anesthesia and surgery in such settings and in natural disaster situations when there are limited resources is not a simple task, and ketamine can offer considerable benefits. Although it carries a risk of misuse for recreational purposes, ketamine must

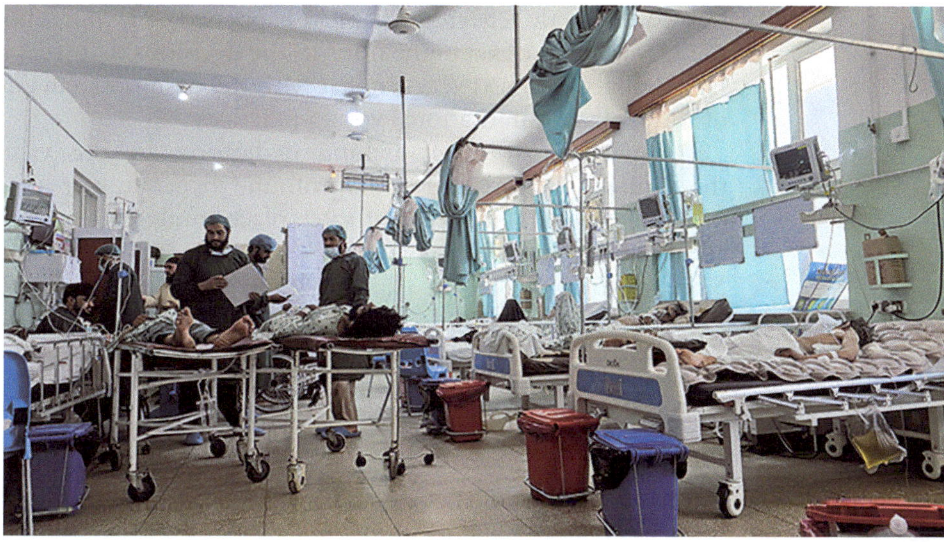

Fig. 4 Photo showing the surgical ward in Mirwais Hospital, Kandahar province, Afghanistan, in November 2023. Thanks to the hybrid solar energy system installed by the International Red Cross (ICRC), the hospital can operate up to 23 hours a day (provided by and with permission from the ICRC, Geneva, Switzerland, without publication restrictions; photographer: Mohammad Masoud Samimi, May 22, 2024, ICRC Number V-P-AF-E-0299).

remain available with easy access for health professionals around the world.

In Memory of Jannicke Mellin-Olsen

Very sadly, Dr. Jannicke Mellin-Olsen passed away only a few weeks after we completed this chapter. She was a former Secretary of the European Society of Anaesthesiology and Intensive Care, former President of the WFSA, first non-American member of the Patient Safety Movement Foundation Board, and active member of many international non-for-profit organizations.

For four decades, she inspired anesthesiologists and colleagues from all disciplines with her commitment for patient safety, playing an instrumental role in the 2010 Helsinki Declaration. Dr. Mellin-Olsen was also a strong advocate for equitable access to healthcare worldwide, which profoundly influenced anesthesia practice on a global scale. And she was a staunch proponent of ketamine, having made good experiences with the substance during humanitarian missions for the UNIFIL, ICRC, and in other tasks such as air ambulance flights and remote delivery of medical care. When we were working on this chapter, she even referred to her old emails from that time when some countries attempted to schedule ketamine at the United Nations (she insisted that every word we used in this book exactly reflected what had happened). Having lobbied at the UN in favor of ketamine prior to the decisions that were made, Dr. Mellin-Olsen was very familiar with the cosmopolitan customs of the UN and international voting practices.

Jannicke, may your lifelong dedication to the values of the United Nations and your profound belief in our ability to make the world a better and safer place guide us, in the times ahead, and forever. Thank you.

S. Himmelseher

Conflicts of Interest

The authors declare no conflicts of interest.

References

1. Falovic R, Graves M, Hendel S (2024) Anaesthesia safety in war zones. In: Lundgren-Steele-C, Warner M (eds) World Federation of Anaesthesiologists 2024, Update in anaesthesia. Dean Bradley House, London, pp 21–25
2. WHO (2023) Health topics. Building peace in fragile and conflict settings through health. Available at https://www.who.int/activities/building-peace-in-fragile-and-conflict-settings-through-health#:~:text=While%20health%20often%20deteriorates%20in%20conflict%2C%20health%20can,such%20as%20vaccinating%20children%20or%20healing%20the%20wounded. Accessed 27 April 2025
3. Lockett C, Naylor JF, Fisher AD et al (2023) A comparison of injury patterns and interventions among US military special operations versus conventional forces combatants. Med J (Ft Sam Houst Tex) Jan–Mar:64–69
4. Mion G (2017) History of anaesthesia: the ketamine story—past, present, and future. Eur J Anaesthesiol 35:571–578
5. Mercer SJ (2009) "The drug of war" a historical review of the use of ketamine in military conflicts. J R Nav Med Serv 95:145–150
6. Lyon RF, Schwan C, Zeal J et al (2018) Successful use of ketamine as a prehospital analgesic by pararescuemen during operation enduring freedom. J Spec Oper Med 18:70–73
7. De Rocquignay G, Dubecq C, Martinez T et al (2020) Use of ketamine for prehospital pain control on the battlefield: a systematic review. Trauma Acute Care Surg 88:180–185
8. Leslie E, Pittman E, Drew B et al (2021) Ketamine use in operation enduring freedom. Mil Med 186:e720
9. Cohen B, Tamly T, Gelikas S et al (2022) Opioid sparing effect of ketamine in military prehospital pain management—a retrospective study. J Trauma Acute Care Surg 93(2S Suppl 1):S71–S77
10. Dubecq C, Montagnon R, Morand G et al (2023) Combat casualties treated with intranasal ketamine for prehospital analgesia: a case series. J Spec Oper Med 23:84–87
11. Hudson IL, Staudt AM, Burgess M et al (2023) Patterns of palliation: a review of casualties that received pain management before reaching role 2 in Afghanistan. Mil Med 188:108–116
12. Westcott SL, Wojahn A, Morrison TC et al (2023) Ketamine decreased opiate use in US military combat operations from 2010 to 2019. BMJ Mil Health 27:e002291
13. Trelles Centurion M, Van Den Bergh R, Gray H (2017) Anaesthesia provision in disasters and armed conflicts. Curr Anaesthesiol Rep 7: 1–7
14. Watso JC, Huang M, Hendrix JM et al (2023) Comparing the effects of low-dose ketamine, fentanyl, and morphine on hemorrhagic tolerance and analgesia in humans. Prehosp Emerg Care 27:600–612
15. Maddry JK, Mora AG, Savell S et al (2016) Combat MEDEVAC: a comparison of care by provider type for en route trauma care in theater and 30-day patient outcomes. J Trauma Acute Care Surg 81(5 Suppl 2):S104–S110
16. Schauer SG, Naylor JF, Davis WT et al (2022) An analysis of prolonged, continuous ketamine infusions. Mil Med 187:e547–e553

17. Schauer SG, April MD, Aden AK et al (2019) Impact of ketamine infusion versus alternative regimens on mortality among burn intensive care unit patients: implications for prolonged field care. J Spec Oper Med 19:77–80
18. Maani CV, Hoffman HG, Fowler M et al (2011) Combining ketamine and virtual reality pain control during severe burn wound care: one military and one civilian patient. Pain Med 12:673–678
19. Cheong SH, Lee KM, Lim SH et al (2011) The effect of suggestion on unpleasant dreams induced by ketamine administration. Anesth Analg 112:1082–1085
20. Garel N, Levesque JT, Sandra DA et al (2023) Imprinting: expanding the extra-pharmacological model of psychedelic drug action to incorporate delayed influences of sets and settings. Front Hum Neurosci 17: 1200393
21. McGhee LL, Maani CV, Garza TH et al (2014) The intraoperative administration of ketamine to burned U.S. service members does not increase the incidence of post-traumatic stress disorder. Mil Med 179:41–46
22. Mion G, Le Masson J, Granier C, Hoffmann C (2017) A retrospective study of ketamine administration and the development of acute or post-traumatic stress disorder in 274 war-wounded soldiers. Anaesthesia 72: 1476–1483
23. Highland KB, Soumoff AA, Spinks EA et al (2020) Ketamine administration during hospitalization is not associated with posttraumatic stress disorder in military combat casualties: a matched cohort study. Anesth Analg 130:402–408
24. Petrucci E, Cofini V, Pizzi B et al (2023) Health status perception and psychological sequelae in buried victims: an observational study on survivors of the earthquake in Amatrice (Italy), three years later. Prehospital Disaster Med 38:193–198
25. Evers AG, Murrough JW, Charney DS, Costi S (2022) Ketamine as a prophylactic resilience-enhancing agent. Front Psych 13:833259
26. Parsaei M, Hasehmi SM, Seyedmirzaei H et al (2024) Perioperative esketamine administration for the prevention of postpartum depression after the cesarean section: a systematic review and meta-analysis. J Affect Disord 361: 564–580
27. Liu JJW, Ein N, Gervasio J (2024) Ketamine in the effective management of chronic pain, depression, and posttraumatic stress disorder for veterans: a meta-analysis and systematic review. Front Psych 15:1338581
28. Meara JG, Leather AJ, Hagander L, Alkire BC, Alonso N, Ameh EA et al (2015) Global surgery 2030: evidence and solutions for achieving health, welfare, and economic development. Lancet 386:569–624
29. Law TJ, Lipnick MS, Morriss W, Gelb A, Mellin-Olsen J et al (2024) The global anaesthesia workforce survey: updates and trends in the anaesthesia workforce. Anesth Analg 139: 15–24
30. Khan IA, Karim HMR (2023) Anaesthesia services in low- and middle-income countries: the fragile point for safe surgery and patient safety. Cureus 15:e43174
31. Dong TT, Mellin-Olsen J, Gelb AW (2015) Ketamine: a growing global health-care need. Br J Anaesth 115:491–493
32. Lohman D, Barrett D (2020) Scheduling medicines as controlled substances: addressing normative and democratic gaps through human rights-based analysis. BMC Int Health Hum Rights 20:10
33. Downing RA (2022) The One Health Vision requires veterinarians and physicians to collaborate to prevent bioethical challenges created by restricted access to ketamine in low- and middle-income countries. J Am Vet Med Assoc 260:1445–1450
34. Ooms GI, Usman MA, Reed T, van den Ham HA, Mantel-Teeuwisse AK (2024) The impact of scheduling ketamine as an internationally controlled substance on anaesthesia care in sub-Saharan Africa: a case study and key informant interviews. BMC Health Serv Res 24:598
35. WHO Expert Committee on Drug Dependence (2016) Thirty-seventh report. WHO technical report series 998. ISBN: 978-92-4-120998-4, 34 pages. Available at https://www.who.int/publications/i/item/9789241209984. Accessed 27 April 2025
36. United Nations Office on Drug and Crime. Ketamine scheduling. Available at https://www.unodc.org/unodc/site-search.html?q=mission+and+documents+ketamine+scheduling. Accessed 27 April 2025
37. Mohan C, Bansal BL (2005) Narcotic Drugs & Psychotropic Substances Act, 1985, 1985–2005. Cee Emm Publishers, Chandigarh
38. Rajagopal MR (2024) India: the principle of balance to make opioids accessible for palliative care. UNODC South Asia. Available at https://www.unodc.org/southasia/frontpage/2011/april/interview-m-r-rajagopal-access-to-opioids-for-palliative-care.html. Accessed 27 April 2025

Chapter 23

Ketamine and Esketamine for the Treatment of Refractory Status Epilepticus Induced by Organophosphorus Compounds

Frederic Dorandeu, Karine Thibault, and Greg Dal-Bo

Abstract

Organophosphorus nerve agents (OPNA) are the most dangerous chemical warfare agents known. They are an immense threat acting as potent irreversible cholinesterase inhibitors in the nervous system. OPNA induce an immediate hypercholinergic activity responsible for most of the initial toxidrome. During intoxications with high sublethal to lethal doses of OPNA, the excess of acetylcholine in the brain triggers seizures and leads to secondary recruitment of the excitatory glutamate system. Through overstimulation of its receptors, including the ionotropic N-methyl-D-aspartate receptors (NMDAR), glutamate is thought to play a prominent role in the long-lasting maintenance of seizure activity, status epilepticus (SE), and the development of seizure-related brain damage. Ketamine is a NMDAR antagonist clinically used in many medical areas. Despite numerous preclinical and clinical publications reporting ketamine's promising properties, the agent is still not widely recognized as a treatment for epileptic conditions. Indeed, when combined with other antiepileptic drugs, it may terminate seizures and prevent neuronal damage and neurobehavioral deficits even when seizures have evolved into refractory SE. In this narrative review, we discuss the data on OPNA-related SE and treatment approaches with ketamine published since 2017. Some standardization of the preclinical models used to evaluate the efficacy of ketamine in OPNA poisoning is recommended along with the use of non-rodent models. Because OPNA-induced SE finally shares many similarities with SE of other origins, randomized clinical trials on ketamine for treatment of SE may facilitate its promotion as a therapeutic approach combined with other drugs to improve the management of casualties with OPNA intoxication.

Key words Ketamine, S+ ketamine (esketamine), Organophosphorus compounds, Status epilepticus, Refractory status epilepticus

1 Introduction

Working for the French national defense, our laboratory has been and is still more specifically interested in the medical management of victims poisoned with organophosphorus compounds (OP), especially the OP nerve agents (OPNA), a military and terrorist chemical threat. Treatment of seizures, status epilepticus (SE),

Sabine Himmelseher (ed.), *Ketamine: Pain Management, Critical Care, Depression, and Resource-Poor Settings Volume II*, Neuromethods, vol. 225, https://doi.org/10.1007/978-1-0716-4599-4_23,

refractory SE (RSE), and neuroprotection has been and is our main focus. Because of the clear involvement of glutamatergic excitatory transmission in the pathophysiology of the OP-induced central nervous system insults which has been demonstrated since the early 1990s (e.g., [1]), our team has worked on experimental treatment approaches to such SE with antagonists of the alpha-amino-3-hydroxy-5-methyl-4-isoxazolepropionic acid receptors (AMPAR) such as 2,3-dihydroxy-6-nitro-7-sulfamoyl-benzo(F)-quinoxaline NBQX [2] or of the N-methyl-D-aspartate acid receptor (NMDAR) such as N-[1-(2-thienyl)cyclohexyl]-piperidine TCP [3] or gacyclidine (GK-11) [4]. Most of our initial research and that of others was concentrated on stopping the seizures as early as possible to provide neuroprotection but without really assessing potential seizure recurrence and epileptogenesis. In the late 1990s, we then started to work on ketamine, a NMDAR antagonist that was interestingly licensed for clinical use compared to the other substances studied, and we looked more specifically at therapeutic regimens that would provide neuroprotection, reduce recurrence of seizures, and decrease risks for epileptogenesis. We could then demonstrate the excellent potential of ketamine in association with atropine, an anticholinergic agent that will always be part of the treatment [5]. The comparable efficacy of the S+ enantiomer, S+ ketamine or esketamine, at lower doses strongly suggested that antagonism at the NMDAR level was indeed part of the mechanisms of ketamine's neuroprotective action [6]. After our seminal findings in a Guinea pig model of soman intoxication, we kept on studying the neuroprotective properties of such combinations in rat and mouse models of soman-induced SE. More than 10 years ago, we wrote a paper reviewing the data available to support a wider use of ketamine in treating seizures and SE, especially when they become refractory to first-line treatment. This was usually performed with benzodiazepines for the specific situations of military interest [7]. We could also bring forth information that the ketamine-atropine combination did not impair heat tolerance and was not deleterious in hot environments, such as encountered by our forces at that time [8, 9]. We were also interested in more common situations when seizures/SE/RSE had a different etiology. Indeed, OPNA-induced SE shares many characteristics with other forms of SE [10]. The conclusions of our 2013 review [11] were that ketamine was still underused and suffered from misconceptions and lack of knowledge of its value among clinicians. Four years later, the situation was not significantly different when we were asked to review ketamine again [12]. Unfortunately, ketamine was still not in the mind of many practitioners, sometimes used, but mostly only at a very late stage when first- and second-line treatments had failed. Often patients were then debilitated by continuous seizures and had aggressively been treated with pharmacological polytherapy. For the well-being of patients, we will therefore look for positive developments in the use of racemic ketamine and/or its more potent enantiomer, S+ ketamine (esketamine) for therapy of SE. To keep this analysis concise, we will focus on papers published since 2017, refer the reader to recent reviews, and only highlight the most interesting information after a brief description of OPNA-induced SE and brain lesions. We will not discuss the use of ketamine for seizures or SE of different etiologies. The reader is kindly invited to read other chapters of this textbook such as that about ketamine/esketamine for treatment of refractory seizures for more clinical details (*please see* chapter 9 by ElSayed et al).

2 Organophosphorus Compounds, Status Epilepticus, and Risks for Permanent Brain Lesions

In the chemical warfare arsenal, OPNA are the most dangerous agents known. The compounds have been used in the past such as in the Iran–Iraq war (1980–1988) or in Syria (2013) [13]. They are much feared as terrorist attacks. Terrorists were capable of acquiring these substances, for example, in Japan (1994–1995), and OPNA have recently been used for assassination purposes in three known occasions, i.e., against the half-brother of the North Korean leader (agent VX, 2017) and against Russian citizens (agent Novichok, 2018 and 2020 [14, 15]). OPNA act as potent irreversible inhibitors of cholinesterases (ChEs) in both central and peripheral nervous systems and induce an immediate hypercholinergic activity responsible for most of the initial toxidrome. Several OPNA are known, but soman (pinacolyl methyl phosphonofluoridate) was considered to be the major threat in the 1980s/1990s because of the existence of a stockpile at that time and the difficulty in treating poisonings owing to the very quick aging of the OPNA–cholinesterase complex (aging is a modification that prevents reactivation of the enzyme). Other OPNA such as sarin have been and are studied because of the fact that they remain an immense threat, with sarin having repeatedly been used in the Middle East in recent years. Depending upon the dose, exposure to OPNA can produce hypersalivation, lacrimation, relaxation of sphincters, respiratory distress, cardiovascular dysfunction, coma, and rapid death [16]. In intoxications with high sublethal to lethal doses, the excess of acetylcholine in the brain triggers seizures and leads to secondary recruitment of the excitatory glutamate (Glu) system [1]. Through overstimulation of its receptors, including the ionotropic NMDAR and AMPAR, Glu is thought to play a prominent role in the long-lasting maintenance of seizure activity and SE and in the development of seizure-related brain damage (SRBD) [17] (Fig. 1).

In various models of soman-induced seizures and SE/RSE, we have demonstrated that the administration of ketamine during SE in combination with atropine sulfate (with or without a

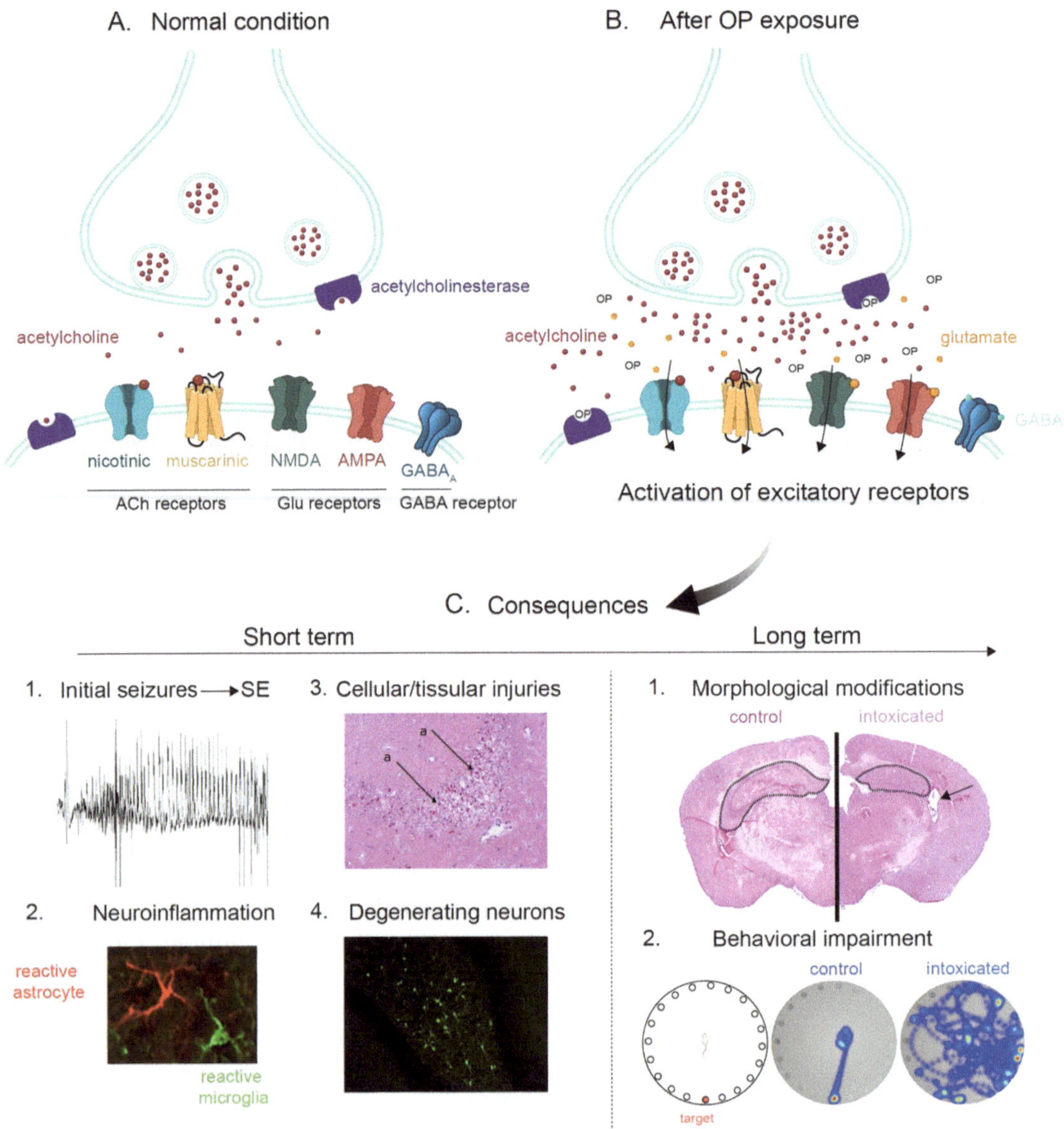

Fig. 1 Illustration of the mechanism of organophosphorus compound poisoning and its consequences. (**a**) Representation of the synaptic elements under normal conditions: acetylcholine (ACh) activates cholinergic receptors (nicotinic and muscarinic), and its signaling is stopped when ACh is degraded by acetylcholinesterase (AChE). (**b**) After exposure to organophosphorus compounds (OP), inhibited AChE induces synaptic ACh accumulation leading to cholinergic hyperactivity. Excess ACh leads to secondary recruitment of the glutamate (Glu) excitatory system and overstimulation of Glu receptors, including ionotropic receptors such as *N*-methyl-D-aspartate (NMDA) and α-amino-3-hydroxy-5-methyl-4-isoxazolepropionic acid (AMPA), resulting in cytotoxicity. Glu increase will also disrupt gamma-aminobutyric acid (GABA) signaling. (**c**) Short- and long-term consequences of OP exposure: (*1*) Glu plays a prominent role in the initiation and maintenance of seizures and status epilepticus (SE). In the days following OP exposure, several changes are observed: (*2*) neuroinflammation with glial activation (astrocytes and microglia), (*3*) cellular and tissular injury (shown by H&P hematoxylin and phloxine (H&P) staining) 48 h after soman intoxication, revealing acidophilic cells (**a**) and significant edema, and (*4*) finally neuronal degeneration (revealed by Fluoro-Jade C staining). In the longer term, (*1*) morphological changes are observed, with shrinkage of the hippocampus and enlargement of the lateral ventricles (arrow); (*2*) the cellular damage induced by OP exposure results in behavioral impairment, including

benzodiazepine) considerably attenuates neuroinflammation, confers neuroprotection with histological and functional evidence, and positively influences soman-induced perturbations in brain metabolism [18–21]. Treatment goals are to abate seizures, prevent their recurrence and epileptogenesis, and prevent or treat neurological damage and deficits which are sequelae of prolonged SE (Fig. 2).

Especially, we aspired to extend the therapeutic window for an effective treatment of OPNA-induced RSE: field conditions may preclude the medical chain to properly manage surviving OPNA casualties before at least 1 h, and early experimental findings gained in our laboratory tended to suggest that NMDAR were involved at a slightly later stage compared to AMPAR. In the US Department of Defense (US DoD), most of the recent work has set an earlier time for administration, i.e., within 40 min after seizure onset (see Table 1).

3 Therapeutic Associations with Ketamine for the Treatment of Status Epilepticus and Super-Refractory Status Epilepticus

The therapeutic efficacy of benzodiazepine decreases drastically approx. 20 min after onset of an epileptic seizure. This progressive pharmacoresistance has extensively been studied by Wasterlain and his collaborators [10] and by a team working for the US DoD (Lumley and coworkers) (for a recent review, see [22]).

Owing to the neurochemical changes in progress during seizures/SE in the brain and thus pharmacological drug efficacy, many research teams considered polytherapy as the path forward. Indeed the chance to find a single molecule with sufficient effects, a "magic bullet," is extremely scarce despite the fact that some molecules with both anticholinergic and antiglutamatergic activities showed promising properties ([22] and references therein). However, when administering such polytherapy, drug doses, drug pharmacokinetics, interactions, and the therapeutic time window based on neuropharmacology have first to be assessed. Drug pharmacokinetics are one of the key considerations, especially when drug effects wane as fast as those of ketamine. This is one of the reasons why we [7] and others (e.g., [23]) recommend repeated ketamine administrations. For translation to animal experiments, the dose of the drug treatment already efficiently used in humans must be determined with

Fig. 1 (continued) cognitive dysfunction, as illustrated by the animals' performance in the Barnes maze test to find the escape box (red dot). Performance was measured by the number of errors the rodents made and the latency to find the target (illustrations **a** and **b** have been created by the authors and A. Novell (2024)in BioRender.com/n18g019; the left part of illustration **c** (consequences, short-term) is from Dhote et al. [18] with permission from Elsevier conveyed via Copyright Clearance Center, Inc., and the right part of illustration **c** (consequences, long-term, all figures) is from the authors' own unpublished results)

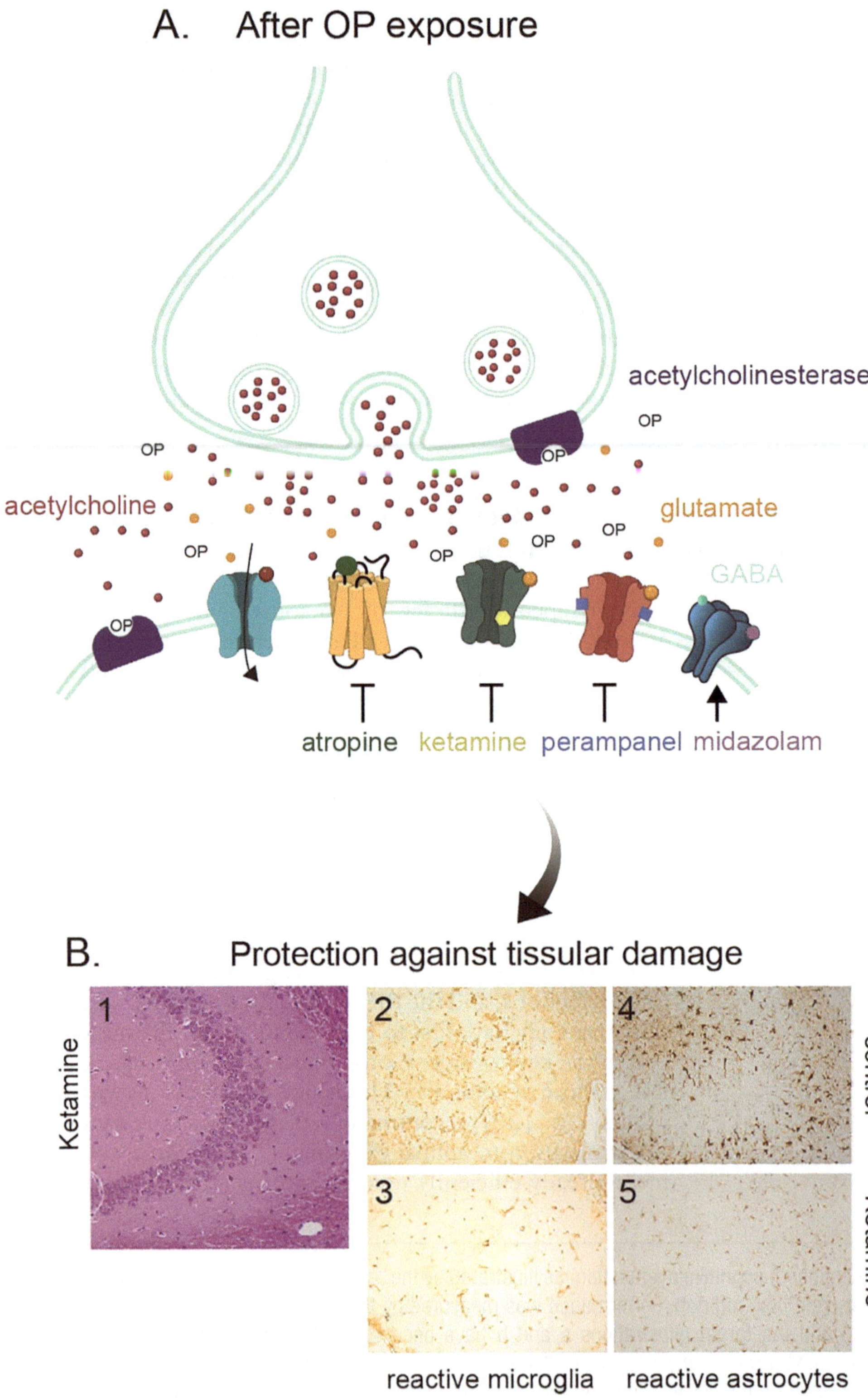

Fig. 2 Illustration of the therapeutic strategy after organophosphorus (OP) compound exposure. (**a**) Acetylcholine antagonists such as atropine act on the muscarinic receptor system, thus treating cholinergic-muscarinic effects of organophosphorus compound (OP) toxicity. The efficacy of benzodiazepines, such as midazolam,

an appropriate method (e.g., [24]). Not doing this may lead to doses that are too low or too high in the experimental model, thus being insufficient, misleading, out-of-reasonable clinical range, or at the very least not optimal. Nonetheless, experiments with unusual drug doses may also bring forth interesting results, for example, pointing to the lowest effective ketamine dose. This is the case, for example, in the study by Lewine and coworkers [23], which applied 7.5 mg/kg IM ketamine in a rat model. This dose cannot be considered as high dosage, but repeated administration of this low dose along with midazolam improved outcomes (Table 1).

The characteristics of the model are also very important, including toxicant properties (nature and dose); nature, dose, and route of administration of the initial drug treatment; and nature, dose, and route of administration of the drugs applied as therapeutic approach against the epileptic seizures. When reading Table 1, it is obvious that it is very difficult or sometimes almost impossible to directly compare some of the recent research findings. However, this does not change the overall conclusion that combinations including ketamine do have beneficial effects. Among the recent papers in Table 1, only one study did not show efficacy of ketamine [25], but its aim was to test the present standard clinical treatment guidelines to SRSE/RSE where ketamine is kept as a third-line treatment after others have failed. All rats that received a third-line therapy in form of IV ketamine or propofol died. This study is therefore not a demonstration of a lack of efficacy of ketamine per se. Moreover, ketamine was not combined to other drugs, whereas others showed that the combination of antiepileptic drugs was more efficient than their sequential administration [26]. As outlined, we mainly tested ketamine combined with atropine in our seminal studies because the latter will always be part of a treatment scheme in OPNA-induced seizures, and atropine reduces some of ketamine's side effects. In our attempt to determine how long after the initiation of seizures delayed ketamine could still be beneficial, we could also provide preliminary data on positive therapeutic

Fig. 2 (continued) decreases dramatically with time after the onset of an epileptic seizure. Glutamatergic antagonism by ketamine at the *N*-methyl-D-aspartate (NMDA) receptor and perampanel at the α-amino-3-hydroxy-5-methyl-4-isoxazolepropionic acid (AMPA) receptor limit postsynaptic glutamate signaling, thereby avoiding cytotoxicity. (**b**) The use of ketamine protects against the consequences of OP exposure, including tissue damage. (*1*) Example of a mouse treated with repeated administration of ketamine 25 mg/kg and atropine sulfate (KET/AS) showing no damage or detectable cell loss with hematoxylin and phloxine (H&P) staining. (*2, 3*) Isolectin *B4* of *Griffonia simplicifolia* (GSA-I-B4) and (*4, 5*) glial fibrillary acidic protein (GFAP) staining show a reduction in microglial and astrocyte reactivity after KET/AS (*3, 5*) compared to a soman-intoxicated animal without treatment (*2, 4*) (illustration **a** has been created by the authors and A. Novell (2024) in BioRender.com/n18g019 and illustration **b** is from Dhote et al. [18] with permission from Elsevier conveyed via Copyright Clearance Center, Inc.). *GABA* gamma-aminobutyric acid

Table 1
Efficacy of ketamine used in therapeutic combinations against organophosphorus nerve agent-induced status epilepticus

Organophosphorus nerve agent	Model (species, dose of agent µg/kg, route of poisoning)	Initial treatment (timing and drugs)	Duration of seizures before ketamine treatment (min)	Drug treatment combinations Drug dose (mg/kg)/route of use	Efficacy	References
Sarin	Rat 105 SC	1 min after exposure: oxime (2-PAM) and nitrate of methylatropine	ca. 45	MDZ (0.66, IM) or "high" KET (7.5, IM, 5 injections 90 min apart) or MDZ (0.66, IM) + "low" KET (2.5, IM, 5 injections 90 min apart) or MDZ (0.66, IM) + "high" KET (7.5, IM, 5 injections 90 min apart)	MDZ + "high" dose KET: reduction of seizures and neuroprotection (behavioral and histological evidence)	[23]
Sarin	Rat 103–115 IM	1 min after exposure: oxime (TMB4) and atropine sulfate	ca. 25 or 55	MDZ (3, IM) or "triple, low dose": MDZ (3, IM) + KET (10, IM) + VA ([illegible]0, IP) Or "triple high dose": MDZ (3, IM) + KET (30, IM) + VA ([illegible]0, IP)	Reduction of seizures and neuroprotection (behavioral and histological evidence)	[29]

Sarin	Rat 90 IM	1 min after exposure: oxime (TMB4) and atropine sulfate	ca. 25	MDZ (3, IM) or MDZ (3, IM) + KET (30, IM) or MDZ (3, IM) + KET (30, IM) + VA (90, IP) or MDZ (3, IM) + KET (30, IM) + LEV (500, IP)	Reduction of mortality, seizures, and neuronal damage. Both triple therapies completely prevented brain injury	[31]
Soman	Rat 132 SC	1 min after exposure: oxime (HI-6 dichloride) and atropine sulfate	40	MDZ (3, IP) or KET (10-30-60-90, IP) or MDZ (3, IP) + KET (30, IP)	KET 60: increased survival (not 10 or 90) KET 30 + MDZ: increased survival (not 60 or 90) and reduce seizures, neuroprotection with improved spatial memory acquisition performance	[27, 30]
Soman	Rat NM SC	NM	40	MDZ (3, NM) or MDZ (3, NM) + KET (30, NM) or MDZ (3, NM) + KET (30, NM) + VA (90, NM)	Provision of better seizure cessation and neuroprotection compared to MDZ alone	[28]

(continued)

Table 1
(continued)

Organophosphorus nerve agent	Model (species, dose of agent μg/kg, route of poisoning)	Initial treatment (timing and drugs)	Duration of seizures before ketamine treatment (min)	Drug treatment combinations Drug dose (mg/kg), route of use	Efficacy	References
Soman	Rat 118 SC	1 min after exposure: oxime (HI-6 dichloride) and atropine sulfate	40	MDZ (3, IP) or PHE (30, IP) or MDZ (3, IP) + PHE (30, IP) or KET (30, IP) + PHE (30, IP) or MDZ (3, IP) + KET (30, IP) + PHE (30, IP)	Provision of seizure cessation, neuroprotection, and prevention of epileptogenesis. Triple therapy results in better outcomes	[33]
Soman	Rat 118 SC	1 min after exposure: oxime (HI-6 dichloride) and atropine sulfate	40	MDZ (3, IP) or MDZ (3, IP)+ LACO (45, IP) or MDZ (3, IP) + RUFI (50, IP) or MDZ (3, IP) + LACO (45, IP) + KET (30, IP) or MDZ (3, IP) + RUFI (50, IP) + KET (30, IP)	Triple therapy (MDZ-LACO-KET) is better than the other combinations, and all combinations are better than MDZ alone	[32]

Soman	Rat 118 SC	1 min after exposure: oxime (HI-6 dichloride) and atropine sulfate	40	MDZ (3, IP) or MDZ (3, IP) + ALLO (6 or 12, IP) or MDZ (3, IP) + ALLO (6/12, IP) + KET (30, IP)	Triple therapy (higher dose of ALLO) reduced severity of toxic signs, recurrence of spontaneous seizures, and neuropathology (beneficial effect of ketamine addition)	[34]
Soman	Rat 150 SC	Pretreatment: HI-6 Post poisoning: nitrate of methylatropine then 20 min after onset of SE MDZ (1.8, IM) + atropine sulfate + 2-PAM then atropine every 3 h	More than 60	KET (low dose 4.5 mg/kg, IV) as third-line treatment after MDZ, lorazepam, PHE, or VA failed KET administered every 15 min until SE stopped, then infusion (30 mg/ kg/ h) for 18 h. Comparison with propofol	Seizure control but all rats died with either ketamine or propofol	[25]
Soman	Es1−/− mouse 80 SC	1 min after exposure: oxime (HI-6 dimethanesulfonate) and atropine sulfate	40	MDZ (3, IP) or MDZ (3, IP) + KET (30, IP)	Increased survival/reduction in seizure severity/ reduction in neuropathology with dual therapy	[36]
Soman	KIKO mouse 80 SC	1 min after exposure: oxime (HI-6 dimethanesulfonate) and atropine sulfate	15	MDZ (3, SC) + KET (30, IP)	Increased survival/reduction in seizure severity/ reduction in neuropathology with dual therapy	[37]

The most recent references are presented here. For work published before 2017, the reader is invited to see [11, 22, 26]. For protocols that involved other types of cholinergic starters for seizures (e.g., pilocarpine or lithium-pilocarpine models), please look at [22, 26]. The indicated experimental groups are the most relevant ones for comparisons, but some other may have been tested, and the reader is invited to look at the reference. Abbreviations: Drugs: *ALLO* allopregnanolone, *HI-6* asoxime, *KET* ketamine, *LACO* lacosamide, *LEV* levetiracetam, *MDZ* midazolam, *2-PAM* pralidoxime, *PHE* phenobarbital, *RUFI* rufinamide, *TMB4* dipyroxime or trimedoxime, VA valproic acid. Routes: *SC* subcutaneous injection, *IM* intramuscular injection, *IP* intraperitoneal injection. Other abbreviations: *Es1−/−* plasma carboxylesterase knockout mouse, ***min*** minute, *NM* not mentioned in the paper, *KIKO* human acetylcholinesterase knock-in carboxylesterase knockout mouse effects of combining ketamine with midazolam and atropine [5]. It is noteworthy that the combination of ketamine with a benzodiazepine already represented a "standard" approach in anesthesiology.

Although we focus on the treatment of OPNA-induced SE/RSE in this work, we would like to briefly mention some results obtained in other seizure models with a "cholinergic starter": in a high-dose lithium pilocarpine rat model, Niquet and coworkers found that a triple combination of midazolam-ketamine-valproate was superior in terminating RSE compared to midazolam-fosphenytoin-valproate or to sequential drug administration (reviewed in [27]). The combination midazolam-ketamine-valproate has also been successfully tested in OPNA-related seizure models [28–31].

Other triple combinations against soman-induced RSE were recently examined by Lumley and co-workers: in a set of experiments, they added one of the two third-generation antiseizure medications, i.e., lacosamide or rufinamide, to the already described ketamine and midazolam combination. The triple combinations were more effective than midazolam alone with lacosamide addition being more efficient than rufinamide [32]. They also tested a combination with phenobarbital that appeared efficient [33]. Lazar et al. successfully studied a triple therapy with midazolam-ketamine-levetiracetam [31]. Finally, a triple therapy including allopregnanolone also proved beneficial [34]. It is worth noting that brexanolone (SAGE-547), a synthetic form of allopregnanolone, has been examined in clinical trials published in 2017 for the treatment of SRSE with a high rate of successful third-line treatment weaning [35]. Genetically engineered mice (either plasma carboxylesterase knockout, Es1−/− mice, or human acetylcholinesterase knock-in serum carboxylesterase knockout, KIKO mice) have also been used to test the dual therapy of midazolam-ketamine with similarly good results [36, 37].

Unfortunately, many research groups are more interested in improving therapeutic approaches with midazolam, so that there is a lack of tests of ketamine combinations with new antiepileptics such as lacosamide-ketamine, rufinamide-ketamine, or other agents that are already clinically used, such as dexmedetomidine-ketamine. From a clinical perspective, the management of double drug combinations is of course better to handle and easier to implement than triple combinations.

4 Do Nowadays' Clinical Trials Better Inform About Ketamine's Value for Treatment of Status Epilepticus Induced by Organophosphorus Compounds?

There is an urgent need for high-quality clinical trials on the treatment of SE that would bring results that are relevant for the specific management of patients with OP poisoning. Unfortunately, among the 1662 clinical trials involving ketamine or esketamine referenced in the website "clinicaltrials.org," only 7 were related to the treatment of SE or epilepsy (according to database browsed with the keywords "ketamine + seizures," "ketamine + status," and "ketamine + epilepsy" at the end of August 2024).

Only one trial has been terminated with results with the number NCT02431663 carrying the name and title "KETASER01: IV administration of ketamine in the treatment of refractory convulsive SE in children compared to administration of midazolam at high doses, thiopental, and/or propofol". However, the KETASER01 study was prematurely halted due to low eligibility of patients and no successful recruitment [38, 39]. Two trials were withdrawn due to lack of sufficient numbers of patients to be enrolled, one (NCT02726867) with the title "evaluation of efficacy and safety of levetiracetam, lacosamide and ketamine as adjunctive treatment of refractory status epilepticus" and the other one (NCT03115489) carrying the title "the efficacy of ketamine infusion compared with traditional anti-epileptic agents in RSE." They might have brought very valuable data. Four trials are currently recruiting, but their results will likely not be significant for medical management of OPNA-related emergencies (NCT05019885, testing the efficacy of subanesthetic dose IV ketamine in the treatment of drug resistant epilepsy, where patients will receive 0.5 mg/kg IV racemic ketamine over 40 min thrice weekly for two consecutive weeks; NCT05779657, examining combined ketamine-midazolam for generalized convulsive SE, where investigators aim to evaluate the efficacy and safety of ketamine plus midazolam versus midazolam alone as first-line therapy in pediatric generalized convulsive SE; NCT05756621, assessing a dual anti-glutamate therapy in SRSE after cardiac arrest (SUPER-CAT) through evaluation of ketamine combined with perampanel; and the not yet recruiting NCT06549426 studying the treatment of electroencephalographic status epilepticus after cardiopulmonary resuscitation-2-(TELSTAR-2) including ketamine use as third-line agent).

From a clinical standpoint regarding advanced therapy of RSE/SRSE with ketamine, the situation we have described in 2017 has not really evolved further [12], despite the fact that some clinicians are trying to advocate for an earlier use of ketamine in SE (e.g., [40, 41]). The lack of randomized prospective studies is a major hindrance in incorporating ketamine and/or esketamine in guidelines and recommendations for therapy of OPNA-induced SE (for more details on ketamine in clinical treatment of RSE/SRSE, see others chapters in this textbook).

5 Conclusions

Taken together, we draw the following conclusions:

- Recent preclinical data confirm the excellent therapeutic potential of ketamine without doubt and propose promising triple drug combinations. Because their efficacy has been demonstrated mostly only at 40 min after seizure onset, it would be valuable to study these drug combinations at a more delayed time point, after a SE of longer duration.
- Most of the recent work investigated endeavors to increase the efficacy of benzodiazepines and other pharmaceutical agents through the adjunctive use of ketamine. Reversing the point of view, we propose to focus on ketamine and examine combinations of two drugs (instead of triple therapies). It is of great interest to test ketamine combined with drugs that are already clinically used in combination therapies, either with or without direct effects on glutamatergic transmission such as dexmedetomidine or those acting on other ionotropic Glu receptors. For instance, LY293558, an AMPA/GluK1 receptor antagonist, alone or in combination with NMDAR antagonists such as caramiphen proved beneficial in young adult and aged rats even if administered at 1 h post-noxious exposure [42]. This is a logical approach that we previously followed in our laboratory [2, 17], and combinations such as ketamine and perampanel are also clinically investigated [43].
- The subtle but multifaceted differences in experimental conditions, circumstances, and protocols studying ketamine's efficacy render a direct study comparison often difficult and/or sometimes almost impossible. This should encourage international teams to collaborate and share models and data. Only then a consistent body of evidence and data foundation can be laid in favor of ketamine demonstrating its clear benefits within polytherapy of OPNA-induced SE, RSE, and SRSE.
- Esketamine remains grossly understudied in the context of OPNA-induced SE. We could not find one single new study since our work from 2007 [6] or a recent trial reporting its effects in therapy of SE of other etiologies. However, this may change soon because IV esketamine has recently been approved for clinical use in France and China.
- Overall, treatment with ketamine is still limited in clinical settings, mostly because of a lack of knowledge about its skillful use, "irrational fears" of side effects, or persistent misconceptions. We hope that our work along with the others in this textbook will help clinicians to discover and take into account the multiple benefits of ketamine.

Acknowledgments

We would like to thank Anthony Novell for participating in developing the figures.

Conflicts of Interest

The authors do not have any conflicts of interest.

References

1. Lallement G et al (1991) Effects of soman-induced seizures on different extracellular amino acid levels and on glutamate uptake in rat hippocampus. Brain Res 563(1-2):234–240
2. Lallement G et al (1994) Coadministration of atropine, NBQX and TCP against soman-induced seizures. Neuroreport 5(9): 1113–1117
3. Carpentier P et al (2001) Effects of thienylphencyclidine (TCP) on seizure activity and brain damage produced by soman in Guinea-pigs: ECoG correlates of neurotoxicity. Neurotoxicology 22(1):13–28
4. Lallement G et al (1999) Review of the value of gacyclidine (GK-11) as adjuvant medication to conventional treatments of organophosphate poisoning: primate experiments mimicking various scenarios of military or terrorist attack by soman. Neurotoxicology 20(4):675–684
5. Dorandeu F et al (2005) Efficacy of the ketamine-atropine combination in the delayed treatment of soman-induced status epilepticus. Brain Res 1051(1-2):164–175
6. Dorandeu F et al (2007) Protective effects of S + ketamine and atropine against lethality and brain damage during soman-induced status epilepticus in Guinea-pigs. Toxicology 234(3):185–193
7. Dorandeu F et al (2013) Ketamine combinations for the field treatment of soman-induced self-sustaining status epilepticus. Review of current data and perspectives. Chem Biol Interact 203(1):154–159
8. Barbier L et al (2015) Beneficial effects of a ketamine/atropine combination in soman-poisoned rats under a neutral thermal environment. Neurotoxicology 50:10–19
9. Barbier L et al (2012) Ketamine does not impair heat tolerance in rats. Eur J Pharmacol 691(1-3):77–85
10. Chen JW, Wasterlain CG (2006) Status epilepticus: pathophysiology and management in adults. Lancet Neurol 5(3):246–256
11. Dorandeu F et al (2013) Treatment of status epilepticus with ketamine, are we there yet? CNS Neurosci Ther 19(6):411–427
12. Dorandeu F (2017) Ketamine for the treatment of (super) refractory status epilepticus? Not quite yet. Expert Rev Neurother 17(5): 419–421
13. Rosman Y et al (2014) Lessons learned from the Syrian sarin attack: evaluation of a clinical syndrome through social media. Ann Intern Med 160(9):644–648
14. Stone R (2018) U.K. attack puts nerve agent in the spotlight. Science 359(6382):1314–1315
15. Steindl D et al (2021) Novichok nerve agent poisoning. Lancet 397(10270):249–252
16. Newmark J (2007) Nerve agents. Neurologist 13(1):20–32
17. Lallement G et al (1993) Neuroprotective activity of glutamate receptor antagonists against soman-induced hippocampal damage: quantification with an omega 3 site ligand. Brain Res 618(2):227–237
18. Dhote F et al (2012) Combinations of ketamine and atropine are neuroprotective and reduce neuroinflammation after a toxic status epilepticus in mice. Toxicol Appl Pharmacol 259(2):195–209
19. Dhote F et al (2007) Prolonged inflammatory gene response following soman-induced seizures in mice. Toxicology 238(2-3):166–176
20. Fauvelle F et al (2012) Prediction of neuroprotective treatment efficiency using a HRMAS NMR-based statistical model of refractory status epilepticus on mouse: a metabolomic approach supported by histology. J Proteome Res 11(7):3782–3795
21. Fauvelle F et al (2010) Changes in mouse brain metabolism following a convulsive dose of

soman: a proton HRMAS NMR study. Toxicology 267(1-3):99–111
22. Niquet J et al (2023) Treatment of cholinergic-induced status epilepticus with polytherapy targeting GABA and glutamate receptors. Epilepsia Open 8 Suppl 1(Suppl 1):S117–S140
23. Lewine JD et al (2018) Addition of ketamine to standard-of-care countermeasures for acute organophosphate poisoning improves neurobiological outcomes. Neurotoxicology 69:37–46
24. Food and Drug Administration, Guidance for Industry (2005). Estimating the maximum safe starting dose in initial clinical trials for therapeutics in adult healthy volunteers. FDA. Accessed 5 Oct 2024. https://www.fda.gov/regulatory-information/search-fda-guidance-documents/estimating-maximum-safe-starting-dose-initial-clinical-trials-therapeutics-adult-healthy-volunteers
25. Morgan JE et al (2021) Refractory and super refractory status epilepticus in nerve agent-poisoned rats following application of standard clinical treatment guidelines. Front Neurosci 15:732213
26. Lumley L et al (2021) Treatment of acetylcholinesterase inhibitor-induced seizures with polytherapy targeting GABA and glutamate receptors. Neuropharmacology 185:108444
27. Niquet J et al (2020) Rational polytherapy in the treatment of cholinergic seizures. Neurobiol Dis 133:104537
28. Niquet J et al (2019) Early polytherapy for benzodiazepine-refractory status epilepticus. Epilepsy Behav 101(Pt B):106367
29. Gore A et al (2021) Neuroprotection by delayed triple therapy following sarin nerve agent insult in the rat. Toxicol Appl Pharmacol 419:115519
30. Lumley LA et al (2019) Dataset of EEG power integral, spontaneous recurrent seizure and behavioral responses following combination drug therapy in soman-exposed rats. Data Brief 27:104629
31. Lazar S et al (2024) Efficacy of a combined anti-seizure treatment against cholinergic established status epilepticus following a sarin nerve agent insult in rats. Toxicol Appl Pharmacol 484:116870
32. Lumley LA et al (2024) Efficacy of lacosamide and rufinamide as adjuncts to midazolam-ketamine treatment against cholinergic-induced status epilepticus in rats. J Pharmacol Exp Ther 388(2):347–357
33. Lumley LA et al (2021) Combination of anti-seizure medications phenobarbital, ketamine, and midazolam reduces soman-induced epileptogenesis and brain pathology in rats. Epilepsia Open 6(4):757–769
34. Nguyen DA et al (2024) Evaluation of midazolam-ketamine-allopregnanolone combination therapy against cholinergic-induced status epilepticus in rats. J Pharmacol Exp Ther 388(2):376–385
35. Rosenthal ES et al (2017) Brexanolone as adjunctive therapy in super-refractory status epilepticus. Ann Neurol 82(3):342–352
36. Marrero-Rosado BM et al (2020) Ketamine as adjunct to midazolam treatment following soman-induced status epilepticus reduces seizure severity, epileptogenesis, and brain pathology in plasma carboxylesterase knockout mice. Epilepsy Behav 111:107229
37. Marrero-Rosado BM et al (2021) Novel genetically modified mouse model to assess Soman-induced toxicity and medical countermeasure efficacy: human acetylcholinesterase Knock-in serum carboxylesterase knockout mice. Int J Mol Sci 22(4):1893
38. Rosati AS, De Masi S, Guerrini R (2018) Ketamine for refractory status epilepticus: a systematic review. CNS Drugs 32(11):997–1009
39. Rosati A et al (2022) KETASER01 protocol: what went right and what went wrong. Epilepsia Open 7(3):532–540
40. Buratti S et al (2023) Ketamine as advanced second-line treatment in benzodiazepine-refractory convulsive status epilepticus in children. Epilepsia 64(4):797–810
41. Choi JW, Shin JW (2021) Early combination therapy of ketamine and midazolam in patients with refractory status epilepticus in hemodynamic unstable state. J Epilepsy Res 11(2):150–153
42. Aroniadou-Anderjaska V et al (2020) Targeting the glutamatergic system to counteract organophosphate poisoning: a novel therapeutic strategy. Neurobiol Dis 133:104406
43. Souidan H et al (2022) Double antiglutamatergic treatment in patients with status epilepticus: a case series. Epilepsy Behav 137 (Pt A):108954

Part V

Ketamine in Exciting Uses

Chapter 24

Ketamine for Emergency Medicine in the Mountains and at High Altitude

Marc Moritz Berger, Richard Steffen, Hülya Kursun, and Jürgen Knapp

Abstract

In austere environments, the provision of effective and safe analgesia and anesthesia is frequently challenging due to difficult terrestrial conditions and limited medical infrastructure. For mountain regions and at high altitude, the use of ketamine as a strong analgesic and anesthetic has been recommended by various experts and recent guidelines. In this chapter, the pharmacological properties of ketamine that are relevant for emergency medical care in the mountains and at high altitude are discussed. A particular focus is set on the hemodynamic and respiratory effects of ketamine as well as ketamine's influence on the hypoxic ventilatory response (HVR), hypoxic pulmonary vasoconstriction (HVR), and alveolar fluid clearance (AFC).

Key words Ketamine, Mountain medicine, High altitude, Pain therapy, Analgesia, Risk, Austere environment, Adverse effects, Hypoxic pulmonary vasoconstriction, Hypoxic ventilatory response, Alveolar fluid clearance, Anesthesia

1 Introduction

In the mountains and at high altitude, numerous specifics have to be considered when providing out-of-hospital emergency medical care. This is especially challenging in remote regions, where medical and all other infrastructure is limited. Treatment often takes place under difficult terrestrial conditions with restricted patient access, lack of (monitoring) equipment, supplemental oxygen, hindrances for re-evaluation during transport, and limited means to respond to complications of any type. In addition, a patient in this environment often requires complex rescue measures in case of injury and/or disease and may have long transport times during which clinical monitoring cannot be performed well. For example, if the medical team has to focus on the difficult terrain and the descent

Sabine Himmelseher (ed.), *Ketamine: Pain Management, Critical Care, Depression, and Resource-Poor Settings Volume II*, Neuromethods, vol. 225, https://doi.org/10.1007/978-1-0716-4599-4_24,

from the mountain itself or has to help with patient evacuation, much attention has to be directed toward managing external circumstances and the transport in addition to medical care (Fig. 1).

Treatment of pain in an austere environment therefore requires a thoughtful step-by-step approach that minimizes potential harm but aims at providing best possible patient care without neglecting the needs of the medical and rescue team. Other than in the lowlands, a patient's cooperation in the evacuation process is much more often needed. At high altitude, oxygen deficiency is an additional possibly critical stressor, affecting both patient's and rescuer's physical, psychological, and neurocognitive functions and states.

2 General Considerations: Analgesia in Austere Environments

Even in remote regions of the world, the provision of effective and safe analgesia is a core maxim of prehospital emergency medicine. Nevertheless, this principle is associated with considerable risks such as cardiorespiratory and neurocognitive depression related to the use of strong analgesics. However, under all conditions and circumstances, pain control is a secondary consideration after management of immediate threats to life. Primary therapeutic efforts need to treat possible compromise of airway, breathing, and circulation, secondary to secure the scene for patient and rescuer safety. When the safety of the scene has been established and a primary survey of the patient has been performed, a secondary survey should follow to evaluate causes and severity of possible pain. Before an analgesic is administered in remote, austere environments, particular attention must be paid to its possible adverse effects because even minor patient problems can become very hazardous during, for example, an evacuation [1].

An ideal medication for austere settings would be compact, lightweight, and durable against crush, abrasion, heat, and cold. It would be non-sedating with a large therapeutic index, work via multiple application routes, and be associated with high circulatory stability and as little respiratory depression as possible. Additionally, it would not have potential for rapid tolerance, dependence, or abuse [2, 3]. Unfortunately, no such medication exists. Medical care providers must therefore carefully weigh advantages and disadvantages of each (analgesic) medication before use. In general, opioids play a pivotal role in the treatment of severe pain. However, opioids have serious side effects. If manifesting in hostile environments where medical equipment and medication is not sufficiently available, opioid-related adverse effects are not only difficult to treat but may exaggerate severe health problems dramatically (e.g., respiratory or circulatory depression).

Fig. 1 (**top**) Evacuation of a patient in a rescue bag by helicopter hoist operation in the Swiss Alps; (**bottom**) Emergency medical care for a seriously injured patient in a high alpine environment with steep terrain and limited human and material resources (author photos)

3 Key Considerations for Ketamine in Austere Environments

For decades, ketamine has been recommended by expert groups and recently in new guidelines (e.g., [1]) as analgesic and anesthetic for use in medical mountain rescue [4], in wilderness medicine [5], and for medical care at high altitudes [6, 7]. Specifically, the use of subanesthetic doses of ketamine for analgesia and sedation with preservation of spontaneous ventilation and laryngeal protective reflexes seems advantageous in environments where advanced airway management is often impossible and severe hypoxia an ubiquitous, threatening risk [6]. Ketamine produces an increase in blood pressure, stroke volume, and heart rate while maintaining systemic vascular resistance in conditions without exhausted volume depletion or cardiovascular reserve. It has a rapid onset of action and is relatively short acting and titratable [8]. At subanesthetic doses, ketamine may induce psychoactive events, which are usually short-lived and resolve by themselves. Another characteristic that favors ketamine use for such conditions are its various application routes, such as intramuscular (IM), intravenous (IV), intraosseous, nasal, oral, sublingual, or rectal [8]. In wilderness medicine, the IM, nasal, IV, or intraosseous route is most often used [1]. Ketamine's good safety profile combined with ease of storage makes it additionally ideal for use in areas where refrigerators, complex monitoring, electricity, and oxygen may all be lacking, be in short supply, or unreliable [9]. For example, no pharmacological degradation was observed after ketamine storage at +40 °C and − 15 °C [10].

When administering medication in austere environments, it has to be noted that the external circumstances like heat and cold much stronger affect patients than most in-hospital settings. Cold can lead to a relevant vasoconstriction with blood volume centralization, while heat can cause an opposite reaction with corresponding changes in plasma concentrations of hydrophilic and lipophilic substances. This complicates caregivers' quest for adequate dosing of medications and makes a patient–individual, response-orientated dose titration all the more important [11].

4 Ketamine's Favorable Profile in Alpine Environments and High-Altitude Settings

4.1 Alpine Environments

Ketamine is well known to preserve ventilation and maintain protective airway reflexes [9]. These properties are advantageous if a patient needs strong analgesia in an environment where advanced airway management is impractical and spontaneous breathing needs to be maintained. This may be the case, for example, in situations where a patient must be evacuated from a steep rock face, crevasse, or ice slope. Respiratory depressant analgesics such as opioids (e.g., fentanyl) are associated with the risk of having to secure the airway up to intubation, which may not be feasible. Any application of a respiratory depressant drug in such situations requires careful consideration. In case of hypovolemia, for example, due to bleeding, ketamine's respiratory and cardiovascular stimulating effects may be beneficial in comparison to other analgesic and anesthetic agents.

A recent study surveyed emergency doctors of the Air Glaciers, which is an alpine helicopter emergency medical service (HEMS) in the French-speaking part of Switzerland. The 39 participating physicians considered ketamine to be useful or even irreplaceable in out-of-hospital settings [4]. An important criterion for its appreciation was the better safety profile as compared to, for example, fentanyl. The significance of ketamine in professional alpine rescue has also been demonstrated in an observational study of physician-staffed helicopter bases in Austria from a 12-year time period: ketamine and opioids alone or in combination were the analgesics of choice [12]. Another study analyzed rescue missions conducted by Air Zermatt, a commercial rescue service in the high-alpine region in Switzerland. This analysis showed that the most common medical intervention on-site was analgesia or analgosedation, with fentanyl and ketamine being the most commonly used analgesic drugs [13]. It has been proven since decades that skillful out-of-hospital injection of subanesthetic doses of ketamine alone or in combination with opioids is safe and does not cause loss of consciousness, a clinically relevant decrease in oxygen saturation, or decline in cognitive functions [12, 14–16].

The high relevance of ketamine for out-of-hospital analgesia or analgosedation is also reflected in the 2024 update of the Wilderness Medical Society (WMS) clinical practice guidelines for the treatment of acute pain in austere environments [1], which recommend ketamine for the control of acute, severe pain. The guidelines suggest a dose range of 0.1–0.5 mg/kg for IV and 1–2 mg/kg for IM racemic ketamine administration. Nevertheless, the authors also point out that ketamine providers should have experiences with ketamine and its effects and should be equipped to manage airway complications in case they arise. These prerequisites are based on strong expert recommendations and clinical evidence of moderate quality [1]. To reduce the anyway low risk of psychomimetic adverse events, ketamine sometimes is combined with a benzodiazepine such as midazolam. This combination is not mentioned in the current WMS guideline, and in our opinion, it is not necessary.

4.2 High-altitude Settings

Data on ketamine administration at high altitude are scarce. An analysis of 11 consecutive cases documented the use of ketamine for anesthesia in a remote hospital at an altitude of 3900 m by primary care physicians without specialist anesthesia training [7]. The study found no significant medical complications or an increased incidence of hypoxemia associated with ketamine use. Another study compared the combination of ketamine-propofol with fentanyl-propofol for short surgical procedures at an altitude of 2514 m [17]: ketamine-propofol was associated with significantly fewer drops in oxygen saturation requiring positive pressure ventilation than fentanyl-propofol. Successful use of ketamine was also reported in a woman presenting with postpartum hemorrhagic shock at 4243 m requiring anesthesia to treat the bleeding [6]. Taken together, compared to other strong analgesics and anesthetics, ketamine's typical pharmacological profile makes its administration at high altitude advantageous.

4.2.1 Hypoxic Ventilatory Response

At sea level, the spontaneous ventilation in humans is mainly modulated by central chemoreceptors which are located in the ventrolateral surface of the medulla. They are sensitive to changes in partial pressure of CO_2 (PCO_2) and proton (H^+) ion concentration in the extracellular fluid of the brain. The blood–brain barrier is highly permeable to dissolved CO_2, and an increase in arterial PCO_2 stimulates ventilation [18]. At high altitude, the contribution of peripheral chemoreceptors to the ventilatory drive increases. Peripheral chemoreceptors are located in the carotid and aortic bodies. With increasing decrease in arterial partial pressure of O_2 (PO_2), hypoxemia becomes the primary breathing stimulus [19]. A hypoxia-induced increase in ventilation starts as the PO_2 falls [20], becoming more prominent once the inspired PO_2 is below 100 mmHg [21]. This corresponds to an altitude of approximately 3000 m or an arterial hypoxic $PO_2 \leq 60$ mmHg [18]. The magnitude of the hypoxic ventilatory response (HVR) is inter-individually highly variable, with a strong HVR being favorable for ventilatory acclimatization.

Experimental data from humans or dogs have shown that ketamine does not depress the HVR, while other anesthetics and analgesics, including barbiturates and opioids, depress the ventilatory response to hypoxia [22, 23] (Fig. 2).

When hypoxemia is the primary breathing stimulus, the maintenance of the HVR may be of clinical importance, especially if spontaneous breathing must be preserved (e.g., during rescue from technically challenging terrain). At least from a theoretical point of view, agents that reduce the HVR such as opioids are therefore more likely to be associated with respiratory depression at altitudes >3000 m than substances such as ketamine which do not affect the HVR. As a consequence, the need for iatrogenic airway protection may be greater with HVR depressant agents.

4.2.2 Hypoxic Pulmonary Vasoconstriction

Ascent to altitudes above 1600–2500 m causes hypoxic pulmonary vasoconstriction (HPV) [24–26]. HPV is beneficial in situations where alveolar hypoxia occurs on a local basis, as seen, for example, in patients with pneumonia. In this situation, blood flow is diverted away from inadequate to better ventilated lung regions, thus

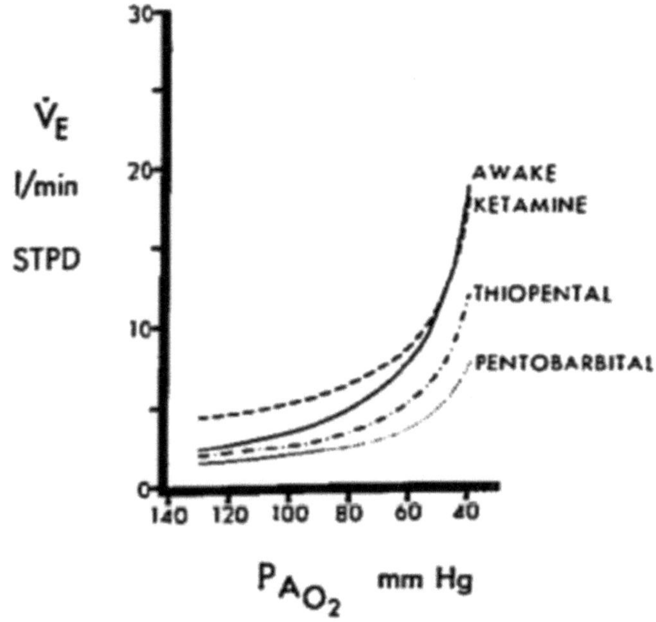

Fig. 2 Mean plots of minute ventilation and increasing hypoxia in seven dogs, showing changes of the ventilatory response to isocapnic hypoxia during IV anesthesia with pentobarbital (30 mg/kg), thiopental (15 mg/kg), or ketamine (10 mg/kg) relative to the awake state. V_E, expiratory minute ventilation in L/min; STPD, standard temperature, pressure, and dry; P_A0_2, partial alveolar pressure of oxygen (from Hirshman et al. [23] with permission from Wolters Kluwer Health, Inc. conveyed through Copyright Clearance Center, Inc.)

optimizing ventilation–perfusion matching and alveolar gas exchange. However, at high altitude, alveolar hypoxia occurs on a global basis, resulting in more widespread HPV. This leads to an increase in pulmonary vascular resistance and right ventricular afterload, without improving pulmonary gas exchange. Exaggerated HPV plays a key role in the pathophysiology of high altitude pulmonary edema (HAPE) [27, 28], a non-cardiogenic form of pulmonary edema seen in healthy individuals who ascend too quickly to altitudes above 3000 m [29]. Measures that increase the pulmonary arterial pressure therefore favor the development of HAPE, whereas interventions that reduce the pulmonary artery pressure are protective (e.g., administration of drugs like nifedipine, sildenafil, or tadalafil [27, 30, 31]).

Data on the effect of ketamine on the HPV of humans at high altitude are lacking. However, ketamine has been found to cause pulmonary vasodilation under conditions of hypoxia-induced pulmonary hypertension in rats and mice [32–34]. It has also been shown to be a safe anesthetic option in children with pre-existing pulmonary arterial hypertension [35]. Based on these few (weak) data, which do not allow to draw a concrete conclusion about the effects of ketamine on pulmonary arterial pressure at high altitude, one may nevertheless speculate that ketamine may improve (or at least not adversely affect) the function of the pulmonary vasculature during hypoxia. In contrast, various IV analgesics and anesthetics including fentanyl and propofol have been shown to not reduce [36] or even potentiate [37] the HPV.

4.2.3 Alveolar Fluid Clearance

Reabsorption of alveolar edema fluid is essential for the resolution of pulmonary edema, including HAPE. The removal of edema fluid from the alveolar space depends on active transport of sodium (Na^+) and chloride (Cl^-) across the alveolar epithelium [38]. Na^+ enters the cell via various Na^+ transporters in the apical plasma membrane of epithelial type I and II cells and is extruded on the basolateral side of the cell by Na^+/K^+ pumps. Water follows passively across the epithelium according to the resulting osmotic gradient. Interstitial lung fluid is primarily cleared via the lymphatics. Hypoxia decreases the transepithelial Na^+ transport by reducing expression and activity of both the epithelial sodium channel (ENaC) and the Na^+/K^+-ATPase. A decreased capacity to clear fluid from the alveolar space may contribute to HAPE [39–41]. Prophylactic stimulation of the alveolar fluid clearance (AFC) has been shown to decrease the incidence of HAPE in susceptible subjects during exposure to 4559 m [42].

Ketamine is a known inhibitor of voltage-gated Na^+ channels [43]. Their quaternary structure is similar to that of the alveolar ENaC [44]. It appears therefore plausible that ketamine may impair the AFC and may facilitate HAPE formation or may inhibit its clearance. Indeed, experimental data have shown that ketamine inhibits the AFC in rats and humans when it is administered intratracheally [45, 46]. After IV administration, however, ketamine did not inhibit the AFC in rats [46]. This was even valid in septic rats with acute lung injury [47]. If these findings also apply to humans, IV ketamine should not exert adverse effects on the AFC and would be safe for use in patients with HAPE. However, a definite assessment of ketamine's safety in patients with HAPE is not possible at present.

5 Conclusion

Even in austere environments, the provision of effective and safe analgesia remains a core maxim of out-of-hospital emergency medicine (Fig. 3).

For mountain regions and at high altitude, the use of ketamine as a strong analgesic and anesthetic has been recommended by various expert groups and recent guidelines. Specifically, ketamine's hemodynamically stabilizing effects as well as preservation of

Fig. 3 Medical care in cold, austere environments in the Swiss Alps (author photo)

spontaneous ventilation and laryngeal protective reflexes highlight advantages over other strong, but cardiorespiratory depressant, analgesics and anesthetics. Ketamine has become an indispensable drug in the field of alpine emergency medicine and is likely to offer many beneficial effects when compared with various other drugs for analgesia and anesthesia at high altitudes.

Conflict of Interest

None of the authors has a conflict of interest to declare.

References

1. Fink PB, Wheeler AR 3rd, Smith WR, Brant-Zawadzki G, Lieberman JR, McIntosh SE et al (2024) Wilderness medical society clinical practice guidelines for the treatment of acute pain in austere environments: 2024 update. Wilderness Environ Med 35(2):198–218
2. Surrett G, Franklin J, Wedmore I (2015) Pain control in austere settings. Curr Sports Med Rep 14(2):117–122
3. Wedmore IS, Johnson T, Czarnik J, Hendrix S (2005) Pain management in the wilderness and operational setting. Emerg Med Clin North Am 23(2):585–601, xi–xii
4. Vanolli K, Hugli O, Eidenbenz D, Suter MR, Pasquier M (2020) Prehospital use of ketamine in mountain rescue: a survey of emergency physicians of a single-center alpine helicopter-based emergency service. Wilderness Environ Med 31(4):385–393
5. Auerbach PS (2001) Wilderness medicine, 4th edn. Mosby Inc., St. Louis

6. Grocott MP, Johannson L (2007) Ketamine for emergency anaesthesia at very high altitude (4243 m above sea-level). Anaesthesia 62(9): 959–962
7. Bishop RA, Litch JA, Stanton JM (2000) Ketamine anesthesia at high altitude. High Alt Med Biol 1(2):111–114
8. Balzer N, McLeod SL, Walsh C, Grewal K (2021) Low-dose ketamine for acute pain control in the emergency department: a systematic review and meta-analysis. Acad Emerg Med 28(4):444–454
9. Craven R (2007) Ketamine. Anaesthesia 62 (Suppl 1):48–53
10. Castner T (2000) Temperaturbelastung von Medikamenten in der präklinischen Notfallmedizin. Dissertation, University of Ulm, Ulm
11. Schäfer L, Schmidt P, Schiefer LM, Sareban M, Macholz F, Berger MM (2020) Pharmakologische Eigenschaften von Notfallmedikamenten unter Extrembedingungen. Notfall Rettungsmed 23:270–275
12. Rugg C, Woyke S, Voelckel W, Paal P, Strohle M (2021) Analgesia in adult trauma patients in physician-staffed Austrian helicopter rescue: a 12-year registry analysis. Scand J Trauma Resusc Emerg Med 29(1):28
13. Pietsch U, Strapazzon G, Ambuhl D, Lischke V, Rauch S, Knapp J (2019) Challenges of helicopter mountain rescue missions by human external cargo: need for physicians onsite and comprehensive training. Scand J Trauma Resusc Emerg Med 27(1):17
14. Bredmose PP, Lockey DJ, Grier G, Watts B, Davies G (2009) Pre-hospital use of ketamine for analgesia and procedural sedation. Emerg Med J 26(1):62–64
15. Losvik OK, Murad MK, Skjerve E, Husum H (2015) Ketamine for prehospital trauma analgesia in a low-resource rural trauma system: a retrospective comparative study of ketamine and opioid analgesia in a ten-year cohort in Iraq. Scand J Trauma Resusc Emerg Med 23: 94
16. Johansson P, Kongstad P, Johansson A (2009) The effect of combined treatment with morphine sulphate and low-dose ketamine in a prehospital setting. Scand J Trauma Resusc Emerg Med 17:61
17. Bhattarai R, Hamal PK (2021) Comparison of fentanyl-propofol and ketamine-propofol combination in induction and maintenance with intravenous anesthesia for short surgical procedures at moderate elevations. J Nepal Health Res Counc 18(4):769–775
18. Berger MM, Luks AM (2023) High altitude. Semin Respir Crit Care Med 44(5):681–695
19. Powell FL, Milsom WK, Mitchell GS (1998) Time domains of the hypoxic ventilatory response. Respir Physiol 112(2):123–134
20. Luks AM, Ainslie PN, Lawley JS, Roach RC, Simonson TS (2021) Ward, Milledge and West's high altitude medicine and physiology, 6th edn. Taylor & Francis Ltd., London
21. Rahn H, Otis AB (1949) Man's respiratory response during and after acclimatization to high altitude. Am J Phys 157(3):445–462
22. Cartwright CR, Henson LC, Ward DS (1998) Effects of alfentanil on the ventilatory response to sustained hypoxia. Anesthesiology 89(3): 612–619
23. Hirshman CA, McCullough RE, Cohen PJ, Weil JV (1975) Hypoxic ventilatory drive in dogs during thiopental, ketamine, or pentobarbital anesthesia. Anesthesiology 43(6): 628–634
24. Levine BD, Zuckerman JH, deFilippi CR (1997) Effect of high-altitude exposure in the elderly: the Tenth Mountain Division study. Circulation 96(4):1224–1232
25. Smith TG, Talbot NP, Chang RW, Wilkinson E, Nickol AH, Newman DG et al (2012) Pulmonary artery pressure increases during commercial air travel in healthy passengers. Aviat Space Environ Med 83(7):673–676
26. Swenson ER (2013) Hypoxic pulmonary vasoconstriction. High Alt Med Biol 14(2): 101–110
27. Bartsch P, Maggiorini M, Ritter M, Noti C, Vock P, Oelz O (1991) Prevention of high-altitude pulmonary edema by nifedipine. N Engl J Med 325(18):1284–1289
28. Berger MM, Hesse C, Dehnert C, Siedler H, Kleinbongard P, Bardenheuer HJ et al (2005) Hypoxia impairs systemic endothelial function in individuals prone to high-altitude pulmonary edema. Am J Respir Crit Care Med 172(6):763–767
29. Bartsch P, Swenson ER (2013) Clinical practice: acute high-altitude illnesses. N Engl J Med 368(24):2294–2302
30. Maggiorini M, Brunner-La Rocca HP, Peth S, Fischler M, Bohm T, Bernheim A et al (2006) Both tadalafil and dexamethasone may reduce the incidence of high-altitude pulmonary edema: a randomized trial. Ann Intern Med 145(7):497–506
31. Berger MM, Sareban M, Schiefer LM, Swenson KE, Treff F, Schafer L et al (2022) Effects of acetazolamide on pulmonary artery pressure and prevention of high-altitude pulmonary edema after rapid active ascent to 4,559 m. J Appl Physiol (1985) 132(6):1361–1369

32. Kaye AD, Banister RE, Anwar M, Feng CJ, Kadowitz PJ, Nossaman BD (1998) Pulmonary vasodilation by ketamine is mediated in part by L-type calcium channels. Anesth Analg 87(4):956–962
33. Busch CJ, Spohr FA, Motsch J, Gebhard MM, Martin EO, Weimann J (2010) Effects of ketamine on hypoxic pulmonary vasoconstriction in the isolated perfused lungs of endotoxaemic mice. Eur J Anaesthesiol 27(1):61–66
34. Maruyama K, Maruyama J, Yokochi A, Muneyuki M, Miyasaka K (1995) Vasodilatory effects of ketamine on pulmonary arteries in rats with chronic hypoxic pulmonary hypertension. Anesth Analg 80(4):786–792
35. Williams GD, Maan H, Ramamoorthy C, Kamra K, Bratton SL, Bair E et al (2010) Perioperative complications in children with pulmonary hypertension undergoing general anesthesia with ketamine. Paediatr Anaesth 20(1):28–37
36. Bjertnaes LJ (1977) Hypoxia-induced vasoconstriction in isolated perfused lungs exposed to injectable or inhalation anesthetics. Acta Anaesthesiol Scand 21(2):133–147
37. Nakayama M, Murray PA (1999) Ketamine preserves and propofol potentiates hypoxic pulmonary vasoconstriction compared with the conscious state in chronically instrumented dogs. Anesthesiology 91(3):760–771
38. Matthay MA, Clerici C, Saumon G (2002) Invited review: active fluid clearance from the distal air spaces of the lung. J Appl Physiol 93(4):1533–1541
39. Berger MM, Schiefer LM, Treff G, Sareban M, Swenson ER, Bartsch P (2020) Acute high-altitude illness: updated principles of pathophysiology, prevention, and treatment. Dtsch Z Sportmed 71:267–274
40. Mairbaurl H (2006) Role of alveolar epithelial sodium transport in high altitude pulmonary edema (HAPE). Respir Physiol Neurobiol 151(2–3):178–191
41. Sartori C, Duplain H, Lepori M, Egli M, Maggiorini M, Nicod P et al (2004) High altitude impairs nasal transepithelial sodium transport in HAPE-prone subjects. Eur Respir J 23(6):916–920
42. Sartori C, Allemann Y, Duplain H, Lepori M, Egli M, Lipp E et al (2002) Salmeterol for the prevention of high-altitude pulmonary edema. N Engl J Med 346(21):1631–1636
43. Haeseler G, Tetzlaff D, Bufler J, Dengler R, Munte S, Hecker H et al (2003) Blockade of voltage-operated neuronal and skeletal muscle sodium channels by S(+)- and R(−)-ketamine. Anesth Analg 96(4):1019–1026
44. Kosari F, Sheng S, Li J, Mak DO, Foskett JK, Kleyman TR (1998) Subunit stoichiometry of the epithelial sodium channel. J Biol Chem 273(22):13469–13474
45. Cui Y, Nie H, Ma H, Chen L, Zhang L, Wang J et al (2011) Ketamine inhibits lung fluid clearance through reducing alveolar sodium transport. J Biomed Biotechnol 2011:460596
46. Berger MM, Pitzer B, Zugel S, Wieland CW, Vlaar AP, Schultz MJ et al (2010) Alveolar but not intravenous S-ketamine inhibits alveolar sodium transport and lung fluid clearance in rats. Anesth Analg 111(1):164–170
47. Fastner C, Mairbaurl H, Weber NC, van der Sluijs K, Hackl F, Hotz L et al (2014) Intravenous S-ketamine does not inhibit alveolar fluid clearance in a septic rat model. PLoS One 9(11):e112622

Chapter 25

Unrealized Capital: Potential Roles for Ketamine, Hypnotherapy, and Virtual Reality in Cardiac Anesthesia

Sabine Himmelseher and Christopher V. Maani

Abstract

Postoperative pain, cognitive decline, and mental health issues continue to represent significant challenges following cardiac surgery. In addition to baseline morbidity and local and systemic responses to surgical trauma, perioperative emotional and psychological stressors are exceptional for cardiac surgery patients. Human beings connect emotions with their heart. Nevertheless, there is a paucity of data and guidance on innovative approaches to improving the outcomes of patients who have undergone cardiac surgery. This review will focus on the available evidence of ketamine's potential to advance the perioperative and long-term therapy of cardiac surgery patients in a multicomponent framework as a future prospect. We will put forth a combination therapy comprising ketamine-mediated hypnotherapy as a mind–body-based approach, behavioral psychotherapy, and neurocognitive training with the use of virtual reality (VR) to augment interventions.

Key words Cardiac anesthesia, Esketamine, Postoperative neurocognitive disorder, Delirium, Depression, Pain, Hypnotherapy, Virtual reality, Cognitive behavioral therapy, Psychotherapy

1 Specifics of Care with Heart Surgery Patients and Ketamine

There is a paucity of data and guidance on innovative approaches to improve the incidence and severity of postoperative neurocognitive decline and delirium, as well as depression, anxiety, and pain in the setting of cardiac and general surgery [1–6]. In a population-based study including more than 490,000 participants from the UK Biobank published September 2024 in the Lancet, data corroborated that surgeries are generally safe but cumulatively associated with postoperative cognitive decline and neurodegeneration, particularly in older and vulnerable patients undergoing multiple procedures [1]. After cardiac surgery, pain, cognition, and mental health issues remain problems of great clinical significance. Inadequate pain management is associated with mood disturbances, functional restrictions, and poor recovery, which is valid vice versa [4, 6–8]. Preoperative depression is an independent risk factor for

Sabine Himmelseher (ed.), *Ketamine: Pain Management, Critical Care, Depression, and Resource-Poor Settings Volume II*, Neuromethods, vol. 225, https://doi.org/10.1007/978-1-0716-4599-4_25,

postoperative delirium, hospital readmission, decreased postsurgical independence and delayed return to normal daily activities and the workforce [3, 9, 10]. The long-term sequelae of chronic pain and depression include an elevated risk for further major cardiovascular events [11, 12]. It is predicted that these morbidities will increase in prevalence as the population continues to age and as the incidence of operable cardiovascular diseases rises [13].

In addition to local and systemic responses to surgical trauma, some patients anticipate and mentally experience cardiac surgery as a traumatic, lack-of-control event with tremendously distressing and disabling feelings [14]. As human beings connect emotions with their heart, perioperative emotional and psychological stressors are particularly impactful for cardiac surgery patients, further straining the already vulnerable cardiovascular system.

A patient-tailored combined pharmacological and biopsychological approach may therefore prove beneficial in enhancing the perioperative care and outcomes of heart surgery patients. On the one hand, medical professional societies have acknowledged the "mind–heart–body connection" and recommended psychological management of patients with cardiovascular disease through the implementation of in-hospital mental health support interventions [14, 15]. On the other hand, ketamine and, in particular, esketamine have been effectively employed to reduce postoperative pain and decrease the development of chronic postsurgical pain [16–18]. Furthermore, both racemic ketamine and the S-isomer have been demonstrated to reduce postoperative depression and negative emotions [16, 17, 19–22], as well as mitigate renal and myocardial injury [23, 24]. Additionally, both have been found to decrease postoperative delirium [25, 26], with esketamine exhibiting particular efficacy in improving postoperative neurocognitive disorders in elderly patients [26–28] and preventing delirium after heart surgery [29].

This review will focus on the potential of ketamine to improve outcomes for cardiac surgery patients within a multicomponent therapeutic framework as a future prospect. The proposed combination therapy comprises perioperative ketamine-mediated hypnotherapy which may be considered a mind–body-based approach and (behavioral) psychotherapy and neurocognitive training with the use of virtual reality (VR) to augment interventions.

2 Hypnotherapy, Virtual Reality, and Neurocognitive Interventions in Heart Surgery Patients

The term "medical hypnosis" has been defined as a treatment that causes a state of consciousness involving focused attention on oneself and a reduction in environmental awareness. This is achieved by redirecting attention and providing suggestions [30]. The concept of hypnotizability or hypnotic susceptibility has been developed as a stable personality trait which describes the response to suggestions given under hypnosis. It is regarded as a pivotal element in elucidating the discrepancies in the consequences of hypnotic suggestions. Virtual reality (VR) is defined as a three-dimensional environmental simulation generated by computer technology, frequently observed through a head-mounted display. It permits a spectrum of engagement, from "non-immersion" to "complete immersion," wherein individuals may transition from their actual physical surroundings into VR environments with multisensory engagement [30].

The application of medical hypnosis with or without virtual reality (VR) techniques in cardiac surgery patients has yielded inconclusive results with respect to the conferral of benefits. For example, when patients were randomly assigned to the control group, which received usual care, or to one of three experimental groups, which received hypnotherapy, virtual reality (VR), or a combination of both, on the day before and after surgery, no significant between-group differences were observed in physiological measures, anxiety, pain, opioid use, and fatigue after each session [31]. In a perioperative, daily goal-oriented, smartphone-based model, developed and evaluated by Indonesian nurses with the support of information technology experts and experienced cardiovascular nurses, a multifaceted approach ranging from prayer, education, and exercise to listening to specific music was successfully applied [32]. Despite the presence of baseline or chronic pain in some patients, the intervention group showed improved postoperative pain, anxiety, and mobility outcomes. Nevertheless, the time to early mobilization remained unchanged. A scoping review on the role of hypnotherapy in postoperative cardiac surgical patients concluded that further evidence is required "to justify its incorporation into routine" management practices [33]. This is in contrast to the clinical experience gained with medical hypnosis as an adjunctive therapy in a large German center performing more than 4000 major heart surgeries per year [34]. In this setting, supervised psychologists who have undergone training in hypnotherapy over a 2-year period provide expert hypnotic interventions as part of routine inpatient care for patients undergoing heart transplantation, mechanical circulatory support, and other major procedures. The experts report an enhanced sense of patient well-being, a reduction in depressive and anxiety symptoms, decreased blood pressure, stabilized pulse, diminished muscle tension, and a decreased reliance on analgesics, anxiolytics, benzodiazepines, and tranquilizers. The beneficial effects of medical hypnosis have also been observed in major oncological surgeries, including the provision of opioid-sparing pain relief and the maintenance of heart rate variability [35, 36]. Additionally, studies have demonstrated the

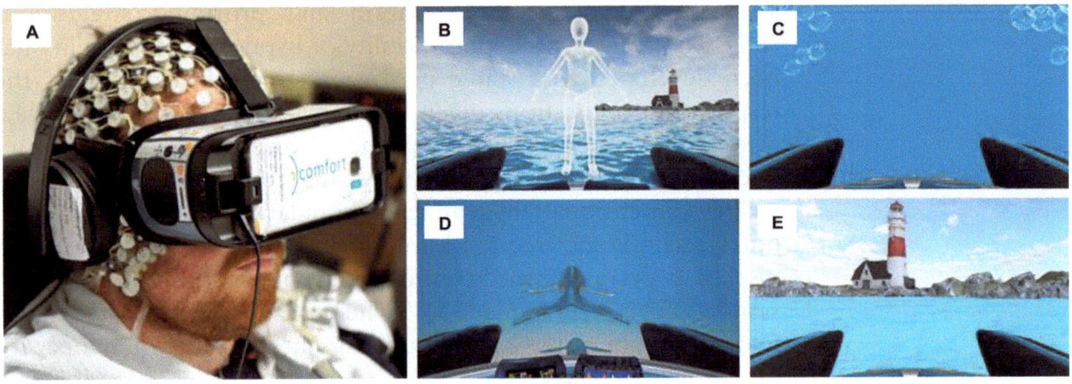

Fig. 1 Virtual reality (VR) kit: (**a**) VR hypnosis device and (**b–e**) hypnosis software images (Oncomfort's Aqua^{R}, HypnoVR, Gosselies, Charleroi, Belgium). (**b**) Session starts with an induction phase based on hypnotic techniques, (**c**) then a guidance phase follows where participants dive in an underwater world, and (**d**) a deepening phase ensues during which participants follow a whale in an underwater journey; (**e**) finally, a realtering phase during which participants are brought back from the deep sea occurs (from Rousseaux et al. [40] with permission from John Wiley and Sons conveyed via Copyright Clearance Center, Inc.)

efficacy of hypnosis in the treatment of depression in sessions with one expert and "one-in-person" patient [37].

The ongoing progress in radiological and cardiological interventional techniques has resulted in a dramatic surge in the number of procedures performed over the past decade. A study conducted by experienced therapists found that when hypnosis was used during atrial flutter ablation in 56 awake patients, there was a reduction in pain and morphine requirements compared to a control group of 57 patients who received non-hypnotic relaxation suggestions and white noise [38]. In another randomized controlled trial, a total of 118 patients without preoperative neurocognitive deficit and a mean age of 80 years undergoing transaortic valvular replacement (TAVR) under general anesthesia with sevoflurane and remifentanil were examined [39]. The administration of positive suggestions and music via headphones and an MP3 player was employed as a therapeutic intervention during the procedure. However, this approach with "suggestions" did not result in any discernible impact on the incidence of postoperative delirium, pain, analgesic requirements, or the prevalence of nausea and vomiting [39].

Recent (systematic) reviews and evidence report that VR technologies can be an effective tool for pain management across a range of medical procedures, with an increasing number of applications in both acute and chronic pain conditions [30, 40–42] (Fig. 1).

However, there is still a considerable statistical heterogeneity in the trials. Within the context of burn wound care, the use of immersive VR for distraction-based analgesia has been shown to improve ratings for "worst pain" and "better mood." Contrary to expectations, interactive computer-generated VR scenarios did not prove to be more effective than still nature pictures in reducing pain [43].

When an early version of VR hypnosis program was used in patients experiencing acute physical trauma pain, it became rapidly clear that a fast-paced, chaotic, highly stimulating trauma care emergency room may not have been the best environment to approach background pain, that the hypnotic intervention may not have been strong enough for such a setting, and that these trauma patients were likely significantly limited in their ability to absorb hypnotic suggestions [44]. Participants had been asked to keep their eyes open for the program. However, recent advances in hardware and software capabilities coupled with enhanced affordability have led to the development of innovative VR interventions that incorporate biopsychosocial, individual patient data for the delivery of cognitive and behavioral therapy. This is based on a mechanism-oriented, goal-directed strategy [42]. Such endeavors have yielded initial evidence of improvements, for instance, in patients with chronic back pain [45]. Furthermore, VR-based mentoring intelligence such as avatars are promising tools that can be operationalized to specifically address individual needs, such as for self-management of chronic pain [42, 46]. This progress offers also new perspectives for managing post-cardiac surgical pain.

Furthermore, neurocognitive deterioration is a common occurrence following cardiac surgery, with the potential to have a deleterious impact on patient quality of life [47]. Nonetheless, both inpatient and outpatient rehabilitation programs have been shown to enhance neurocognition and daily living capabilities [48, 49]. The current array of innovative VR-based techniques holds considerable promise for the provision of individually tailored neurocognitive training, which takes into account both the patients' biopsychosocial determinants and their actual functional and physical status.

3 Ketamine in Cardiac Anesthesia

In addition to its hemodynamically stabilizing effects, the pathophysiological rationale behind the use of ketamine in cardiac anesthesia with regard to the brain is based on its modular effects on (neuro-) inflammation and anti-excitotoxic impact [50]. Both of these play a significant role in postoperative cognitive disorders and delirium (POD) with an age-related increase [51, 52] (Fig. 2).

In recent years, advanced research has indicated that racemic ketamine has no adverse but rather beneficial impact on cognitive function and delirium after cardiac surgery when used for anesthetic induction or during cardiopulmonary bypass (CBP) [23, 53]. A Bayesian network meta-analysis published in 2023

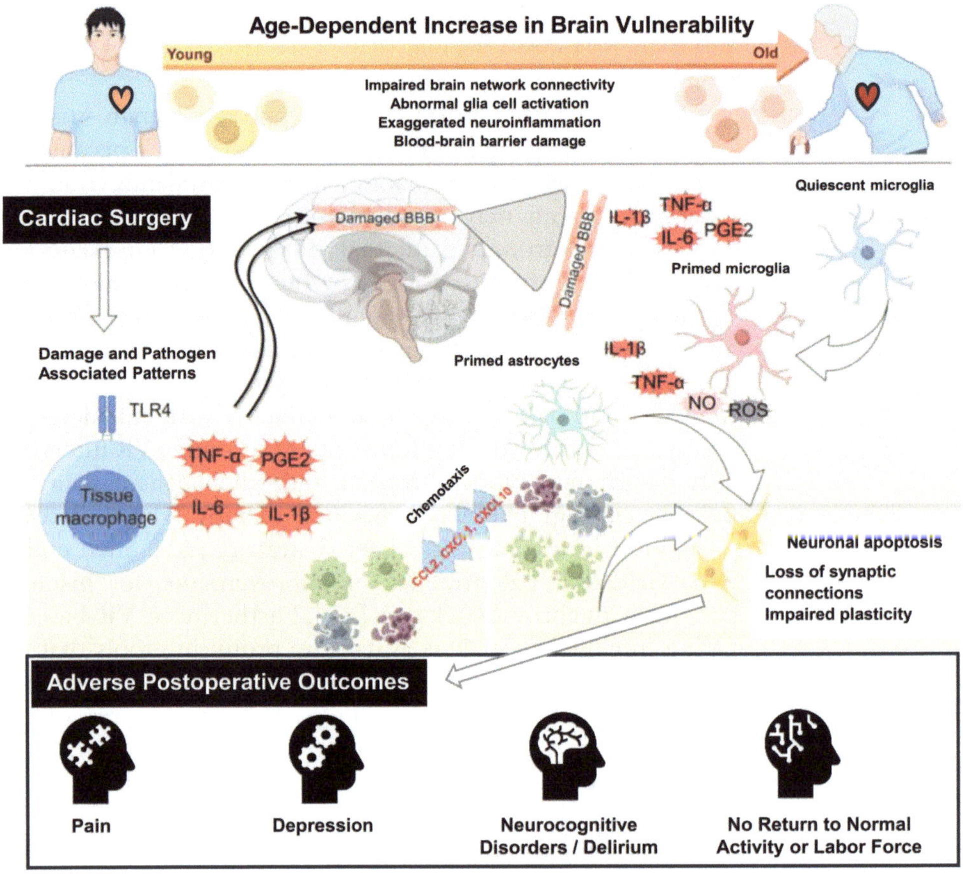

Fig. 2 Important pathophysiological changes assumed to underly postoperative neurocognitive disorders and delirium caused by neuro-inflammation, leading to adverse patient outcomes with an age-dependent increase in brain vulnerability after cardiac surgery: Surgery and other procedures induce damage- and pathogen-associated inflammatory patterns which mediate the secretion of interleukins IL-1β and IL-6, tumor-necrosis-factor-α (TNF-α), and prostaglandin e2 (PGE2) by tissue macrophages through binding of pattern recognition receptors such as toll-like receptor 4 (TLR4). These mediators enter the brain parenchyma through a damaged blood–brain barrier (BBB). They also activate quiescent microglia. These release large quantities of inflammatory mediators including reactive oxygen species (ROS) and nitric oxides (NO). Activated astrocytes release chemokines such as CCL2, CXCL1, CXC10, and CXCL10 to draw monocytes and other immune cell populations into the brain parenchyma. Neuroinflammation can lead to neuronal apoptosis, impaired synaptic plasticity, and dysfunctional synaptic connections. This results in adverse postoperative outcomes, such as postoperative neurocognitive disorders or depression (modified from Zhu et al. [52] under a commercial CC-BY 4.0 international license; https:/creativecommons.org/licenses/by/4.0/)

demonstrated that a preoperative, subanesthetic dose of ketamine can effectively reduce the incidence of postoperative delirium [54]. A sequential trial analysis including all eligible cardiac and non-cardiac surgery study patients receiving various doses of perioperative ketamine indicated that the number of patients required to detect or reject a 20% relative risk reduction for POD with ketamine was 6494 patients [55]. The authors therefore stated that the lack of adequate study power precluded any definitive conclusions regarding the therapeutic effects of ketamine on POD.

Furthermore, a recent report has indicated that a single preoperative adjunctive injection of low-dose esketamine (0.25 mg/kg) has the potential to reduce the incidence of POD following cardiac valve replacement with CBP in relatively young patients with a mean age of 52 years [29]. When compared to sufentanil in total intravenous anesthesia (TIVA), 1–3 mg/kg esketamine used for anesthesia induction followed by a 2–4 mg/kg/h infusion attenuated the pro-inflammatory response during and after CBP, without any negative impact on cardiac troponin T levels [56, 57]. However, a study published two decades ago did not identify superior neurocognitive test outcomes following TIVA with esketamine/propofol in comparison to remifentanil/propofol 10 weeks after open-heart surgery [58]. Unfortunately, this study administered exceptionally high doses of esketamine, reaching approximately 7.5 mg/kg/h during anesthesia. Indeed, it was previously assumed that only high doses of ketamine are neuroprotective. Conversely, it is now understood that the neuroprotective effects of esketamine and ketamine are subject to tight regulation within a specific dose range and—depending on the (potentially) harmful context—need to be adapted properly. It is not beneficial to over- or underdose racemic or esketamine.

Moreover, in a rat model of cardiac surgery, chronic postsurgical unpredictable stress resulted in significant alterations in metabolic pathways of specific amino acids [59]. This was evidenced by integrative proteomic and metabolomic analyses, indicating, for instance, exhaustion [59]. Findings from HV and patients show that the administration of subanesthetic and anesthetic dose ketamine and esketamine exerts a comprehensive influence on metabolomics [60, 61], potentially offering a means of counteracting various postsurgical disease states, including fatigue [62].

4 Ketamine, Depression, and Psychotherapy

As previously outlined in detail in several preceding chapters of this book, ketamine and esketamine have been validated as efficacious antidepressants in the treatment of severe depressive disorders [63, 64]. In perioperative settings, especially esketamine has been demonstrated to alleviate postoperative depression and anxiety, reduce pain, and improve sleep and quality of life [65]. These characteristics are similarly crucial in the context of cardiac surgery and postsurgical time periods.

Nevertheless, an analysis of ketamine trials in psychiatric disorders reveals a distinct predominance of the "only-medical approach" and a paucity of research on biopsychological interventions to enhance and prolong the pharmacological effects of ketamine [64]. Furthermore, although ongoing ketamine treatment has been demonstrated to prevent depressive episodes, the "optimal" long-term therapeutic approach for ketamine therapy remains uncertain. Concerns pertaining to both long-term remission and relapse persist. Following ketamine discontinuation and particularly after hospital discharge, there is an elevated risk of suicide in patients with suicidal ideation. Therefore, there is a pressing need for strategies to maintain treatment gains. A systematic review published in 2023 determined that comprehensive recommendations are not yet feasible, but that the inclusion of psychotherapy in ketamine treatments may prove advantageous despite the considerable trial heterogeneity [66]. Furthermore, a proof-of-concept, randomized study has indicated that cognitive behavioral therapy (CBT) may be an effective method of sustaining the antidepressant effects of IV ketamine and improving outcomes in treatment-resistant depression [67]. Given the efficacy of CBT in preventing relapse and reducing suicide attempts, an ongoing trial is examining the combination of intranasal esketamine (administered twice weekly for 4 weeks) and a 16-week course of CBT, versus esketamine alone, with a 6-month follow-up period [68]. A separate trial is investigating the efficacy of oral ketamine at an initial dose of 1 mg/kg, administered twice per week, followed by variable dosing according to response over a period of 8 weeks. Behavioral activation therapy is conducted on a twice-weekly basis for a period of 4 weeks, followed by once-weekly sessions for an additional 4 weeks, in comparison to oral ketamine treatment alone [69]. These data are of significant importance as a basis for the design of doses and timelines, for example, in a study on the use of ketamine or esketamine in perioperative and post-cardiac surgery patients.

5 Combining Ketamine and Medical Hypnosis as Mind–Body-Based Therapy: Opportunities and Challenges

As a psychophysiological trait, high hypnotizability, or hypnotic susceptibility, has frequently been linked to the efficacy of hypnotherapy, whereas low hypnotizability has been regarded as a hindrance. However, it seems that hypnotizability can be influenced by a number of factors, including individual experiences [70] and the use of psychoactive drugs. As predicted by the authors, an infusion of ketamine at a low dose of 0.21 mg/kg/h was observed to increase the hypnotizability of healthy volunteers (HV) who had scored "low" on the Stanford Clinical Hypnotizability Scale prior to ketamine administration [71]. Conversely, there was a tendency for ketamine to reduce the hypnotizability of HV who had been classified as medium hypnotizable individuals. While the full clinical relevance of this phenomenon is not yet clear, all HV reported subjective psychoactive and dissociative effects related to ketamine. Preparation, positive suggestions before and during such events can prevent or at least reduce their unwanted quality and decreases patient disenrollment [72, 73]. As it is standard practice in high-volume antidepressant treatment centers, positive education, supportive guidance through transient psychoactive events, and debriefing by experienced mental health professionals is of course recommendable for potential ketamine use in perioperative, awake cardiac surgery patients. Nevertheless, there are psychiatrists who advocate for the incorporation of psychedelic experiences during ketamine treatment as a form of therapy. They posit that such experiences may facilitate amplified consciousness, personal growth, inner healing, and enhanced clarity in the recipient [74]. In such settings, psychoactive experiences have to be imbedded into the therapeutic concept. However, the mental effects that follow ketamine treatment in post-cardiac surgery patients are poorly defined.

Some researchers are convinced that the antidepressant effects of ketamine/esketamine are largely influenced by placebo effects. A meta-analysis published in 2023 underscores this notion [75]. In an ongoing trial, the same researchers are now evaluating esketamine versus placebo with verbal induction of "high" or "low" expectations for treatment success in patients with major depressive disorder [76]. The study of resting state functional connectivity in magnetic resonance imaging (rs-fcMRI) with emotion and reward processing paradigms may elucidate the psychological and neural mechanisms underlying expectation effects and their interaction with esketamine's pharmacological impact.

Furthermore, it has been proposed that the phenomenology of psychoactive events and outcomes of antidepressant ketamine therapy are markedly influenced by "imprinted" patient's experiences made in the days preceding treatment sessions [77]. A review of the current literature reveals that the potential for imprinted mental experiences to exert delayed influences in postoperative ketamine therapy has yet to be examined. The potential impact of surgical and anesthetic procedures on the delayed experiential effects of ketamine remains unclear. However, it seems prudent to consider the possibility of such effects when ketamine is used in perioperative and postsurgical care, particularly in the context of heart surgery. It is similarly reasonable to educate patients about such phenomena in order to help avoid any unwanted, deeply "impressive" negative experiences before ketamine sessions. It is noteworthy that the literature on this topic is inconclusive regarding the question of whether a state of high pre-treatment anxiety can impact the response to ketamine's antidepressant effects, necessitating higher or prolonged doses.

6 Future Prospects and Conclusions

In light of the current evidence pertaining to the pharmacological potential and hypnotic, suggestion, and psychotherapeutic-enhancing effects of esketamine, especially this compound appears to be a promising candidate offering multifactorial benefits for advanced care of cardiac surgery patients. Esketamine has the potential to stabilize hemodynamics during and after anesthesia while simultaneously inducing antidepressant, analgesic, anti-inflammatory, and neuroplastic effects. It may also be administered as a resilience-enhancing agent, used before surgery to treat and prevent depressive disorders. Following acute surgery, its use may be extended into longer-term postsurgical times. However, it is unlikely that postsurgical "stand-alone" ketamine therapy is sufficient for heart surgery patients.

To provide the shortest effective and lowest drug dose use schedule possible, the combination with non-pharmacological interventions is promising. These include postsurgical medical hypnosis, (behavioral) psychotherapy, and longer-term neurocognitive training. It has also been demonstrated that ketamine can enhance suggestibility and that ketamine-assisted or augmented mental health therapy can be more effective. In a comparison of the effects of hypnosis, meditation, and two psychedelics, psilocybin and lysergic acid diethylamide (LSD) in rs-fMRI study, all modalities induced distinct connectivity patterns in neural correlates, but no network change reached statistical significance [78]. This may be different in a comprehensive ketamine-assisted therapy with synergistic effects of different modalities (not only in cardiac surgery patients).

The advent of new, affordable digital technologies and the recent, rapid developments of VR technology have created a cost-effective opportunity for the realization of such therapeutic combinations, whereby multiple pieces of information can be integrated into a single virtual space [42, 79]. As a result of continued treatment, VR may serve as an alternative to in-patient sessions, thereby reducing costs for resources and therapeutic personnel. At best, VR techniques are personalized, offering interventions which have been programmed according to the individual patient's biopsychosocial and physical conditions, including assessment of efficacy of therapies and patient progression (Fig. 3).

The use of virtual reality (VR) also allows for concurrent electroencephalography (EEG) and functional near-infrared spectroscopy (NIRS) measurements which may be of additional value after cardiac surgery. Of course, disadvantages and pitfalls, such as sensory overstimulation, "cyber sickness," or distraction from the desired experiences, need to be prevented.

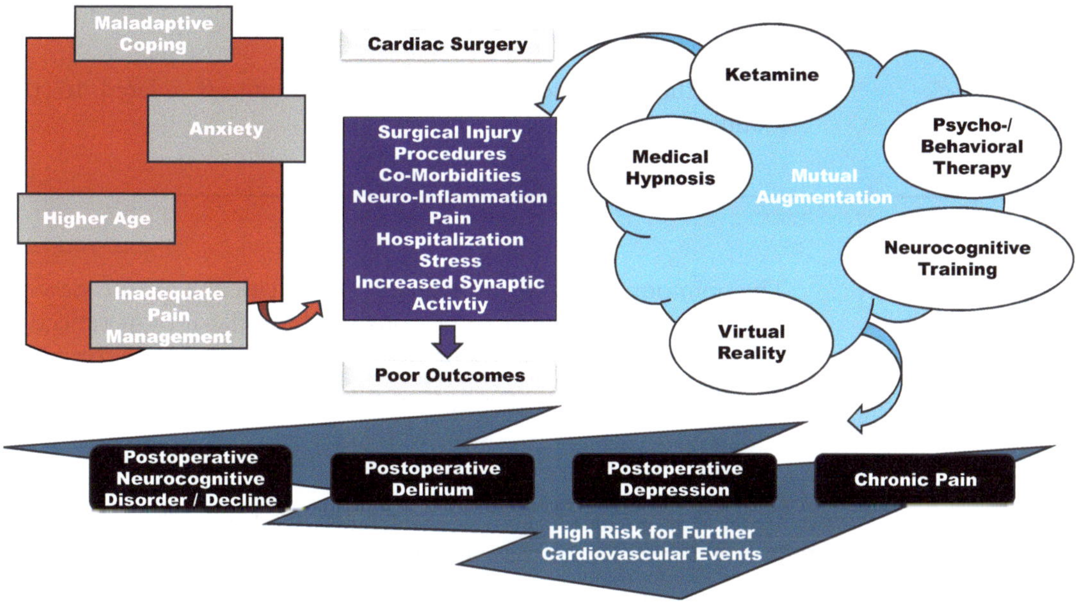

Fig. 3 Factors and events which can cause poor patient outcomes after cardiac surgery and an innovative cloud framework proposal for combining treatments as new therapeutic approaches to improve outcomes (created by S. Himmelseher in Power Point, Microsoft Corp., Redmond, WA, USA)

In conclusion, a combination of repeated ketamine administration, hypnosis, and mental health interventions including the use of VR as a vehicle for therapy delivery at the appropriate time represents a promising framework for enhancing perioperative care and outcomes for cardiac surgery patients. In order to more fully exploit the neuroplastic capabilities of ketamine, it is essential to conduct comprehensive studies on the interaction effects of cardiac surgery, CPB, and ketamine-induced brain changes. It is essential to elucidate those time periods that may offer a "window of neuroplastic opportunity" for targeted neurocognitive interventions. In the absence of surgery, the neuroplastic effects of ketamine reach a peak at approximately 1–2 days after its administration in antidepressant therapy [63, 64]. This may serve as a preliminary indicator for a potential timeline for behavioral therapy and neurocognitive training in a ketamine combination regimen. Initial steps have been taken, and preliminary investigations on hypnotherapy and ketamine in heart surgery patients are underway. In each subsequent research endeavor, it would be beneficial to aim to maximize the synergistic potential of the proposed therapeutic components. For the benefit of our patients, may this prospect chapter inspire future interdisciplinary research!

Conflict of Interest

The authors do not have conflicts of interest related to this manuscript.

Disclaimer

The opinions or assertions contained herein are the private views of the authors and are not to be construed as official or as reflecting the views of the Department of the Army or the Department of Defense of the USA.

References

1. Taylor J, Robledo KP, Medel V et al (2024) Association between surgical admissions, cognition, and neurodegeneration in older people: a population-based study from the UK biobank. Lancet Healthy Longev 5:100623
2. Whitlock EI, Diaz-Ramirez G, Smith AK et al (2021) Association of coronary artery bypass grafting vs percutaneous coronary intervention with memory decline in older adults undergoing coronary revascularization. JAMA 325: 1955–1964
3. Mattimore D, Fischl A, Christophides A et al (2023) Delirium after cardiac surgery – a narrative review. Brain Sci 13:1682
4. Chen N, Mo YC, Xu M et al (2024) Risk factors for postoperative delirium in elderly patients undergoing heart valve surgery. J Cardiothorac Surg 19:106
5. O'Gara B, Espinosa Leon JP, Robinson K et al (2023) New onset postoperative depression after major surgery: an analysis from a national claims database. BJA Open 8(C):100223
6. Xiao MZX, Kann JS, Dana E et al (2023) Prevalence and risk factors for chronic postsurgical pain after cardiac surgery: a single-center prospective cohort study. Anesthesiology 139: 309–320
7. Kelly I, Fields K, Sarin P et al (2024) Identifying patients vulnerable to inadequate pain resolution after cardiac surgery. Semin Thorac Cardiovasc 36:182–194
8. Sun T, Fan M, Peng D et al (2024) Association of chronic poststernotomy pain and health-related quality of life: a prospective cohort study. Can J Anaesth 71:579–589
9. Falk A, Kählin J, Nymark C et al (2022) Depression is associated with delirium after cardiac surgery – a population-based cohort study. Interact Cardiovasc Thorac Surg 35: ivac151
10. Posen I, Grosman-Rimon L, Caspi-Avissar N et al (2024) Depression predicts delayed return to daily activities in patients post-cardiac surgery: a prospective observational study. J Cardiothorac Surg 19:504
11. Chouairi F, Fuery MA, Mullan CW et al (2021) The impact of depression on outcomes in patients with heart failure and reduced ejection fraction treated in the GUIDE-IT trial. J Card Fail 12:1359–1366
12. Reynolds CA, Minic Z (2023) Chronic pain-associated cardiovascular disease: the role of sympathetic nerve activity. Int J Mol Sci 24: 5378
13. Thang AB, Diaz-Ramirez IG, Smith AK (2023) Preoperative factors predict memory decline after coronary artery bypass grafting or percutaneous coronary intervention in an epidemiological cohort of older adults. J Am Heart Assoc 12:e027849
14. Tigges-Limmer K, Sitzer M, Gummert J (2021) Perioperative psychological interventions in heart surgery. Dtsch Ärzteblatt Int 118:339–345
15. Levine GN, Cohen BE, Commodore-Mensah Y et al (2021) Psychological health, well-being, and the mind-heart-body connection: a scientific statement from the American Heart Association. Circulation 143:e763–e783
16. Hung KC, Kao CL, Ho CN et al (2024) The impact of perioperative ketamine or esketamine on the subjective quality of recovery after surgery: a meta-analysis of randomised controlled trials. Br J Anaesth 132(6):1293–1303
17. Huang C, Yang R, Xie X et al (2024) Effect of small dose esketamine on perioperative neurocognitive disorder and postoperative depressive

symptoms in elderly patients undergoing major elective noncardiac surgery for malignant tumors: a randomized clinical trial. Medicine (Baltimore) 103:e40028
18. Abouarab AH, Brülle R, Aboukilila MY, Weibel S, Schnabel A (2024) Efficacy and safety of perioperative ketamine for the prevention of chronic postsurgical pain: a meta-analysis. Pain Pract 24:553–566
19. Zhou Y, Sun W, Zhang G et al (2021) Ketamine alleviates depressive symptoms in patients undergoing intracranial tumor resection: a randomized controlled trial. Anesth Analg 133: 1588–1597
20. Lou XJ, Ren ZY, Hashimoto K et al (2024) Efficacy and safety of esketamine for perioperative depression in patients undergoing elective surgery: a meta-analysis of randomized controlled trials. Asian J Psychiatr 95:103997
21. Wen W, Wenjing Z, Xia X et al (2024) Efficacy of ketamine versus esketamine in the treatment of perioperative depression: a review. Pharmacol Biochem Behav 242:173773
22. Luo T, Deng Z, Ren Q et al (2024) Effects of esketamine on postoperative negative emotions and early cognitive disorders in patients undergoing non-cardiac thoracic surgery: a randomized controlled trial. J Clin Anesth 95: 111447
23. Wittwer ED, Ceerhan JH, Schroeder DT et al (2023) Impact of ketamine versus propofol for anesthetic induction on cognitive dysfunction, delirium, and acute kidney injury following cardiac surgery in elderly, high-risk patients. Ann Card Anaesth 26:274–280
24. Bi Z, Kong L, Zhao J et al (2024) Positive effects of low dose S-ketamine on preventing myocardial injury after thoracoscopic lobectomy in patients aged 70 to 85. BMC Anesthesiol 24: 103
25. Abd Ellatif SE, Mowafy SMS, Shahin MA (2024) Ketofol versus dexmedetomidine for preventing postoperaive delirium in elderly patients undergoing intestinal obstruction surgeries: a randomized controlled study. BMC Anesthesiol 24:1
26. Wang Y, Ma B, Wang C et al (2024) The influence of low-dose s-ketamine on postoperative delirium and cognitive function in older adults undergoing thoracic surgery. J Cardiothorac Surg 19:324
27. Tu W, Yuan H, Zhang S et al (2021) Influence of anesthetic induction of propofol combined with esketamine on perioperative stress and inflammatory responses and postoperative cognition of elderly surgical patients. Am J Transl Res 13:1701–1709
28. Han C, Ji H, Guo Y et al (2023) Effect of subanesthetic dose of esketamine on perioperative neurocognitive disorders in elderly undergoing gastrointestinal surgery: a randomized controlled trial. Drug Des Devel Ther 17: 863–873
29. Xiong X, Shao Y, Chen D et al (2024) Effect of esketamine on postoperative delirium in patients undergoing cardiac valve replacement with cardiopulmonary bypass: a randomized controlled trial. Anesth Analg 139:743–753
30. Louras M, Vanhaudenhuyse A, Panda R et al (2024) Virtual reality combined with mind body therapies for the management of pain: a scoping review. Int J Clin Exp Hypn 72:435–471
31. Rousseaux F, Dardenne N, Massion PB et al (2022) Virtual reality and hypnosis for anxiety and pain management in intensive care units: a prospective randomised trial among cardiac surgery patients. Eur J Anaesthesiol 39:58–66
32. Awaludin S, Nurachmah E, Wisesa Soetisna T, Umar J (2022) The effect of smartphone-based perioperative nursing intervention: prayer, education, exercise therapy, hypnosis, and music toward pain, anxiety, and early mobilization on cardiac surgery. J Public Health Res 11: 2742
33. Ahmad B, Gill I, Liblik K et al (2023) The role of hypnotherapy in postoperative cardiac surgical patients, a scoping review of current literature. Curr Probl Cardiol 48:101787
34. Tigges-Limmer K, Brocks Y, Winkler Y et al (2024) Clinical experience with medical hypnosis as an adjunctive therapy in heart surgery. Front Psychol 5:1356392
35. Azam MA, Weinrib AZ, Slepian PM et al (2024) Effects of perioperative clinical hypnosis on heart rate variability in patients undergoing oncologic surgery: secondary outcomes of a randomized controlled trial. Front Pain Res (Lausanne) 5:1354015
36. Rosenbloom BN, Slepian PM, Azam MA et al (2024) A randomized controlled trial of clinical hypnosis as an opioid-sparing adjunct treatment for pain relief in adults undergoing major oncologic surgery. J Pain Res 17:45–59
37. Chen X, Yang R, Ge L et al (2017) Hypnosis in the treatment of major depression: an analysis of heart rate variability. Int J Clin Exp Hypn 65: 52–63
38. Garcia R, Bouleti C, Li A et al (2020) Hypnosis versus placebo during atrial flutter ablation. The PAINLESS study: a randomized controlled trial. JACC Clin Electrophysiol 6: 11551–11560

39. Kaufmann C, Zech N, Brandt F et al (2023) Intraoperative suggestions to prevent postoperative delirium in patients undergoing transaortic valvular replacement: a randomized placebo-controlled trial. Aging Clin Exp Res 35:1865–1872
40. Rousseaux F, Panda R, Toussaint C et al (2023) Virtual reality hypnosis in the management of pain: self-reported and neurophysiological measures in healthy subjects. Eur J Pain 27:148–162
41. Teh JJ, Pascoe DJ, Parchure R et al (2024) Efficacy of virtual reality for pain relief in medical procedures: a systematic review and meta-analysis. BMC Med 22:64
42. Medina S, Clarke S, Hughes S (2024) Virtual reality-based analgesia: towards a novel framework for the biopsychosocial management of chronic pain. Br J Anaesth 133:486–490
43. Patterson DR, Drever S, Soltani M et al (2023) A comparison of interactive immersive virtual reality and still nature pictures as distraction-based analgesia in burn wound care. Burns 49: 182–192
44. Wiechmann SA, Jensen MP, Sharer SR et al (2022) The impact of virtual reality hypnosis on pain and anxiety caused by trauma: lessons learned from a clinical trial. Int J Clin Exp Hypn 70:156–173
45. Eccleston C, Fisher E, Liikaanen S et al (2022) A prospective, double-blind, pilot, randomized, controlled trial of an embodied virtual reality intervention for adults with low back pain. Pain 163:1700–1715
46. Barlett J, Fisher E, Liikkanen S et al (2022) The design and development of an embodied semi-autonomous mentoring intelligence (SAMI) for use in virtual reality interventions, operationalized for the self-management of chronic pain. Front Virtual Real 3:882980
47. Greaves B, Psaltis PJ, Ross TJ (2019) Cognitive outcomes following coronary artery bypass grafting: a systematic review and meta-analysis of 91,829 patients. Int J Cardiol 289:43–49
48. Ajtahed SS (2019) Efficacy of neurocognitive rehabilitation after coronary artery bypass graft surgery in improving quality of life. Front Psychol 10:1759
49. Butz M, Gerriets T, Sammer G et al (2023) Twelve-month follow-up effects of cognitive training after heart valve surgery on cognitive functions and health-related quality of life: a randomised clinical trial. Open Heart 10: e002411
50. Mion G, Himmelseher S (2024) Esketamine: less drowsiness, more analgesia. Anesth Analg 139:78–91
51. Zhu Y, Zhou M, Jia X et al (2023) Inflammation disrupts the brain network of executive function after cardiac surgery. Ann Surg 277: e689–e698
52. Zhu Y, Feng W, Kong Q et al (2023) Evaluating the effects of S-ketamine on postoperative delirium in elderly patients following total hip or knee arthroplasty under intraspinal anesthesia: a single-center randomized, double-blind, placebo-controlled, pragmatic study protocol. Front Aging Neurosci 15:1298661
53. Siripoonyothai S, Sindhvananda W (2021) Comparison of postoperative delirium within 24 hours between ketamine and propofol infusion during cardiopulmonary bypass machine: a randomized controlled trial. Ann Card Anaesth 24:294–301
54. Meco M, Giustiniano E, Cecconi M et al (2023) Pharmacological prevention of postoperative delirium in patients undergoing cardiac surgery: a Bayesian network meta-analysis. J Anesth 37:294–310
55. Hung KC, Chen IW, Liu PH (2024) Impact of ketamine on postoperative delirium: insights from trial sequential analysis. Int J Surg 110: 1834–1835
56. Neuhäuser C, Preiss V, Feurer MK et al (2008) Comparison of S(+)-ketamine- with sufentanil-based anaesthesia for elective coronary artery bypass graft surgery: effect on troponin T levels. Br Anaesth 100:765–771
57. Welters ID, Feurer MK, Preiss V et al (2011) Continuous S-(+)-ketamine administration during elective coronary artery bypass graft surgery attenuates pro-inflammatory cytokine response during and after cardiopulmonary bypass. Br J Anaesth 106:172–179
58. Nagels W, Demeyere R, van Hemelrijk J et al (2004) Evaluation of the neuroprotective effects of S(+)-ketamine during open-heart surgery. Anesth Analg 98:1595–1603
59. Yan H, Wang H, Chen W et al (2024) Integrative proteomics and metabolomics data analysis exploring the mechanism of brain injury after cardiac surgery in chronic stress rats. BMC Anesthesiol 24:111
60. Cavaleri D, Riboldi I, Crocoma C et al (2024) Evidence from preclinical and clinical metabolomics studies on the antidepressant effects of ketamine and esketamine. Neurosci Lett 831: 137791
61. Nummela AJ, Laaksonen LT, Laitio TT et al (2022) Effects of dexmedetomidine, propofol, sevoflurane and S-ketamine on the human metabolome: a randomised trial using nuclear magnetic resonance spectroscopy. Eur J Anaesthesiol 39:521–532

62. Sun L, Zhao Y, Li Y et al (2023) Effect of continuous subanesthetic esketamine infusion on postoperative fatigue in patients undergoing laparoscopic radical resection for colorectal cancer: a randomized controlled trial. Am J Cancer Res 13:2554–2563
63. Johnston JN, Henter ID, Zarate CA (2023) The antidepressant action of ketamine and its enantiomers. Pharmacol Ther 246:108431
64. Gutierrez G, Swainson J, Ravindran N et al (2024) IN esketamine and IV ketamine: results of a multisite observational study assessing the effectiveness and tolerability of two novel therapies for treatment-resistant depression. Psychiatry Res 340:116125
65. Niu G, Zheng X, Deng B et al (2024) The effects of prophylactic use of esketamine on postoperative depression and quality of life: a meta-analysis. Minerva Anestesiol 90:321–329
66. Kew BM, Porter RJ, Douglas KM et al (2023) Ketamine and psychotherapy for the treatment of psychiatric disorders; systematic review. BJPsych Open 9:e79
67. Wilkinson ST, Rhee TG, Joormann J et al (2021) Cognitive behavioral therapy to sustain the antidepressant effects of ketamine in treatment-resistant depression: a randomized clinical trial. Psychother Psychosom 90:318–327
68. Kitay BM, Murohy E, Macaluso M et al (2023) Cognitive behavioral therapy following esketamine for major depression and suicidal ideation for relapse prevention: the CBT-ENDURE randomized clinical trial study protocol. Psychiatry Res 330:115585
69. Beaglehole B, Porter R, Douglas K et al (2024) Protocol for a randomized controlled trial of ketamine versus ketamine and behavioral activation therapy for adults with treatment-resistant depression in the community. BMJ Open 14:e084844
70. Rasch B, Cordi MJ (2024) The influence of experience and modality of presentation (online vs. offline) on hypnotizability. Front Psychol 14:1293070
71. Patterson DR, Hoffer C, Jensen MP et al (2018) Ketamine as a possible moderator of hypnotizability: a feasibility study. Int J Clin Exp Hypn 66:298–307
72. Sklar GS, Zukin SR, Reilly TA (1981) Adverse reactions to ketamine anesthesia. Anaesthesia 36:183–187
73. Cheong SH, Lee KM, Lim SH et al (2011) The effect of suggestion on unpleasant dreams induced by ketamine administration. Anesth Analg 112:1082–1085
74. Wolfson P, Vaid G (2024) Ketamine-assisted psychotherapy, psychedelic methodologies, and the impregnable value of the subjective—a new and evolving approach. Front Psych 15: 1209419
75. Matsingos A, Wilhlem M, Noor L et al (2024) Hype or hope? High placebo response in major depression treatment with ketamine and esketamine: a systematic review and meta-analysis. Front Psych 15:1346697
76. Falkenberg I, Bitsch F, Liu W et al (2023) The effects of esketamine and treatment expectation in acute major depressive disorder (expect): study protocol for a pharmacological fMRI study using a balanced placebo design. Trials 24:514
77. Garel N, Levesque JT, Sandra DA et al (2023) Imprinting: expanding the extra-pharmacological model of psychedelic drug action to incorporate delayed influences of sets and settings. Front Hum Neurosci 17: 1200393
78. Moujaes F, Rieser NM, Philipps C et al (2024) Comparing neural correlates of consciousness: from psychedelics to hypnosis and meditation. Biol Psychiatry Cogn Neurosci Neuroimaging 9:533–543
79. Sekula AD, Downey L, Puspanathan P (2022) Virtual reality as a moderator of psychedelic-assisted psychotherapy. Front Psychol 13: 813746

Chapter 26

Ketamine in Microgravity

Lydia Johnson Kolaparambil Varghese, Tobias Warnecke, and Jochen Hinkelbein

Abstract

This paper examines the potential of ketamine as a vital medication for space missions, addressing the unique challenges of providing medical care in microgravity environments with limited resources and significant communication delays. Ketamine's fast-acting effects, flexible ways of being administered, and strong safety profile—especially its ability to avoid respiratory depression—make it well suited for space, where airway management can be complex. Its stability under extreme conditions implies that it could also be stored safely for long time periods on missions to the Moon, Mars, and beyond. Including ketamine in emergency protocols gives astronauts the ability to handle a variety of medical emergencies on their own, helping ensure safety and self-reliance as we venture into space. More research is needed to comprehend the behavior of ketamine in space.

Key words Ketamine, Microgravity, Emergency medicine, Space missions, Airway management

1 Introduction

In an era of advanced space exploration, with plans for settlements on Mars, a lunar gateway around the Moon, and the continuous habitation of the International Space Station (ISS), the medical care of space travelers has become increasingly complex and demanding. Ensuring greater autonomy and reliability in onboard medical treatments is crucial. Alongside the challenges of microgravity and radiation, communication latency is a significant hurdle. Telemedicine becomes problematic when delays can reach up to 45 min, making real-time medical care virtually impossible [1]. Moreover, the microgravity environment induces significant physiological changes, including muscle atrophy, bone density loss, fluid redistribution, and alterations in the cardiovascular and immune systems [2–4]. These changes can affect how medications work in the body, making pharmacological interventions more complex than on Earth.

Sabine Himmelseher (ed.), *Ketamine: Pain Management, Critical Care, Depression, and Resource-Poor Settings Volume II*, Neuromethods, vol. 225, https://doi.org/10.1007/978-1-0716-4599-4_26,

Astronauts, despite their extensive technical and operational training, typically receive only basic medical training, and the presence of a fully trained physician is not feasible on every mission [5]. The absence of a dedicated medical professional on many missions further complicates matters, leaving astronauts to manage potentially complex medical situations with minimal support. In this environment, the need for a versatile, safe, and easy-to-administer drug is vital. A drug that stands out for its wide range of applications and uniqueness is ketamine.

Ketamine is an anesthetic, developed in the twentieth century, which antagonizes N-methyl-D-aspartate (NMDA) receptors. It also interacts with multiple other systems, including opioid receptors (μ, δ, and κ) and monoaminergic, cholinergic, muscarinic, and nicotinic receptors. It produces analgesia, sedation, dissociation, and anaesthesia with increasing doses [6]. Nowadays, it is a cornerstone in emergency and trauma care due to its versatility and reliability.

One of ketamine's key pharmacodynamic features is its rapid onset of action, along with the possibility to be administered via various routes, including oral, intramuscular, subcutaneous, and intravenous application.

The administration of ketamine is thus adaptable to a variety of clinical situations. This plasticity has made it invaluable for pain management and anesthesia, and more recently, it has been recognized for its significance in treating refractory depression [7].

Ketamine is frequently the preferred choice among non-anesthesiologist clinicians in multiple settings, owing to its excellent safety profile with lack of cardiocirculatory depression and preservation of spontaneous respiration within a large dose range. Its high therapeutic index and cost-effectiveness makes ketamine an indispensable tool in both resource-limited and high-stakes medical environments [8–10].

2 Advantages of Ketamine in Microgravity

Ketamine's effectiveness has been well-demonstrated in extreme settings, notably during the War in Vietnam, where it was widely used as an anesthetic in resource-limited environments [11]. Its strong safety profile and adaptability in such austere conditions establish a compelling case for its use in the demanding context of space exploration.

The key characteristic that makes ketamine very suitable as medication for space environments is that it does not cause respiratory depression [5–7]. This feature stands in sharp contrast to opioids and other sedatives. In the microgravity setting where airway management can be exceedingly challenging, maintaining spontaneous ventilation is crucial. The absence of respiratory compromise and the non-impairment of protective airway reflexes with ketamine ensures that astronauts can undergo necessary medical procedures without requiring complex respiratory support, which would be and is difficult to implement in space [12]. Komorowski and co-workers recommend ketamine as the drug of choice for airway management in austere environments and for future space exploration missions [9].

Ketamine also provides significant hemodynamic stability, as it does not induce hypotension and can be particularly advantageous in patients with shock or sepsis [13–15]. Its ability to promote constriction and dilation may explain its favorable profile in hypovolemic patients [12]. Due to its sympathomimetic effects on the cardiovascular system, ketamine is often the anesthetic of choice for rapidly stabilizing patients in shock or hypotension when emergency anesthesia is required [15]. This feature is invaluable in space missions, where fluid management and circulatory support are of course limited, and those astronaut caregivers or "crew medical officers" administering ketamine must rely on drugs that provide a reliable and safe physiological response.

One of ketamine's key advantages is its minimal effects on renal and hepatic function without repeated uses. This may be particularly important during long-duration space missions, where maintaining optimal organ function is essential. Any compromise to kidney or liver health could threaten an astronaut's ability to perform mission-critical tasks or may even become a matter "of life and death." Ketamine's well-established safety profile allows for its use without adding undue stress to vital organs in short-term administration [16]. Furthermore, unlike many anesthetics, ketamine does not increase the intracranial pressure if normocapnia is kept. Together with its neuroprotective properties, ketamine may therefore be the option for managing traumatic brain injuries in space, which are of significant concern. Actually, the risk of head injuries from floating debris or accidental impacts is definitely elevated under microgravity conditions [17, 18].

Far beyond, the isolation and confinement of space missions can lead to psychological issues such as depression and anxiety [19]. Clinical evidence has shown ketamine's rapid antidepressant effects, and the substance has meanwhile been established as antidepressant therapy in medical centers and community programs in many countries of the world (please see other chapters in this textbook). Ketamine's antidepressant effects at low dose could be beneficial for maintaining crew mental health during extended missions.

3 Medical Uses of Ketamine in Space

In several medical scenarios that may arise in space, ketamine may well be used: for instance, while suturing lacerations, ketamine can provide both effective sedation and pain relief, facilitating these procedures with minimal discomfort [5]. Small and greater injuries after various accidents are likely to occur more frequently in microgravity. During extravehicular activities (EVAs), astronauts are at risk of traumatic injuries from micrometeoroid impacts or equipment malfunctions [20, 21]. In such scenarios, ketamine can provide rapid analgesia and sedation, facilitating emergency interventions while maintaining cardiorespiratory stability. The combination of esketamine with lidocaine, as studied in terrestrial applications, could provide enhanced pain management in space settings, such as during minor surgical procedures or injury care [22]. Multimodal approaches like these might help reduce the required dosages of each agent, a valuable advantage in environments where medical resources are limited. Although further research is needed to validate its use in space, this approach aligns well with the demands of long-duration missions, where efficient, low-risk pain management strategies are vital.

In case of burn injury management, ketamine offers reliable analgesia without compromising respiratory function, which is particularly important if frequent dressing changes or grafting procedures are required [23, 24]. All of these are critical factors in the confined environment of space missions [12].

In terms of pain management, ketamine has proven its efficacy for treatment of acute and chronic pain including oral and intranasal use [16, 25]. These application routes may be very advantageous because they may allow to get along without insertion of an intravenous line. Ketamine also is a valuable alternative to opioids, which have adverse effects which are very undesirable under microgravity conditions, such as respiratory depression, nausea, vomiting, and constipation [26].

Ketamine's bronchodilator properties also make it an excellent option for managing asthma exacerbations [27]. Such a condition could be triggered by the unique atmospheric conditions of a space cabin.

In addition, especially esketamine's ability to provide short-term analgesia and sedation with a rapid recovery period would be ideal for minor but distressing procedures of all kinds, ensuring that crew members can quickly recover.

4 Ketamine in Mars Habitats

As humanity prepares to establish habitats on Mars, providing effective emergency medical care in this remote environment becomes a critical challenge. Mars missions present unique risks, including injuries during spacewalks, radiation exposure, and accidents within habitats due to equipment malfunctions or environmental hazards like dust storms.

Ketamine may serve as an essential tool for these missions. Its potent analgesic and dissociative anaesthetic properties do not compromise respiratory function—a crucial advantage when advanced airway management is difficult or impossible in reduced gravity. Its chemical stability and resistance to cosmic radiation make it suitable for long-term storage on Mars. Moreover, ketamine's ease of use by crew members without specialized anesthesiology training and minimal equipment requirements may enhance its practicality.

By integrating ketamine into emergency medical protocols for Mars missions, we can equip crews to handle a wide range of acute medical situations autonomously. This proactive approach addresses the critical needs for anaesthesia and pain management where immediate evacuation is impossible. Embracing ketamine as part of our medical preparedness is a vital step toward ensuring the safety and self-sufficiency of future human exploration endeavors on the Red Planet.

5 Future Perspectives

Ketamine offers clear advantages over traditional opioids, providing effective analgesia and sedation with small risks of respiratory depression or hypotension with skilled use. Opioid-related adverse effects are especially worrying in space. Ketamine's versatility is perhaps its greatest strength for treatment approaches in microgravity: it addresses a broad spectrum of medical needs from trauma care and pain management to short-term sedation, status asthmatic therapy, and depression treatment. Combined with ketamine's low cost and decades of safe use and experience in diverse environments on earth, ketamine's multifaceted impact positions the substance as one of the first choices for future space missions.

As space exploration pushes further from Earth, the need for safe, reliable, and multipurpose medications will only increase. Ketamine's well-established safety profile, ease of administration, and broad clinical applications make it an ideal candidate for these demanding conditions. Moving forward, further research into ketamine's pharmacokinetics and pharmacodynamics in microgravity will help optimize its use on long-duration missions. With its proven efficacy and wide range of applications, ketamine is poised to become a cornerstone of medical care in our continuing exploration of space.

While ketamine shows great promise, future research into its pharmacology under conditions of microgravity is necessary to optimize dosing and administration protocols for space missions.

Conflicts of Interest

The authors declare no conflicts of interest.

In Memory of Jochen Hinkelbein

We are deeply saddened and still shocked by the sudden death of Professor Dr. Jochen Hinkelbein, who died in a tragic accident in March 2025. He was Full Professor of Emergency Medicine, and Chair and Director of the Department of Anaesthesiology, Intensive Care, Emergency and Pain Medicine, at the Johannes Wesling Klinikum Minden, University Hospital Ruhr-University Bochum, Minden, and Director of the Department of Anesthesiology, Intensive Care, and Pain Therapy at the Bad Oeynhausen Hospital and the Auguste Victoria Clinic, Bad Oeynhausen, Germany.

Professor Hinkelbein was an outstanding and highly respected expert in emergency and aerospace medicine. Among the many positions he held in professional societies, he was a past Chairman of the ESAIC Subcommittee on Resuscitation, Emergency Medicine, and Trauma, and served as President of the German Society of Aerospace Medicine (DGLRM) and as President of the European Society of Aerospace Medicine (ESAM) [28]. He was a Diplomate and Fellow of several associations such as the Aerospace Medical Association (FAsMA). Until his appointment in Minden in 2023, Jochen Hinkelbein was one of the most important senior physicians at the Department of Anesthesiology and Operative Intensive Care at the University Hospital of Cologne for more than a decade. During this time, he did not only establish the European Trauma Course (ETC), but also initiated a collaboration with the European Space Agency (ESA) (28). As a result, the Cologne training laboratory was able to prepare all European astronauts for emergency scenarios on board the ISS. One of his favorites was working with the German Aerospace Center (DLR) on weightlessness research, with a focus on resuscitation techniques tested during parabolic flights.

When discussing the content of this chapter as an outlook on emergency medicine in microgravity, he always emphasized that "resuscitation as we know today is almost impossible in microgravity". We came to the conclusion that – although we both wanted to be "visionary" about possible Mars and other space habits - we had to stick to the "reality" and the available evidence. Nevertheless, he was convinced that ketamine's proven efficacy and wide range of applications make it a really good candidate for medical care in (future) space exploration endeavors.

Together with many others, we wish to honor a dedicated physician and teacher, and a kind and humble man, Jochen Hinkelbein.

S. Himmelseher

References

1. Vidal V, Boyer L, Luciani A, Tacher V, Kobeiter H, Longere B et al (2023) Bringing interventional radiology to Mars! Cardiovasc Intervent Radiol 46(4):425–427
2. Sy MR, Keefe JA, Sutton JP, Wehrens XHT (2023) Cardiac function, structural, and electrical remodeling by microgravity exposure. Am J Physiol Heart Circ Physiol 324(1):H1–H13
3. Tanaka K, Nishimura N, Kawai Y (2017) Adaptation to microgravity, deconditioning, and countermeasures. J Physiol Sci 67(2):271–281
4. Kim DS, Vaquer S, Mazzolai L, Roberts LN, Pavela J, Watanabe M et al (2021) The effect of microgravity on the human venous system and blood coagulation: a systematic review. Exp Physiol 106(5):1149–1158
5. Silverman GL, McCartney CJ (2008) Regional anesthesia for the management of limb injuries in space. Aviat Space Environ Med 79(6): 620–625
6. Zorumski CF, Izumi Y, Mennerick S (2016) Ketamine: NMDA receptors and beyond. J Neurosci 36(44):11158–11164
7. Nowacka A, Borczyk M (2019) Ketamine applications beyond anesthesia—a literature review. Eur J Pharmacol 860:172547
8. Chesters A, Webb T (2015) Ketamine for procedural sedation by a doctor-paramedic prehospital care team: a 4-year description of practice. Eur J Emerg Med 22(6):401–406
9. Komorowski M, Fleming S, Mawkin M, Hinkelbein J (2018) Anaesthesia in austere environments: literature review and considerations for future space exploration missions. NPJ Microgravity 4:5
10. Guldner GT, Petinaux B, Clemens P, Foster S, Antoine S (2006) Ketamine for procedural sedation and analgesia by nonanesthesiologists in the field: a review for military health care providers. Mil Med 171(6):484–490
11. 'Mercer SJ (2009) The drug of war'—a historical review of the use of ketamine in military conflicts. J R Nav Med Serv 95(3):145–150
12. Komorowski M, Watkins SD, Lebuffe G, Clark JB (2013) Potential anesthesia protocols for space exploration missions. Aviat Space Environ Med 84(3):226–233
13. Morris C, Perris A, Klein J, Mahoney P (2009) Anaesthesia in haemodynamically compromised emergency patients: does ketamine represent the best choice of induction agent? Anaesthesia 64(5):532–539
14. Burke T, Manglani Y, Altawil Z, Dickson A, Clark R, Okelo S et al (2015 Sep) A safe-Anesthesia innovation for emergency and life-improving surgeries when no anesthetist is available: a descriptive review of 193 consecutive surgeries. World J Surg 39(9):2147–2152
15. Srivilaithon W, Bumrungphanithaworn A, Daorattanachai K, Limjindaporn C, Amnuaypattanapon K, Imsuwan I et al (2023) Clinical outcomes after a single induction dose of etomidate versus ketamine for emergency department sepsis intubation: a randomized controlled trial. Sci Rep 13(1):6362
16. Peltoniemi MA, Hagelberg NM, Olkkola KT, Saari TI (2016) Ketamine: a review of clinical pharmacokinetics and pharmacodynamics in anesthesia and pain therapy. Clin Pharmacokinet 55(9):1059–1077
17. Godoy DA, Badenes R, Pelosi P, Robba C (2021) Ketamine in acute phase of severe traumatic brain injury 'an old drug for new uses?'. Crit Care 25(1):19
18. Laws JC, Vance EH, Betters KA, Anderson JJ, Fleishman S, Bonfield CM et al (2023) Acute effects of ketamine on intracranial pressure in children with severe traumatic brain injury. Crit Care Med 51(5):563–572
19. Yin Y, Liu J, Fan Q, Zhao S, Wu X, Wang J et al (2023) Long-term spaceflight composite stress induces depression and cognitive impairment

in astronauts-insights from neuroplasticity. Transl Psychiatry 13(1):342

20. Poveda L, Devane K, Lalwala M, Gayzik FS, Stitzel JD, Weaver AA (2024) Injury risk predictions in lunar terrain vehicle (LTV) extravehicular activities (EVAs): a pilot study. Ann Biomed Eng 52(9):2534–2545
21. Scheuring RA, Mathers CH, Jones JA, Wear ML (2009) Musculoskeletal injuries and minor trauma in space: incidence and injury mechanisms in U.S. astronauts. Aviat Space Environ Med 80(2):117–124
22. Zhou L, Lv L, Wu R, Mang W, Hu L (2024) Improvement in pain by using lidocaine combined with esketamine in elderly patients receiving local anaesthesia for percutaneous kyphoplasty: a randomized controlled study. BMC Anesthesiol 24(1):384
23. Galeiras R, Mourelo M, Pértega S, López ME, Esmoris I (2018) Procedural sedation and analgesia during enzymatic debridement of burn patients. Ann Burns Fire Disasters 31(3): 223–227
24. Barrett W, Buxhoeveden M, Dhillon S (2020) Ketamine: a versatile tool for anesthesia and analgesia. Curr Opin Anaesthesiol 33(5): 633–638
25. Andolfatto G, Willman E, Joo D, Miller P, Wong WB, Koehn M et al (2013) Intranasal ketamine for analgesia in the emergency department: a prospective observational series. Acad Emerg Med 20(10):1050–1054
26. Schug SA, Zech D, Grond S (1992) Adverse effects of systemic opioid analgesics. Drug-Safety 7(3):200–213
27. Agrawal A, Goyal S (2013) Ketamine in status asthmaticus: a review. Indian J Crit Care Med 17(3):154–161
28. Böttiger BW (2025) In Memoriam: Prof. Dr. med. Jochen Hinkelbein. European Resuscitation Council Newsletter. Available at HYPERLINK "https://www.erc.edu/in-memoriam-prof-dr-med-jochen-hinkelbein" ERC | Bringing resuscitation to the world Accessed 06 May 2025

INDEX

Sabine Himmelseher (ed.), *Ketamine: Pain Management, Critical Care, Depression, and Resource-Poor Settings Volume II*, Neuromethods, vol. 225, https://doi.org/10.1007/978-1-0716-4599-4,

B

E

F

G

H

I

N

O

P

Q

R

S

T

U

V

W

Y

MIX
Papier aus verantwortungsvollen Quellen
Paper from responsible sources
FSC® C105338
FSC
www.fsc.org

If you have any concerns about our products,
you can contact us on
ProductSafety@springernature.com

In case Publisher is established outside the EU,
the EU authorized representative is:
Springer Nature Customer Service Center GmbH
Europaplatz 3, 69115 Heidelberg, Germany

Printed by Libri Plureos GmbH
in Hamburg, Germany